高等院校海洋科学专业规划教材

MARINE ECOTOXICOLOGY
海洋生态毒理学

主编
（西）朱利安·布拉斯科(Julián Blasco)
（加）彼得·M.查普曼(Peter M. Chapman)
（英）奥利维亚·坎帕纳(Olivia Campana)
（西）玛丽安·汉佩尔(Mirian Hampel)

译 孙 显

中山大学出版社
SUN YAT-SEN UNIVERSITY PRESS
·广州·

SunYat-sen University Press is authorized to publish and distribute exclusively the Chinese (Sim-plified Characters) language edition. This edition is authorized for sale throughout People's Republic of China. No part of the publication may be reproduced or distributed by any means, or stored in a database or retrieval system, without the prior written permission of the publisher.

图书在版编目（CIP）数据

海洋生态毒理学/（西）朱利安·布拉斯科等主编；孙显译. —广州：中山大学出版社，2024.7

（高等院校海洋科学专业规划教材）

书名原文：Marine Ecotoxicology

ISBN 978-7-306-08049-3

Ⅰ.①海… Ⅱ.①朱… ②孙… Ⅲ.①海洋污染—毒理学—高等学校—教材 Ⅳ.①X55

中国国家版本馆 CIP 数据核字（2024）第 043666 号

HAIYANG SHENGTAI DULIXUE

出 版 人：王天琪
策划编辑：李　文
责任编辑：李　文
封面设计：曾　斌
责任校对：石玉珍
责任技编：靳晓虹
出版发行：中山大学出版社
电　　话：编辑部 020-84110283，84113349，84111997，84110779，84110776
发行部 020-84111998，84111981，84111160
地　　址：广州市新港西路 135 号
邮　　编：510275　传　真：020-84036565
网　　址：http://www.zsup.com.cn　E-mail：zdcbs@mail.sysu.edu.cn
印 刷 者：佛山市浩文彩色印刷有限公司
规　　格：787mm×1092mm　1/16　24.25 印张　600 千字
版次印次：2024 年 7 月第 1 版　2024 年 7 月第 1 次印刷
定　　价：108.00 元

如发现本书因印装质量影响阅读，请与出版社发行部联系调换

《高等院校海洋科学专业规划教材》
编审委员会

主　　任　王东晓　李春荣　陈省平　赵　俊

委　　员　（以姓氏笔画排序）

万志峰　王天霖　王东晓　王江海
卢建国　刘　岚　刘维亮　苏　明
李　雁　李春荣　李朝政　来志刚
吴玉萍　吴加学　吴景峰　邹世春
陈省平　陈保卫　邱春华　易梅生
罗一鸣　赵　俊　郭长军　胡　湛
贾坤同　龚　骏　龚文平　谢　伟
翟　伟

总　　序

海洋与国家安全和权益维护、人类生存和可持续发展、全球气候变化、油气和某些金属矿产等战略性资源保障等休戚相关。贯彻落实"海洋强国"建设和"一带一路"倡议，不仅需要高端人才的持续汇集，实现关键技术的突破和超越，而且需要培养一大批了解海洋知识、掌握海洋科技、精通海洋事务的卓越拔尖人才。

海洋科学涉及的领域极为宽广，几乎涵盖了传统所熟知的"陆地学科"。当前，海洋科学更加强调整体观、系统观的研究思路，从单一学科向多学科交叉融合的发展趋势十分明显。海洋科学本科人才培养中，处理好"广博"与"专深"的关系，十分关键。基于此，我们本着"博学专长"的理念，按"243"思路来构建"学科大类→专业方向→综合提升"的专业课程体系。其中，学科大类板块设置了基础和核心两类课程，以拓宽学生知识面，助其掌握海洋科学理论基础和核心知识；专业方向板块从本科第四学期开始，按海洋生物、海洋地质、物理海洋和海洋化学四个方向对学生进行"四选一"分流，以帮助学生掌握扎实的专业知识；综合提升板块则设置选修课、实践课和毕业论文三个模块，以推动学生更自主、个性化、综合性地学习，养成专业素养。

相对于数学、物理学、化学、生物学、地质学等专业，海洋科学专业开设时间较短，教材积累相对欠缺，部分课程尚无正式教材，部分课程虽有教材但专业适用性不理想或知识内容较为陈旧。我们基于"243"课程体系，固化课程内容，从以下三个方面建设海洋科学专业系列教材：一是引进、翻译和出版 Descriptive Physical Oceanography: An Introduction, 6ed（《物理海洋学·第6版》）、Chemical Oceanography, 4ed（《化学海洋学·第4版》）、Biological Oceanography, 2ed（《生物海洋学·第2版》）、Introduction to Satellite Oceanography（《卫星海洋学》）、Coastal Storms: Processes and Impacts（《海岸风暴：过程与作用》）、Marine Ecotoxicology（《海洋生态毒理学》）等原版教材；二是编著、出版《海洋植物学》《海洋仪器分析》《海岸动力地貌学》《海洋地图与测量学》《海洋污染与毒理》《海洋气象学》《海洋观测技术》

《海洋油气地质学》等理论课教材；三是编著、出版《海洋沉积动力学实验》《海洋化学实验》《海洋动物学实验》《海洋生态学实验》《海洋微生物学实验》《海洋科学专业实习》《海洋科学综合实习》等实验教材或实习指导书，预计最终将出版40多部系列性教材。

教材建设是高校的基本建设，对于实现人才培养目标起着重要作用。在教育部、广东省和中山大学等教学质量工程项目的支持下，我们以教师为主体、以学生为中心，及时地把本学科发展的新成果引入教材，使教学内容更具针对性和适用性。谨此对所有参与系列教材建设的教师和学生表示感谢。

系列教材建设是一项长期持续的工作，我们致力于突出前沿性、科学性和适用性，并强调内容的衔接，以形成完整的知识体系。

因时间仓促，教材中难免有不足和疏漏之处，敬请不吝指正。

《高等院校海洋科学专业规划教材》编审委员会

目 录

序	(1)
1 海洋污染物	(1)
1.1 引言	(2)
1.2 来源和属性	(2)
1.3 分析方法	(9)
1.4 海洋环境中的赋存	(16)
1.5 海洋环境中的微塑料	(21)
1.6 未来趋势	(22)
1.7 结论	(23)
参考文献	(23)
2 现代统计设计与分析方法	(43)
2.1 引言	(44)
2.2 毒性检测	(44)
2.3 设计考虑	(46)
2.4 数据处理	(50)
2.5 估计和推断	(55)
2.6 浓度–响应（C-R）模型	(58)
2.7 物种敏感性分布法模型	(65)
2.8 生态毒理学统计软件工具	(67)
2.9 展望	(73)
参考文献	(74)
3 化学物质吸收和效应的动态模型	(83)
3.1 引言	(84)
3.2 建模基本原则	(85)
3.3 毒物代谢动力学	(89)
3.4 毒物效应动力学概述	(94)
3.5 对生存的影响	(97)
3.6 对亚致死终点的影响	(100)
3.7 种群及以上水平	(105)
3.8 未来的可能性	(106)

参考文献 (108)

4 生物富集和生物监测 (113)
4.1 生物富集的通则 (114)
4.2 生物富集模型 (117)
4.3 动力学参数 (120)
4.4 生物富集模型的应用 (126)
4.5 生物监测 (129)
4.6 生物监测原理及注意事项 (130)
4.7 展望 (132)
参考文献 (132)

5 生物标记物和效应 (137)
5.1 引言 (138)
5.2 生物标记物 (138)
5.3 高通量筛选技术或"组学" (157)
5.4 展望 (164)
参考文献 (165)

6 海水毒性试验 (189)
6.1 引言 (190)
6.2 专业术语 (190)
6.3 一般海水毒性试验方法及规程 (191)
参考文献 (214)

7 沉积物毒性试验 (227)
7.1 引言 (228)
7.2 如何进行沉积物毒性试验 (242)
7.3 全沉积物毒性鉴定评价 (248)
7.4 展望 (249)
参考文献 (254)

8 海洋环境围隔与实地毒性试验 (271)
8.1 引言 (272)
8.2 围隔——保护海洋生态系统的管理工具 (274)
8.3 水生围隔设计通则 (276)
8.4 拓展:为大型鱼类开发系统 (280)
8.5 海洋围隔设计的思考 (283)
8.6 展望 (283)

8.7　结论：围隔试验在海洋生态毒理学中的应用 …………………………………… (284)
参考文献 ………………………………………………………………………………… (284)

9　生态风险和证据权重评价 …………………………………………………………… (293)
9.1　引言 ………………………………………………………………………………… (294)
9.2　生态风险评价 ……………………………………………………………………… (294)
9.3　证据权重 …………………………………………………………………………… (298)
9.4　生态风险评价中的海洋生态毒理学和证据权重 ………………………………… (302)
9.5　展望 ………………………………………………………………………………… (303)
参考文献 ………………………………………………………………………………… (305)

10　全球变化 ……………………………………………………………………………… (311)
10.1　引言 ……………………………………………………………………………… (312)
10.2　流域土地利用变化 ……………………………………………………………… (312)
10.3　港口和工业相关变化 …………………………………………………………… (321)
10.4　气候变化 ………………………………………………………………………… (332)
10.5　展望 ……………………………………………………………………………… (340)
10.6　总结 ……………………………………………………………………………… (343)
参考文献 ………………………………………………………………………………… (344)

索引 …………………………………………………………………………………………… (355)

后记 …………………………………………………………………………………………… (375)

序

　　海洋、沿海地区和过渡水域（比如河口和沿海潟湖）是生产力较高的区域，对地球生态系统的整体功能至关重要。目前，它们正在遭受各种压力源所带来的不利影响，这些压力源包括气候和生境变化、入侵/引进物种、富营养化（包括有害赤潮）和化学污染物。

　　生态毒理学是研究上述压力因素影响的一门较新的学科。"生态毒理学"一词最初是由 Truhaut 在 1969 年创造（Truhaut，1977），被定义为预测潜在有毒物质或其他压力源对自然生态系统和目标物种影响的科学（Hoffman et al.，2003）。

　　以前关于这一主题的书主要关注淡水而非海洋生态系统。因此，我们认为有必要为本科生和研究生提供一本专门研究海洋生态毒理学的书。本书提供的参考信息对高级研究人员也有益处，尤其是计划从事相关研究的人员。

　　本书分十章，由一些在相关领域较资深的研究人员合作编写而成。本书内容大致如下：污染物、统计学、污染物建模、生物积累和生物监测、生物标记、海水毒性试验、沉积物毒性试验、围隔试验和现场试验、生态风险和参数的权重、全球变化。

　　作为主编，我们从编写这十章的作者身上学到了很多。我们希望读者也能学到很多，并利用书中所提供的知识进一步推进海洋生态毒理学的发展。

朱利安·布拉斯科
彼得·M. 查普曼
奥利维亚·坎帕纳
玛丽安·汉佩尔

1 海洋污染物

D. Álvarez-Muñoz[①], **M. Llorca**[②], **J. Blasco**[③], **D. Barceló**[①,②]

---

① 西班牙加泰罗尼亚水研究所
② 西班牙 IDAEA-CSIC
③ 西班牙安达卢西亚海洋科学研究所（CSIC）

1.1 引　　言

　　大量化学品因人类活动输入了海洋环境。特别是工业化国家，每天都在向环境排放各种生活、商业和工业生产活动产生的污染物。这些污染物在海洋环境中形成复杂的、有害的混合物，不仅对野生生物构成潜在风险，还会通过污染海产品危害人类健康。本章汇总了常见的海洋污染物，包括金属、持久性和新兴有机污染物；综述了它们的主要污染源、理化特性以及常用的定量检测分析技术；评估了它们在海水、沉积物和海洋生物中的含量，为后续海洋野生生物污染物暴露的毒性研究提供依据。此外，受科学界和社会日益关注的一类污染物——微塑料，也将在本章被提及。此类污染物的分析测量方法预示非靶向技术（nontarget techniques）的应用将是未来环境监测的研究方向。

1.2 来源和属性

1.2.1 金属

　　在生态毒理学中，"重金属"与环境污染物有关，而"痕量金属"仅指基质中浓度为痕量（0.01%）的金属。通常这两个词所指相同。Nieboer（1980）提出一种基于金属离子路易斯酸性质的化学分类方法：根据金属酸或碱的"软""硬"度将其分为A类、B类和临界类。虽然金属化学特性的数据很好地支持了此分类法，但"重金属"和"痕量金属"这两个词仍在科学文献中被广泛使用。本章将用"金属"一词来表示生态毒理学中的金属污染物，如银（Ag）、砷（As）、金（Au）、铋（Bi）、镉（Cd）、钴（Co）、铬（Cr）、铜（Cu）、铁（Fe）、镓（Ga）、汞（Hg）、铟（In）、铱（Ir）、锰（Mn）、钼（Mo）、镍（Ni）、铅（Pb）、钯（Pd）、铂（Pt）、铑（Rh）、锑（Sb）、硒（Se）、锡（Sn）、钛（Ti）、钒（V）和锌（Zn）。

　　金属存在于土壤和岩石中，经侵蚀风化过程被释放。在水体中，它们以溶解或颗粒形态通过溪流和河流进行迁移；在大气中，有些金属结合在颗粒物、气溶胶上，或以蒸汽形式（如汞）进行迁移。除自然来源外，人类活动是金属输入环境的主要来源。金属通过地表径流、大气传输、热液排放、地下水渗漏、沉积物扩散和外层空间输入等方式进入海洋，而前三个途径的通量为主要输入。通量和含量的平衡关系可用于建立金属源的联系。人源金属进入海洋环境的主要途径是风积输入。Libes（2009）总结了自然和人源金属通量，经计算得知大气/河流输入至海洋的通量比值介于3.6（As）～300（Pb）。与其他污染物相比，金属为非生物降解污染物。它们的分布、迁移、赋存和毒性受海洋环境的物理和化学特性影响。金属可被水体中的悬浮固体基质吸附（吸附过程是指溶质在海水或沉积物中与固体基质发生可逆结合的过程）。其中部分被水洗脱，最终沉淀于沉积物中。水体中的胶体颗粒（黏土、铁和锰的氧化物或氢氧化物、碳酸钙和有机颗粒）可提供与金属结合的活性表面。然而，金属形态是决定其赋存、反应性、毒性和生态风险的主要因素。许多溶解在海水中的金属能以各种无机络合物的形式存在。例如，Hg 主要以 $HgCl_4^{2-}$ 的形式存在，有些金属主要以氯化物的形式存在（如 $AgCl_3^{2-}$ 和 $AuCl_2^-$），有些金属则以碳酸盐或氢氧根离子的形式存在。我们可以通过计算机程序解决金属在复杂系统中的形态问题，例如 MINEQL、MINTEQ、MINEQLt、WHAM & CHESS。这些程序可用于检测金属的不同反应过程以及化学平衡计算，如溶解、沉淀、氧化还原转化和吸附过程（详见第 4 章）。

1.2.2 持久性有机污染物

1.2.2.1 多环芳烃

多环芳烃（polycyclic aromatic hydrocarbons，PAHs）是最重要的环境污染物之一，主要为燃烧过程产生的副产物。在自然界，森林大火和火山喷发可产生 PAHs，但大多数 PAHs 是因工业和其他人类活动（如煤炭和原油加工、天然气燃烧、取暖、垃圾燃烧、交通尾气排放、烹饪和吸烟）产生（WHO，2000）。PAHs 分子结构完全由碳和氢原子构成，由两个或多个稠苯环排列组成。PAHs 在水中的溶解度较低，且当分子质量增加时，溶解度几乎呈对数下降（Johnsen et al.，2005）。相比之下，PAHs 具有高度亲脂性，可以吸附在悬浮颗粒物上，并在水体中逐渐沉降，最终富集在沉积物中（Karickhoff et al.，1979）。此外，大多数 PAHs 的蒸气压较低。在大气中，大于 4 环的 PAHs 更易吸附在颗粒物上；而少于 4 环的 PAHs 则常以气体形式存在，直至通过降水被去除（Skupinska et al.，2004）。因此，PAHs 进入环境的方式有两种：吸附在大气和水生环境中的颗粒物上进行长距离迁移；或者当地来源的溶解（Douben，2003）。第二种方式通常发生在双环 PAHs 和少量的三环 PAHs 中。因为它们易溶于水体，从而可被水生生物吸收利用（Mackay & Callcott，1998）。

1.2.2.2 表面活性剂

表面活性剂是有机化合物中最普遍的家族之一。尽管表面活性剂的主要用途是配制家用和工业洗涤剂，但也被用于制造其他产品，如化妆品、油漆、纺织品、染料、聚合物、农用化学品和石油。表面活性剂的分子结构由亲水基团和疏水基团两部分组成。每个分子中极性和非极性基团的存在促进它们在水溶液中形成胶束，这对表面活性剂的去污力和增溶作用至关重要。污物在胶束中被溶解，避免再沉积，从而完成洗涤过程。根据溶液中离子的电荷性质，可将表面活性剂分为阴离子表面活性剂（带负电荷）、阳离子表面活性剂（带正电荷）、非离子表面活性剂（不带电荷）和两性表面活性剂（基于介质的 pH 而带正电荷或带负电荷）。这种分类是基于分子的亲水部分，因为疏水部分是非离子的，且通常由烷基链组成。因具通用性和实用性，阴离子和非离子表面活性剂是最重要的表面活性剂。常见的阴离子表面活性剂有直链烷基苯磺酸盐（linearalkyl-benzene sulfonic acid，LAS）、烷基硫酸盐（alkyl sulfates，AS）、烷基醚硫酸盐（alkyl ether sulfates，AES）和肥皂。非离子表面活性剂的主要种类有醇聚氧乙烯酯（alcohol polyethoxylates，AEO）和烷基酚聚氧乙烯（alkylphenol polyethoxylates，APEO）等。

表面活性剂是水溶性化合物，主要污染源是污水处理厂排放的废水（WWTPs）。它们的溶解度随烷基链长度的增加而下降。不同类型表面活性剂的溶解度也会随着亲水基团（位置或数量）的不同而变化。除可在大气中检测到的壬基酚外，其他表面活性剂都是低挥发性化合物（Van Ry et al.，2000）。表面活性剂的辛醇-水分配系数（K_{ow}）差异取决于其类型，其 K_{ow} 会随着分子疏水性的增加而增加。此外，早期研究已发现化合物的 K_{ow} 与其在生物中的富集度直接相关。表面活性剂的水-沉积物分配系数（K_d）值通常高于 K_{ow}，这表明表面活性剂对沉积物的亲和力更高。

1.2.2.3 多氯联苯

20 世纪 30—70 年代中期，多氯联苯（polychlorinated biphenyls，PCBs）是在全球范围内生产和使用的合成化合物（Solaun et al.，2015）。PCBs 的结构特征是联苯结构上连接氯原子（数量为 1～10），化学通式为 $C_{12}H_{10-x}Cl_x$。PCBs 包含 209 个结构相似的同系物（Solaun et al.，2015），其中有 130 种不同结构的 PCBs 被用作商业用途（United Nations，1999）。PCBs 因具有化学稳定性高、热容量大、易燃性低和绝缘性好等属性，而被广泛应用于工业和商业领域。例如，作为变压器和电容器的液体介质，以及用于印刷墨水、油漆、除尘剂、农药、液压油、增塑剂、黏合剂、阻燃剂和润滑剂的生产（Erickson & Robert，2011；Solaun et al.，2015）。虽然《关于持久性有机污染物的斯德哥尔摩公约》已禁止 PCBs 的生产，但 PCBs 仍能通过大型电子设备的处理和废物排放进入环境（WHO，2010）。此外，《欧盟水框架指令》也要求对 PCBs 环境浓度进行监控（European Commission，2002）。海洋环境中 PCBs 的主要来源是沿海城市的生活污水、工业废弃物排放以及大气沉降（Dorr & Liebezeit，2009；Solaun et al.，2015；UNEP，2002）。PCBs 具有亲脂性，不会稳定存在于水体中，而易富集在富含脂质的生物组织和沉积物中（Erickson & Robert，2011；Solaun et al.，2015）。PCBs 具有生殖毒性及免疫毒性（OSPAR Commission，2010）。PCBs 中生物毒性最强的一类具有与二噁英和呋喃化学性质相似的平面结构（OSPAR Commission，2010），其中 12 种 PCBs 在某些条件下可部分氧化形成二苯并呋喃类化合物，故而被称为类二噁英-PCBs（DL-PCBs）（Tuomisto，2011）。

1.2.2.4 农药

农药的作用是杀死、减少或驱除昆虫、杂草、啮齿动物、真菌和其他可能威胁公众健康或经济的生物。农药的作用方式为靶向攻击目标生物体内的系统（生殖、免疫、内分泌等）和酶。而这些系统和酶与人体的系统和酶相似或相同，从而导致农药对人类健康和环境也构成了威胁。农药可以按用途分为杀虫剂、杀真菌剂、灭鼠剂、杀螨剂；或按化学特征分为拟除虫菊酯类、有机磷类、有机氯类、氨基甲酸酯类、草甘膦类和三唑类。例如，有机磷类农药都是磷酸衍生物；有机氯类农药是由碳、氢和氯元素组成的化合物。农药制剂包括具有活性的主要成分和呈惰性或增强杀虫特性的其他成分。典型农药制剂的剂型为水剂、粉剂、可湿性粉剂和乳油。这些制剂用于保护农业、林业和园艺业植物，或作为生物灭杀产品（包括木材浸渍、船龙骨涂料和杀黏菌产品）。农药主要随地表径流或"喷雾漂移"，由农田进入地表水，随着排水系统进入河流，最后汇入海洋。此外，土壤中的农药通过淋溶作用可进入地下水系统。因此，农药的水溶性和挥发性是农药污染相关研究所关注的两个主要化学特征。农药的水溶性越高，以地表径流和淋溶方式迁移的潜力则越大；挥发性越大，以喷雾漂移方式迁移的潜力则越大。在室温（20 ℃或25 ℃）下，农药活性成分的水溶性通常以每升水的所含溶质毫克数（mg/L）表示，蒸气压以毫米汞柱（mmHg）表示。蒸气压低的农药不易挥发，若溶于水，则有可能在水中累积；若不溶于水，则会根据其辛醇-水分配系数（K_{ow}）在土壤或生物体中累积。

1.2.2.5 二噁英

二噁英化合物包括多氯代联二苯-对-二噁英（polychlorinated dibenzodioxins, PCDDs）、10种多氯二苯并呋喃（polychlorinated dibenzofurans, PCDFs）和12种PCBs（DL-PCBs，可通过部分氧化形成二苯并呋喃的同类物）。虽然二噁英有其特定构象，但因PCDDS、PCDFs & DL-PCBs含有以1,4-二噁英为中心环的二苯并-1,4-二噁英骨架结构，故也被称为二噁英类。PCDDs & PCDFs是工业生产过程，包括氯酚和氯代苯氧除草剂的生产、纸浆的氯气漂白以及冶炼过程产生的副产品。它们也可由火山爆发和森林火灾等自然事件产生（WHO，2010）。另外，PCDFs也是PCB商用混合物中常见的污染物（WHO，2010）。二噁英的主要来源是焚烧不充分的废气、受污染的土壤和沉积物（WHO，2010）。二噁英和类二噁英化合物一旦进入环境，很难通过化学和微生物方式降解。二噁英呈高亲脂性，易累积在脂肪组织中，并通过食物链实现生物富集（WHO，2010）。此外，二噁英中的氯原子数量较多，促使其具有更高的持久性和生物富集能力。海洋食物链的生物放大作用，增加了它们对人类和动物的潜在危害（WHO，2010）。其中，2,3,7,8-四氯二苯并-对-二噁英（2,3,7,8-tetrachlorodibenzo-p-dioxin, 2,3,7,8-TCDD）是目前所有已知的二噁英类中毒性最强的单体，其毒性当量因子（toxic eguivalency factor, TEF）为1（Van den Berg et al., 2006）。

1.2.3 新兴有机污染物

1.2.3.1 全氟和多氟烷基

全氟和多氟烷基物质（perfluoroalkyl substances, PFASs）为人工合成化合物，已有60多年生产历史，主要分子结构为$C_nF_{2n+1}^-$（Buck et al., 2011）。PFASs化合物结构各异，故而表现出不同的特性、赋存和毒性。但它们都具有碳链键（自然界中键能最强的一种），因此普遍具有高稳定性（Llorca, 2012）。全氟烷烃具有疏水性和疏油性，与水混合时，会形成三个不混溶的相态。这类化合物常用于耐火材料添加剂以及防油、防污、防脂和防水剂的生产（Llorca, 2012）。PFASs被应用于厨具上的不粘层、服装防水和透气膜，以及航空航天、汽车、建筑/建造、化工、电子、半导体和纺织业等诸多领域（EPA，2009；Llorca，2012）。PFASs的主要来源分为两类：直接来源，如泡沫灭火剂、商业和工业产品生产过程的废水和废气排放；间接来源，如全氟烷基产品的降解、造纸工业、污水处理厂和烹饪（Llorca，2012）。大多数PFASs具有物理、化学和生物的高度稳定性，广泛分布于环境中（Llorca，2012）。因此，PFASs被归为持久性有机污染物（persistent organic pollutants，POPs）。例如，全氟辛烷磺酸盐（perfluorooctane sulphonate，PFOS）已被列入《关于持久性有机污染物的斯德哥尔摩公约》（USEPA，2006）。

1.2.3.2 溴化阻燃剂

阻燃剂是添加到可燃材料中或与可燃材料发生反应以提高材料耐火性的一类化学物质（WHO，1997）。它们被添加到可燃材料中，如塑料、木材、纸张和纺织品等。阻燃聚合物广泛应用于汽车、消费类电子产品、计算机、电气设备和建筑材料等领域。阻燃剂通常出现在制造生产过程所产生的废水中，在产品的制造或使用过程中，它们通过泡沫产品的分解、设备的处理等途径挥发和浸出而进入环境；或通过垃圾填埋场浸出和回收废物而燃烧，吸附到粉尘颗粒上（Alaee et al.，2003；Murphy，2001）。阻燃剂分为四大类：无机阻燃剂类、有机磷阻燃剂类、卤化有机阻燃剂类和含氮阻燃剂类（Alaee & Wenning，2002）。卤化有机阻燃剂又分为含氯和含溴有机阻燃剂，约25%的阻燃剂都含有溴（Andersso et al.，2006）。根据阻燃剂结合到聚合物的方式，可将其分为不同的亚组。添加型溴化阻燃剂（brominated flame retardants，BFRs）可和其他化合物混合形成聚合物，这类BFRs含有多溴联苯（polybrominated biphenyls，PBBs）、多溴二苯醚（poly brominated diphenyl ethers，PBDEs）和六溴环十二烷（hexabromocyclododecane，HBCD）等化合物，具有低蒸汽压和低水溶性的特点。反应型溴化阻燃剂能与塑料发生化学键合，如四溴双酚A（tetrabromobisphenol A，TBBPA）。聚合型溴化阻燃剂，如溴化聚苯乙烯（brominated polystyrene，BPS），其特征为溴原子结合在聚合物的主链上，形成更为稳定的化学构型，使其具有高分子量以及低挥发性、低生物利用度和低毒性等特点（Guerra et al.，2011）。通常溴化阻燃剂的挥发性随着溴原子数的增加而降低。因此，多溴化合物不易挥发，而低溴化合物在水中的流动性更强，更易从水体表层挥发。此外，低溴化合物（如TBBPA）的log K_{ow}值为4.5，而高溴化合物如十溴二苯醚（deca-BDE）则为10，这表明TBBPA在水生生物中具有更高的生物累积能力。

1.2.3.3 内分泌干扰物

内分泌干扰物（endocrine disrupting chemicals，EDCs）是一种能干扰生物内分泌系统并改变生物正常生长发育的化学物质。EDCs对人类和动物的生殖、神经和免疫系统具有毒性效应。EDCs具有多种作用途径，包括雌激素、抗雄激素、甲状腺激素、过氧化物酶体增殖物激活受体γ（peroxisome proliferator-activated receptor γ，PPARγ）、维甲酸、其他核受体作用、类固醇生成酶、神经递质受体和系统，以及许多在动物和人类中高度保守的途径（Diamanti-Kandarakis et al.，2009）。大剂量内分泌干扰物会导致内分泌系统紊乱。EDCs的结构、理化性质和来源主要取决于污染物的类型。EDCs包括杀虫剂、杀菌剂、除草剂、药品和工业污染物等（Hotchkiss et al.，2008）。双酚、烷基酚、雌激素和全氟类化合物与其他 EDCs 相比更普遍存在于水生生态系统中（Vandermeersch et al.，2015）。

1.2.3.4 药品和个人护理品

药物的活性化合物在维护人和动物健康方面起着重要作用。它们是有不同功能、物理化学和生物学性质的复杂化合物。在环境中，它们通常呈碱性或酸性，可分为中性离子、阳离子、阴离子或两性离子四种类型。通常按治疗目的可将药物分为不同的组或家族，如抗生素、精神病药物、止痛药、抗炎药、镇静剂、激素、β-受体阻滞剂和利尿剂。根据药物溶解度和肠壁渗透性的不同可将药物分为四类：Ⅰ类为高溶高渗，Ⅱ类为低溶高渗，Ⅲ类为高溶低渗，Ⅳ类为低溶低渗。所有药物都必须以水溶液的形式存在于吸收部位，才能被生物吸收利用。因此会采用不同的技术来提高难溶性药物的溶解度，如粒径缩小、晶体工程、成盐、固体分散、表面活性剂的使用和络合等。一旦药物被吸收，生物体可以将其转化为更多的极性分子。然而，生物体往往不能将药物完全代谢（排泄率为 0～100%）。因此，生活污水是药物进入环境的主要来源之一。药物的其他重要来源包括废物处理、水产养殖、畜牧业和园艺业（Gaw et al.，2014）。这些来源使药物不断地被输入到环境中。虽然药物具有很高的转化或去除率，但持续地排放导致水环境中的药物含量不断上升。

个人护理品包括种类繁多的化合物，如消毒剂（如三氯苯氧氯酚）、芳香剂（如麝香）、驱虫剂（如避蚊胺）、防腐剂（如对羟基苯甲酸酯）和防晒霜（如甲基亚苄亚基樟脑）。这些产品主要用于人体外部，包括化妆品、凝胶和肥皂等（Brausch & Rand，2011）。外部身体应用的要求限制了这些个人护理品产生任何代谢变化。因此这些物质得以未经转化就进入环境（Ternes et al.，2004）。一些研究表明，这些化合物在环境中具有环境持久性、生物活性及生物累积性（Brausch & Rand，2011）。

1.2.3.5 海洋生物毒素

海洋生物毒素是由海洋微藻（浮游植物或底栖微藻）合成的天然化学物质。在特定气候和水体条件下，浮游植物具有很高的增殖率，从而形成高密度的藻云（即藻华）（Gerssen et al.，2010）。虽然在特定情况下，有些藻华有益于水产养殖和海洋生物。迄今为止，已知有 40 种甲藻和硅藻在藻华形成期间会产生海洋毒素（被命名为藻毒素）（Gerssen et al.，2010）。尽管藻毒素均具环状和脂肪链结构，且含有 C、H、O 原子（有时具 N 原子），但它们的分子结构在大小和构象上仍有差异。学者已对藻毒素的结构进行了大量的研究（Gerssen et al.，2010）。藻毒素可以在鱼类、贝类和其他海洋生物体内累积，但不表现出毒性作用。然而当人类食用被藻毒素污染的贝类时，则会引发一系列胃肠道、心脏甚至神经系统的疾病（EURLMB，2016；Hallegraeff et al.，1995）。人们主要关注的亲水性藻毒素包括：①软骨藻酸引起的记忆丧失性贝毒（amnesic shellfish poison，ASP）；②蛤蚌素引起的麻痹性贝毒（paralytic shellfish poisoning，PSP）（Gerssen et al.，2010）。亲脂性藻毒素包括：①神经性贝毒（neurotoxic shellfish poisoning，NSP）中的短裸甲藻毒素；②腹泻性贝毒（diarrhetic shellfish poisoning，DSP）中的甲基软海绵酸、鳍藻毒素和扇贝毒素;③氮杂螺环酸贝毒(azaspiracid poisoning,AZP)中的原多甲藻酸(Gerssen et al.,2010);④与多种胃肠道、神经或心血管病症有关的雪卡鱼毒家族中的雪卡毒素(Vandermeersch et al.,2015)。有害、有毒藻华的爆发与扩散是当今备受瞩目的研究方向之一(Van Dolah,2000)。基于藻华扩散及其对人类的影响,欧洲食品安全局(European Food Safety Authority,EFSA)已经对藻毒素给出了不同的吸收限值(Alexander et al.,2008a,2008b,2008c,2009a,2009b,2009c)。

1.2.3.6 纳米材料

纳米材料(nanometer materials,NMS)通常分为碳基纳米材料(carbon-based nano-materials,CNMs)、金属氧化物纳米颗粒(metal oxide nanoparticles,NPs)、金属纳米颗粒、聚合物纳米材料和量子点(Sanchís,2015)。本节将重点介绍碳基纳米材料。纳米材料通过自然来源(自然偶发)和人为活动(纳米技术应用)进入环境(Sanchís,2015)。CNMs 的自然来源包括燃烧和其他高能过程,如火山爆发(Buseckhe & Adachi 2008;Jehlička et al.,2003)、闪电(Daly et al.,1993)、沙尘暴(Gu et al.,2003)和野火(Sanchís,2015)。人为来源中,人为燃烧是 CNMs 的主要来源,其次是机械摩擦和采矿,最后是纳米技术应用(Sanchís,2015)。CNMs 包括非晶碳 NPS 和不同碳同素异形体(Hirsch,2010),在纳米尺度具有碳键 sp^2 和 sp^3 杂化的物理化学特征(Sanchís,2015)。对 CNMs 研究最多的是富勒烯——由二十到数百个碳原子排列成为空心多面体型的不饱和的未官能化碳材料(Astefanei et al.,2015;Maruyama et al.,1991;Prinzbach et al.,2000;Sanchís,2015)。

1.2.3.7 聚二甲基硅氧烷

聚二甲基硅氧烷(polydimethylsiloxanes,PDMS)主要作为有机硅生产的基本物料,也可作为清洁剂、汽车蜡和抛光剂中的添加剂。PDMS 还可以直接应用于卫生和个人护理产品(如化妆品、除臭剂、护发素、护肤霜)、生物医学用品(如用于整容手术的植入物)和家用产品。PDMS 具有一个聚合物的分子结构,包含硅原子和氧原子交替构成的骨架以及与硅原子相连的有机侧基(如甲基、苯基或乙烯基),基本结构是 $[SiO(CH_3)_2]_n$,其中 n 表示重复单体的数量,数值范围从零到几千。通过调整—SiO—链长度、功能性侧基和分子链之间的交联,有机硅可用来合成多种具有不同化学性质和性能特征的材料。此外,PDMS 分子结构既有线性的,也有环形的,后者在环境中难以降解。通常 PDMS 是难溶性化合物,但其具有高亨利常数,易挥发(Hamelink et al.,1996)。90% 的 PDMS 蒸发到大气中,主要分解成羟基自由基(Dewil et al.,2007);仅有 10% 的 PDMS 进入污水处理厂(Allen et al.,1997)。虽然进入污水处理厂的 PDMS 也会挥发,但 PDMS 的水-沉积物分配系数(K_d 范围为 $2.2 \sim 5.0$)表明其对污泥的吸附能力较强,因而并不易挥发(David et al.,2000;Whelan et al.,2009)。利用污泥作为肥料是 PDMS 进入环境(尤其土壤)的主要途径。虽然在废水中检测到的 PDMS 浓度较低,但它们是水生环境的长期污染源。PDMS 一旦被释放到水中,往往会与颗粒有机物相结合/溶解并沉降下来。因此,最高浓度的 PDMS 通常出现在废水排放区附近的沉积物中(Sparham et al.,2011)。

1.3 分析方法

1.3.1 金属

许多金属在海水中浓度较低，通常存在于由盐和其他物质形成的复杂基质（有机和无机）中。此外，在取样、保存和前处理过程中，外来污染会增加检测难度。基于上述原因，学者对1975年以前的海水金属检测数据存疑。超净技术的应用使我们能够有效收集高价值的海洋金属数据。例如，微量元素和同位素生物地球化学循环（GEO-TRACES）计划已经成功地获得了有关金属生物地球化学循环的数据资料。当前获取海洋痕量金属数据的过程都需要防止污染问题。为避免采样过程中微量元素的污染，研究人员会设计特定的取样装置，并使用特定的塑料器皿（Kremling，2002）。有关超净技术的资料、处理方法和其他步骤的具体信息不在本章细述，可从其他资料获得（Tovar-Sánchez，2012）。海水中金属元素的检测方法主要有石墨炉原子吸收光谱法（graphite furnace atomic absorption spectrometry，GF-AAS）、电感耦合等离子体原子发射光谱法（inductively coupled plasma-atomic emission，ICP-AE）、电感耦合等离子体质谱法（ICP-mass spectrometry，ICP-MS）、通过灵敏检测技术（分光光度计或荧光计）的流动分析体系法（如流动注射分析法、顺序注射分析法）、系统的灵敏检测技术顺序注射分析法（荧光分光光度计或荧光计）、伏安法（阳极溶出伏安法，cationic stripping voltammetry，ASV；阴极溶出伏安法，cationic stripping voltammetry，CSV）。ICP-AE和ICP-MS的多元素特性使其成为最常用的两种分析方法。但这两种方法不能直接分析海水样品，即使检测稀释后的海水样品，检出限仍不理想。虽然在新型ICP-MS设备中使用气体稀释法是一种很有前景的替代方案，但至今为止，它在超痕量水平上的效果仍不能令人满意。为解决这些问题，可以增加样品预浓缩步骤和采用多种选择性萃取技术。例如，使用螯合剂（如二硫代氨基甲酸盐或螯合阳离子树脂Chelex 100）进行液相萃取（Kremling，2002）。

1.3.2 持久性有机污染物

1.3.2.1 多环芳烃

由于多环芳烃（PAHs）具有高疏水性，所以PAHs主要在海洋环境中的生物体、颗粒物和沉积物中积累。因水体中PAHs含量较低，所以在分析水样之前必须对样品进行浓缩（Li et al.，2013）。不同的萃取和预浓缩技术已被广泛应用，如液液萃取（liquid-liquid extraction，LLE）、搅拌棒吸附萃取（stir bar sorptive extraction，SBSE）、固相萃取（solid phase extraction，SPE）和固相微萃取（solid phase microextraction，SPME）

(Kruger et al., 2011; Li et al., 2015; Robles-Molina et al., 2013)。此外，抓斗式采样体积限制是海洋 PAHs 研究的限制因素之一。目前通过被动采样器的应用，已解决了这一问题。被动采样器已被应用于沿海地区溶解态 PAHs 的时空变化趋势研究（Alvarez et al., 2014）。沉积物和生物中 PAHs 的萃取，往往采用非极性溶剂和各种不同的方法，如索氏提取法（De Boer & Law, 2003）、超临界流体萃取（supercritical fluid extraction, SFE）（Berg et al., 1999）、加压液体萃取（pressurized liquid extraction, PLE；也称为加速溶剂提取, accelerated solvent extraction, ASE）（Burkhardt et al., 2005）、亚临界水萃取（subcritical water extraction, SWE）（Ramos et al., 2002）、微波辅助萃取（microwave-assisted extraction, MAE）（Banjoo & Nelson, 2005; Pena et al., 2006）、超声波辅助提取（ultrasonic-assisted extraction, USE）（Banjoo & Nelson, 2005）& QuEChERS（quick, easy, cheap, effective, rugged and Safe）（Johnson, 2012）等。研究表明，超声波浴的使用在以上方法中占主导地位（Navarro et al., 2009）。在为 PAHs 萃取开发的数种清洗程序中，基于 SPE 的清洗程序最为方便；其中，二氧化硅（Pena et al., 2006）和弗罗里硅土（Burkhardt et al., 2005; Navarro et al., 2009）的清洗效果获得了广泛认可。相较于液相色谱（liquid chromatography, LC），气相色谱（gas chromatography, GC）通常是分离、鉴定和定量分析 PAHs 的首选方法。这主要是因为 GC 通常比 LC 具有更高选择性、分辨效能和高灵敏度（Poster et al., 2006）。

1.3.2.2 表面活性剂

商业化的表面活性剂通常为复杂混合物，由数百种具有不同物理化学性质的不同异构体、同系物和/或乙氧基-酶组成。因此，分析方法的差异主要取决于是否分析表面活性剂的某一类型（阳离子、阴离子、非离子）或不同类别化合物的总含量。第一种情况的操作相对简单，只需根据基质类型进行固相萃取或液相萃取（Roslan et al., 2010）。第二种情况则需将萃取后的分析物再进行分离和/或预浓缩。对于水样的处理，通常在水样中加入生物杀灭剂（如甲醛）后放入冰箱保存，从而最大限度地抑制水样中生物的降解（Lara-Martin et al., 2006a），并在短时间内（48 h）进行分析。液液萃取技术曾被广泛用于表面活性剂的分析。目前，十八烷基硅胶（C18）固相萃取技术是从液体和固体样品中提取和纯化表面活性剂的最常用技术。C18 可通过与阴离子表面活性剂的强阴离子交换（strong anionic exchange, SAX）而实现结合（Leon et al., 2000; Matthijs et al., 1999）。对于非离子表面活性剂，可使用石墨化炭黑（graphitized black carbon, GBC）（Houde et al., 2002）和硅胶柱（C2 – C18）（Petrovic et al., 2001），通过强阳离子交换（strong cationic exchange, SCX）和强阴离子交换实现结合（Dunphy et al., 2001）。目前大多数研究使用 C18 和 GBC。因为它们可在一个操作流程中同时提取和分离多种表面活性剂，包括阴离子、非离子及其代谢产物。对于固体样本（如沉积物和生物体），通常先用加热器干燥或冷冻干燥（Alvarez-Munoz et al., 2004; Lara-Martin et al., 2006a），然后使用不同的方法提取目标物，如索氏提取（Lara-Martin et al., 2006b; Saez et al., 2000）、超声波萃取（Lara-Martin et al., 2011; Versteeg & Rawlings, 2003）、PLE（Alvarez-Munoz et al., 2007; Lara-Martin et al., 2006a）、微波辅助溶剂萃取（microwave-assisted solvent extraction, MASE）（Croce et al., 2003）、基质固相分散萃取（matrix solid phase dispersion，MSPD，适用于从生物体中提取表面活性剂）（Tolls et al.，1999）。关于溶剂，首选甲醇来萃取 LAS、AEO 和 AES 及其主要降解产物；而对于 APEO 及其降解产物，则用其他极性较小的溶剂（如己烷或二氯甲烷）取代甲醇（Shang et al.，1999）。净化阶段的方法与水样处理相似，由表面活性剂的类型决定。仪器分析主要通过液相或气相色谱（衍生化后）与质谱联用进行（Lara-Martin et al.，2006a；Reiser et al.，1997）。但也使用其他检测设备，如紫外（UV）或荧光检测器（fluorescence detector，FLD）（Alvarez-Munoz et al.，2004；Croce et al.，2003）。

1.3.2.3 多氯联苯

多氯联苯（PCBs）具有高疏水性，不存于水体。因此，对海洋环境中 PCBs 的研究主要集中于沉积物和生物体。分析过程主要包括对冷冻干燥后的样品进行提取和净化处理。此外，需要使用内标物（internal standards，IS）来检测是否准确提取了目标化合物。IS 通常是同位素标记化合物，属于研究所关注的一类污染物。在分析时，可以使用自然界中未检测到的 PCBs 同系物作为 IS。沉积物中 PCBs 通常使用固液萃取法进行提取。首先使用正己烷和二氯甲烷（体积比 1∶1）进行索氏提取 16 h（Castells et al.，2008）。尽管索氏提取法仍在使用，但更快的提取方法如 PLE 等也已被应用。PLE 主要基于高压(1000～2000 psi)和高温（通常高于溶剂蒸气温度）提取目标分析物；以二氯甲烷作为溶剂萃取剂，用于分析沉积物中 PCBs（Barakat et al.，2013）。随后使用氧化铝－二氧化硅或弗罗里硅土进行清洗，然后用戊烷和二氯甲烷（体积比 1∶1）洗脱（Barakat et al.，2013）。最后在用仪器分析之前，通过氮吹和加热将洗脱液还原。

生物样本的提取方法和沉积物相似。例如，在 100 ℃ 和 1750 psi（1 kPa = 0.145 psi）压力下使用 PLE、戊烷和二氯甲烷（体积比 1∶1）提取 PCBs（Solaun et al.，2015）。萃取过程中使用弗罗里硅土和 Na_2SO_4 对样本进行预净化以及水分去除，通过氮吹对样品进行预浓缩。在使用凝胶渗透色谱柱（gel permeation chromatography，GPC）净化样品前，用二氯甲烷复溶，并过 0.45 mm 滤膜。随后将洗脱液旋蒸，并用异辛烷溶解，使用 Na_2SO_4 净化，再进行离心，最后分析有机相。PCBs 的仪器分析基于气相色谱－电子捕获检测器（electron capture detector，ECD）或质谱分析仪（mass spectrometer，MS）。

1.3.2.4 农药

20 世纪 70 年代，人们已开始对农药残留进行分析研究。但因基质中不同类别化合物（物理化学性质具有较大差异）的含量较低，所以对农药残留的检测分析至今仍然是一个挑战。分离和浓缩农药残留是样品处理的重要步骤，往往占用大部分分析时间。固相萃取（SPE）通常为水样处理的首选方法（Hernandez et al.，2012；Masia et al.，2013a，2013b）。但其他一些方法如 LLE（液液萃取）（Jiang et al.，2013）和被动取样技术（Martínez Bueno et al.，2009）也被应用。目前已发现几种类型的 SPE 吸附剂适用于从水样中提取农药，如 GCB、Oasis HLB、Strata-X、Strata C18。对于高极性农药的提取，C18 键合硅胶和苯乙烯－二乙烯基苯共聚物则最为常用（Primel et al.，2012）。

QuEChERS 是从生物或沉积物等固体样品中提取农药的最常用技术（Andreu & Pico，2012；Bruzzoniti et al.，2014）。QuEChERS 根据农药的类型（pH 的影响、降解度）和分析目的，为样品提供了一个多功能的处理平台（Masia et al.，2014）。其他非水基质中农药的研究是基于单独或组合形式技术（Masia et al.，2014）。这些替代技术包括 LLE、膜辅助 MSPD、涡流色谱技术、MAE、PLE、SPME、SBSE、SPE、高通量平面固相萃取（HTpSPE）或样品稀释和直接进样（详情可见 Masia et al.，2014）。

如今 SPE 和/或 QuEChERS 与 LC-MS 结合应用是分析水或固体样品中农药的最常用技术。农药极性强、热稳定性差和不易挥发，所以气相色谱法检测效果较差（Masia et al.，2014）。因此，液相色谱法优于气相色谱法。在线 SPE 技术被认为是传统离线 SPE-LC-MS 检测的理想替代方法。它提供一种自动化的样品预处理方式，可直接自动分析批量样品。

1.3.2.5 二噁英

二噁英具有高疏水性，不存于水体。因此，对二噁英的分析主要以沉积物和生物体为对象。对于沉积物中二噁英和 DL-PCBs 的提取，主要以甲苯或其他适宜溶剂组合为提取液，通过索氏提取法抽提 24 h 完成（Castro-Jiménez et al.，2013）。然后通过酸消化，再用硅胶色谱柱进行纯化。随后使用基于连续多层二氧化硅膜、碱性氧化铝和 PX－21 碳吸附剂的 PowerPrep 自动化系统净化样品。生物体二噁英的提取则以甲苯和环己烷（体积比 1∶1）为提取液，通过索氏提取法抽提 24 h 完成（Castro-Jiménez et al.，2013；Parera et al.，2013）。净化的方法与上述沉积物的处理方法相似，也可以采用 Parera 等人所述的更有效的净化方法（Parera et al.，2013）。随后对索氏提取法抽提的溶液进行旋蒸，使用烘箱 105 ℃ 烘干，以消除用于重量法测定脂肪含量的溶剂残留；将脂肪残余物溶解在正己烷中，再使用经硫酸（质量分数 44%）改性后的硅胶柱去除；再使用 PowerPrep 自动化系统进行净化，通过旋蒸和氮吹将洗脱液浓缩至近干，最后使用壬烷复溶（Parera et al.，2013）。二噁英的分析仪器主要是气相色谱－高分辨率质谱仪（GC-high-resolution mass spectrometer，GC-HRMS）（Castro-Jiménnez et al.，2013；Cloutier et al.，2014；Parera et al.，2013）。

1.3.3 新兴有机污染物

1.3.3.1 全氟和多氟烷基物质

PFASs 的分析面临实验材料和仪器分析系统交叉污染的问题，因此样品处理分析过程中需要空白提取样品作为对照。通常用玻璃纤维过滤器过滤水样以去除颗粒物（Ahrens et al.，2009，2010a，2010b）。加入内标物后，通过 SPE 进行提取。根据 PFAS 的特性，通常使用配有阴离子交换固定相的小柱。随后将样品洗脱液氮吹至近干，用甲醇复溶，最后进行高效液相色谱分析。

对于沉积物的分析，先用稀酸进行固－液萃取，然后在甲醇中加入氢氧化铵进行超声波浴萃取（Long et al.，2013）。其他研究已证明，用甲醇进行超声波辅助萃取 1 h 是一种有效的提取方法（Llorca et al.，2014）。使用阴离子 SPE 处理或高纯碳粉净化离心后的上清液（Llorca et al.，2014；Long et al.，2013）。洗脱液的处理方法与水样相似。

对于生物样品来说，脂质沉淀是第一步必要处理步骤。例如，使用甲醇（NaOH，10 mmol/L）（Llorca，2012）。随后样品通过轨道消化器提取 1 h 后离心。所得上清液使用阴离子 SPE 纯化。洗脱液的处理与上述水样及沉积物样处理方法相同。仪器分析基于液相串联质谱联用（LC-MS/MS）。不过，目前也采用 GC-MS/MS 联用技术对挥发性 PFASs 进行分析。

1.3.3.2 溴化阻燃剂

PBBs、PBDEs、TBBPA & HBCD 具有疏水性，在水体中浓度较低。因此，通常需要大体积样品（最高 1000 mL）以确保对这些化合物进行准确的检测。在提取过程中通常使用非极性溶剂。液液萃取（LLE）处理需要大量的溶剂，所以通常被固相萃取（SPE）所取代（Covaci et al.，2007）。C18 可同时保留非极性和中等极性化合物，是目前应用最广泛的吸附剂，尤其对 PBDEs 适用（Fulara & Czaplicka，2012）。其他相如聚苯乙烯 – 二乙烯基苯共聚物已成功应用于中等极性和水溶性的新型 BFRs 提取（Lopez et al.，2009）。此外，固相微萃取（SPME）（Polo et al.，2006）、搅拌棒吸附萃取（SBSE）（Quintana et al.，2007）、分散液相微萃取（dispersive liquid-liquid microextraction，DLLME）（Li et al.，2007）和浊点萃取法（cloud point extraction，CPE）（Fontana et al.，2009）等其他技术也被应用于水样的 BFRs 测定。在对沉积物或生物等固体样品进行萃取前，通常将样品与硫酸钠混合或冷冻干燥，以除去样本中水分。溴化阻燃剂的提取可以通过索氏抽提、加压流体萃取（PLE）（Lacorte et al.，2010）、SPME（Montes et al.，2010）和微波固相分散（Covaci et al.，2007）等方法完成。由于提取技术的选择性较少，且样品基质复杂，因此必须对样品进行纯化处理。沉积物样品的净化过程主要是脱硫，生物样品则需要去除脂肪。最佳的方法是用凝胶色谱法和用硅胶、氧化铝或不同活化程度的弗罗里硅土做固定相的吸附色谱法分馏（Fulara & Czaplicka，2012）。GC-MS 和 GC-ECD 是最常用的仪器分析技术（Fontana et al.，2009；Rezaee et al.，2010）。当然，LC-MS 也有应用（Quintana et al.，2007）。

1.3.3.3 内分泌干扰物

如前文所述，EDCs 的分子异质性和物理化学性质差异使其分析方法的开发变得十分困难。尽管已实现对多种残留的分析方法的应用，但大多数已发表的研究都只是针对少量化合物（Jakimska et al.，2013）。通常在样品（水、沉积物或生物样品）中添加内标物是前处理分析的第一步。C18、—NH$_2$ 或—CN 改性二氧化硅固相萃取是水样提取最常用的方法（Petrovic et al.，2002a）。通过温和氮吹将洗脱液浓缩至近干，加入适当溶剂复溶后，使用液相色谱（LC）或气相色谱（GC）进行分析（Petrovic et al.，2002a）。对于沉积物和生物等固体样品，在提取前通常需先完成冷冻干燥预处理后，再进行下一步分析。沉积物的提取采用超声波萃取或加压流体萃取（PLE），再用固 – 液吸附色谱的层析柱（弗罗里硅土、氧化铝和不同类型的碳等），或者 C18、—NH$_2$ 或—CN 改性二氧化硅固相萃取完成净化（Petrovic et al.，2002a）。对于冻干生物样品的提取可采用不同的技术，如 PLE（Al-Ansari et al.，2010；Rudel et al.，2013；Schmitz-

Afonso et al.，2003）、超声波萃取（Pojana et al.，2007）、高速溶剂萃取（High Speed Solvent Extraction，HSSE）（Kim et al.，2011）和微波辅助萃取（MAE）（Liu et al.，2011）。这些方法都需要消耗大量的溶剂和时间。其他更快捷的技术，如 QuEChERS 也已成功应用于生物样品处理（Jakimska et al.，2013）。目前最常用的净化方法包括弗罗里硅土 SPE 吸附柱（沉积物样品处理）（Pojana et al.，2007）、C18 固相萃取小柱（Schmitz-Afonso et al.，2003）、GPC（Navarro et al.，2010）等。样品净化后的处理与水样处理方法相似。固体样本的仪器分析是基于 LC-MS/MS 或 GC-MS，具体取决于所测定 EDCs 的性质（Navarro et al.，2010；Petrovic et al.，2002a；Schmitz-Afonso et al.，2003）。

1.3.3.4 药品和个人护理品

多残留分析方法是检测海洋环境中药物和个人护理产品（pharmaceuticals and personal care products，PPCP）最常用的方法（Rodríguez-Mozaz et al.，2015）。这种方法可同时分析具有不同治疗目的的化合物（以药物为例），如抗生素、精神药物、镇痛药/消炎药、镇定剂、激素、β 受体阻滞剂、利尿剂等。由于目标化合物的理化性质不同，需要协调实验条件以实现所有目标化合物回收率在可接受范围。此外，通常使用含同位素标记的标准品来避免不准确定量和补偿基质效应。对于海水样本，虽然以被动取样器作为同时取样和提取的方法也被使用（Martínez Bueno et al.，2009；Munaron et al.，2012；Tertuliani et al.，2008），但固相萃取净化、液相色谱与质谱联用是首选的分析方法（Borecka et al.，2015；Gros et al.，2009，2013；Jiang et al.，2014；Loos et al.，2013；Yang et al.，2011）。对于沉积物和生物样本，第一步是从基质中通过不同的方法提取目标分析物并加入内标物。由于操作简单，基于有机溶剂的振荡、涡旋和超声波萃取都是最常见的提取方法（Klosterhaus et al.，2013；Kwon et al.，2009；Na et al.，2013）。加压流体萃取和微波辅助萃取也被应用于沉积物和生物样本的分析（Alvarez-Muñoz et al.，2015a，2015b；Berrada et al.，2008；Hibberd et al.，2009；Jelic et al.，2009；McEneff et al.，2013；Wille et al.，2011）。虽然采用 Oasis 混合型阴离子交换反相吸附剂小柱（Oasis HLB）已成为固相萃取（SPE）的首要选择（Azzouz et al.，2011；Dodder et al.，2014；Kwon et al.，2009；Samanidou & Evaggelopoulou 2007），但有些研究人员也将 QuEChERS 用于生物样本的提取和净化（Martinez Bueno et al.，2013，2014；Villar-Pulido et al.，2011）。有时生物样品需要使用 GPC 完成更进一步的净化，如富含脂肪的样品（Huerta et al.，2013；Tanou et al.，2014）。现有以沉积物为对象的研究中，都是以液相色谱与质谱联用作为药品和个人护理品污染物分析的首选方法（Klosterhau et al.，2013；Kwon et al.，2009）。对于生物样本，液相色谱与质谱联用也是最常用的技术。但也有使用其他类型的检测器，如荧光检测器（FLD）、紫外吸收检测器（UV）或二极管阵列检测器（DAD）（Cueva-Mestanza et al.，2008；Fernandez-Torres et al.，2011；He et al.，2012）。其他如酶联免疫吸附试验（enzyme linked immunosorbent assay，ELISA）和时间分辨荧光免疫分析（time-resolved fluoroimmunoassay，TR-FIA）技术也被用于生物样本的分析（Chafer-Pericas et al.，2010a，2010b）。

1.3.3.5 海洋生物毒素

因其分子结构特异性，海洋毒素难以被人工合成，导致了标准品（尤其是亲脂性毒素）的缺乏，从而阻碍了对海水、沉积物和生物样本中海洋毒素的分析操作。在上述背景下，必须对藻类中生物大分子进行纯化。目前官方的生物样品（被人类消费的海产食品是海洋生物毒素研究最有代表性的基质）分析方法是基于小鼠或大鼠的体内实验（Gerssen et al.，2010）。但新的体外试验、生化（免疫化学）和化学方法正在兴起（FAO，2004；Gerssen et al.，2010）。根据分子类型和结构，化学分析方法主要采用液相色谱-串联质谱（LC-MS/MS）或液相色谱（LC）荧光分析仪。

1.3.3.6 纳米材料

对于纳米材料，应分为两种不同类型进行分析，即定量分析（用于浓度测定）和定性分析（如粒径大小）。本节内容主要讨论水、沉积物和生物体中纳米材料的定量分析。与其他污染物分析一样，第一步是加入内标物质。

虽然水体中检测不到游离的 CNM，但 CNM 能在水中形成悬浮的聚集物。水样中 CNM 的提取遵循不同的策略：①基于甲苯的液体萃取；②以甲苯为洗脱剂的 C18 柱固相萃取（Xiao et al.，2011）；③基于甲苯的过滤和颗粒物提取（尽管尚未在海水中进行试验）（Sanchís et al.，2012）。根据流动相的初始条件进行分析，对甲苯悬浮液进行旋蒸浓缩，随后在甲苯或甲醇中复溶。

CNMs 的定量分析主要基于液相色谱-紫外线光谱（LC-UV/Visible）和液相色谱-串联质谱（LC-MS/MS）（现在研究者开始使用高分辨质谱法 HRMS），但也可以使用透射电子显微镜（transmission electron microscopy，TEM）和纳米颗粒跟踪分析（nanoparticle tracking analysis，NTA）。对于液相色谱，最常用的色谱分离柱是 C18 键合硅胶柱和丙基酰胺键合胶柱。如前所述，可使用紫外线光谱仪（UV-Visible）或质谱仪（MS）进行分析。与电喷雾电离（electrospray ionization，ESI）源相比，大气压化学电离（atmospheric pressure chemical ionization，APCI）和大气压光电离（atmospheric pressure photoionization，APPI）更有效（Astefanei et al.，2015；Núñez et al.，2012；Sanchís，2015）。

1.3.3.7 聚二甲基硅氧烷

不同环境基质中硅氧烷的分析方法仍十分有限。二甲基硅氧烷在水体中浓度较低，且样品处理过程中存在高污染风险（因此类化合物具有挥发性），使此类化合物分析方法的开发具有较大挑战性。水样分析主要针对地表水、废水和河水。对水样萃取可采用不同的方法，如吹扫捕集（Kaj et al.，2005）、顶空萃取（Sparham et al.，2008）、顶空固相微萃取（Companioni-Damas et al.，2012）、液液萃取（Sanchis et al.，2013）、膜辅助溶剂萃取、超声辅助分散液相微萃取等方法（Cortada et al.，2014）。大多数关于沉积物的研究往往采用固液萃取技术（Sparham et al.，2011；Warner et al.，2010；Zhang et al.，2011），其他技术如 PLE 和超声波辅助有机溶剂萃取（ultrasound-assisted solvent extraction，USASE）也被应用于此类研究（Sanchís et al.，2013；Sparham et al.，

2011）。鱼类组织的提取可采用固相支持液液萃取（SLE）（Kaj et al., 2005；Warner et al., 2010）、USASE（Sanchis et al., 2016）、吹扫捕集法（Kierkegaard et al., 2010）。大多数研究都是以正己烷作为萃取溶剂，也有部分研究使用戊烷（Wang et al., 2013）和乙酸乙酯（Sparham et al., 2011）。为避免因化合物的蒸发而影响萃取，通常不对样本进行净化处理。而对于复杂的基质（如鱼类组织），通过简单的离心即可澄清正己烷萃取液（Sanchis et al., 2016）。由于硅氧烷低分子量和低极性的特点，气相色谱-质谱（GC-MS）分析已成为大多数分析方法中的首选技术。但也有少量的研究采用气相色谱（GC）与其他检测器（如火焰离子化检测器）耦合进行分析（Huppmann et al., 1996；Dewil et al., 2007）。

1.4 海洋环境中的赋存

1.4.1 金属

净化技术的应用提高了人们对海水中金属分布和行为的认识。海水中的金属浓度范围在 nM（10^{-9} mol/L）到 pM（10^{-12} mol/L）数量级之间，Au 的含量则是 fM（10^{-15} mol/L）数量级（Donat, Bruland, 1995）。2014 年，痕量元素和同位素海洋生物地球化学循环（GEOTRACES）研究小组发布了相关数据，汇编了 796 个站位的数据，包括水文参数、溶解态和颗粒微量元素（如 Al、Ba、Cd、Cu、Fe、Mn、Mo、Ni、Pb）、稀土元素（REEs）、稳定和放射性同位素（The GEOTRACES group, 2015）。海水中金属的水平和垂直分布是其去除和补给之间平衡后的结果。微量元素的垂直分布可分为：营养盐型、保守型和清除型，当然也可能发生混合。营养型分布由生物限制性元素分布表示，呈现出表层水低、深层水高的特点，且受生物代谢过程调控。保守行为型元素呈现出与盐度有关的分布特征，不呈梯度趋势，随着深度均匀变化。清除型元素分布呈垂直分布，溶解浓度随深度的增加而降低，如 Mn、Pb & Co。这类元素的去除是由沉淀或吸附过程介导。Nozaki（1997）收集了不同元素的垂直剖面分布数据。海水中的金属分布不均匀，沿海和河口地区的金属浓度比远海高。历史上，研究发现市政、工业和污水排放的水域都被受金属污染，如纽约湾、波士顿港和泰晤士河河口等（Kennish, 1997）。Luoma & Rainbow（2008）综合分析了英国河口、半封闭海湾、沿海水域和未受干扰沿海水域的溶解态金属浓度，结果显示 Cu 的浓度范围为 0.1～4.64 μg/L、Zn 为 0.27～5.1 μg/L、Cd 为 0.003～0.19 μg/L、Pb 为 0.004～0.269 μg/L，河口和海岸地区的浓度差异不超出 10～20 倍（Luoma & Rainbow, 2008）。全球范围内可获得的有关金属溶解浓度的信息是有限的。因此，制定以减轻干扰和改善水质为目的的评估战略，需要对水体污染物的生物地球化学循环和归宿进行长期研究（SañudóWihelmy et al., 2004；TovaSánchez, 2012）。沿海水域中的金属和颗粒物的去除是河口咸淡水交汇作用所导致的。所以这些区域是去除和捕获沉积物中金属的重要场所。溶解态金属（孔隙水和上覆水中）对颗粒物（沉积或悬浮）具有很高的亲和力。这种亲和力可以用分配系数（K_d）来评估。虽然水环境理化条件的变化会导致沉积物中金属的释放，增加了金属的生物可利用度。但是颗粒物中金属浓度比溶解态金属浓度高几个数量级。这一事实表明沉积物起着金属汇的作用。当比较不同站位的金属含量时，应考虑归一化标准。因为影响金属含量的颗粒物粒径大小是一个混淆因素。较细的颗粒具有更高的单位质量比表面积，增加了金属结合位点。归一化方法包括化学萃取、细粒沉积物的物理分离、对粒度敏感的天然成分归一化和统计技术（Luom & Rainbow，2008）。Kennish（2000）总结了其选定的美国河口、沿海海洋系统以及 19 个英国河口地区的金属浓度，用干重表示金属含量（Kennish，2000）。一些生态系统的金属含量表现出高水平，如美国旧金山湾和英国的 Restronguet Creek。然而该研究中沉积物重金属总含量不能提供生物利用度信息。可以考虑使用不同的方法以获得生物利用度数据，但这不在本章的讨论范围内，更多内容请参阅 Parsons et al.（2007）。

水体或沉积物中的生物可以通过不同途径（如水和悬浮物）蓄积金属。这种机制被称为生物富集。生物的生理机能和生存方式等都可以调节这一过程。本书将在第 4 章对这一主题进行深入论述。

1.4.2　持久性有机污染物

1.4.2.1　多环芳烃

如前所述，PAHs 在水中的溶解度很低。因此溶解态的 PAHs 浓度较低，范围为 pg/L 至 ng/L 数量级。低分子量的 PAHs 更易溶于水。水中最常检测到的 PAHs 大多数为 2～4 个苯环，如芘烯、萘和苯蒽等。中国东海和南海 PAHs 总含量范围为 30.40～120.29 ng/L，多为常见的 PAHs 化合物类型（Ren et al.，2010）。PAHs 的疏水性导致沉积物成为该污染物的"汇"。沉积物中 PAHs 的检出浓度高达数千 ng/g 水平。Baumard et al.（1998）研究表明地中海巴塞罗那港采集的沉积物中 PAHs 含量高达 8400 ng/g（干重）（Baumard et al.，1998）。Leon et al.（2014）研究发现在上述同一地区沉积物中 PAHs 的总浓度为 1006 ng/g。PAHs 具有亲脂性，易富集在水生生物体内，尤其是软体动物中，因为脊椎动物具有更强的 PAHs 代谢和排泄能力（Meador et al.，1995）。例如，地中海西部地区贻贝中 PAHs 的含量高达 80 ng/g 干重（Baumard et al.，1998），而鲻鱼中 PAHs 含量为 40 ng/g（干重）（Leon et al.，2014）。

1.4.2.2　表面活性剂

表面活性剂及其降解代谢产物在各种类型的环境中均有检出。它们广泛存在于地表水中，含量范围通常为 1 ng/L 到数百 μg/L 之间。检出最多的化合物分别是作为阴离子和非离子表面活性剂的 LAS & NPEOs。意大利威尼斯潟湖中检出的 LAS 浓度范围为 1～296 μg/L（Stalmans et al.，1991）；西班牙塔拉戈纳河口 NPEOs 含量为 1～37 μg/L（Petrovic et al.，2002b）。因为表面活性剂具有中等至高等的吸附能力，所以沉积物中表面活性剂的含量比水体高出几个数量级。沉积物中 LAS 和 NPEO$_s$ 的浓度范围为低

mg/kg 到数百 mg/kg（Traverso-Soto et al.，2012）。表面活性剂在海洋生物中的赋存也有报道，尤其是那些表现出内分泌干扰活性的化合物（如壬基酚）。台湾沿海地区的牡蛎和蜗牛中烷基酚浓度范围为 20～5190 ng/g（Chin-Yuan et al.，2006）。

1.4.2.3 多氯联苯

海洋环境中 PCBs 的赋存特点是易富集在生物体内，尤其是鱼类。PCBs 的生物富集特性可导致该类污染物通过海洋食物链产生生物放大作用。Lu et al.（2014）对坎伯兰湾的格陵兰鲨鱼及其北极海洋食物网中手性多氯联苯的研究证实了这个观点。PCBs 在浮游动物中的含量为 153 ng/g lw（脂重）（McKinney et al.，2012），鲱鱼为 109～561 ng/g lw（McKinney et al.，2012），毛鳞鱼为 438 ng/g lw（Lu et al.，2014），杜父鱼为 867 ng/g lw（McKinney et al.，2012），格陵兰比目鱼为 221 ng/g lw（McKinney et al.，2012），鲑鱼为 76～4810 ng/g lw（McKinney et al.，2012）；同一地区鲨鱼肝脏中则高达 4600 ng/g lw（Lu et al.，2014）。通常 PCBs 同系物具有高亲脂性，肝脏或血浆中 PCBs 浓度高于肌肉（Lu et al.，2014）。埃及海岸地区沉积物中 PCBs 含量为 2.29～377 ng/g dw（干重）（Barakat et al.，2013）。地中海沿岸沉积物的 PCBs 含量为 2.33～44.00 ng/g dw；西班牙贝索斯河近海区域沉积物的 PCBs 为 22.34～37.74 ng/g dw（Castells et al.，2008）。

1.4.2.4 杀虫剂

因为主要的农业地区位于沿海平原和河谷，所以杀虫剂存在于沿海环境中并不让人意外。尽管一些杀虫剂（如 DDT）在全球范围内已被禁止使用。但由于它们很难降解，因此仍存在于世界各地的海洋环境中。例如，一项对全球环航所收集的海洋浮油研究表明（Menzies et al.，2013），巴拿马运河 Gatun Locks 水域 DDT 及其代谢物的含量高达 96.6 ng/L，东萨摩亚的帕果帕果氯丹和相关化合物含量高达 285.1 ng/L，法属社会群岛莫雷阿岛库克斯湾的氯苯含量高达 1213.1 ng/L。河口地区的沉积物中也检出了 ng/g 级水平的杀虫剂。Zheng et al.（2016）研究发现中国九龙河口地区沉积物的主要污染物为 DDT、DDD 和 DDE 等有机氯农药，其中 DDE 含量高达 311 ng/g dw。虽然表层水中杀虫剂的含量通常低于野生动物的致死暴露浓度，但杀虫剂可积累在水生生物体内，从而引起亚致死等不良反应。例如，同样在 Zheng et al.（2016）的研究中，海水的腐霉利为 3094 ng/L。进一步的实验研究证明，这种化合物会干扰河口鱼类中卵黄蛋白原的表达（Zheng et al.，2016）。生物体中杀虫剂的含量水平为 ng/g 级。Zhou et al.（2014）对中国长江三角洲沿海地区采集的 11 种软体动物进行分析，得出 DDTs、六氯环己烷和氯丹的含量分别为 6.22～398.19 ng/g（湿重）、0.66～7.11 ng/g 和 0.14～4.08 ng/g（Zhou et al.，2014）。

1.4.2.5 二噁英

二噁英在海洋环境中的赋存受人类活动影响。例如，对埃布罗河三角洲地区（西班牙地中海海岸）的贻贝、地毯蛤、骨螺、康吉鳗、金头鲷、比目鱼和沙丁鱼等生物进行研究，得出总 PCDD/F 的含量为 0.29～2.29 pg/g ww，总 DL-PCBs 为 24.6～

5503 pg/g ww（Parera et al.，2013）。在地中海布拉内斯海底峡谷（西北地中海）的甲壳类和鱼类的研究中，得出总 PCDD/F 的含量为 110～795 pg/g lw（Castro-Jiménez et al.，2013）。时间趋势的研究表明，生物体内二噁英的浓度呈轻微下降特征（Parera et al.，2013）。

在沉积物相关的研究中，布拉内斯海底峡谷（西北地中海）深海沉积物中二噁英含量为 102～680 pg/g dw（Castro-Jiménez et al.，2013）。

1.4.3 新兴有机污染物

1.4.3.1 全氟和多氟烷基物质

PFASs 具有远距离迁移的特点，广泛存在于海洋环境中，在北极和南极大陆等偏远地区均有检出。Ahrens et al.（2010b）对北海、波罗的海和挪威海域表层水中最难降解的 PFASs（PFOA & PFOS）、C6 羧酸和 C8 磺胺（FOSA）进行定量分析，发现这些污染物的含量范围为 0.02～6.16 ng/L。在另一项研究中，南极绕极流区域 PFOS 检出的浓度范围为 11～51 pg/L（Ahrens et al.，2010a）。

对海洋沉积物中 PFASs 的分析表明，西北太平洋（美国华盛顿普吉湾）沉积物中 C4 羧酸、PFOS & FOSA 的含量范围为 0.13～1.50 ng/g（Long et al.，2013）；高浓度的 PFOA、PFOS、C10 羧酸和磺酸盐化合物在地中海的希腊附近水域沉积物中被检出，范围为 8.2～146 ng/kg（Llorca et al.，2014）；地中海西部克雷乌斯角沉积物中 PFOS、PFOA、C4、C6 & C9 羧酸的浓度范围为 120～11650 ng/kg（Sanchez-Vidal et al.，2015）。

关于海洋生物的研究，如对北卡罗来纳州历史上 PFAS 输入源附近的蓝鳃太阳鱼（*Lepomis macrochirus*）肌肉的研究发现，鱼体内含有高浓度的 PFOS，达 2.08～275 ng/g dw（Delinsky et al.，2009）。对西班牙产品市场中鳕鱼籽、旗鱼、乌贼、幼鳕和凤尾鱼的 PFASs 含量进行分析，结果表明 C5、C9 & C10 羧酸、C4 & C10 磺酸盐、PFOA & PFOS 的浓度范围为 0.09～50 ng/g dw（Llorca et al.，2009）。对于欧洲海产品市场的相似研究发现，双壳类、鲱鱼、石首鱼、白鲑、黑鳕、鲑鱼、鳕鱼和金枪鱼中 C4、C5、C6 & C9 羧酸、PFOA、PFOS & FOSA 的浓度范围为 0.09～54 ng/g dw（Llorca，2012）。

1.4.3.2 溴化阻燃剂

BFRs 具有远距离大气迁移的潜力，在远离来源地区域的沉积物和鱼类中均有检出 BFRs，这表明它们是广泛存在于环境中的污染物。加拿大北部的鱼类、螃蟹，北极环斑海豹和其他海洋哺乳动物，以及格陵兰岛的鱼类和贻贝中均发现了高浓度的 PBDEs（Ikonomou et al.，2002；Christensen et al.，2002）。此外，挪威海峡远离已知点源样带沉积物中 HBCD 的浓度为 35～9000 μg/kg（Haukas et al.，2009）。对已知污染源附近水域进行分析，发现东京湾、苏格兰的克莱德河口和厦门近海地区等沉积物的岩芯中 BFRs 含量通常为低 μg/kg 到数千 μg/kg。

1.4.3.3 内分泌干扰物

威尼斯潟湖（意大利亚得里亚海湾）水体中 EDCs 浓度为 2.8～211 ng/L，沉积物中相似化合物的含量范围为 3.1～289 ng/g dw（Pojana et al.，2007）。此 EDCs 浓度略低于加利西亚海岸沉积物中的 EDCs 浓度（20.1～1409 ng/g dw），主要为壬基酚和辛基酚（Salgueiro-González et al.，2014）。海洋生物中不同同系物 EDCs 的浓度不等。例如，中国水域鱼类中对羟基苯甲酸酯的平均浓度范围为 0.005 ng/g～1.45 ng/g ww（Liao et al.，2013）；葡萄牙塔霍河口鱼体内对羟基苯甲酸酯含量最高为 98.4 ng/g dw（Alvarez-Muñoz et al.，2015b）；意大利菲乌米奇诺虾类中总烷基酚类化合物含量最高为 1255 ng/g ww（Ferrara et al.，2008）。

1.4.3.4 药品和个人护理产品

药品和个人护理产品在全世界沿海地区中被广泛检出，主要分布在河口、港口、潟湖、封闭或半封闭水域，详见 Rodríguez-Mozaz et al.（2015）。至少在海水中检出 100 多种药物。其中抗生素是研究最多的药物，其次是精神病治疗药物、止痛药和消炎药。在海水中最常检出的化合物是乙酰氨基酚、布洛芬、双氯芬酸、红霉素、克拉霉素、磺胺甲恶唑、甲氧苄啶、卡马西平、吉非罗齐和阿替洛尔。通常在水生环境中检出的药物水平高低不等。例如，夏季地中海海水中检出的普萘洛尔为 0.5 ng/L，水杨酸为 11.8 ng/L，红霉素为 40.7 ng/L，磺胺甲恶唑为 64.8 ng/L（Gonzalez et al.，2013）。中国渤海湾水体中检出抗生素的浓度范围为 2.3 ng/L（磺胺甲恶唑片）至 6800 ng/L（诺氟沙星）（Zhou et al.，2011）。沉积物是水体许多污染物的天然储存库。就药物而言，沉积物已被确定为抗生素主要的"汇"（Kim & Carlson，2007）。沉积物中检出频率最高的药物是抗生素，如四环素、土霉素、诺氟沙星、氧氟沙星、恩诺沙星和环丙沙星（Rodríguez-Mozaz et al.，2015）。中国长江口沉积物中的氧氟沙星浓度高达 458 ng/g（Shi et al.，2014）。生物中检出最多的药物是卡马西平和土霉素。研究结果表明，药物浓度通常在低 ng/g 的数量级水平，少有例外。例如，意大利波河三角洲地区双壳类中文拉法辛和阿奇霉素浓度分别为 36.1 ng/g 和 13.3 ng/g（Alvarez-Muñoz et al.，2015b）。中国渤海湾野生软体动物中诺氟沙星高达 370 ng/g dw，氧氟沙星为 242 ng/g dw，环丙沙星为 208 ng/g dw，恩诺沙星为 147 ng/g dw（Li et al.，2012）。

关于海洋环境中 PcPS 类化合物的研究较少。水生生物体内已检出麝香香料（Kannan et al.，2005）；不同地区的海豚中检出防晒霜（Gago-Ferrero et al.，2013）。此外，部分研究在北海海水检出了驱虫剂（Weigel et al.，2002）。

1.4.3.5 海洋生物毒素

大量研究表明，地中海和黑海贻贝中原多甲藻酸浓度为 6～900 ng/g（Elgarch et al.，2008；Taleb et al.，2006）；大西洋东北部贻贝和牡蛎中原多甲藻酸的含量略高，可达 1.4～4200 ng/g（Furey et al.，2002；James et al.，2002；Magdalena et al.，2003；Vandermeersch et al.，2015）。对于雪卡毒素，太平洋中东部和大西洋中西部鱼

类、珊瑚鳕鱼和无鳔石首鱼中发现的浓度范围为 0.1~52.9 ng/g（Dickey 2008；Stewart et al., 2010；Vandermeersch et al., 2015）。

1.4.3.6 纳米材料

就目前而言，尚无研究报道海洋环境中（包括水、沉积物和生物）纳米材料的存在。然而，Sanchí et al.（2011）在地中海地区大气中检出气溶胶结合富勒烯（Sanchís et al., 2011）。

1.4.3.7 硅氧烷类

如前所述，由于具有低溶解度和高蒸气压的特点，硅氧烷在水中的浓度为痕量水平（Sparham et al., 2008）。然而由于硅氧烷具有高疏水性，它们在沉积物和生物中的含量则较高。例如，硅氧烷在奥斯陆峡湾内的沉积物中检出的浓度高达 920 ng/g dw；Bekkelaget 污水处理厂附近检出含量最多、浓度最高的化合物为十甲基环五硅氧烷（D5）；亨伯河口地区 D5 的检出浓度为 60~260 ng/g dw（Kierkegaard et al., 2011）。在人口较少的地区（如北极），沉积物中硅氧烷类的含量有所下降，浓度为 2 ng/g dw（Sparham et al., 2011）。对于生物中的硅氧烷类，八甲基环四硅氧烷（D4）和 D5 是西班牙巴塞罗那市场的海鲜中检出率最高的化合物，浓度范围为低浓度 pg/g~30 ng/g ww（Sanchís et al., 2015）。此外，也有研究表明，野生贻贝、大菱鲆和鳕鱼中六甲基环三硅氧烷（D3）浓度为 50~321.3 ng/g，D4 为 1.3~134.4 ng/g，D5 为 3.3~2200 ng/g，十二甲基环六硅氧烷（D6）为 0.9~151.5 ng/g（Schlabach et al., 2007）。

1.5 海洋环境中的微塑料

除本章所述海洋环境中存在的多种化合物外，微塑料（Microplastics，MP）是另一类新兴的污染物。虽然塑料垃圾和颗粒不是一个新问题，但是人们越来越关注环境中的微塑料及其引起的环境破坏问题，并对此问题的认知日新月异（Hartl et al., 2015）。在一般环境条件下，较大的塑料制品会降解为所谓的微塑料，通常为直径小于 5 mm 的碎片（Wagner et al., 2014）。MP 具有高流动性，广泛分布于世界海洋环境中。其中，大型海洋环流是塑料污染的"热点区域"。大多数研究调查了中性和中上水层 MP（Wagner et al., 2014）。然而，沉积物中也能检出 MP，浓度范围为 1~100 个/kg（Hidalgo-Ruz et al., 2012）。MP 可以被水生生物吸收，并在整个食物网中累积（Wright et al., 2013）。海洋生物，如贻贝（Browne et al., 2008）、海龟（Stamper et al., 2009）和鱼类（Boerger et al., 2010）等，已被证明吸收了微塑料。MP 可以在生物的消化道累积，导致生物的营养障碍和机能下降（Browne et al., 2008）；MP 还可以作为其他有机污染物（如 PCBs、DDT & PAHs）的载体（Mato et al., 2001；Rios et al., 2007），成为野生动植物接触这些污染物的媒介（Oehlmann et al., 2009；Teuten et al., 2009）。

关于监测方法，尚未有完善的 MP 采样规范。目前最常见的分离技术是高密度浮选技术，然后在显微镜下检查碎片。但该技术可能导致一些高密度聚合物（如 PVC 或聚酯）的密度被低估。

迄今为止，关于海洋环境中 MP 研究的重点是其丰度及其对环境的影响。需要进行更多的研究来了解它们的潜在危害，包括揭示其在海洋环境中的来源和归宿、开发合适的生物标志物、明确 MP 对人类健康的影响（Hartl et al., 2015）；此外，MP 与其他污染物的相互作用、化学负荷、吸收/解吸动力学，以及从塑料到生物的转移过程等研究也同样值得深入探索。

1.6 未来趋势

值得强调的是，本章中介绍的许多污染物往往在海洋环境（尤其是沿海和河口地区）中同时存在。这些污染物形成的有害"混合物"对野生生物和人类健康带来了潜在风险。即使较低的浓度（ng/L 或 pg/L）也可能造成毒性效应。此外，同时存在的不同污染物浓度总和会导致较高的污染水平。有科学证据表明，当多种不同的化学物质同时接触生物体时，它们可能通过附加、拮抗、叠加和协同等共同作用而影响毒性的整体水平。

特定污染物的传统靶标分析如本章所述，需遵循一定的方法。目前在环境评估中，系统地开发混合污染处理方法（European Commission，2012）以及确定需优先关注的潜在混合污染物是主要的挑战。靶标分析具有高灵敏度和可靠性。但它有一个明显的缺点，即无法测得非靶标污染物。这些无法检测的"未知"化合物浓度远高于可测化合物浓度，它们可以更好地解释某特定样品的毒性。非靶标分析技术可以填补该领域的不足。它是一种无需事先选择目标化合物即可识别环境污染物的有力工具。此外，痕量分析方法的改进可以帮助识别含量较少的污染物（Chetwynd et al., 2014；David et al., 2014）；高分辨率质谱仪可用于回顾性分析，查找此前未知的污染物（Acuña et al., 2015）。因此，非靶标分析是未来环境化学的发展趋势。迄今为止，它主要用于水样分析，包括废水（Gomez Ramos et al., 2011）、河水（Ruff et al., 2015）、地表水（Ibañez et al., 2008）、地下水（Ibañez et al., 2011）和饮用水（Muller et al., 2011）。由于目标分析物的复杂性，尤其是生物样本（Simon et al., 2015），非靶标分析方法在沉积物和生物体中的应用仅有较小进展（Alvarez-Munoz et al., 2015；Rostkowski et al., 2011）。

1.7 结 论

海洋环境（河口和沿海地区）的化学污染是一个非常复杂的问题。海洋生态系统中可以同时存在多达15种不同污染水平的污染物，包括金属、持久性有机污染物和新兴有机污染物。这些污染物形成的有害"混合物"，可能会对水生环境、人类健康（食用受污染的海产品）以及相关沿海活动造成负面影响，如渔业、水产养殖或娱乐活动等。

本章节对金属、PAHs、PCBs、表面活性剂、杀虫剂、二噁英、PPCPs、EDCs、PFASs、BFRs、NMs、海洋生物毒素、PDMS & MP 等污染物进行了综述。它们的主要理化性质与其对应分子结构直接相关。它们的主要污染来源已被确定，大多是通过人类活动进入水生环境。本章还介绍了海洋环境污染物最常用的检测和定量分析技术，其中 SPE 是液体样品最广泛应用的分析技术。对海水、沉积物和海洋生物中的污染物检测表明，海洋污染物具有广泛的地理分布特征。

参考文献

ACUÑA J, STAMPACHIACCHIERE S, PÉREZ S, et al., 2015. Advances in liquid chromatography-high resolution mass spectrometry for quantitative and qualitative environmental analysis. Anal. Bioanal. Chem., 407: 6289 – 6299.

AHRENS L, BARBER J L, XIE Z, et al., 2009. Longitudinal and latitudinal distribution of perfluoroalkyl compounds in the surface water of the Atlantic Ocean. Environ. Sci. Technol., 43: 3122 – 3127.

AHRENS L, XIE Z, EBINGHAUS R, 2010a. Distribution of perfluoroalkyl compounds in seawater from Northern Europe, Atlantic Ocean, and Southern Ocean. Chemosphere, 78: 1011 – 1016.

AHRENS L, GERWINSKI W, THEOBALD N, et al., 2010b. Sources of polyfluoroalkyl compounds in the North Sea, Baltic Sea and Norwegian Sea: evidence from their spatial distribution in surface water. Mar. Pollut. Bull., 60: 255 – 260.

ALAEE M, WENNING R J, 2002. The significance of brominated flame retardants in the environment: current understanding, issues and challenges. Chemosphere, 46: 579 – 582.

ALAEE M, ARIAS P, SJODIN A, et al., 2003. An overview of commercially used brominated flame retardants, their applications, their use patterns in different countries/regions and possible modes of release. Environ. Int., 29: 683 – 689.

AL-ANSARI A M, SALEEM A, KIMPE L E, et al., 2010. Bioaccumulation of the pharmaceutical 17alpha-ethinylestradiol in shorthead redhorse suckers (*Moxostoma macrolepidotum*) from the St. Clair River, Canada. Environ. Pollut., 158: 2566 – 2571.

ALEXANDER J, BENFORD D, COCKBURN A, et al., 2008a. Marine biotoxins in shellfish: yessotoxin group. EFSA J., 7 (2): 907: 1-62.

ALEXANDER J, BENFORD D, COCKBURN A, et al., 2008b. Marine biotoxins in shellfish: azaspiracid group. EFSA J., 6 (10): 723: 1-52.

ALEXANDER J, AUDUNSSON G A, BENFORD D, et al., 2008c. Marine biotoxins in shellfish: okadaic acid and analogues. EFSA J., 6 (1): 589: 1-62.

ALEXANDER J, BENFORD D, COCKBURN A, et al., 2009a. Marine biotoxins in shellfish: saxitoxin group. EFSA J., 7 (4): 1019.

ALEXANDER J, BENFORD D, COCKBURN A, et al., 2009b. Marine biotoxins in shellfish: domoic acid. EFSA J., 7 (7): 1181.

ALEXANDER J, BENFORD D, COCKBURN A, et al., 2009c. Marine biotoxins in shellfish: pectenotoxin group. EFSA J., 7 (6): 1109: 1-47.

ALLEN R B, KOCHS P, CHANDRA G, 1997. Organosilicon materials // Handbook of environmental chemistry. Springer-Verlag.

ALVAREZ D A, MARUYA K A, DODDER N G, et al., 2014. Occurrence of contaminants of emerging concern along the California coast (2009-10) using passive sampling devices. Mar. Pollut. Bull., 81: 347-354.

ALVAREZ-MUNOZ D, SAEZ M, LARA-MARTIN P A, et al., 2004. New extraction method for the analysis of linear alkylbenzene sulfonates in marine organisms: pressurized liquid extraction versus Soxhlet extraction. J. Chromatogr. A, 1052: 33-38.

ALVAREZ-MUNOZ D, GOMEZ-PARRA A, GONZALEZ-MAZO E, 2007. Testing organic solvents for the extraction from fish of sulfophenylcarboxylic acids, prior to determination by liquid chromatography-mass spectrometry. Anal. Bioanal. Chem., 388: 1013-1019.

ALVAREZ-MUNOZ D, INDIVERI P, ROSTKOWSKI P, et al., 2015. Widespread contamination of coastal sediments in the Transmanche Channel with anti-androgenic compounds. Mar. Pollut. Bull., 95: 590-597.

ALVAREZ-MUÑOZ D, HUERTA B, FERNANDEZ-TEJEDOR M, et al., 2015a. Multi-residue method for the analysis of pharmaceuticals and some of their metabolites in bivalves. Talanta, 136: 174-182.

ALVAREZ-MUÑOZ D, RODRÍGUEZ-MOZAZ S, MAULVAULT A L, et al., 2015b. Occurrence of pharmaceuticals and endocrine disrupting compounds in macroalgaes, bivalves, and fish from coastal areas in Europe. Environ. Res., 143: 56-64.

ANDERSSON P L, OBERG K, ORN U, 2006. Chemical characterization of brominated flame retardants and identification of structurally representative compounds. Environ. Toxicol. Chem., 25: 1275-1282.

ANDREU V, PICO Y, 2012. Determination of currently used pesticides in biota. Anal. Bioanal. Chem., 404: 2659-2681.

ASTEFANEI A, NÚÑEZ O, GALCERAN M T, 2015. Characterisation and determination of fullerenes: a critical review. Anal. Chim. Acta, 882: 1-21.

AZZOUZ A, SOUHAIL B, BALLESTEROS E, 2011. Determination of residual pharmaceuticals in edible animal tissues by continuous solid-phase extraction and gas chromatography-mass spectrometry. Talanta, 84: 820 – 828.

BANJOO D R, NELSON P K, 2005. Improved ultrasonic extraction procedure for the determination of polycyclic aromatic hydrocarbons in sediments. J. Chromatogr. A, 1066: 9 – 18.

BARAKAT A O, MOSTAFA A, WADE T L, et al., 2013. Distribution and ecological risk of organochlorine pesticides and polychlorinated biphenyls in sediments from the Mediterranean coastal environment of Egypt. Chemosphere, 93: 545 – 554.

BAUMARD P, BUDZINSKI H, MICHON Q, et al., 1998. Origin and bioavailability of PAHs in the Mediterranean Sea from mussel and sediment records. Estuar. Coast. Shelf Sci., 47: 77 – 90.

BERG B E, LUND H S, KRINGSTAD A, et al., 1999. Routine analysis of hydrocarbons, PCB and PAH in marine sediments using supercritical CO_2 extraction. Chemosphere, 38: 587 – 599.

BERRADA H, BORRULL F, FONT G, et al., 2008. Determination of macrolide antibiotics in meat and fish using pressurized liquid extraction and liquid chromatography-mass spectrometry. J. Chromatogr. A, 1208: 83 – 89.

BOERGER C M, LATTIN G L, MOORE S L, et al., 2010. Plastic ingestion by planktivorous fishes in the North Pacific Central Gyre. Mar. Pollut. Bull., 60: 2275 – 2278.

BORECKA M, SIEDLEWICZ G, HALINSKI L P, et al., 2015. Contamination of the southern Baltic Sea waters by the residues of selected pharmaceuticals: method development and field studies. Mar. Pollut. Bull., 94: 62 – 71.

BRAUSCH J M, RAND G M, 2011. A review of personal care products in the aquatic environment: environmental concentrations and toxicity. Chemosphere, 82: 1518 – 1532.

BROWNE M A, DISSANAYAKE A, GALLOWAY T S, et al., 2008. Ingested microscopic plastic translocates to the circulatory system of the mussel, *Mytilus edulis* (L). Environ. Sci. Technol., 42: 5026 – 5031.

BRUZZONITI M C, CHECCHINI L, DE CARLO R M, et al., 2014. QuEChERS sample preparation for the determination of pesticides and other organic residues in environmental matrices: a critical review. Anal. Bioanal. Chem. 406: 4089 – 4116.

BUCK R C, FRANKLIN J, BERGER U, et al., 2011. Perfluoroalkyl and polyfluoroalkyl substances in the environment: terminology, classification, and origins. Integr. Environ. Assess. Manag., 7: 513 – 541.

BURKHARDT M R, ZAUGG S D, BURBANK T L, et al., 2005. Pressurized liquid extraction using water/isopropanol coupled with solid phase extraction cleanup for semivolatile organic compounds, polycyclic aromatic hydrocarbons (PAH) and alkylated PAH homolog groups in sediment. Anal. Chim. Acta, 549: 104 – 116.

BUSECK P R, ADACHI K, 2008. Nanoparticles in the atmosphere. Elements, 4: 389 – 394.

CASTELLS P, PARERA J, SANTOS F J, et al., 2008. Occurrence of polychlorinated naphthalenes, polychlorinated biphenyls and short-chain chlorinated paraffins in marine sediments from Barcelona (Spain). Chemosphere, 70: 1552-1562.

CASTRO-JIMÉNEZ J, ROTLLANT G, ÁBALOS M, et al., 2013. Accumulation of dioxins in deep-sea crustaceans, fish and sediments from a submarine canyon (NW Mediterranean). Prog. Oceanogr., 118: 260-272.

CHAFER-PERICAS C, MAQUIEIRA A, PUCHADES R, et al., 2010a. Fast screening immunoassay of sulfonamides in commercial fish samples. Anal. Bioanal. Chem., 396: 911-921.

CHAFER-PERICAS C, MAQUEIRA A, PUCHADES R, et al., 2010b. Immunochemical determination of oxytetracycline in fish: comparison between enzymatic and time-resolved fluorometric assays. Anal. Chim. Acta, 662: 177-185.

CHETWYND A J, DAVID A, HILL E M, et al., 2014. Evaluation of analytical performance and reliability of direct nanoLC-nanoESI-high resolution mass spectrometry for profiling the (xeno) metabolome. J. Mass Spectrom., 49: 1063-1069.

CHIN-YUAN C, LI-LIAN L, WANG-HSIEN D, 2006. Occurrence and seasonal variation of alkylphenols in marine organisms from the coast of Taiwan. Chemosphere, 65: 2152-2159.

CHRISTENSEN J H, GLASIUS M, PECSELI M, et al., 2002. Polybrominated diphenyl ethers (PBDEs) in marine fish and blue mussels from southern Greenland. Chemosphere, 47: 631-638.

CLOUTIER P, FORTIN F, FOURNIER M, et al., 2014. Development of an analytical method for the determination of low-level of dioxin and furans in marine and freshwater species. J. Xenobiot., 4: 73-75.

COMPANIONI-DAMAS E Y, SANTOS F J, GALCERAN M T, 2012. Analysis of linear and cyclic methylsiloxanes in water by headspace-solid phase microextraction and gas chromatography-mass spectrometry. Talanta, 89: 63-69.

CORTADA C, DOS REIS L C, VIDAL L, et al., 2014. Determination of cyclic and linear siloxanes in wastewater samples by ultrasound-assisted dispersive liquid-liquid microextraction followed by gas chromatography-mass spectrometry. Talanta, 120: 191-197.

COVACI A, VOORSPOELS S, RAMOS L, et al., 2007. Recent developments in the analysis of brominated flame retardants and brominated natural compounds. J. Chromatogr. A, 1153: 145-171.

CROCE V, PAGGIO S, PAGNONI A, et al., 2003. Determination of 4-nonylphenol and 4-nonylphenol ethoxylates in river sediments by microwave assisted solvent extraction. Ann. Chim., 93: 297-304.

CUEVA-MESTANZA R, TORRES-PADRON M E, SOSA-FERRERA Z, et al., 2008. Microwave-assisted micellar extraction coupled with solid-phase extraction for preconcentration of pharmaceuticals in molluscs prior to determination by HPLC. Biomed. Chromatogr., 22: 1115-1122.

DALY T K, BUSECK P R, WILLIAMS P, et al., 1993. Fullerenes from a fulgurite.

Science, 259: 1599 - 1601.

DAVID M D, FENDINGER N J, HAND V C, 2000. Determination of Henry's law constants for organosilicones in actual and simulated wastewater. Environ. Sci. Technol., 34: 4554 - 4559.

DAVID A, ABDUL-SADA A, LANGE A, et al., 2014. A new approach for plasma (xeno) metabolomics based on solid-phase extraction and nanoflow liquid chromatography-nanoelectrospray ionisation mass spectrometry. J. Chromatogr. A, 1365: 72 - 85.

DE BOER J, LAW R J, 2003. Developments in the use of chromatographic techniques in marine laboratories for the determination of halogenated contaminants and polycyclic aromatic hydrocarbons. J. Chromatogr. A, 1000: 223 - 251.

DELINSKY A D, STRYNAR M J, NAKAYAMA S F, et al., 2009. Determination of ten perfluorinated compounds in bluegill sunfish (*Lepomis macrochirus*) fillets. Environ. Res., 109: 975 - 984.

DEWIL R, APPELS L, BAEYENS J, et al., 2007. The analysis of volatile siloxanes in waste activated sludge. Talanta, 74 (1): 14 - 19.

DIAMANTI-KANDARAKIS E, BOURGUIGNON J P, GIUDICE L C, et al., 2009. Endocrine-disrupting chemicals: an Endocrine Society scientific statement. Endocr. Rev., 30: 293 - 342.

DICKEY R W, 2008. Ciguatera toxins: chemistry, toxicology, and detection. Food Sci. Technol., New York, Marcel Dekker: 173 - 479.

DODDER N G, MARUYA K A, LEE FERGUSON P, et al., 2014. Occurrence of contaminants of emerging concern in mussels (*Mytilus* spp.) along the California coast and the influence of land use, storm water discharge, and treated wastewater effluent. Mar. Pollut. Bull., 81: 340 - 346.

DONAT J R, BRULAND K W, 1995. Trace elements in the oceans // SALBU B, STEINESS E (Eds.). Trace elements in natural waters. CRC-Press, Boca Raton, FL: 247 - 281.

DÖRR B, LIEBEZEIT G, 2009. Organochlorine compounds in blue mussels, *Mytilus edulis*, and Pacific oysters, *Crassostrea gigas*, from seven sites in the Lower Saxonian Wadden Sea, Southern North Sea. Bull. Environ. Contam. Toxicol., 83: 874 - 879.

DOUBEN P E T, 2003. PAHs: an ecotoxicological perspective. John Wiley & Sons.

DUNPHY J C, PESSLER D G, MORRALL S W, et al., 2001. Derivatization LC/MS for the simultaneous determination of fatty alcohol and alcohol ethoxylate surfactants in water and wastewater samples. Environ. Sci. Technol., 35: 1223 - 1230.

ELGARCH A, VALE P, RIFAI S, et al., 2008. Detection of diarrheic shellfish poisoning and azaspiracids toxins in Moroccan mussels: comparison of LC-MS method with the commercial immunoassay kit. Mar. Drugs, 6: 587 - 594.

EPA, 2009. Provisional health advisories for perfluorooctanoic acid (PFOA) and perfluorooctane sulfonate (PFOS).

ERICKSON M D, ROBERT G K I, 2011. Applications of polychlorinated biphenyls. Environ. Sci. Pollut. Res., 18: 135 - 151.

EURLMB, 2016. European union reference laboratory for marine biotoxins. http://aesan.msssi.gob.es/en/CRLMB/web/faqs/biotoxinas.shtml.

European_Commission, 2002. Water framework directive: directive 2000/60/EC of the European parliament and of the council establishing a framework for the community action in the field of water policy. http://ec.europa.eu/environment/water/water-framework/index_en.html.

European_Commission, 2012. The combination effects of chemicals. Chemical mixtures. Communication from the Commission to the Council, 252.

FAO, 2004. Food and agriculture organization of the united nations: marine biotoxins.

FARRE M, PICO Y, BARCELO D, 2014. Application of ultra-high pressure liquid chromatography linear ion-trap orbitrap to qualitative and quantitative assessment of pesticide residues. J. Chromatogr. A, 1328: 66 - 79.

FERNANDEZ-TORRES R, BELLO LOPEZ M A, OLIAS CONSENTINO M, et al., 2011. Simultaneous determination of selected veterinary antibiotics and their main metabolites in fish and mussel samples by highperformance liquid chromatography with diode arrayfluorescence (HPLC-DAD-FLD) detection. Anal. Lett., 44: 2357 - 2372.

FERRARA F, ADEMOLLO N, DELISE M, et al., 2008. Alkylphenols and their ethoxylates in seafood from the Tyrrhenian Sea. Chemosphere, 72 (9): 1279 - 1285.

FONTANA A R, SILVA M F, MARTINEZ L D, et al., 2009. Determination of polybrominated diphenyl ethers in water and soil samples by cloud point extraction-ultrasound-assisted back-extraction-gas chromatography-mass spectrometry. J. Chromatogr. A, 1216: 4339 - 4346.

FULARA I, CZAPLICKA M, 2012. Methods for determination of polybrominated diphenyl ethers in environmental samples: review. J. Sep. Sci., 35 (16): 2075 - 2087.

FUREY A, BRAÑA-MAGDALENA A, LEHANE M, et al., 2002. Determination of azaspiracids in shellfish using liquid chromatography/tandem electrospray mass spectrometry. Rapid Commun. Mass Spectrom., 16: 238 - 242.

GAGO-FERRERO P, ALONSO M B, BERTOZZI C P, et al., 2013. First determination of UV filters in marine mammals. Octocrylene levels in Franciscana dolphins. Environ. Sci. Technol., 47 (11): 5619 - 5625.

GAW S, THOMAS K V, HUTCHINSON T H, 2014. Sources, impacts and trends of pharmaceuticals in the marine and coastal environment. Philos. Trans. R. Soc. Lond. B Biol. Sci., 369 (1656).

GERSSEN A, POL-HOFSTAD I E, POELMAN M, et al., 2010. Marine toxins: chemistry, toxicity, occurrence and detection, with special reference to the Dutch situation. Toxins (Basel), 2 (4): 878 - 904.

GOMEZ-RAMOS M D M, PÉREZ-PARADA A, GARCÍA-REYES J F, et al., 2011. Use of an accurate-mass database for the systematic identification of transformation products of organic contaminants in wastewater effluents. J. Chromatogr. A, 1218 (44): 8002 – 8012.

GROS M, PETROVIC M, BARCELO D, 2009. Tracing pharmaceutical residues of different therapeutic classes in environmental waters by using liquid chromatography/quadrupole-linear ion trap mass spectrometry and automated library searching. Anal. Chem., 81: 898 – 912.

GROS M, RODRIGUEZ-MOZAZ S, BARCELO D, 2013. Rapid analysis of multiclass antibiotic residues and some of their metabolites in hospital, urban wastewater and river water by ultra-high-performance liquid chromatography coupled to quadrupole-linear ion trap tandem mass spectrometry. J. Chromatogr. A, 1292: 173 – 188.

GU Y X, ROSE W I, BLUTH G J S, 2003. Retrieval of mass and sizes of particles in sandstorms using two MODIS IR bands: a case study of April 7, 2001 sandstorm in China. Geophys. Res. Lett., 30 (15): 1805.

GUERRA P, ALAEE M, ELJARRAT E, et al., 2011. Brominated flame retardants: chapter 1: Introduction to brominated flame retardants: commercially products, applications, and physicochemical properties. Springer.

HALLEGRAEFF G, MCCAUSLAND M, BROWN R, 1995. Early warning of toxic dinoflagellate blooms of *Gymnodinium catenatum* in southern Tasmanian waters. J. Plankton Res., 17: 1163 – 1176.

HAMELINK J L, SIMON P B, SILBERHORN E M, 1996. Henry's law constant, volatilization rate and aquatic half-life of octamethylcylotetrasiloxanes. Environ. Sci. Technol., 30: 1946 – 1952.

HARTL M G J, GUBBINS E, GUTIERREZ T, et al., 2015. Review of existing knowledge: emerging contaminants: focus on nanomaterials and microplastics in waters. CSsCoEf, ed. Edinburgh. www.crew.ac.uk/pulications.

HAUKAS M, HYLLAND K, BERGE J A, et al., 2009. Spatial diastereomer patterns of hexabromocyclododecane (HBCD) in a Norwegian fjord. Sci. Total Environ., 407: 5907 – 5913.

HE X, WANG Z, NIE X, et al., 2012. Residues of fluoroquinolonas in marine aquaculture environment of the Pearl River Delta, South China. Environ. Geochem. Health, 34: 323 – 335.

HERNANDEZ F, PORTOLES T, IBANEZ M, et al., 2012. Use of time-of-flight mass spectrometry for large screening of organic pollutants in surface waters and soils from a rice production area in Colombia. Sci. Total Environ., 439: 249 – 259.

HIBBERD A, MASKAOUI K, ZHANG Z, et al., 2009. An improved method for the simultaneous analysis of phenolic and steroidal estrogens in water and sediment. Talanta, 77: 1315 – 1321.

HIDALGO-RUZ V, GUTOW L, THOMPSON R C, et al., 2012. Microplastics in the marine environment: a review of the methods used for identification and quantification. Environ. Sci. Technol., 46: 3060-3075.

HIRSCH A, 2010. The era of carbon allotropes. Nat. Mater., 9: 868-871.

HOTCHKISS A K, RIDER C V, BLYSTONE C R, et al., 2008. Fifteen years after "Wingspread": environmental endocrine disrupters and human and wildlife health: where we are today and where we need to go. Toxicol. Sci., 105 (2): 235-259.

HOUDE F, DEBLOIS C, BERRYMAN D, 2002. Liquid chromatographic-tandem mass spectrometric determination of nonylphenol polyethoxylates and nonylphenol carboxylic acids in surface water. J. Chromatogr. A, 961: 245-256.

HUERTA B, JAKIMSKA A, GROS M, et al., 2013. Analysis of multi-class pharmaceuticals in fish tissues by ultra-high-performance liquid chromatography tandem mass spectrometry. J. Chromatogr. A, 1288: 63-72.

HUPPMANN R, LOHOFF H W, SCHRÖDER H F, 1996. Cyclic siloxanes in the biological waste water treatment process: determination, quantification and possibilities of elimination. Fres. J. Anal. Chem., 354 (1): 66-71.

IBÁÑEZ M, SANCHO J V, HERNÁNDEZ F, et al., 2008. Rapid non-target screening of organic pollutants in water by ultraperformance liquid chromatography coupled to time-of-light mass spectrometry. Trends Anal. Chem., 27 (5): 481-489.

IBAÑEZ M, SANCHO J V, POZO J O, et al., 2011. Use of quadrupole time-of-flight mass spectrometry to determine proposed structures of transformation products of the herbicide bromacil after water chlorination. Rapid Commun. Mass Spectrom., 25: 3103-3113.

IKONOMOU M G, RAYNE S, FISCHER M, et al., 2002. Occurrence and congener profiles of polybrominated diphenyl ethers (PBDEs) in environmental samples from coastal British Columbia, Canada. Chemosphere, 46: 649-663.

JAKIMSKA A, HUERTA B, BARGANSKA Z, et al., 2013. Development of a liquid chromatography-tandem mass spectrometry procedure for determination of endocrine disrupting compounds in fish from Mediterranean rivers. J. Chromatogr. A, 1306: 44-58.

JAMES K J, FUREY A, LEHANE M, et al., 2002. First evidence of an extensive northern European distribution of azaspiracid poisoning (AZP) toxins in shellfish. Toxicon, 40: 909-915.

JEHLIČKA J, SVATOŠ A, FRANK O, et al., 2003. Evidence for fullerenes in solid bitumen from pillow lavas of Proterozoic age from Mítov (Bohemian Massif, Czech Republic). Geochim. Cosmochim. Acta, 67: 1495-1506.

JELIC A, PETROVIC M, BARCELO D, 2009. Multi-residue method for trace level determination of pharmaceuticals in solid samples using pressurized liquid extraction followed by liquid chromatography/quadrupole-linear ion trap mass spectrometry. Talanta, 80: 363-371.

JIANG H, ZHANG Y, CHEN X, et al., 2013. Simultaneous determination of pentachlorophenol, niclosamide and fenpropathrin in fishpond water using an LC-MS/MS method for forensic investigation. Anal. Methods, 5: 111-115.

JIANG J J, LEE C L, FANG M D, 2014. Emerging organic contaminants in coastal waters: anthropogenic impact, environmental release and ecological risk. Mar. Pollut. Bull., 85: 391-399.

JOHNSEN A R, WICK L Y, HARMS H, 2005. Principles of microbial PAH-degradation in soil. Environ. Pollut., 133: 71-84.

JOHNSON Y S, 2012. Determination of polycyclic aromatic hydrocarbons in edible seafood by QuEChERS - based extraction and gas chromatography-tandem mass spectrometry. J. Food Sci., 77: T131-T137.

KAJ L, SCHLABACH M, ANDERSSON J, et al., 2005. Siloxanes in the Nordic environment.

KANNAN K, REINER J L, YUN S H, et al., 2005. Polycyclic musk compounds in higher trophic level aquatic organisms and humans from the United States. Chemosphere, 61: 693-700.

KARICKHOFF C W, BROWN D S, SCOTT T A, 1979. Sorption of hydrophobic pollutants on natural sediments. Water Res., 13: 241-248.

KENNISH M J, 1997. Practical handbook of estuarine and marine pollution. CRC Press, Boca Raton, FL.

KENNISH M J, 2000. Practical handbook of marine science. CRC Press, Boca Raton, FL.

KIERKEGAARD A, ADOLFSSON-ERICI M, MCLACHLAN M S, 2010. Determination of cyclic volatile methylsiloxanes in biotawith a purge and trap method. Anal. Chem., 82: 9573-9578.

KIERKEGAARD A, VAN EGMOND R, MCLACHLAN M S, 2011. Cyclic volatile methylsiloxane bioaccumulation in flounder and ragworm in the Humber estuary. Environ. Sci. Technol., 45: 5936-5942.

KIM S C, CARLSON K, 2007. Temporal and spatial trends in the occurrence of human and veterinary antibiotics in aqueous and river sediment matrices. Environ. Sci. Technol., 41: 50-57.

KIM J W, RAMASWAMY B R, CHANG K H, et al., 2011. Multiresidue analytical method for the determination of antimicrobials, preservatives, benzotriazole UV stabilizers, flame retardants and plasticizers in fish using ultra high performance liquid chromatography coupled with tandem mass spectrometry. J. Chromatogr. A, 1218: 3511-3520.

KLOSTERHAUS S L, GRACE R, HAMILTON M C, et al., 2013. Method validation and reconnaissance of pharmaceuticals, personal care products, and alkylphenols in surface waters, sediments, and mussels in an urban estuary. Environ. Int., 54: 92-99.

KREMLING K, 2002. Determination of trace elements // GRASSHOFF K, KREMLING K, EHRHARDT M (Eds.). Methods of seawater analysis. Wiley-VCH, Weinheim: 253-273.

KRUGER O, CHRISTOPH G, KALBE U, et al., 2011. Comparison of stir bar sorptive extraction (SBSE) and liquid-liquid extraction (LLE) for the analysis of polycyclic aromatic hydrocarbons (PAH) in complex aqueous matrices. Talanta, 85: 1428 – 1434.

KWON J W, ARMBRUST K L, VIDAL-DORSCH D, et al., 2009. Determination of 17alpha-ethynylestradiol, carbamazepine, diazepam, simvastatin, and oxybenzone in fish livers. J. AOAC Int., 92: 359 – 369.

LACORTE S, IKONOMOU M G, FISCHER M, 2010. A comprehensive gas chromatography coupled to high resolution mass spectrometry based method for the determination of polybrominated diphenyl ethers and their hydroxylated and methoxylated metabolites in environmental samples. J. Chromatogr. A, 1217: 337 – 347.

LARA-MARTIN P A, GOMEZ-PARRA A, GONZALEZ-MAZO E, 2006a. Development of a method for the simultaneous analysis of anionic and non-ionic surfactants and their carboxylated metabolites in environmental samples by mixed-mode liquid chromatography-mass spectrometry. J. Chromatogr. A, 1137: 188 – 197.

LARA-MARTIN P A, GOMEZ-PARRA A, GONZALEZ-MAZO E, 2006b. Simultaneous extraction and determination of anionic surfactants in waters and sediments. J. Chromatogr. A, 1114: 205 – 210.

LARA-MARTIN P A, GONZALEZ-MAZO E, BROWNAWELL B J, 2011. Multi-residue method for the analysis of synthetic surfactants and their degradation metabolites in aquatic systems by liquid chromatography-time-of-flight-mass spectrometry. J. Chromatogr. A, 1218: 4799 – 4807.

LEON V M, GONZALEZ-MAZO E, GOMEZ-PARRA A, 2000. Handling of marine and estuarine samples for the determination of linear alkylbenzene sulfonates and sulfophenylcarboxylic acids. J. Chromatogr. A, 889: 211 – 219.

LEON V M, GARCIA I, MARTINEZ-GOMEZ C, et al., 2014. Heterogeneous distribution of polycyclic aromatic hydrocarbons in surface sediments and red mullet along the Spanish Mediterranean coast. Mar. Pollut. Bull., 87: 352 – 363.

Li Y, Wei G, Wang X, 2007. Determination of decabromodiphenyl ether in water samples by single-drop microextraction and RP-HPLC. J. Sep. Sci., 30: 2698 – 2702.

LI W, SHI Y, GAO L, et al., 2012. Investigation of antibiotics in mollusks from coastal waters in the Bohai Sea of China. Environ. Pollut., 162: 56 – 62.

LI N, QI L, SHEN Y, et al., 2013. Amphiphilic block copolymer modified magnetic nanoparticles for microwave-assisted extraction of polycyclic aromatic hydrocarbons in environmental water. J. Chromatogr. A, 1316: 1 – 7.

LI J Y, CUI Y, SU L, et al., 2015. Polycyclic aromatic hydrocarbons in the largest deepwater port of East China Sea: impact of port construction and operation. Environ. Sci. Pollut. Res., 22: 12355 – 12365.

LIAO C, CHEN L, KANNAN K, 2013. Occurrence of parabens in foodstuffs from China and its implications for human dietary exposure. Environ. Int., 57 – 58, 68 – 74.

LIBES S M, 2009. Introduction to marine biogeochemistry. Academic Press, Burlington.

LIU J, WANG R, HUANG B, et al., 2011. Distribution and bioaccumulation of steroidal and phenolic endocrine disrupting chemicals in wild fish species from Dianchi Lake, China. Environ. Pollut., 159: 2815 - 2822.

LLORCA M, FARRÉ M, PICÓ Y, et al., 2009. Development and validation of a pressurized liquid extraction liquid chromatography-tandem mass spectrometry method for perfluorinated compounds determination in fish. J. Chromatogr. A, 1216: 7195 - 7204

LLORCA M, FARRÉ M, KARAPANAGIOTI H K, et al., 2014. Levels and fate of perfluoroalkyl substances in beached plastic pellets and sediments collected from Greece. Mar. Pollut. Bull., 87: 286 - 291.

LLORCA M, 2012. Analysis of perfluoroalkyl substances in food and environmental matrices (Bachelor thesis). University of Barcelona, Department of Environmental Chemistry, IDAEA-CSIC.

LONG E R, DUTCH M, WEAKLAND S, et al., 2013. Quantification of pharmaceuticals, personal care products, and perfluoroalkyl substances in the marine sediments of Puget Sound, Washington, USA. Environ. Toxicol. Chem., 32: 1701 - 1710.

LOOS R, TAVAZZI S, PARACCHINI B, et al., 2013. Analysis of polar organic contaminants in surface water of the northern Adriatic Sea by solid-phase extraction followed by ultrahigh-pressure liquid chromatography-QTRAP® MS using a hybrid triple-quadrupole linear ion trap instrument. Anal. Bioanal. Chem., 405: 5875 - 5885.

LOPEZ P, BRANDSMA S A, LEONARDS P E, et al., 2009. Methods for the determination of phenolic brominated flame retardants, and by-products, formulation intermediates and decomposition products of brominated flame retardants in water. J. Chromatogr. A, 1216: 334 - 345.

LOSADA S, ROACH A, ROOSENS L, et al., 2009. Biomagnification of anthropogenic and naturally-produced organobrominated compounds in a marine food web from Sydney Harbour, Australia. Environ. Int., 35: 1142 - 1149.

LU Z, FISK A T, KOVACS K M, et al., 2014. Temporal and spatial variation in polychlorinated biphenyl chiral signatures of the Greenland shark (*Somniosus microcephalus*) and its arctic marine food web. Environ. Pollut., 186: 216 - 225.

LUOMA S N, RAINBOW P S, 2008. Metal contamination in aquatic environments: science and lateral management. Cambridge University Press, Cambridge.

MACKAY D, CALLCOTT D, 1998. Partitioning and physical chemical properties of PAHs. Springer, Berlin Heidelberg.

MAGDALENA A B, LEHANE M, KRYS S, et al., 2003. The first identification of azaspiracids in shellfish from France and Spain. Toxicon, 42: 105 - 108.

MARTÍNEZ BUENO M J, HERNANDO M D, AGÜERA A, et al., 2009. Application of passive sampling devices for screening of micro-pollutants in marine aquaculture using LC-

MS/MS. Talanta, 77: 1518 – 1527.

MARTÍNEZ BUENO M J, BOILLOT C, FENET H, et al., 2013. Fast and easy extraction combined with high resolution-mass spectrometry for residue analysis of two anticonvulsants and their transformation products inmarinemussels. J. Chromatogr. A, 1305: 27 – 34.

MARTÍNEZ BUENO M J, BOILLOT C, MUNARON D, et al., 2014. Occurrence of venlafaxine residues and its metabolites in marine mussels at trace levels: development of analytical method and a monitoring program. Anal. Bioanal. Chem., 406: 601 – 610.

MARUYAMA S, LEE M Y, HAUFLER R E, et al., 1991. Thermionic emission from giant fullerenes. Z. Phys. D Atoms Mol. Clusters, 19: 409 – 412.

MASIA A, CAMPO J, VAZQUEZ-ROIG P, et al., 2013a. Screening of currently used pesticides in water, sediments and biota of the Guadalquivir River Basin (Spain). J. Hazard. Mater., 263 (15 Pt 1): 95 – 104.

MASIA A, IBANEZ M, BLASCO C, et al., 2013b. Combined use of liquid chromatography triple quadrupole mass spectrometry and liquid chromatography quadrupole time-of-flight mass spectrometry in systematic screening of pesticides and other contaminants in water samples. Anal. Chim. Acta, 761: 117 – 127.

MASIA A, BLASCO C, PICÓ Y, 2014. Last trends in pesticide residue determination by liquid chromatography-mass spectrometry. Trends Anal. Chem., 2: 11 – 24.

MASON R P, 2013. Trace metals in aquatic systems. Wiley Blackwell, Chichester.

MATO Y, ISOBE T, TAKADA H, et al., 2001. Plastic resin pellets as a transport medium for toxic chemicals in the marine environment. Environ. Sci. Technol., 35: 318 – 324.

MATTHIJS E, H M S, KIEWIET A, et al., 1999. Environmental monitoring of linear alkylbenzene sulfonates, alcohol ethoxylate, alcohol ethoxy sulfate, alcohol sulfate and soap. Environ. Toxicol. Chem., 18: 2634 – 2644.

MCENEFF G, BARRON L, KELLEHER B, et al., 2013. The determination of pharmaceutical residues in cooked and uncooked marine bivalves using pressurized liquid extraction, solid-phase extraction and liquid chromatography-tandem mass spectrometry. Anal. Bioanal. Chem., 405: 9509 – 9521.

MCKINNEY M A, MCMEANS B C, TOMY G T, et al., 2012. Trophic transfer of contaminants in a changing Arctic marine food web: Cumberland Sound, Nunavut, Canada. Environ. Sci. Technol., 46: 9914 – 9922.

MEADOR J P, STEIN J E, REICHERT W L, et al., 1995. Bioaccumulation of polycyclic aromatic hydrocarbons by marine organisms. Rev. Environ. Contam. Toxicol., 143: 79 – 165.

MENZIES R, SOARES QUINETE N, GARDINALI P, et al., 2013. Baseline occurrence of organochlorine pesticides and other xenobiotics in the marine environment: Caribbean and Pacific collections. Mar. Pollut. Bull., 70: 289 – 295.

MONTES R, RODRIGUEZ I, CELA R, 2010. Solid-phase microextraction with simultaneous oxidative sample treatment for the sensitive determination of tetra-to hexabrominated diphenyl ethers in sediments. J. Chromatogr. A, 1217: 14 – 21.

MORENO-GONZALEZ R, RODRIGUEZ-MOZAZ S, GROS M, et al., 2015. Seasonal distribution of pharmaceuticals in marine water and sediment from a Mediterranean coastal lagoon (SE Spain). Environ. Res., 138: 326 – 344.

MULLER A, SCHULZ W, RUCK W K, et al., 2011. A new approach to data evaluation in the non-target screening of organic trace substances in water analysis. Chemosphere, 85: 1211 – 1219.

MUNARON D, TAPIE N, BUDZINSKI H, et al., 2012. Pharmaceuticals, alkylphenols and pesticides in Mediterranean coastal waters: results from a pilot survey using passive samplers. Estuar. Coast. Shelf Sci., 114: 82 – 92.

MURPHY J, 2001. Modifying specific properties: flammability-flame retardants // Additives for plastics handbook, 115 – 140.

NA G, FANG X, CAI Y, et al., 2013. Occurrence, distribution, and bioaccumulation of antibiotics in coastal environment of Dalian, China. Mar. Pollut. Bull., 69: 233 – 237.

NAVARRO P, ETXEBARRIA N, ARANA G, 2009. Development of a focused ultrasonic-assisted extraction of polycyclic aromatic hydrocarbons in marine sediment and mussel samples. Anal. Chim. Acta, 648: 178 – 182.

NAVARRO P, BUSTAMANTE J, VALLEJO A, et al., 2010. Determination of alkylphenols and 17beta-estradiol in fish homogenate. Extraction and clean-up strategies. J. Chromatogr. A, 1217: 5890 – 5895.

NIEBOER E R D, 1980. The replacement of the nondescript term "heavy metal" by abiologically and chemically significant classification of metal ions. Environ. Pollut., 1: 3 – 26.

NOZAKI Y, 1997. A fresh look at element distribution in the North Pacific Ocean. Eos, 78: 221 – 223.

NÚÑEZ Ó, GALLART-AYALA H, MARTINS C P B, et al., 2012. Atmospheric pressure photoionization mass spectrometry of fullerenes. Anal. Chem., 84: 5316 – 5326.

OEHLMANN J, SCHULTE-OEHLMANN U, KLOAS W, et al., 2009. A critical analysis of the biological impacts of plasticizers on wildlife. Philos. Trans. R. Soc. Lond. B Biol. Sci., 364: 2047 – 2062.

OSPAR_Commission, 2010. Polychlorinated byphenyls. Status and trend in marine chemical pollution.

PARERA J, ÁBALOS M, SANTOS F J, et al., 2013. Polychlorinated dibenzo-p-dioxins, dibenzofurans, biphenyls, paraffins and polybrominated diphenyl ethers in marine fish species from Ebro River Delta (Spain). Chemosphere, 93 (3): 499 – 505.

PARSONS J, BELZUNCE-SEGARRA M J, CORNELISSEN G, et al., 2007. Characterisation of contaminants in sediments: effects of bioavailability on impact // BARCELÓ D, PETROVIC M (Eds.). Sustainable management of sediment resources: sediment quality and impact assessment of pollutants. Elsevier, Amsterdam: 35 – 60.

PENA T, PENSADO L, CASAIS C, et al., 2006. Optimization of a microwave-assisted extraction method for the analysis of polycyclic aromatic hydrocarbons from fish samples. J. Chromatogr. A, 1121 (2): 163-169.

PETROVIC M, DIAZ A, VENTURA F, et al., 2001. Simultaneous determination of halogenated derivatives of alkylphenol ethoxylates and their metabolites in sludges, river sediments, and surface, drinking, and wastewaters by liquid chromatography-mass spectrometry. Anal. Chem., 73: 5886-5895.

PETROVIC M, ELJARRAT E, LÓPEZ DE ALDA M J, et al., 2002a. Recent advances in the mass spectrometric analysis related to endocrine disrupting compounds in aquatic environmental samples. J. Chromatogr. A, 974: 23-51.

PETROVIC M, FERNANDEZ-ALBA A R, BORRULL F, et al., 2002b. Occurrence and distribution of nonionic surfactants, their degradation products, and linear alkylbenzene sulfonates in coastal waters and sediments in Spain. Environ. Toxicol. Chem., 21: 37-46.

POJANA G, GOMIERO A, JONKERS N, et al., 2007. Natural and synthetic endocrine disrupting compounds (EDCs) in water, sediment and biota of a coastal lagoon. Environ. Int., 33: 929-936.

POLO M, LLOMPART M, GARCIA-JARES C, et al., 2006. Development of a solid-phase microextraction method for the analysis of phenolic flame retardants in water samples. J. Chromatogr. A, 1124: 11-21.

POSTER D L, SCHANTZ M M, SANDER L C, et al., 2006. Analysis of polycyclic aromatic hydrocarbons (PAHs) in environmental samples: a critical review of gas chromatographic (GC) methods. Anal. Bioanal. Chem., 386: 859-881.

PRIMEL E G, CALDAS S S, ESCARRONE A L V, 2012. Multiresidue analytical methods for the determination of pesticides and PPCPs in water by LC-MS/MS: a review. Cent. Eur. J. Chem., 10: 876-899.

PRINZBACH H, WEILER A, LANDENBERGER P, et al., 2000. Gas-phase production and photoelectron spectroscopy of the smallest fullerene, C20. Nature, 407: 60-63.

QUINTANA J B, RODIL R, MUNIATEGUI-LORENZO S, et al., 2007. Multiresidue analysis of acidic and polar organic contaminants in water samples by stir-bar sorptive extraction-liquid desorption-gas chromatography-mass spectrometry. J. Chromatogr. A, 1174: 27-39.

RAMOS L, KRISTENSON E M, BRINKMAN U A, 2002. Current use of pressurised liquid extraction and subcritical water extraction in environmental analysis. J. Chromatogr. A, 975: 3-29.

REISER R, TOLJANDER H O, GIGER W, 1997. Determination of alkylbenzenesulfonates in recent sediments by gas chromatography/mass spectrometry. Anal. Chem., 69: 4923-4930.

REN H, KAWAGOE T, JIA H, et al., 2010. Continuous surface seawater surveillance on poly aromatic hydrocarbons (PAHs) and mutagenicity of East and South China Seas. Estuar. Coast. Shelf Sci. , 86: 395 - 400.

REZAEE M, YAMINI Y, FARAJI M, 2010. Evolution of dispersive liquid-liquid microextraction method. J. Chromatogr. A, 1217: 2342 - 2357.

RIOS L M, MOORE C, JONES P R, 2007. Persistent organic pollutants carried by synthetic polymers in the ocean environment. Mar. Pollut. Bull. , 54: 1230 - 1237.

ROBLES-MOLINA J, GILBERT-LOPEZ B, GARCIA-REYES J F, et al. , 2013. Comparative evaluation of liquid-liquid extraction, solid-phase extraction and solid-phase microextraction for the gas chromatography-mass spectrometry determination of multiclass priority organic contaminants in wastewater. Talanta, 117 (15): 382 - 391.

RODRÍGUEZ-MOZAZ S, ÁLVAREZ-MUÑOZ D, BARCELÓ D, 2015. Environmental problems in marine biology: methodological aspects and applications. chapter: pharmaceuticals in marine environment: analytical techniques and applications. CRC Press, Taylor & Francis.

ROSLAN R N, HANIF N M, OTHMAN M R, et al. , 2010. Surfactants in the sea-surface microlayer and their contribution to atmospheric aerosols around coastal areas of the Malaysian peninsula. Mar. Pollut. Bull. , 60: 1584 - 1590.

ROSTKOWSKI P, HORWOOD J, SHEARS J A, et al. , 2011. Bioassay-directed identification of novel antiandrogenic compounds in bile of fish exposed to wastewater effluents. Environ. Sci. Technol. , 45 (24): 10660 - 10667.

RUDEL H, BOHMER W, MULLER M, et al. , 2013. Retrospective study of triclosan and methyl-triclosan residues in fish and suspended particulate matter: results from the German Environmental Specimen Bank. Chemosphere, 91: 1517 - 1524.

RUFF M, MUELLER M S, LOOS M, et al. , 2015. Quantitative target and systematic non-target analysis of polar organic micro-pollutants along the River Rhine using high-resolution mass-spectrometry identification of unknown sources and compounds. Water Res. , 87: 145 - 154.

SAEZ M, LEON V M, GOMEZ-PARRA A, et al. , 2000. Extraction and isolation of linear alkylbenzene sulfonates and their intermediate metabolites from various marine organisms. J. Chromatogr. A, 889: 99 - 104.

SALGUEIRO-GONZÁLEZ N, TURNES-CAROU I, MUNIATEGUI-LORENZO S, et al. , 2014. Analysis of endocrine disruptor compounds in marine sediments by in cell clean up-pressurized liquid extraction-liquid chromatography tandem mass spectrometry determination. Anal. Chim. Acta, 852: 112 - 120.

SAMANIDOU V F, EVAGGELOPOULOU E N, 2007. Analytical strategies to determine antibiotic residues in fish. J. Sep. Sci. , 30: 2549 - 2569.

SANCHEZ-VIDAL A, LLORCA M, FARRÉ M, et al. , 2015. Delivery of unprecedented amounts of perfluoroalkyl substances towards the deep-sea. Sci. Total Environ. , 526: 41 - 48.

SANCHÍS J, BERROJALBIZ N, CABALLERO G, et al. , 2011. Occurrence of aerosol-bound fullerenes in the Mediterranean Sea atmosphere. Environ. Sci. Technol., 46: 1335 – 1343.

SANCHÍS J, FARRÉ M, BARCELÓ D, 2012. Analysis and fate of organic nanomaterials in environmental samples. Comprehensive Analytical Chemistry, 59: 131 – 168.

SANCHÍS J, MARTÍNEZ E, GINEBREDA A, et al., 2013. Occurrence of linear and cyclic volatile methylsiloxanes in wastewater, surface water and sediments from Catalonia. Sci. Total Environ., 443: 530 – 538.

SANCHÍS P, LLORCA M, FARRÉ M, et al., 2015. Volatile methyl siloxanes in market seafood and fish from the Xuquer River, Spain. Sci. Total Environ., 443: 530 – 538.

SANCHÍS J, LLORCA M, PICO Y, et al., 2016. Volatile dimethylsiloxanes in market seafood and freshwater fish from the Xuquer River, Spain. Sci. Total Environ., 545 – 546: 236 – 243.

SANCHÍS J, 2015. Analysis of nanomaterials and nanostructures in the environment (Bachelor thesis). Universitat de Barcelona, Department of Environmental Chemistry, IDAEA-CSIC.

SAÑUDO-WIHELMY S A, TOVAR-SÁNCHEZ A, FISHER N S, et al., 2004. Examining dissolved toxic metals in U.S. estuaries. Environ. Sci. Technol., 38 (2): 34A-38A.

SCHLABACH M, ANDERSEN M S, GREEN N, et al., 2007. Siloxanes in the environment of the inner Oslofjord. NILU OR, 27/2007.

SCHMITZ-AFONSO I, LOYO-ROSALES J E, DE LA PAZ AVILES M, et al., 2003. Determination of alkylphenol and alkylphenolethoxylates in biota by liquid chromatographywith detection by tandemmass spectrometry and fluorescence spectroscopy. J. Chromatogr. A, 1010: 25 – 35.

SHANG D Y, IKONOMOU M G, MACDONALD R W, 1999. Quantitative determination of nonylphenol polyethoxylated surfactants in marine sediment using normal-phase liquid chromatography-electrospray mass spectrometry. J. Chromatogr. A, 849: 467 – 482.

SHI H, YANG Y, LIU M, et al., 2014. Occurrence and distribution of antibiotics in the surface sediments of the Yangtze Estuary and nearby coastal areas. Mar. Pollut. Bull., 83: 317 – 323.

SIMON E, LAMOREE M, HAMERS T, et al., 2015. Challenges in effect-directed analysis with a focus on biological samples. Trends Anal. Chem., 67: 179 – 191.

SKUPINSKA K, MISIEWICZ I, KASPRZYCKA-GUTTMAN T, 2004. Polycyclic aromatic hydrocarbons: physiochemical properties, environmental appearance and impact on living organisms. Acta Pol. Pharm., 61: 233 – 240.

SOLAUN O, RODRÍGUEZ J, BORJA A, et al., 2015. Relationships between polychlorinated biphenyls in molluscs, hydrological characteristics and human pressures, within Basque estuaries (northern Spain). Chemosphere, 118: 130 – 135.

SPARHAM C, VAN EGMOND R, O'CONNOR S, et al., 2008. Determination of decamethylcyclopentasiloxane in river water and final effluent by headspace gas chromatography/

mass spectrometry. J. Chromatogr. A, 1212: 124 – 129.

SPARHAM C, VAN EGMOND R, HASTIE C, et al., 2011. Determination of decamethylcyclopentasiloxane in river and estuarine sediments in the UK. J. Chromatogr. A, 1218: 817 – 823.

STALMANS M, MATTHIJS E, DE OUDE N T, 1991. Fate and effects of detergent chemicals in the marine and estuarine environment. Water Sci. Technol., 24: 115 – 126.

STAMPER M A, SPICER C W, NEIFFER D L, et al., 2009. Morbidity in a juvenile green sea turtle (*Chelonia mydas*) due to ocean-borne plastic. J. Zoo Wildl. Med., 40: 196 – 198.

STEWART I, EAGLESHAM G. K, POOLE S, et al., 2010. Establishing a public health analytical service based on chemical methods for detecting and quantifying Pacific ciguatoxin in fish samples. Toxicon, 56: 804 – 812.

TALEB H, VALE P, AMANHIR R, et al., 2006. First detection of azaspiracids in mussels in north west Africa. J. Shellfish Res., 25: 1067 – 1070.

TANOUE R, NOMIYAMA K, NAKAMURA H, et al., 2014. Simultaneous determination of polar pharmaceuticals and personal care products in biological organs and tissues. J. Chromatogr. A, 1355: 193 – 205.

TERNES T A, JOSS A, SIEGRIST H, 2004. Peer reviewed: scrutinizing pharmaceuticals and personal care products in wastewater treatment. Environ. Sci. Technol., 38: 392A-399A.

TERTULIANI J S, ALVAREZ D A, FURLONG E T, et al., 2008. Occurrence of organic wastewater compounds in the Tinkers Creek watershed and two other tributaries to the Cuyahoga River, Northeast Ohio//U. S. geological survey scientific investigations report, 5173: 60.

TEUTEN E L, SAQUING J M, KNAPPE D R, et al., 2009. Transport and release of chemicals from plastics to the environment and to wildlife. Philos. Trans. R. Soc. Lond. B Biol. Sci., 364: 2027 – 2045.

TOLLS J, HALLER M, SIJM D T, 1999. Extraction and isolation of linear alkylbenzenesulfonate and its sulfophenylcarboxylic acid metabolites from fish samples. Anal. Chem., 71: 5242 – 5247.

TOVAR-SÁNCHEZ A, 2012. Sampling approaches for trace element determination in seawater//PAWLISZYN J (Eds.). Comprehensive sampling and sample preparation: analytical techniques for scientists. vol. 1. Elsevier-AP, Amsterdam: 318 – 334.

TRAVERSO-SOTO J M, GONZÁLEZ-MAZO E, LARA-MARTÍN P A, 2012. Analysis of surfactants in environmental samples by chromatographic techniques.

TUOMISTO J, 2011. Synopsis on dioxins and PCBs. Report/National Institute for Health and Welfare (THL) ¼ Raportti/Terveyden ja hyvinvoinnin laitos: 14/2011.

UNEP, 2002. PCB transformers and capacitors: from management to reclassification and disposal. Geneva, United Nations Environment Programme, UNEP Chemicals.

United_Nations, 1999. Guidelines for the Identification of PCBs and Materials Containing

PCBs.

USEPA, 2006. 2010/15 Stewardship program. http://www.epa.gov/oppt/pfoa/pubs/stewardship/index.html.

VAN DEN BERG M, BIRNBAUM L S, DENISON M, et al., 2006. The 2005 World Health Organization re-evaluation of human and mammalian toxic equivalency factors for dioxins and dioxin-like compounds. Toxicol. Sci., 93: 223-241.

VAN DOLAH F M, 2000. Marine algal toxins: origins, health effects, and their increased occurrence. Environ. Health Perspect., 108: 133.

VAN RY D A, DACHS J, GIGLIOTTI C L, et al., 2000. Atmospheric seasonal trends and environmental fate of alkylphenols in the Lower Hudson River Estuary. Environ. Sci. Technol., 34: 2410-2417.

VANDERMEERSCH G, LOURENÇO H M, ALVAREZ-MUÑOZ D, et al., 2015. Environmental contaminants of emerging concern in seafood: European database on contaminants levels. Environ. Res., 143: 29-45.

VERSTEEG D J, RAWLINGS J M, 2003. Bioconcentration and toxicity of dodecylbenzene sulfonate (C12LAS) to aquatic organisms exposed in experimental streams. Arch. Environ. Contam. Toxicol., 44: 237-246.

VILLAR-PULIDO M, GILBERT-LOPEZ B, GARCIA-REYES REYES J F, et al., 2011. Multiclass detection and quantitation of antibiotics and veterinary drugs in shrimps by fast liquid chromatography time-of-flight mass spectrometry. Talanta, 85: 1419-1427.

WAGNER M, SCHERER C, ALVAREZ-MUÑOZ D, et al., 2014. Microplastics in freshwater ecosystems: what we know and what we need to know. Environ. Sci. Eur., 26: 12.

WANG D G, STEER H, TAIT T, et al., 2013. Concentrations of cyclic volatile methylsiloxanes in biosolid amended soil, influent, effluent, receiving water, and sediment of wastewater treatment plants in Canada. Chemosphere, 93: 766-773.

WARNER N A, EVENSET A, CHRISTENSEN G, et al., 2010. Volatile siloxanes in the European Arctic: assessment of sources and spatial distribution. Environ. Sci. Technol., 44: 7705-7710.

WEIGEL S, KUHLMANN J, HÜHNERFUSS H, 2002. Drugs and personal care products as ubiquitous pollutants: occurrence and distribution of clofibric acid, caffeine and DEET in the North Sea. Sci. Total Environ., 295: 131-141.

WHELAN M J, SANDERS D, VAN EGMOND R, 2009. Effect of Aldrich humic acid on watereatmosphere transfer of decamethylcyclopentasiloxane. Chemosphere, 74: 1111-1116.

WHO, 1997. Flame retardants: a general introduction. environmental health criteria geneva. World Health Organization.

WHO, 2000. Air quality guidelines for Europe. 2nd ed. WHO Regional Publications (European series).

WHO, 2010. Exposure to dioxins and dioxin-like substances: a major public health concern preventing disease through healthy environments.

WILLE K, KIEBOOMS J A L, CLAESSENS M, et al., 2011. Development of analytical strategies using U-HPLC-MS/MS and LC-ToF-MS for the quantification of micropollutants in marine organisms. Anal. Bioanal. Chem., 400: 1459 – 1472.

WRIGHT S L, THOMPSON R C, GALLOWAY T S, 2013. The physical impacts of microplastics on marine organisms: a review. Environ. Pollut., 178: 483 – 492.

XIAO Y, CHAE S-R, WIESNER M R, 2011. Quantification of fullerene (C60) in aqueous samples and use of C70 as surrogate standard. Chem. Eng. J., 170: 555 – 561.

YANG Y, FU J, PENG H, et al., 2011. Occurrence and phase distribution of selected pharmaceuticals in the Yangtze Estuary and its coastal zone. J. Hazard. Mater., 190: 588 – 596.

ZHANG Z, QI H, REN N, et al., 2011. Survey of cyclic and linear siloxanes in sediment from the Songhua River and in sewage sludge from wastewater treatment plants, Northeastern China. Arch. Environ. Contam. Toxicol., 60: 204 – 211.

ZHENG S, CHEN B, QIU X, et al., 2016. Distribution and risk assessment of 82 pesticides in Jiulong River and estuary in South China. Chemosphere, 144: 1177 – 1192.

ZHOU S, TANG Q, JIN M, et al., 2014. Residues and chiral signatures of organochlorine pesticides in mollusks from the coastal regions of the Yangtze River Delta: source and health risk implication. Chemosphere, 114: 40 – 50.

ZHOU L J, YING G G, ZHAO J L, et al., 2011. Trends in the occurrence of human and veterinary antibiotics in the sediments of the Yellow River, Hai River and Liao River in northern China. Environ. Pollut., 159: 1877 – 1885.

2　现代统计设计与分析方法

D. R. Fox[①]

[①] 澳大利亚环境测量所；墨尔本大学。

2.1 引　言

本章尽量避免赘述大学统计学基础课程中所涉及的常见统计学概念。现在不仅有大量教科书对经典统计的理论和实践进行了论述（Cox & Snel，2000；Bickel & Doksum，2016；Kutner et al.，2016），还积累了大量的指南材料，其中包含由世界各国及地区（如澳大利亚、新西兰、美国、加拿大和欧盟等）环境司法管辖部门发布的统计指导文件。因此本章的重点不在于介绍现有统计方法，而着力于未来发展。例如，不再使用方差分析（analysis of variance，ANOVA）计算无观察效应浓度（no observed effect concentrations，NOECs），得到"在 x 浓度下无观察效应"这种陈词滥调的结论，而是将资源重新分配，根据观察到的浓度–响应（concentration-response，C-R）现象，进行数学建模。

本章从各方面对现代生态毒理学中的统计分析进行介绍。涉及的内容并不完整和全面，仅反映了作者的偏向和兴趣。例如，未提及污染物吸收、迁移和归宿的模型，也未讨论种群动态的随机模型。但是，我们相信本章所涉及的内容是生态毒理学家进行可持续生态系统保护研究时，所面对的代表性统计问题和挑战。

2.2 毒性检测

为了监测和管理海洋环境中的污染物（潜在有毒物质），需要检测其在水体中的浓度和对生态系统中特定对象的"效应"。效应分为急性和慢性，主要由剂量决定，而剂量是浓度和时间的乘积。急性效应是短期（相对于生物体的寿命）暴露于高浓度引起的效应，慢性效应则是长期（相对于生物体的寿命）暴露于相对较低浓度引起的效应。

毒性是指物质对生物体产生急性或慢性效应的能力。什么是"效应"？这是一个关键但不易回答的问题。为此，生态毒理学家选择多种的毒性终点，从亚致死（如生殖能力受损）到致死（死亡），将其定义为 C-R 实验中的因变量（详见第 2.6 节）。

通过检查毒物浓度变化时生物的响应特征（特定的毒性终点），建立 C-R 曲线，是毒性量化的主要方法。

毒性指标的发展发轫于人类对毒性的研究，可以追溯到 20 世纪初。1908 年，Theodore Cash 在《英国医学杂志》上发表了一篇论文，描述了早期的剂量–响应实验（Cash，1908）。他在该文中谈道"药物产生作用的最小剂量是最适合建立数学关系的毒性指标"。这一药物剂量也被其称为"Grenzdose"或"limit dose"。此外，他还提出了附加术语"最小有效剂量"和"最大无效剂量"。这些术语后来被半数致死剂量

(median lethal dose) 或 LD_{50} 代替 (Trevan, 1927)。需要注意的是, *LC* (lethal concentration) 是指生物组织体外浓度, *LD* 指生物体内浓度。

新的毒性指标不断地大量出现。在生态毒理学中常用的有:
- LD_x 或 LC_x, 导致 $x\%$ 的生物体致死的剂量（浓度）;
- EC_x（效应浓度）, 使 $x\%$ 的生物体产生相应"效应"的浓度;
- IC_x（抑制浓度）, 发生 $x\%$ "损伤"的浓度;
- *NEC*（无效应浓度）, 不产生效应时的最高浓度;
- *NOEC/NOAEL*［无可见（有害）作用浓度］, 在一定时间内, 受试生物没有产生统计显著性有害效应的最大毒物浓度;
- *LOEC/LOAEL*［最低可见（有害）作用浓度］, 试验组与对照组的响应产生统计显著性差异时的最低浓度;
- *MATC*（最大可接受毒物浓度）, 全生命周期或部分生命周期的慢性毒性实验得出的 *NOEC* 和 *LOEC* 之间的毒物浓度几何平均值;
- BEC_x（边界效应浓度）, 对被测物种的效应不超过 $x\%$（置信度95%）时的最高浓度（Hoekstra & van Ewijk, 1993）。

为对这一系列监管参数指标的有效性和其组合应用于生态系统保护的效果进行评估, 学者们进行了大量研究 (Crane & Newman, 2000; Hose & Van den Brink, 2004; Payet, 2004; Shieh et al., 2001; van der Hoeven et al., 1997)。在实际中, 特定毒性检测指标的选择很大程度上取决于标准检测协议的要求。但不幸的是, 由于指标选择的灵活性, 从业人员常会通过简单的缩放将这些指标视为可转换或相关的。后者最常见的例子是, 为了使慢性和急性毒性浓度相协调, 将急性浓度按一定数量级任意缩放。虽然 Fox (2006) 试图通过确定急性和慢性比率 (acute-to-chronic ratio, ACR) 的"最优值", 尽量消除 ACR 的随机性。但从统计学来看, 其方法不合理。

毒性指标的选用规范是混乱的。长期以来, *NOEC* 和 *LOEC* 是毒性检测的首选指标, 但是它们一直受到质疑 (Chapman et al., 1996; Jager, 2012; Fox, 2008; Fox et al., 2012)。甚至有人呼吁禁止它们的使用 (van Dam et al., 2012; Warne & van Dam, 2008; Landis & Chapman, 2011)。然而, 这些反对观点也受到其他研究者的质疑。他们认为禁止使用 *NOEC* 是错误的, 是统计思维简化的结果 (Green et al., 2013)。

世界各地的环境司法管辖部门都发布了各种指南文件, 提出了首选策略, 但实行的具体细节由研究人员自行决定。例如, 经济合作与发展组织 (Organization for Economic Cooperation and Development, OECD) 发布的指南中建议基于模型进行毒性测量, 但在统计细节上, 就"由读者自行决定"(OECD, 2014)。最新修订的《澳大利亚和新西兰淡水和海水水质指南》采取了更规范的方法 (Warne et al., 2013)。对于慢性毒性检测, 该指南给出了以下毒性指标的优先序列:

1. *NEC*。
2. $EC_x/IC_x/LC_x$, 其中 $x \leq 10$（注: 三者排序不分先后）。
3. BEC_{10}。
4. EC_x 或 LC_x, 其中 $15 \leq x \leq 20$。
5. *NOEC* 或由 *MATC*、*LOEC* 或 LC_{50} 值计算出的 *NOEC*。

在慢性毒性数据不足时，OECD（2014）指南建议将急性 EC_{50}、IC_{50} & LC_{50} 转换为"慢性等效数据"，以推导出一个保护浓度。

随着毒性指标研究的发展，如贝叶斯分析方法的应用（Link & Albers，2007；Fox，2010；Cliffroy et al.，2013；Zhang et al.，2012；Grist et al.，2006；Jaworska et al.，2010），监管机构和从业人员面临着挑战，即确保毒性数据和得到数据的统计方法是：①与生态学相关；②适用于研究目的；③适度违背假设时有稳定性；④经得起科学推敲；⑤在统计上可信的。我们将在第2.7节涉及物种敏感度分布（species sensitivity distributions，SSDs）相关统计问题时，详细讨论这些挑战。在当前阶段，有以下重要的统计因素需要考虑：

（1）毒性终点的性质决定了随机变量的类型和毒性数据相应的统计处理方式。随机变量可以是离散的，也可以是连续的。根据经验，如果毒性终点是通过观察（如死亡状态）而不是通过测量来判断，那么随机变量通常是通过计数（生物的死亡数量）获得的离散数据。相反，如果毒性终点需要对某些属性（如鱼体长）进行物理测量，则获得的是连续数据（如增长率）。这不仅仅是学术上的区别，也是满足上述第②点的必要（但不充分）条件。但许多生态毒理学研究中缺乏对这一点的认识，将不相容的数据用于不恰当的统计检验和/或模型。例如，对少量样本（样本量<20）中获得的二元（死亡/存活）数据进行 t 检验和方差分析。

（2）假设毒性数据呈一般性分布（正如 SSD 建模），通过对毒性指标任意地缩放和组合，产生"足量"数据的做法，在生态学和统计学上都是不合理的。它隐含一个错误的假设：多物种单毒性终点和单物种多毒性终点的数据百分位数具有相同的潜在统计分布。只要考虑到不同毒性终点的数据存在离散和连续的区别，这一假设即不成立。

（3）在选择使用毒性指标之前，需要对其统计特性进行仔细的考察和评估。论证"新"指标、方法或评估策略的采用时，仅依靠直觉是不够的。主观地设置 ACR 等于100，很可能会对"保护浓度"产生一个保守估计，却完全忽略了环境评价中相互竞争的风险和目标。在制定环境标准时，至少有两种可能出错的方式（参见统计假设检验中的第Ⅰ、Ⅱ类错误）：一种是限制太低，未能充分保护环境；另一种是限制太高，禁止了人类活动。

2.3 设计考虑

目前，已有很多研究对一般环境数据和更具体的生态毒理学数据的设计和分析进行了报道（Environment Canada，2005；CCME，2007；European Commission，2011；Newman，2012；ANZECC/ARMCANZ，2000a, b；OECD，2012，2014）。但其中大多数都只是基于入门统计学课程中讲述的基本原则进行的重复性研究（Sparks 2000）。以 OECD（2012）& Environment Canada（2005）的两份文件为例，其主题包括假设检验、Ⅰ/Ⅱ类错误、统计功效、随机化、重复、异常值和数据转换。虽然这些统计概念很重要，但对经典/频率理论统计学的强调并不利于新策略和程序的发展。新策略和程序往往有能力更好地处理生态毒理学中出现的多种假定条件的扰动。随机性假设就是一个例证。可以肯定地说，大多数统计方法都是基于随机性和独立性的联合概念。事实上，统计学理论要求 SSD 模型中采用的毒性数据来自随机物种样本。然而实际情况并非如此（Fox，2015 以及其他参考资料）。值得注意的是，指南虽然强调随机的重要性（如毒性测试在设计和步骤的各个方面都应符合随机性，Environment Canada，2005），但同时也建议步骤中确保样品存在偏向性。例如，修订后的《澳大利亚水质指南》建议毒性数据至少要来自四个分类群中的八个物种（Batley et al.，2014）。这种有目的的抽样与随机性原则不符。尽管在环境监测的大背景下（USEPA，2002，2006），生态毒理学研究中并不总能明确区分概率抽样和判断抽样的差异，但好的建议确实存在。

虽然合理的统计学设计在指导数据收集和分析的过程中一直发挥着重要作用，但实际上其在生态毒理学研究中往往受到以下条件的严重制约：①数据采集成本高；②不能或受限使用随机性、重复性和区组等核心统计学原理；③非一致性。现场采样的物流成本和为获取毒性数据而进行实验室分析测试的昂贵费用，共同导致数据采集成本居高不下。严格的随机性定义意味着目标"种群"中每个"个体"被选入样本的概率相同。但已有指南表明随机性无法被满足《澳大利亚水质指南》，ANZECC/ARMCANZ，2000a，b）。此外，实验室毒性试验的标准协议往往只适用于数量相对较小的动物或生物体，这又产生了另一种非随机选择。重复性虽然提高了评价和推断的准确性，但也受到上述数据采集成本的限制。若通过"区组化"（根据其他外生变量组织的实验单元）控制成本，则只有事先已知无关变异的主要来源才可能采用。非一致性指的是生态毒理学数据的倾向违反了各种指南文件中描述的统计测试和步骤需要的先决条件或假设。其包括但不限于违反以下有关的假设：独立性、分布方式、方差结构、样本量、异常值、删失和响应产生机制。

本节剩余部分致力于探索生态毒理学实验设计中较新的内容，而不是总结教科书（如 Hinkelmann & Kempthorne，2008；Gad，2006）和上文指南文件中很容易习得的标准统计设计理论。

方差分析在生态毒理学数据分析中持续发挥着重要作用。尽管随着科学家们越来越少地选择 NOEC 作为指标，预计方差分析的使用会减少。但方差分析在环境化学物质毒性效应假设的测试中仍然发挥着重要的作用。使用方差分析及相应统计推断工具的关键前提是确定合适的实验设计，至少应该符合以下要求：①能对所有感兴趣的效应进行公正有效的评估；②尽可能控制可能影响实测响应的无关变异源（如 C-R 实验中的温度或盐度影响）；③尽量节约有限的资源。

减少偏差、提高精度和控制无关变异都倾向于增加处理水平的结合数和/或提高重复度，提高了实验成本。基于上述问题，正交分式析因设计在生态毒理中没有得到广泛应用令人诧异（Dey，1985）。虽然本书不可能对这一重要话题进行综合全面的解析，但可以举一个简单的例子来说明正交分式析因设计的潜在优势。

近期的一项研究中，Webb et al.（2014）描述了一项"没有遵从标准实验设计"的毒理学实验。该实验所面临的挑战是以适应独特的物理和逻辑约束的方式来满足上述三点要求。他们的解决方案依赖于先进的数学和计算——细节超出了本书的范围。下文将举出一个虚构的例子来说明正交分式析因设计的使用，该部分参考了 Webb et al.（2014）的研究。

2.3.1 示例

通过毒性实验研究一个拟建海水淡化厂对环境的潜在影响。在诸多测试中，研究人员感兴趣的一个是高盐废水是否对海洋生物具有毒性。其他重要因素还有：一天中接触有毒物质的时间（time of day，TOD）、废水的温度和盐度。在 Webb et al.（2014）的研究中，样本的放置方式可能是另一个需要控制的变异源。在该实验中，可以使用 3 个架子来放置烧杯，每个架子有 4 层。烧杯在架子上的位置（对应着不同剂量、TOD、温度和盐度的组合）非常重要。因为不同位置受到光照强度、与门的距离和热量分层等因素的潜在影响。实验受 4 个因素的直接影响：剂量（添加/不添加）、TOD（am/pm）、温度（15 ℃/25 ℃）和盐度（环境的/高的）；再加上决定烧杯位置的两个因素（架号和层号）。一个"全因子"实验要求对这些因子的 384 种组合至少全部测试一次。这不仅费时费钱，也没有必要。如果实验目的是只研究"主效应"（即各因素单独的影响），其他相互作用的影响忽略不计。那么可以选取完成实验设计的一部分以显著节省实验成本，而这就是分式析因设计。此外，如果选取的部分实验处理是有目的的筛选，则可以评估各独立因素的主效应——该特性显然是可取的，但不能通过随机或主观选择来保证。在统计学中，两个随机变量的独立性具有正交性的几何解释（如彼此呈直角）。因此能对效应独立评估的分式析因设计被称为正交分式析因设计。这种方法的应用最早可以追溯到 Adelman（1961）、Bose & Bush（1952）、Rao（1950）、Kempthorne（1947）等的研究。有一点需要注意，为满足正交性要求，对处理方式和处理数量的选择并不是一件简单的工作，需要很好地理解高层次数学概念，如线性代数、哈达玛矩阵和伽罗瓦域理论。统计软件工具的问世使这项工作变得容易，即使它们往往无法完成较复杂的设计。R（R Development Core Team，2004）是这类工具的唯一免费软件。它拥有由用户贡献的大量且快速增长的函数库，包括用于创建和分析分式析因设计的软件包。有兴趣学习的读者可以查阅 R 官网（CRAN，2015）和相关资料（Lawson，2015；Gad，2006）。

表 2.1 详细列出了通过合理分配节省成本的实验设计。该设计使得实验次数从 384 次减少到仅 25 次。这种设计被称为分辨率Ⅲ，代表主效应的评估是相互独立的，但非交互作用的独立（需要知道或假设相互作用的影响可以忽略不计）。其他一些限制较少的分辨率设计也可能有效，但它们通常需要更多的实验资源（通常以更多处理组合的形式）。

在第 2.6 节中，将讨论进行 C-R 实验时的"最优"设计。

表 2.1　污水毒性研究的正交分式析因设计

次数	层号	架号	剂量	TOD	温度/ ℃	盐度
1	1	3	不添加	am	15	环境
2	2	3	不添加	am	15	环境
3	3	3	不添加	am	15	环境
4	3	3	添加	pm	25	升高
5	4	3	添加	pm	25	升高
6	1	1	不添加	am	25	升高
7	2	1	不添加	am	25	环境
8	3	1	添加	pm	15	环境
9	3	1	添加	am	15	环境
10	4	1	不添加	am	15	升高
11	1	1	不添加	pm	15	升高
12	2	1	添加	am	15	升高
13	3	1	添加	am	25	环境
14	3	1	不添加	am	25	环境
15	4	1	不添加	pm	15	环境
16	1	2	添加	am	25	环境
17	2	2	添加	am	15	升高
18	3	2	不添加	pm	15	升高
19	3	2	不添加	pm	15	环境
20	4	2	不添加	am	25	环境
21	1	2	添加	pm	15	环境
22	2	2	不添加	pm	25	环境
23	3	2	不添加	am	25	升高
24	3	2	不添加	am	15	升高
25	4	2	添加	am	15	环境

2.4 数据处理

大数据的迅速兴起重新激发了人们对数据存储、运算和操作（被一些人称为"统计管理工作"）的兴趣，尽管这一工作被科学家们认为是平凡的，但又必不可少（New York Times，2014）。

根据定义，大数据的特征是"数据集的大小超出了常规软件工具在可接受时间内捕捉、存储、管理和处理数据的能力"（Wikipedia，2015）。产生大数据的科学领域包括天文学、电信、基因学和自然资源管理。虽然这些研发都值得投资，但也有研究者表示担心，认为急于加入大数据浪潮可能会把人们的注意力从同样重要的"小数据"问题上转移（Environmetrics Australia，2014a）。

虽然生态毒理学数据集可能永远不会属于大数据的范畴，但其规模也可以跨越几个数量级。例如，用户可从美国环保局生态数据库（USEPA，2015）下载多达10000条单一化学毒性的记录，而SSDs建模通常只使用5组或6组观测值，一般不超过20组观测值。

就实验设计而言（第2.3节），世界各地环境司法管辖区所发布的指南（或补充材料）通常对常用的数据分析技术进行综合整理。其中大部分是有用的，有些也已过时。例如，在加拿大环境局发布的"加拿大指南"（Environment Canada，2005）中，附录一是用于手工绘制C-R实验结果的空白坐标纸。

所有指南文件的另一个特点是它们对数据统计处理流程图的依赖性很强。事实上，统计新手会觉得流程图有用且便利。然而在进行初步数据分析时，严格遵循高度结构化的方法与探索性数据分析（exploratory data analysis，EDA）的目标背道而驰。事实上，EDA本质上是采用非结构化的方法，通过功能强大的计算机图形和专门的软件工具来梳理出隐藏的结构，揭示数据集的模式、趋势、异常值、相关性、极值和其他重要特征。而这一过程并没有流程图。

准备、组织和操作数据的过程有必要但耗时，有人声称这些工作量占据了数据分析工作的80%（Dasu & Johnson，2003）。缺乏一致性意见或一种标准的业务守则是造成该情况的主要原因。Wickham（2014）指出了数字数据集常见的几个问题：

- 列标题是数值，而不是变量名；
- 多个变量存储于一列；
- 行和列均存储变量；
- 同一表格中存储多类型的观测单元；
- 单一观测单元存储于多表格。

为了解决缺乏指导规范的问题，Wickham（2009）开发了R语言包tidyr、dplyr和ggplot2。前两个包用于数据整理，即将数据转换为下一步分析和查询可接受格式，这通常是一个繁琐过程。ggplot2是一个功能强大的图形软件包，它允许用户通过添加或删除图层来修改基础图,从而以交互方式探索和展示数据。因篇幅限制,本章不对这些软件包进行详细讨论,下面仅简要介绍它们的部分功能。

2.4.1 数据操作:tidyr 包

tidyr 和 dplyr 包是协同使用的。它们被设计的目的是用于清理或处理混乱的数据(Quora,2014)。tidyr 致力于简单和一致的理念,即数据框(R 中的术语,一组矩形数据)的列表示变量,行表示观测值。为了解 tidyr 具体如何工作,以表 2.2 的数据为例。表 2.2 给出了 4 次重复实验中,毒物浓度(%)与生物存活量(100 个样本)的联系。虽然结果的表格展示形式紧凑,但不适合进一步的统计分析。

表2.2 不同毒物浓度(%)与生物存活数(100 个样本)的联系,4 次重复

浓度(%)	1	2	3	4
0	89	92	88	93
3.1	87	93	97	91
6.3	96	91	90	94
12.5	93	89	95	95
25	76	67	73	85
50	0	0	0	0
100	0	0	0	0

函数 gather() 解决了这个问题。下面的 R 代码演示了存储在一个 example_1.csv 文件中的表格数据被 R 读取,然后转换为"标准"格式:

```
dat<-read.csv("example_1.csv") # read in tabular data
names(dat)<-c("concen","1","2","3","4") # assign column names
dat1<-gather(dat,rep,surv,2:5) # convert to standard format
head(dat1) # display first 6 rows of converted data
    concen  rep  surv
1    0.0    1    89
2    3.1    1    87
3    6.3    1    96
4   12.5    1    93
5   25.0    1    76
6   50.0    1    0
```

表 2.3 为行和列同时存在变量的一组数据集示例。这部分数据来自一项更复杂的实

验，该实验通过检测不同浓度的废水（以未稀释废水的百分比计）对大型藻类生长的影响（以配子体长度衡量）来评估海水淡化厂废水的毒性。同时记录了协变量信息（包括 pH、盐度和溶解氧）。

表 2.3 中共有 6 个变量（浓度、长度、pH、盐度、DO 和重复次数）。这些变量在行和列中均存在。先利用函数 gather（）将变量分离，然后利用函数 spread（）根据其他变量的级别创建多个列。

表2.3 用于大型藻类生长试验的配子体长度部分数据

浓度/%	类型	A	B	C	D
0	长度/μm	19.027	23.476	20.908	20.559
0	pH	8	8	8	8
0	盐度/‰	37.3	37.3	37.3	37.3
0	DO/(mg·L^{-1})	98	98	98	98
3.1	长度/μm	20.98	20.581	21.867	19.663
3.1	pH	7.98	7.98	7.98	7.98
3.1	盐度/‰	37.3	37.3	37.3	37.3
3.1	DO/(mg·L^{-1})	103.2	103.2	103.2	103.2

下面的 R 代码演示了存储在一个 example_2.csv 文件中的表格数据如何被 R 读取，然后转换为"标准"格式：

```
Dat<-read.csv("example_2.csv") # read in tabular data
dat2<-gather(dat,type,rep,A:D) # stacks all reps into single col
names(dat2)[3:4]<-c("rep","value") # assign names to cols
dat3<-spread(dat2,type) # creates separate cols from 'type'
str(dat3) # get details of structure of new dataframe

'data.frame':   28 obs. Of  6 variables:
$ concen: num    0 0 0 0 3.1 3.1 3.1 3.1 6.3 6.3 ...
$ rep : Factor w/ 4 leve"s""""""""""""D": 1 2 3 4 1 2 3 4 1 2 ...
$ DO : num    98 98 98 98 103 ...
$ length: num    19 23.5 20.9 20.6 21 ...
$ pH : num    8 8 8 8 7.98 7.98 7.98 7.98 8.01 8.01 ...
$ salin : num   37.3 37.3 37.3 37.3 37.3 37.3 37.3 37.3 36.7 36.7 ...
```

通过以上代码生成标准格式的数据框 dat3，有 6 列（变量）和 28 行（观测值）。

2.4.2 数据可视化：ggplot2 包

按照开发人员的观点，ggplot2 与大多数其他图形包不同，因为它具有一个深层的基本语法（Wickham，2009）。ggplot2 的主要优点是通过交互式添加"图层"来创建可发布图形。这些图形可能由不同数据集或对象的相同变量组成，比如由一个 R 包生成的平滑拟合。

以下示例来自 Eduard Szöcs 的网站（Szöcs，2015a）。示例展示了如何使用 ggplot2 为毒死蜱数据生成带注释的 SSD。数据来自于美国环保局生态数据库（USEPA，2015）。数据存储在名为 df 的 R 数据框中（图 2.1A 和 B）。

```
df <- df[order(df$val), ]          # rearrange toxicity data in ascending order
df$frac <- ppoints(df$val, 0.5)    # use intrinsic function ppoints to compute
                                   # empirical estimates of cumulative
                                   # probabilities
#
# Next use ggplot2 to build up SSD
#
require(ggplot2)                   # load the ggplot2 package
p<-ggplot(data = df)               # sets up base layer of plot and stores as
                                   # an R object

   p<-p+geom_point(aes(x = val, y = frac), size = 5)
                                   # adds a layer of points to base layer and
                                   # stores back into object p

   p                               # plot the object simply by naming it
                                   # see Figure 2.1A

# Next use log-scale for concen, add species and axis labels,
# and change background theme.

   p<-p+geom_text(aes(x = val, y = frac, label = species), hjust = 1.1,
      size = 4) + theme_bw() + scale_x_log10(limits=c(0.0075, max(df$val))) +
      labs(x = expression(pas'e('Concentration of Chlorpyrifos'[ ', m', 'g
      ', L^', ' ]')), y'= 'Fraction of species affec'ed')

   p                    # plot the object simply by naming it
                        # see Figure 2.1B
```

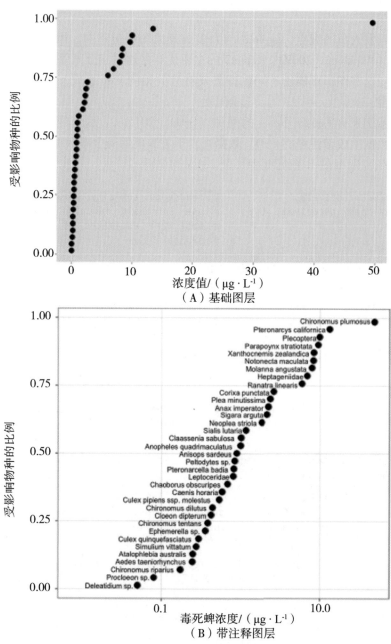

图2.1 使用 ggplot2 创建实证 SSD

2.5 估计和推断

统计推断涉及参数估计和假设检验的对偶问题。虽然它们相互关联，但有细微的差别，目的也不相同（图2.2）。

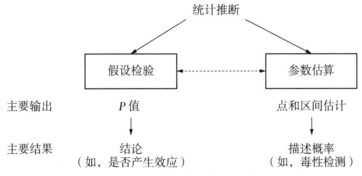

图2.2 生态毒理学中统计推断相关概念示意

（摘自 Fox, D. R., Billoir, E., Charles, S., Delignette-Muller, M. L., Lopes, C., 2012. What to do with NOECs/NOELs-prohibition or innovation? Integr. Environ. Assess. Manag. 8, 764–766.）

估算是指以某种"最佳"方式来量化未知的模型参数。而假设检验程序从一个关于未知参数的陈述开始，利用样本中包含的信息来评估这个陈述的可信性。统计推断存在两种统计范例：频率统计（即第2.1节提到的古典统计）和贝叶斯统计。虽然贝叶斯统计越来越受欢迎，但生态毒理学领域的大多数统计程序是基于频率统计设计的（Evans et al., 2010; Fox, 2010; Billoir et al., 2008）。

这两种范例的主要区别在于，频率统计将模型参数视为未知的固定常数，将数据用于寻找未知参数的"最佳"估值；而贝叶斯统计将模型参数视为以概率分布为特征的随机变量（称为"先验概率"），利用贝叶斯公式将数据用于修正先验概率分布。修正后的分布称为后验概率，是后续参数推断的基础。这两种统计和推断模式在统计学界引起了很大的分歧。值得庆幸的是，这种分歧已在很大程度上消散了，许多科学家都承认这两种方法的合理性，认为在任何研究中采用贝叶斯统计或频率统计，取决于对个人偏好、易用性和有效性的考虑。

随着贝叶斯统计在生态毒理学中的应用稳步增长，监管者面临的挑战是如何适应这种推断模式中始终存在的主观性要素。事实上，频率统计与贝叶斯统计之间的争议大多集中于此。也许正因为频率统计是"数据驱动"的，并且不接受主观概率，所以它在制定环境标准方面享有非常突出的地位。相反，贝叶斯统计的主观因素使许多人认为它提供了更丰富、更翔实的分析。贝叶斯统计吸引人的特点是先验密度可以作为载体将专家知识引入分析。然而，如 Barnett 和 O'Hagan（1997）所指出的，从一个或多个专家那里获取信息的过程"充满了技术上的困难"。最近,使用结构化信息获得技术的研究,试图减少在认知不确定存在时,专家偏差和错误的来源(McBride et al., 2012)。

尽管如 Fox(2015)所指出的那样,使用频率统计进行数据处理也并不像许多人想象的那么客观;但在推断海水和淡水中有毒物质的安全浓度时,贝叶斯统计使用主观信息的问题一直受到人们关注。即使基本方法合理,使用贝叶斯统计推导出的"保护"浓度也难以让人信服。可以预见,贝叶斯统计所得结果的可信度会在许多方面受到挑战。有人认为贝叶斯统计中负责先验概率的"专家"在选择上是存在偏见的;或者贝叶斯统计结果与更传统的频率统计结果进行比较时,不可避免地会产生不一致和/或不可调和的差异。例如,Cliffroy 等(2013)比较了由频率统计和贝叶斯方法得出的 HC_5 估算值(经 SSDs 获得的 5%的危害浓度),发现贝叶斯统计倾向于低估参考值,两者因子的差异从 0.2 到近 6。研究者认为低估倾向是一种积极特征,因为基于低估值,保护水平得到了提高。但在环境规划评估过程中,这一观点不可能得到既得利益相关者的认同。

虽然本节不可能全面介绍统计性估计程序,但广义线性模型值得一提,因为它是许多常用方法的基础,如回归分析、方差分析、协方差分析(analysis of covariance, AN-COVA)(Graybill, 1976)。所有这些方法的通用模型可用式 2.1 表示:

$$Y = X\beta + \varepsilon \quad (2.1)$$

式中,X 是 $(n \times p)$"设计"矩阵,它包含实验因素和(或)协变量 $\{x_1, x_2, \ldots\}$ 的代码;β 是 $(p \times 1)$ 待估算的参数向量;ε 是 $(n \times 1)$ 随机误差向量,假定随机误差服从独立正态分布,均值为 0,方差为 σ_ε^2。式(2.1)的最简单形式对应于 y 对 x 的简单回归,即

$$y_i = a + bx_i + e_i \quad (2.2)$$

式中,y 是 $(n \times 1)$ 响应矢量;$y^T = [y_1 \ y_2 \ \cdots \ y_n]$;$X = [1^T \ x^T]$;$\beta^T = [a \ b]$,当部分或全部的协变量是因子时,相应的 x 是"虚拟"代码向量。例如,当 x 简单地指示毒物的存在与否时,分配给 x 的值可以是 $\{0 \ 1\}$。

无论 X 是否包含所有检测值、所有虚拟代码或两者的组合,模型的参数都使用相同的方程来估算,即

$$\hat{\beta} = (X^TX)^{-1}X^TY \quad (2.3)$$

此外,估算参数的方差-协方差矩阵为

$$\text{Cov}[\hat{\beta}] = \sigma_\varepsilon^2(X^TX)^{-1} \quad (2.4)$$

一个关键之处是理解统计设计如何影响推断的质量。在上述简单线性回归方程中,这很容易由式(2.5)中 X^TX 验证,其逆方程见式(2.6)。

$$X^TX = \begin{bmatrix} n & \sum_{i=1}^{n} x_i \\ \sum_{i=1}^{n} x_i & \sum_{i=1}^{n} x_i^2 \end{bmatrix} \quad (2.5)$$

$$(X^T X)^{-1} = \frac{1}{n\sum_{i=1}^{n} x_i^2 - \left(\sum_{i=1}^{n} x_i\right)^2} \times \begin{bmatrix} \sum_{i=1}^{n} x_i^2 & -\sum_{i=1}^{n} x_i \\ -\sum_{i=1}^{n} x_i & n \end{bmatrix} \quad (2.6)$$

所以，我们从式（2.4）和式（2.6）可以看出，估算参数的方差和协方差完全是 x_i 的函数（对于给定的 σ_ε^2），而与响应 y_i 无关。此外，式（2.6）中非零的非对角项表明，参数 $\{a \ b\}$ 不是独立估算出来的（除非 $\sum_{i=1}^{n} x_i = 0$）。

回顾第 2.3 节表 2.1 中列出的分式析因设计，因子和不同条件数量为层号（4）、架号（3）、剂量（2）、TOD（2）、温度（2）和盐度（2）。不深究细节的情况下，式（2.1）中设计矩阵 X 所需的参数 p 的数量为 $p = \sum_{j}^{k} L_j - k + 1$，其中 k 是因子的数量，L_j 是因子 j 的层数。所以对于表 2.1 的设计，$p = 15 - 6 + 1 = 10$，得到一个（25×10）设计矩阵 X。对应表 2.1 中数据对 X 编码（此处未给出），得到以下 $X^T X$ 矩阵：

$$X^T X = \begin{bmatrix} 25 & 0 & 0 & 0 & 0 & 0 & 0 & 0 & 0 & 0 \\ 0 & 15 & 0 & 0 & 0 & 0 & 0 & 0 & 0 & 0 \\ 0 & 0 & 15 & 0 & 0 & 0 & 0 & 0 & 0 & 0 \\ 0 & 0 & 0 & 15 & 0 & 0 & 0 & 0 & 0 & 0 \\ 0 & 0 & 0 & 0 & 10 & 0 & 0 & 0 & 0 & 0 \\ 0 & 0 & 0 & 0 & 0 & 10 & 0 & 0 & 0 & 0 \\ 0 & 0 & 0 & 0 & 0 & 0 & 25 & 0 & 0 & 0 \\ 0 & 0 & 0 & 0 & 0 & 0 & 0 & 25 & 0 & 0 \\ 0 & 0 & 0 & 0 & 0 & 0 & 0 & 0 & 25 & 0 \\ 0 & 0 & 0 & 0 & 0 & 0 & 0 & 0 & 0 & 25 \end{bmatrix}$$

从 $X^T X$ 中所有非对角项均为零这一事实可以看出分式析因设计的正交性。如前所述，这是可取的。因为所有主效应的估算都是相互独立的。此外，$X^T X$ 的对角结构意味着通过将对角元素替换为它们的倒数，就可以得到逆矩阵。如果对角线项并不完全相同，则意味着模型参数的估算精度是可变的。在理想情况下，编码方案会使对角线项都等于 25。用理想的设计来衡量所提议设计的总体效率，在判断我们的设计优劣时很有用。其中一种方法是检查矩阵 $(X^T X)^{-1}$ 的行列式（Wikipedia，2016）。因此选择"好"分式析因设计的一个标准是使该行列式最小化，这是 D – 最优性的基础，第 2.6 节将进一步讨论。本示例可以证明，表 2.1 中的分式析因设计效率约为全析因设计的 70%。

如本节开头所述，虽然估计和假设检验程序侧重点不同，但存在重叠。估计 SSDs 或复杂 C-R 模型的参数时，通常需要使用复杂的数学和统计工具。而更具挑战的是确定从拟合模型中所得量的标准误差。例如，式（2.7）是 C-R 数据建模中用到的一个四参数逻辑函数：

$$y = \beta_0 + \frac{\beta_1 - \beta_0}{1 + \exp\{\beta_2[\ln x - \ln \beta_3]\}} \quad (2.7)$$

这是一个复杂的非线性模型。不仅浓度 (x) 是非线性的，而且参数 $\{\beta_0, \beta_1, \beta_2, \beta_3\}$ 也是非线性的。没有简单的方法来估算这些参数，必须使用专业计算机软件，如将在第 2.9 节中讨论的 R 软件包。一旦拟合，通过将式（2.1）中的参数替换为估算值 $\{\hat{\beta}_0, \hat{\beta}_1, \hat{\beta}_2, \hat{\beta}_3\}$ 和（或）使用式（2.8）估算给定响应 y_0 时的浓度 $\hat{x}_0$。

$$\hat{x}_0 = \exp\left\{\frac{\ln\left(\frac{\hat{\beta}_1 - y_0}{\hat{\beta}_0 - y_0}\right) + \hat{\beta}_2 \ln \hat{\beta}_3}{\hat{\beta}_2}\right\} \tag{2.8}$$

对给定值进行参数估计时，可在计算机上进行式（2.7）和（2.8）的计算。但得出估计值的标准误差没有"封闭式"的表达式，需要更复杂的方法。这一问题通常使用统计重采样技术（如自举法或刀切法）来解决。这涉及从样本数据中重复抽样（和替换）。对每个样本进行参数估计，并使用式（2.7）或（2.8）计算兴趣量（quantity of interest）。获得的估计值之间的差异提供了统计量的标准误差估值。显然这是一项复杂且重复的任务，非常适合使用计算机来完成。

2.6 浓度-响应（C-R）模型

本节概述了与 C-R 建模相关的一些统计问题，并指出了未来的发展方向。但本节并不会全面介绍 C-R，如使用毒代-毒效动力学（toxicokinetic-toxicodynamic，TKTD）模型评估体单个生物体水平上的毒性。

目前，C-R 实验产生的数据用于设置毒物的"安全"暴露浓度，这是第 2.7 节中 SSDs 的主题。值得注意的是，术语"C-R"常常等同于"剂量-反应"。但严格来讲，它们并不相同。因为剂量是浓度、频率和持续时间（暴露）的函数。

C-R 实验及为分析其数据而开发的数学和统计建模工具具有悠久而卓越的历史，接下来对此进行简要追溯。

动物或生物体的 C-R 曲线可以呈现出各种各样的形状，有简单的线性关系，也有更复杂的毒物兴奋效应（低浓度下的刺激效应）和滞后效应（浓度增加函数的响应轨迹并非与浓度减小函数相反）曲线形式。图 2.3 显示了在理想状态下的 C-R 曲线的两个例子。

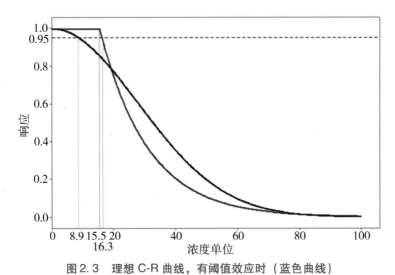

图2.3 理想 C-R 曲线，有阈值效应时（蓝色曲线）
$EC_5 = 16.3$；非阈效应（红色曲线），$EC_5 = 8.9$；无效应浓度（NEC）= 15.5

图 2.3 中表明了两个 EC_5（即预期 5% 效应浓度）和 NEC 的估计值。虽然是假设的，但图 2.3 确实突出了 C-R 建模中一个真实且常见的问题，即派生度量对模型选择和参数化的敏感性。

如果图 2.3 中的反应表示未受影响生物体的比例，则受影响的比例呈现形状如图 2.4 所示。这种细长的"S"形在 C-R 模型和种群增长模型中均有出现。

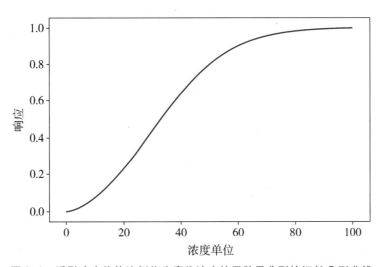

图 2.4 受影响生物体比例作为毒物浓度的函数呈典型的细长 S 形曲线

英国学者托马斯·马尔萨斯（1978）在对种群动态的早期研究中，使用了简单的指数模型来描述种群增长。但这不太符合实际，因为种群并不能无限增长。更真实的生长模型呈现出渐近特性，其中最重要的是 1838 年由弗朗索瓦·韦尔赫斯特（1838）提出的逻辑斯蒂函数。逻辑斯蒂方程一直是一个相对晦涩的数学成果，直到 1920 年才被约翰·霍普金斯大学的雷蒙德·珀尔和洛厄尔·里德重新发掘（Pearl & Reed，1920）。20世纪30年代，著名统计学家切斯特·布利斯和罗纳德·费舍尔采用了略微不同的方法进行生物测定建模。因个体耐受程度不同，将处理刺激（剂量）作为协变量，而将反应作为随机变量（Bliss，1935）。在这个公式中，概率模型与反应相关，反应通常服从正态分布，但这不是必要条件。Bliss创造了术语"probit"作为概率单位（probability unit）的简写（Bliss，1935）。

概率分析很快在实际建模中得到了应用，以描述一个离散的二元结果与一个或多个解释变量的关系。1944年，美国统计学家约瑟夫·伯克森提议使用逻辑斯蒂函数替代概率分析（Berkson，1944），并引入术语"logit"作为比例数学变换的缩写，该变换采用让步比的对数（让步比被宽泛的定义为"成功的"概率/"失败的"概率）。

如第2.2节中所述，ANOVA方法多年来常与C-R实验相结合，来推导NOEC/NOEL毒性指标，而这些指标一直受到质疑。尽管不愿再回顾一遍反对意见，但为了更好地设计C-R实验，浅析这些检测的关键缺陷是有益的。

在C-R实验中，假设检验被用来评估毒性效应陈述的有效性，而估计方法被用于获得毒性效应的测定方法（Fox et al.，2012）。实际上，使用ANOVA方法完成毒性测定是对实验资源的巨大浪费，因为它没有使用任何包含在反应-生成机制中存储的信息（即C-R模型）。此外，ANOVA需要在每个浓度下进行重复实验，这与C-R建模要求相悖。有人认为，如果效应和毒物浓度的关系未知（Newman & Clements，2008）或定义不清（Green，2016）时，假设检验比建模更可取。这正是第2.4节中讨论过的EDA方法的作用。最后，采用ANOVA及相关的多重比较方法推导毒性测定，是将浓度当作因子，而非协变量。这意味着实验室中经仔细测量的浓度数据，不被认为是定量信息，而是仅仅被当作不确定性估计和精度说明的"标签"，这让人难以接受（Fox et al.，2012；Fox & Landis，2016a）。

倡导使用基于模型估计方法的建议可以追溯到20多年前（OECD，1998）。但即使开展了广泛的宣传（Landis & Chapman，2011；Warne & van Dam，2008；van Dam et al.，2012；Fox & Landis，2016b），革新却一直进展缓慢，并被惯性所阻碍（Green et al.，2013；Green，2016）。

为了突出基本难点，图2.5展示了除草剂对黄瓜生长的毒性实验结果。图2.5A为实测浓度对数与反应的曲线关系，方差分析仅用于区分不同的剂量组（图2.5B）。此外，虽然图2.5A表示在$10^{-4} \sim 10^{-3}$区间有"效应"，但使用方差分析得出图2.5B中的剂量组（1—7）与对照组相比无统计显著性。取剂量组7中"最大值"作为效应水平的估计值，剂量组7对应于初始浓度为0.0055单位或$10^{-2.26}$，该浓度刚好在图2.5A的效应范围内。

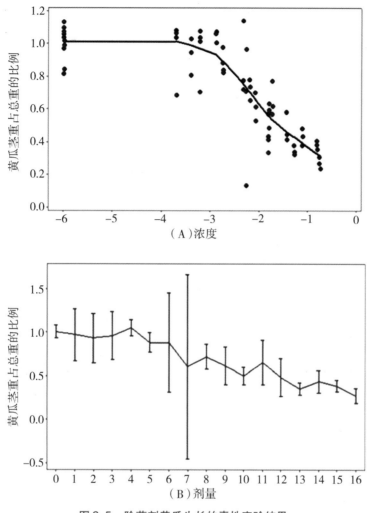

图 2.5 除草剂黄瓜生长的毒性实验结果

图片来源于 Moore, D. R. J., Warren-Hicks, W. J., Qian, S., Fairbrother, A., Aldenberg, T., Barry, T., Luttik, R., Ratte, H. -T., 2010. Uncertainty analysis using classical and Bayesian hierarchical models. In: Warren-Hicks, W. J., Hart, A. (Eds), Application of Uncertainty Analysis to Ecological Risks of Pesticides, CRC Press, Boca Raton FL, USA.

如果我们假设（或确定）剂量组 7 的反应被破坏，可以合理地将其移除，然后重新分析。这样做的结果是 NOEC 增加到 0.0065 单位。而基于模型的分析（本节未显示）表明 NEC 的估计值实际上略有下降，相对不受删除异常数据的影响。虽然最终很容易获得基于模型估计的精确度或置信度结果，但这些结果对基于方差分析的分析毫无意义，也根本无法获得（Fox et al., 2012）。

显然，使用模型进行毒性估计是一种更丰富和全面的推断模式。但这些增益的"代价"通常是计算复杂度的增加。这在过去可能会带来问题，但成熟工具（如 R）和专业软件（drc 包）的可用性，意味着可相对容易进行对复杂 C-R 模型的估计和推断。

如第 2.5 节所述，贝叶斯方法在生态毒理学中的应用越来越普遍，而如何将主观评估正式纳入监管框架仍是一个悬而未决的问题。但毫无疑问，这种统计思维和分析模式会继续发展。以图 2.5A 中的数据为例，遵循 Fox（2010）中的步骤，使用贝叶斯方法可以拟合式（2.9a）和式（2.9b）给出的指数阈值模型。

$$Y_i \stackrel{d}{\sim} g\gamma \tag{2.9a}$$

$$E[Y_i \mid x_i] = \mu_i = \alpha\exp[-\beta(x_i - \gamma)I(x_i - \gamma)] \tag{2.9b}$$

其中，Y_i 表示浓度为 x_i 时的响应，$I(z) = \begin{cases} 1, z > 0 \\ 0, z \leq 0 \end{cases}$。

$E[Y_i \mid x_i]$ 表示以 x_i 为条件时 Y_i 的数学期望值；符号"$\stackrel{d}{\sim}$"表示"分布为"。参数 α 是浓度为 $0 \sim \gamma$ 之间的平均响应；β 控制浓度大于 γ 的响应速率；γ 为 NEC。

尽管看起来很复杂，但是可以借助免费软件，如 OpenBUGS（Openbugs，2009）、JAGS（Sourceforge，2015）或 Stan（Stan Development Team，2015）进行贝叶斯分析，约 10 行代码就可以完成分析。

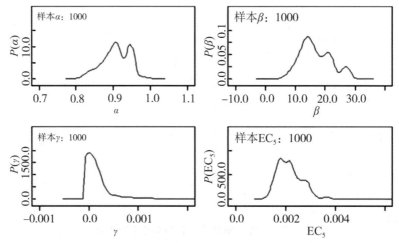

图 2.6　不同的参数以及图 2.5 中数据的 EC_5 对应的经验后验密度

频率统计和贝叶斯统计之间的一个分歧点是频率使用点/区间估计和/或假设检验，而贝叶斯的所有推断都基于后验密度。假设检验是频率统计的概念，贝叶斯理论不存在这一说法。图 2.6 总结了参数 α、β、γ 和 EC_5 的相关后验分布抽样结果。对于这些分布，可以获得平均数和标准差等常规汇总统计数据。而最优贝叶斯点估计是相关后验分布的中位数或众数。贝叶斯统计通过确定后验密度下的某个标称面积值（如 0.95）的最短区间，获得最高后验密度（highest posterior density，HPD）区间，以取代置信区间。参数 γ 的 95% 可信区间为 [0, 0.0014]，与基于图 2.5A 中数据的主观评价一致。值得注意的是，对于该数据，获得与先前确定的 NOEC 相同 γ 值的估计后验概率仅为 0.003，这表明贝叶斯方法和频率方法之间存在不可调和的差异。

C-R 建模另一个日益引人注意的方面是其设计。为了获得和分析 C-R 模型中的实测数据，高昂的成本引起了人们对实验设计的关注。实验设计需要能够从固定样本量中获得对关键参数的最精确估计。在实践中,毒理学家约定俗成的观点是,C-R 实验中所使用的每种浓度至少需要重复 3 次实验。这一惯例无疑是由生成 NOECs 的单向方差分析方法的重复性要求所驱动。如果没有重复实验,方差分析模型中的均方误差项就无法估计,从而缺少了推断(包括 NOEC 的测定)的基础。尽管选择使用建模方法的趋势在上升,但每种浓度进行至少 3 次重复实验的设计并不少见。这不仅没有必要,而且浪费宝贵的实验资源。在统计学中有个简单的公式:

$$数据 = 模型 + 误差$$

上述方程中的"误差"是模型未考虑到的所有数据变化的"集合",因此常被称为"残差"。"误差"的来源多种多样,但随机误差和失拟误差是两个重要的来源。随机误差在一些软件中被称为"纯误差"。它代表来自"期望值"的不可预测的、随机发生的变化,因此只能用概率来描述。失拟误差源于模型的错误设定,并表示由于所选模型无法准确描述反应而产生的差异。图 2.7 是 R 软件 drc 包提供的叶长数据图,其展示了不同误差的区别。因此,如果进行了重复实验,就得到纯误差的独立估计值。若再减去从许多软件包中报告为残差的均方,就获得失拟误差的估计值。

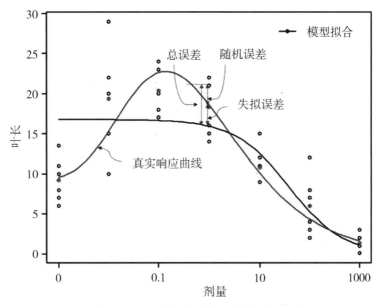

图 2.7　总误差、随机误差和失拟误差间的关系

空心圆圈为实测数据,实心红点是相应浓度下的平均响应。纵轴表示测得叶长(cm),横轴表示甲磺隆浓度(mg/L)

回到设计的问题上,除非需要检验失拟误差的重要性,否则不必进行重复性实验。即使有需要,还有其他"成本较低"的方法来评估替代模型形式的充分性。例如,drc 包具有通过赤池信息或 AIC(akaike information criterion,AIC,与数据相符模型的效用的一种衡量指标)标准或贝叶斯信息准则(类似于 AIC,其计算不需要在贝叶斯分析中指定先验分布)来比较模型列表效用的功能。

正如 OECD(2012)所指出的,剂量水平的数量和剂量梯度的选择不仅对研究目标

（如危害识别或剂量－反应/风险评估）的实现至关重要，对随后的统计分析也很重要。由于不再需要重复性实验，可以将实验资源重新分配，用于增加在 C-R 曲线关键区域附近提取的信息密度（如 NEC）。虽然 C-R 实验的设计（如剂量梯度和时间梯度的选择）是主观地或基于试探规则进行，但可以通过优化某些标准来确定"最佳"实验设计，从而使用更正式的程序。如第 2.5 节中介绍的 D－最优准则。

2.6.1 C-R 实验的局部 D－最优设计

目前，采用数学优化方法设计 C-R 实验的情况并不普遍。但随着生态毒理学家从基于 ANOVA 转向基于模型的估计和推断，这种情况有望得到改善。D－最优准则在第 2.5 节中的分式析因设计部分有所提及，通常使用 ANOVA 方法进行分析。与第 2.5 节中的例子相同，最优准则仅是设计矩阵 X 的函数。X 可以在数据采集之前完成，然后通过构造 X 来计算任何给定设计的效率。但由于评价最优准则需要模型参数的背景信息，所以设计一个非线性 C-R 实验的统计模型不可能实现。因此，我们面临一个"死循环"：设计一个实验来有效估计模型参数，但"最优"实验设计需要先收集数据来估计参数。虽然可以使用参数的"最优猜测值"[例如，Chevre & Brazzale（2008）采用了他们认为的"合理值"]。但此方法显然不是完全令人满意的。由于无法评估猜测值的准确性，因此任何基于猜测值的设计都存在此问题。

对于这一问题存在更为复杂的解决方案。例如，Li & Fu（2013）将贝叶斯方法与设计问题的自适应方法相结合。这个过程很复杂，需要理解高等数学和统计概念。其方法修改版的简化概要如下：

（1）将总样本量的一小部分作为"预实验"，采用主观设计（可能受类似研究的结果启发）。例如，在传统 ANOVA 型的 C-R 实验中，假设总样本量 $n = k \cdot r$，其中 k 是采用的浓度组数，r 是每种浓度需要重复的实验次数。假设预实验中 $n_1 = k$。

（2）利用预实验结果，获得经验后验分布、感兴趣参数向量的 $p(\Theta | \text{data})$、Θ（如何做到这一点取决于所使用的特定软件包）。确定候选设计列表 $\{D_i\}$，$i = 1, \cdots, m$，其中，D_i 是剩余样本量相应的设计浓度，剩余样本量 $n_2 = n - n_1 = k(r-1)$。

（3）对于给定的设计，D_i：①从后验分布 $p(\Theta | \text{data})$ 抽样，得到一个响应预测 $Y = \{Y_1, \ldots, Y_{n_2}\}$，重复 L 次；②对于（a）中的每个样本，求得效用函数 $g(Y)$ [例如，$g(\cdot)$ 可能等于 D－最优标准]；③根据（a）中 L 次重复的平均值，估算 $E[g(Y) | D_i]$。

（4）步骤（c）中平均值最大的设计即为候选设计 $\{D_i\}$ 中的最优设计。

2.7 物种敏感性分布法模型

本节简要概述 SSD 模型的现状，但不涉及 SSD 的计算和推断方法，如有兴趣可参考其他资料（Posthuma et al.，2002；Duboudin et al.，2004）。

从某种意义上说，SSD 是一种推断工具，它使我们能够超越个体毒性指标来判断某种化学品对整个"种群"的毒性。但是种群是被定义的（尽管通常指构成生态系统的所有物种）。

SSD 是一种统计策略。因此在使用和解释它的同时，需要考虑一些重要的统计因素。与所有的推断统计技术一样，种群和样本之间存在重要的区别。对于统计学家来说，种群只是其感兴趣"事物"的最大集合。它可能是有生命的或无生命的。例如，电子工程师可能对推断大量电路板的总体故障率感兴趣；而植物生理学家可能感兴趣的是估计某植物物种的个体在 CO_2 水平升高时的气孔反应。在生态毒理学中，SSD 使我们能够在只观察到一小部分反应的情况下，推断某一物质对特定种群中所有物种的毒性效应。

虽然 SSD 是生态毒理学的一个重要发展方向，但其应用也一直存在问题和争议（Forbes & Forbes，1993；Forbes & Calow，2002；Hickey & Craig，2012；Wheeler et al.，2002a，b；Zajdlik，2006，2015）。至少从概念上来说，借鉴统计推断思维是有意义的。如果我们从定义的种群中随机抽取一个物种样本，就可以将从样本中获得的毒性测定集合进行一个理论概率分布拟合。使用这种理论结构的优点（大概）是可以通过少量样本对毒性进行评估，进而扩展到整个种群，这也是 SSD 有时被称为"外推法"的原因。但 SSD 也存在一些严重的缺陷。即使经过 30 多年的应用和改进，这些缺陷仍然存在，并有可能影响整个方法的可信度。已有许多文章论述了 SSD 方法及相应指标（如 NOECs 等的缺陷）（Newman et al.，2000；Okkerman et al.，1991；Wang et al.，2014）。虽然其中一些缺陷已经通过一些方法得到明显改善，如增加样本量以提高推断质量，但这些方法在实际操作中并不总是具有可行性。而且有些缺陷一直都无法解决。其中最突出的缺陷是：①与其他科学领域（如物理、化学和热力学）不同，其他领域有一些理论可以解释为什么要应用某种特定的函数形式，而 SSDs 却无此基础；②尽管随机性是统计假设的核心要求，但用于获得样本统计分布的物种选择从不是随机的。

关于上文所述缺陷①，目前的做法是使用少量概率模型（特别是对数正态分布、logistic 分布和对数 logistic 分布）作为 SSD 合适的描述符。然而，生态毒理学中没有指导理论来证明一种分布形式优于另一种。对于许多应用来说，这并不是问题。因为所有选择提供的拟合都是合理的。但在生态毒理学中，这种"自由度"是很有问题的。标准拟合优度检验对获得"最佳"分布没有帮助。因为小样本将不可避免地导致低效率的拟合优度检验。这意味着任何看似可信的候选模型都不太可能被否决。因此函数形式选择这一令人困扰的问题不太可能得到解决。难以在 SSD 模型中选择函数形式的另一个原因是人们最终兴趣在于拟合曲线中定义最不明确的部分——即左尾部极值（假设值越大，不利结果越大）。拟合 SSD 的最终目标是估计受指定毒物浓度暴露不利影响的物种比例，或是估计对不超过所有物种 $x\%$ 有害的浓度。后者被定义为 HC_x，在数值上等同于 SSD 的第 x 百分位，通常为 HC_5。

在物种选择问题上，Posthuma et al.（2002）提到"SSD 的一个严重缺陷是假设测试物种的一个非随机样本代表所研究群体"。随机性假设对统计分析至关重要。但令人惊讶的是在生态毒理学领域，几乎没有采取什么措施来克服该问题。世界各国的指南和监管要求实际上通过强制使用选择性抽样而将非随机性嵌入 SSDs 中。例如，修订版《澳大利亚和新西兰水质指南》建议，使用至少来自于 4 个分类群中 8 个物种的毒性数据（Batley et al.，2014；Warne et al.，2014）。随机抽样在 SSD 建模中可能仍然是一种难以实现的理想。其原因至少有两个：①不可能鉴别一个生态系统中的所有物种；②只能选择少数（非随机）物种进行检测。有趣的是，尽管这是一个长期存在且被广泛承认的问题，但直到最近人们才认真尝试量化非随机物种选择对 SSD 拟合和 HCx 估计的影响，并对此进行改善（Fox，2015）。

尽管这一关键问题需要更多的研究，但 Fox（2015）已证明，在为物种选择函数指定一个 β 概率密度函数（probability density function，pdf）的较弱假设下，实际的 SSD 并非假设的 SSD（除非选择过程真的是随机的）。Fox（2015）特别指出当采用 SSD 描述毒性数据（X），并且假设 SSD 服从参数为（α，β）的 log-logistic 分布，X 服从参数为（a，b）的 β 分布时，那么实际分布是修正后的 F 分布，具有以下概率密度函数（pdf）：

$$gx(x;a,b,\alpha,\beta) = \frac{b\beta}{a\alpha}\left(\frac{x}{\alpha}\right)^{\beta-1} dF\left[\frac{b}{a}\left(\frac{x}{\alpha}\right)^{\beta};2a,2b\right] \quad (2.10)$$

符号 $dF(\cdot;v_1,v_2)$ 表示具有 v_1 和 v_2 自由度的标准 F 分布。上述 pdf 所满足的一个要求是，物种选择为真正随机（对应于 β 分布的特殊情况 $a=b=1$），样本数据的实际分布服从假设的 log-logistic 分布。非随机物种选择的影响是深远的——导致 HC_x 的估计误差达到或超过 20 倍（Fox，2015）。幸运的是，HC_x 的非随机性偏差可以通过偏差校正因子（bias correction factor，bcf）来校正。假设一个 log-logistic 的 SSD 模型具有参数为（a，b）的 β 选择函数，以此对（偏差）数据进行拟合得到 HC_x。适用于 HC_x 的 bcf 为

$$bcf(x) = \left[\frac{b}{a\,\xi_x}\frac{x}{100-x}\right]^{\frac{1}{\beta}} \quad (2.11)$$

ξ_x 是标准 F 分布中具有 $2a$ 和 $2b$ 自由度的第 x 百分位数（Fox，2015）。运用式（2.11）的唯一困难在于，需要知道所假设 log-logistic SSD 形状参数 b 的真实值。对此，Fox（2015）建议将式（2.11）中的 β 替换为其样本估算值 $\hat{\beta}$。例如，假设拟合的 log-logistic SSD 参数为 $\hat{\alpha}=5.46$，$\hat{\beta}=1.76$。此外，物种选择函数可以优先选择较敏感的物种，用 $a=0.5$ 和 $b=2.0$ 的 β 密度来描述。使用已发布的具有 1~4 个自由度的 F 分布表，或者更方便地使用 EXCEL 中固有函数 F.INV（），很容易求得当 $x=5$，$\xi_x=0.004453$ 时，$bcf(5)=8.94$。换句话说，根据样本数据拟合 SSD 得出的 HC_5 几乎增加一个数量级，以补偿非随机选择过程产生的偏差。

另一个活跃的研究领域是明确地将时间因素考虑在内，旨在提高 SSD 建模的有效性和适用性，这一直是整个方法中的"缺失维度"。如我们所看到的，采用少量、非随机的毒性数据样本估计理论 SSD 的参数。无论是回归还是基于 ANOVA 分析，这些度量无一例外地通过已具标准化实验室规程的 C-R 实验得到。其中一部分是确定了实验持续暴露时间——通常为 24 h、48 h 或 96 h。因此，从这些实验中检测到的毒性数据实际上仅与某一特定的暴露时间有关。Fox & Billoir（2013）在研究中指出，几乎没有人尝试完全集成 SSD 中的时间组件。在该研究中，只采用将时间引入 SSDs 模型的简单假设，量化该额外维度对诸如 HC_x 等重要数值的影响。当然，HC_x 本身就是 C-R 持续暴露时间的函数。Kon Kam King et al.（2015）使用"毒物动力学"TD 模型在 SSD 中增加了时间维度。TD 模型是人们更熟悉的 TKTD 模型的一个"集成"版本，TKTD 模型用于描述生物体个体的毒物吸收、转化和毒性效应相关的各种生物学过程（Jager et al., 2011）。在 TD 模型中，个体的数量为 N_{ijk}，它表示物种 j 暴露于第 i 浓度水平时，正常存活了时间 t_{k-1}，直至时间 t_k 时死亡，N_{ijk} 服从二项分布，t_k 到 t_{k-1} 区间的存活率是时间和暴露浓度的四参数函数，其中一个参数为 NEC。采用贝叶斯层次模型估计参数并推断 HC_5 的时变特性。Kon Kam King et al.（2015）使用已发表的澳大利亚墨累达令盆地区域 217 种大型无脊椎动物耐盐性数据进行分析，发现 HC_5 对时间存在强烈的依赖性，72 h C-R 实验获得的 HC_5 可能高于更长暴露实验时间获得的 HC_5。从环境角度来看，这一结果令人担忧。因为长期暴露在不超过常规方法确定的 HC_5 盐度环境中，潜在受影响的物种比例可能远远高于假定的 5%。

2.8　生态毒理学统计软件工具

本节概述了生态毒理学中的一些常用软件工具，并非面面俱到，主要聚焦于 C-R 数据分析和 SSD 拟合。前一节中简要提及了 TKTD 模型。虽然它用于获得毒性估计，但重点是通过对个体暴露于某污染物的程度和生物体内部浓度之间的关系建模，来预测一段时间内的毒性效应。因此，模型需要明确地描述控制吸收、消除、体内分布和代谢过程。与 TKTD 模型密切相关的是围绕动态能量预算（dynamic energy budget，DEB）理论发展起来的模型。其中最常见的是 DEBtox，最早由 Kooijman & Bedaux 在 1996 年提出（Kooijman & Bedaux，1996）。DEB 是"完全机械地将毒物的外部浓度与对生活史特征（如生存、生长和繁殖）的影响联系（时变的）起来"的理论（Debtox Information Site，2011）。它通过构建规则来实现这一点。这些规则控制着生物体的能量分配，从而控制各种生命阶段。

信息和通信技术"革命"以及前面所提及的大数据的兴起，促进了计算机软硬件迅速繁盛的发展。这也意味着在生态毒理学中可以使用更复杂的模型。这些模型参数可以通过计算密集型技术进行估计，如马尔科夫链蒙特卡洛方法（Markov Chain Monte Carlo，MCMC）和其他重采样策略（Gamerman，2006；Gilks et al.，1998）。从某种程度上看，生态毒理学指南文件和教科书未与时俱进。

更现代的数据操作、展示、可视化和统计分析方法都利用了专门构建的软件工具。虽然旨在达到单一的结果，如毒性值和/或保护浓度的估计专用软件（如 ToxCalc）(Tidepool Scientific Software, 2016) 已广泛使用多年，但从 2012 年开始，R 统计计算环境在学术界和科学界引起了翻天覆地的变化。只因 R 语言有两大优点：①R 系统是开源的，因此完全免费；②软件自带和用户贡献包的内容无比丰富。为符合生态毒理学趋势的前瞻性，本节的其余部分将重点介绍 R 在生态毒理学中的应用。

2.8.1 webchem 包

Webchem 是一个从网上检索化学信息的 R 包（Szöcs, 2015b）。它可以查询和检索大量网络来源的信息，包括：
- 美国国家癌症研究所的化学识别码解析器；
- 英国皇家化学学会的 ChemSpider；
- 美国国家生物技术信息中心的 PubChem BioAssay 数据库；
- 加州大学的化学转化服务处；
- 杀虫剂行动网的杀虫剂数据库；
- 德国联邦环境局 ETOX 数据库的水生和陆地生态毒理学信息；
- 赫特福德郡大学的农药化学特性、物理化学、人体健康和生态毒理学数据库；
- 美国国立医学图书馆的医学 TOXNET 数据库组，覆盖了化学制品和药物、疾病和环境、环境健康、职业安全和健康、中毒、风险评估和条例、毒理学。

例如，输入命令 get_cid（"Triclosan"），将生成以下列表：

```
#>  [1] "5564"     "131203"   "627458"   "15942656" "16220126" "16220128"
#>  [7] "16220129" "16220130" "18413505" "22947105" "23656593" "24848164"
#> [13] "25023954" "25023955" "25023956" "25023957" "25023958" "25023959"
#> [19] "25023960" "25023961" "25023962" "25023963" "25023964" "25023965"
#> [25] "25023966" "25023967" "25023968" "25023969" "25023970" "25023971"
#> [31] "25023972" "25023973" "45040608" "45040609" "67606151" "71752714"
#> [37] "92024355" "92043149" "92043150"
```

下面的示例演示了 IDs 之间如何进行转换：

```
cts_convert(query = '3380-34-5', from = 'CAS', to = 'PubChem CID')
#> [1] "5564"   "34140"
cts_convert(query = '3380-34-5', from = 'CAS', to = 'ChemSpider')
#> [1] "31465"
(inchk <- cts_convert(query = '50-00-0', from = 'CAS', to = 'inchikey'))
#> [1] "WSFSSNUMVMOOMR-UHFFFAOYSA-N"
```

2.8.2 ggplot2 包

ggplot2 是一个功能丰富的通用图形包。虽然不是专为生态毒理学所开发，但它可以用于生成高度可定制的 SSDs（图 2.8 和图 2.9）。

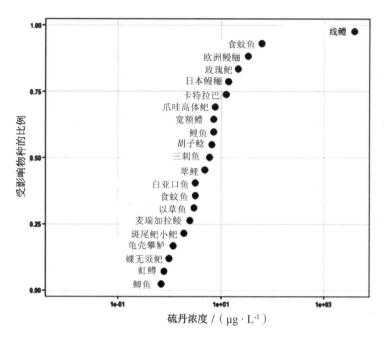

图2.8 硫丹对非澳大利亚鱼类 EC_{50} 数据的经验 SSD

源自 Hose, G. C., Van den Brink, P. J., 2004. Confirming the species-sensitivity distribution concept for endosulfan using laboratory, mesocosm, and field data. Environ. Contam. Toxicol. 47, 511–520.

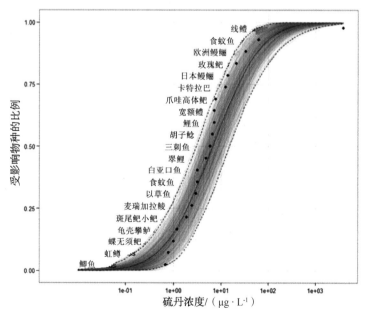

图2.9 基于图2.8中硫丹数据（黑色点）的理论对数正态 SSD（红色曲线），以及经过自举法样本拟合的对数正态分布（蓝色曲线）

置信区间为95%（黑色虚线）。R 源代码改编自 http://bit.ly/1OrFC5n。

2.8.3　drc 包

　　drc 包提供了一套全面的建模和分析工具，用于将复杂模型与 C-R 实验生成的数据拟合。Ritz & Streibig（2005）对 drc 包的特点和使用方法进行了很好的概述。例如，下面的 R 代码可以将 C-R 数据进行四参数和五参数的 log-logistic 函数拟合。使用拟合模型作为参数，anova（）可以快速检查额外的参数值。在这种情况下，报告的 p 值为 0.3742，表明采用五参数模型没有明显改善整体拟合程度。在图形绘制和模型拟合方面，drc 包既便捷又美观，只需简单的代码就可以得到结果（图 2.10）。

```
> head(df)    # inspect first 6 rows of dataframe df
    concen    response
1       0    0.8457565
2       0    0.9741697
3       0    0.9874539
4       0    1.0228782
5       0    1.0405904
6       0    1.0583026
> fit1<-drm(df$response ~ df$concen,fct=LL.4())    # fit 4-parameter log-logistic
> fit2<-drm(df$response ~ df$concen,fct=LL.5())    # fit 5-parameter log-logistic
>
> anova(fit1,fit2)    # compare models
1st model
  fct:   LL.4()
2nd model
  fct:   LL.5()
ANOVA table
           ModelDf     RSS     Df    F value    p value
1st model      59    1.2922
2nd model      58    1.2746    1     0.8019     0.3742

    > # plot results
    >
    > plot(fit1,broken=TRUE,legend=TRUE,legendText="4-param log-logistic",
    + col="red",type="confidence",ylab="fraction",xlab="concen")
    >
    > plot(fit2,add=TRUE,legend=TRUE,legendText="5-param log-
logistic",legendPos=c(0.19,1.2),
    + col="blue",type="confidence",ylab="fraction",xlab="concen")
    >
    > plot(fit2,add=TRUE,type="obs",pch=16,cex=0.6)
```

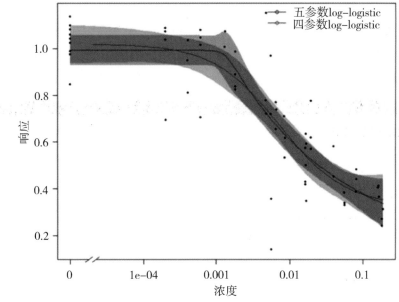

图 2.10　应用 drc 包绘制的数据和模型拟合
彩色波段是平均响应的 95% 置信区间。

2.8.4　函数 fitdistr（）& fitdistrplus 包

二者是通用的分布拟合程序。Fitdistal（）是 MASS 包的一部分，采用最大似然估计拟合单变量概率分布。fitdistrplus 包具有更多的特性，能够使用包括最大似然估计、矩量法、分位数匹配和最大拟合优度在内的各种算法对各种单变量概率模型的删失和非删失数据进行拟合。fitdistrplus 是在线 SSD 工具 MOSAIC 的核心。

使用 fitdistrip（）或 fitdistripplus 可以很容易地获得对图 2.8 中数据对数正态分布拟合的参数估计：

```
> fitdistr(data$EC50,"lognormal")
    meanlog      sdlog
   1.9084029   1.8440993
  (0.4024155) (0.2845507)

> fit<-fitdist(data$EC50, "lnorm",method="mme")
> fit
Fitting of the distribution ' lnorm ' by matching moments
Parameters:
         estimate
meanlog  3.825651
sdlog    1.716488
```

2.8.5　MOSAIC

MOSAIC 是一个与 R 接合的在线工具，可以使用一些预定义的概率模型对经验毒性数据进行拟合。它还使用自举法提供参数估计的标准误差估计和派生的 HC_x 值估计。该工具由里昂大学生物计量和进化生物学实验室的研究人员开发，可在线获取（Biometry and Evolutionary Biology Laboratory，2015）。使用 MOSAIC 对图 2.8 中鱼类数据进行拟合，结果如图 2.12 所示。

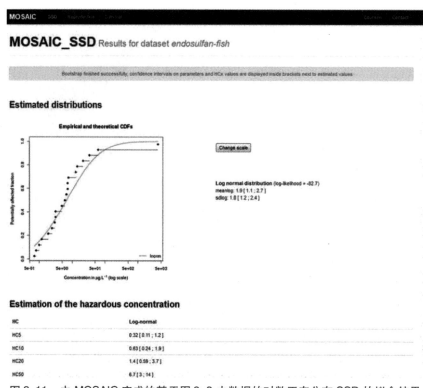

图 2.11　由 MOSAIC 完成的基于图 2.8 中数据的对数正态分布 SSD 的拟合结果

2.8.6　BurrliOZ

BurrliOZ（CSIRO，2015）在概念上与 MOSAIC 相似，也是与 R 接合，用于拟合 log-logistic 或 Burr-type 分布的毒性数据。但与 MOSAIC 不同，BurrliOz 需要下载并在桌面计算机上本地运行。它最初是作为一个独立项目发布，由澳大利亚联邦科学和工业研究组织（CSIRO）的统计学家开发，以支持 2000 年修订的《澳大利亚和新西兰淡水和海洋水质指南》（ANZECC/ARMCANZ，2000a）。虽然提供与 MOSAIC 类似的功能，都采用 R 来完成必要的运算，但 BurrliOz 更复杂，并且其 R 源代码也无法访问。使用 BurrliOz 对图 2.8 中数据进行拟合，结果如图 2.12 所示。

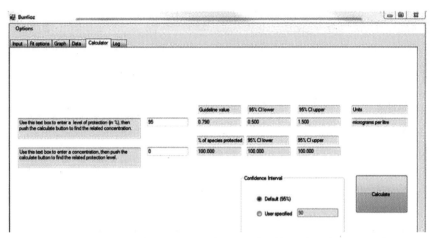

图 2.12　应用 BurrliOz 对图 2.8 中数据进行对数正态分布 SSD 拟合，计算出的 HC_x 结果截图

2.9　展　　望

在过去 40 年里，统计方法的进步在很大程度上是由计算能力和软件开发的进步所推动的。例如，R 的爆发式发展、MCMC 等计算密集型技术的普及以及只需点击鼠标就可生成达到出版质量的图形的能力。因此，统计生态毒理学的机遇和发展也呈现出计算机依赖性。本章中介绍的独立应用程序正在逐渐被在线工具所取代，如 MOSAIC（Biometry and Evolutionary Biology Laboratory，2015）& BurrliOz（CSIRO，2015）。使用 R 语言编写的自定义生态毒理学软件包数量的增加，加上如 R-studio 的 Shiny 软件（RStudio，2016）等"包装"程序和语言的开发，将提高用户友好性，从而鼓励更多人使用这些工具。我们预测这种技术的融合将促进平板电脑、智能手机和 iPads 等移动计算设备上生态毒理学"应用程序"的发展。因此在某种程度上，统计生态毒理学的新前沿将完全数字化。

在理论挑战和机遇方面，Eggen et al.（2004）认为两者相关，并与以下因素有关：
- 低浓度污染物和慢性效应；
- 单一污染物的多重效应；
- 污染物的复杂混合物；
- 多重压力源；
- 生态系统复杂性。

统计工具（如最大似然估计、概率分析和逻辑回归）以及在较小程度上的广义线性模型和贝叶斯方法是当代生态毒理学估计和推断方法的基础。我们所面临的挑战是，如何将生态毒理学的重要方面更紧密地整合到大学课程和专业培训项目中。许多生态毒理学的统计框架都是自然发展而不是按计划发展。令人遗憾的是,在世界各地的自然资源管理机构中,生物统计学家是一个正在消失的"物种"。这将导致生态毒理学研究人员倾向于规避新的、具有挑战性的统计概念,从而倾向于维持生态毒理学统计分析现状。这也部分地导致了关于这类不确定问题的争论。而且我们怀疑生物统计学的发展停滞使得关于这些问题的争论变得毫无意义。未来更具成效的研究将与以下方面相关:

- 对 SSD & HC_x 计算中样本量少和样本非随机性的问题进行修正和处理;
- 发展统计方法,对 TKTD 模型和基于 SSD 方法的预测进行验证;
- 如何以最小成本设计出最大化收益的 C-R 实验;
- 是否有可能为化学物质混合物设置 HC_x;
- 如何将时间因素完美地加入 SSD 建模和 HC_x 估算中,而不是将其忽略;
- 误差传递策略将产生不确定性数据收集过程、不精确的模型规范和数据统计处理;
- 优化模型功能,以进行外部和内部暴露评估;
- 在贝叶斯先验和协议的设定中引入"专家"意见,以便在监管环境下达成共识。

由于 SSD 的重要性和广泛应用,它的方法和相关技术是本章的重点。以上大部分要点都涉及 SSD 方法理论和应用方面的改进。虽然近期许多专家认为,"不要将 SSDs 作为一种纯粹的统计结构应用于对物种敏感性数据知之甚少的情况,而是将 SSDs 作为框架,提供基于过程的方法",但 SSD 将保持其作为建立保护浓度的首选(通常是强制的)方法,甚至是唯一方法(ECETOC,2014)。

SSD 提供了一种一致且可重复的方法来建立水质标准,虽然在分配、参数化和估计技术方面也受到主观选择的影响,但它消除了先前评估因子方法的随意性。这种"统计结构"的问题并不是由统计科学的缺陷造成的,而是违反关键假设和/或不适当的应用而产生的。科学界如何处理这些问题还有待摸索。

参考文献

ADELMANN S, 1961. Irregular fractions of the 2^n factorial experiments. Technometrics, 3 (4): 479-496.

ANZECC/ARMCANZ, 2000a. Australian and New Zealand guidelines for fresh and marine water quality. Australian and New Zealand Environment and Conservation Council/Agricultural and Resource Management Council of Australia and New Zealand, Canberra, ACT, Australia.

ANZECC/ARMCANZ, 2000b. Australian guidelines for water quality monitoring and reporting. Australian and New Zealand Environment and Conservation Council, Agriculture and Resource Management Council of Australia and New Zealand (National Water Quality Management Strategy number 7).

BARNETT V, O'HAGAN A, 1997. Setting environmental standards: the statistical approach to handling uncertainty and variation. Chapman and Hall, London, UK.

BATLEY G E, BRAGA O, VAN DAM R, et al., 2014. Technical rationale for changes to the method for deriving Australian and New Zealand water quality guideline values for toxicants. Council of Australian Government's Standing Council on Environment and Water, Sydney, Australia.

BERKSON J, 1944. Application of the logistic function to bioassay. J. Am. Stat. Assoc., 39: 357 – 365.

BICKEL P J, DOKSUM K A, 2016. Mathematical statistics: basic ideas and selected topics. CRC Press, Boca Raton, FL, USA.

BILLOIR E, DELIGNETTE-MULLER M L, PÉRY A R R, et al., 2008. A Bayesian approach to analyzing ecotoxicological data. Environ. Sci. Technol., 42: 8978 – 8984.

Biometry and Evolutionary Biology Laboratory, 2015. MOSAIC: modelling and statistical tools for ecotoxicology. University of Lyon, France. http://pbil.univ-lyon1.fr/software/mosaic/ssd/.

BLISS C I, 1935. The calculation of the dosage-mortality curve. Ann. Appl. Biol., 22: 134 – 167.

BOSE R C, BUSH K A, 1952. Orthogonal arrays of strength two and three. Ann. Math. Stat., 23: 508 – 524.

CASH J T, 1908. The relationship of action to dose especially with reference to repeated administration of indaconitine. Br. Med. J.: 1213 – 1218.

CCME, 2007. A protocol for the derivation of water quality guidelines for the protection of aquatic life. Canadian Council of Ministers of the Environment, Ottawa, ON, Canada: 37.

CHAPMAN P M, CARDWELL R S, CHAPMAN P F, 1996. A warning: NOECs are inappropriate for regulatory use. Environ. Toxicol. Chem., 15: 77 – 79.

CHÈVRE N, BRAZZALE A R, 2008. Cost-effective experimental design to support modelling of concentration-response functions. Chemosphere, 72: 803 – 810.

CLIFFROY P, KELLER M, PASANISI A, 2013. Estimating hazardous concentrations by an informative Bayesian approach. Environ. Toxicol. Chem., 32: 602 – 611.

COX D R, SNELL E J, 2000. Statistics: principles and examples. Chapman and Hall/CRC, Boca Raton, FL, USA.

CRAN, 2015. Design of experiments (DoE) and analysis of experimental data. https://cran.r-project.org/web/views/ExperimentalDesign.html.

CRANE M, NEWMAN M C, 2000. What level of effect is a no observed effect? Environ. Toxicol. Chem., 19: 516 – 519.

CSIRO, 2015. Burrlioz 2.0. https://research.csiro.au/software/burrlioz/.

DASU T, JOHNSON T, 2003. Exploratory data mining and data cleaning. John Wiley & Sons. http://ca.wiley.com/WileyCDA/WileyTitle/productCd-0471268518,subjectCd-CSB0.html.

DEBtox Information Site, 2011. DEBtox information: making sense of ecotoxicity test results. http://www.debtox.info/about_debtox.html.

DEY A, 1985. Orthogonal fractional factorial designs. John Wiley and Sons, Haslted Press, New York.

DUBOUDIN C, CIFFROY P, MAGAUD H, 2004. Effects of data manipulation and statistical methods on species sensitivity distributions. Environ. Toxicol. Chem., 23: 489 – 499.

ECETOC, 2014. Estimating toxicity thresholds for aquatic ecological communities from sensitivity distributions 11 – 13 February, 2014. Amsterdam Workshop Report No. 28. European Centre for Ecotoxicology and Toxicology of Chemicals.

EGGEN R I L, BEHRA R, BURKHARDT-HOLM P, et al., 2004. Challenges in ecotoxicology. Environ. Sci. Technol., 38: 58A-64A.

Environment Canada, 2005. Guidance document on statistical methods for environmental toxicity tests. EPS 1/RM/46, Ottawa, ON, Canada.

Environmetrics Australia, 2014a. Big data is watching you. http://www.environmetrics.net.au/index.php? news&nid = 81.

Environmetrics Australia, 2014b. The explosive growth of R. http://environmetrics.net.au/index.php? news&nid = 79.

European Commission, 2011. Technical guidance for deriving environmental quality standards// Guidance document no. 27, common implementation strategy for the water framework directive. Brussels, Belgium: 204.

EVANS D A, NEWMAN M C, LAVINE M, et al., 2010. The Bayesian vantage for dealing with uncertainty// WARREN-HICKS J, HART A (Eds.). Application of uncertainty analysis to ecological risks of pesticides. CRC Press, Boca Raton, FL, USA.

FORBES T L, FORBES V E, 1993. A critique of the use of distribution-based extrapolation models in ecotoxicology. Funct. Ecol., 7: 249 – 254.

FORBES V E, CALOW P, 2002. Species sensitivity distributions: a critical appraisal. Hum. Ecol. Risk Assess, 8: 473 – 492.

FOX D R, 2006. Statistical issues in ecological risk assessment. Hum. Ecol. Risk Asses, 12: 120 – 129.

FOX D R, 2008. NECs, NOECs, and the ECx. Australas. J. Ecotoxicol., 14: 7 – 9.

FOX D R, 2010. A Bayesian approach for determining the no effect concentration and hazardous concentration in ecotoxicology. Ecotoxicol. Environ. Saf., 73: 123 – 131.

FOX D R, 2015. Selection bias correction for species sensitivity distribution modelling and hazardous concentration estimation. Environ. Toxicol. Chem., 34: 2555 – 2563.

FOX D R, LANDIS W G, April 18, 2016a. Don't be fooled: a NOEC is no substitute for a poor concentration-response experiment. Environ. Toxicol. Chem., http://dx.doi.org/10.1002/etc.3459.

FOX D R, LANDIS W G, 2016b. Comment on ET&C perspectives November 2015: a holistic view. Environ. Toxicol. Chem., 35: 1337 – 1339.

FOX D R, BILLOIR E, 2013. Time dependent species sensitivity distributions. Environ. Toxicol. Chem., 32: 378 – 383.

FOX D R, BILLOIR E, CHARLES S, et al., 2012. What to do with NOECs/NOELs: prohibition or innovation? Integr. Environ. Assess. Manag., 8: 764 – 766.

GAD S C, 2006. Statistics and experimental design for toxicologists and pharmacologists. 4th ed. Taylor and Francis, Boca Raton, FL, USA.

GAMERMAN D, 2006. Markov chain Monte Carlo: stochastic simulation for Bayesian inference. Chapman and Hall/CRC, Boca Raton, FL, USA.

GILKS W R, RICHARDSON S, SPIEGELHALTER D J, 1998. Markov chain Monte Carlo in practice. Chapman and Hall/CRC, Boca Raton, FL, USA.

GRAYBILL F A, 1976. Theory and application of the linear model. Duxbury Press, North Scituate.

GREEN J W, 2016. Issues with using only regression models for ecotoxicology studies. Integr. Environ. Assess. Manag., 12: 198 – 199.

GREEN J W, SPRINGER T A, STAVELEY J P, 2013. The drive to ban the NOEC/LOEC in favor of ECx is misguided and misinformed. Integr. Environ. Assess. Manag., 9: 12 – 16.

GRIST E P M, O'HAGAN A, CRANE M, et al., 2006. Comparison of frequentist and Bayesian freshwater species sensitivity distributions for chlorpyrifos using time-to-event analysis and expert elicitation. Environ. Sci. Technol., 40: 295 – 301.

HICKEY G L, CRAIG P S, 2012. Competing statistical methods for the fitting of normal species sensitivity distributions: recommendations for practitioners. Risk Anal., 32: 1232 – 1243.

HINKELMANN K, KEMPTHORNE O, 2008. Design and analysis of experiments // Introduction to experimental design. vol. I. Wiley, NJ, USA.

HOEKSTRA J A, VAN EWIJK P H, 1993. The bounded effect concentration as an alternative to the NOEC. Sci. Tot. Environ., 134: 705 – 711.

HOSE G C, VAN DEN BRINK P J, 2004. Confirming the species-sensitivity distribution concept for endosulfan using laboratory, mesocosm, and field data. Environ. Contam. Toxicol., 47: 511 – 520.

JAGER T, 2012. Bad habits die hard: the NOECs persistence reflects poorly on ecotoxicology. Environ. Toxicol. Chem., 31: 228 – 229.

JAGER T, ALBERT C, PREUSS T G, et al., 2011. General unified threshold model of survival: a toxicokinetic-toxicodynamic framework for ecotoxicology. Environ. Sci. Technol., 45: 2529 – 2540.

JAWORSKA J, GABBERT S, ALDENBERG T, 2010. Towards optimization of chemical testing under REACH: a Bayesian network approach to integrated testing strategies. Regul. Toxicol. Pharmacol., 57: 157 – 167.

KEMPTHORNE O, 1947. A simple approach to confounding and fractional replication in factorial experiments. Biometrika, 34: 255-272.

KON KAM KING G, DELIGNETTE-MULLER M L, KEFFORD B J, et al., 2015. Constructing time-resolved species sensitivity distributions using hierarchical toxico-dynamic model. Environ. Sci. Technol., 49 (20): 12465-12473.

KOOIJMAN S, BEDAUX J J M, 1996. The analysis of aquatic toxicity data. VU University Press, Amsterdam, Netherlands.

KUTNER M H, NACHTSHEIM C J, NETER J, et al., 2016. Applied linear statistical models. 5th ed. Irwin/McGraw-Hill, New York, USA.

LANDIS W G, CHAPMAN P M, 2011. Well past time to stop using NOELs and LOELs. Integr. Environ. Assess. Manag., 7 (4): vi-viii.

LAWSON J, 2015. Design and analysis of experiments with R. CRC Press, Boca Raton, FL, USA.

LI J, FU H, 2013. Bayesian adaptive D-optimal design with delayed responses. J. Biopharm. Stat., 23 (3): 559-568.

LINK W A, ALBERS P H, 2007. Bayesian multimodel inference for dose-response studies. Environ. Toxicol. Chem., 26: 1867-1872.

MALTHUS T R, 1798. An essay on the principle of population. J. Johnson, London, UK.

MCBRIDE M F, GARNETT S T, SZABO J K, et al., 2012. Structured elicitation of expert judgments for threatened species assessment: a case study on a continental scale using email. Methods Ecol. Evol., 3: 906-920.

MOORE D R J, WARREN-HICKS W J, QIAN S, et al., 2010. Uncertainty analysis using classical and Bayesian hierarchical models // WARREN-HICKS W J, HART A (Eds.). Application of uncertainty analysis to ecological risks of pesticides. CRC Press, Boca Raton FL, USA.

NEWMAN M C, 2012. Quantitative ecotoxicology. CRC Press, Boca Raton, FL, USA.

NEWMAN M C, CLEMENTS W H, 2008. Ecotoxicology: a comprehensive treatment. CRC Press, Boca Raton, FL, USA.

NEWMAN M C, OWNBY D R, MÉZIN L C A, et al., 2000. Applying species sensitivity distributions in ecological risk assessment: assumptions of distribution type and sufficient numbers of species. Environ. Toxicol. Chem., 19: 508-515.

New York Times, 2014. For big-data scientists, 'janitor work' is key hurdle to insights. http://www.nytimes.com/2014/08/18/technology/for-big-data-scientists-hurdle-to-insights-is-janitor-work.html?_r=2.

OECD (Organization for European Cooperation and Development), 1998. Report of the OECD workshop on statistical analysis of aquatic toxicity data // OECD environmental health and safety publications series on testing and assessment. No. 10. Paris, France.

OECD, 2012. Guidance document 116 on the conduct and design of chronic toxicity and carcinogenicity studies, supporting test guidelines 451, 452, and 453. 2nd ed. // OECD series on testing and assessment. No. 116. Paris, France.

OECD, 2014. Current approaches in the statistical analysis of ecotoxicity data: a guidance to application (annexes to this publication exist as a separate document) // OECD series on testing and assessment. No. 54. Paris, France.

OKKERMAN P C, PLASSCHE E J V D, SLOOFF W, et al., 1991. Ecotoxicological effects assessment: a comparison of several extrapolation procedures. Ecotoxicol. Environ. Saf., 21: 182 – 193.

OpenBUGS, 2009. Bayesian inference using gibbs sampling. http://www.openbugs.net/w/FrontPage.

PAYET J, 2004. Assessing toxic impacts on aquatic ecosystems in life cycle assessment (LCA) (Ph. D. thesis). École Polytechnique Fédérale de Lausanne, Lausanne, Switzerland.

PEARL R, REED L J, 1920. On the rate of growth of the population of the United States since 1790 and its mathematical representation. Proc. Natl. Acad. Sci. USA, 6: 275 – 288.

POSTHUMA L, SUTER G W, TRAAS T P, 2002. Species sensitivity distributions in ecotoxicology. Lewis Publishers, Boca Raton, FL, USA.

QUORA, 2014. What is data munging? https://www.quora.com/What-is-data-munging.

R Development Core Team, 2004. R: a language and environment for statistical computing. R Foundation for Statistical Computing, Vienna, Austria. http://www.R-project.org.

RAO C R, 1950. The theory of fractional replication in factorial experiments. Sankhya, 10: 81 – 86.

RITZ C, STREIBIG J C, 2005. Bioassay analysis using R. J. Stat. Softw., 12: 1 – 22.

RSTUDIO, 2016. Shiny software. https://www.rstudio.com/products/shiny/.

SHIEH J N, CHAO M R, CHEN C Y, 2001. Statistical comparisons of the no-observed-effect concentration and the effective concentration at 10% inhibition (EC10) in algal toxicity tests. Water Sci. Technol., 43: 141 – 146.

Sourceforge, 2015. JAGS: just another gibbs sampler. http://mcmc-jags.sourceforge.net/.

SPARKS T, 2000. Statistics in ecotoxicology. Wiley and Sons, Chichester, UK.

Stan Development Team, 2015. Stan modeling language users guide and reference manual. Version 2.8.0. http://mc-stan.org/documentation/.

SZÖCS E, 2015a. Species sensitivity distributions (SSD) with R. http://edild.github.io/ssd/.

SZÖCS E, 2015b. Introducing the webchem package. http://edild.github.io/webchem/.

Tidepool Scientific Software, 2016. ToxCalc software. https://tidepool-scientific.com/ToxCalc/ToxCalc.html.

TREVAN J W, 1927. The error of determination of toxicity. Proc. R. Soc. B., 101

(712): 483 - 514.

USEPA (U. S. Environmental Protection Agency), 2002. Guidance on choosing a sampling design for environmental data collection for use in developing a quality assurance project plan. EPA QA/G - 5S, Washington, DC, USA.

USEPA, 2006. Data quality assessment: statistical methods for practitioners: EPA QA/G - 9S. Washington, DC, USA.

USEPA, 2015. ECOTOX user guide: ecotoxicology database system. version 4.0. http://www.epa.gov/ecotox/.

VAN DAM R A, HARFORD A J, WARNE M S, 2012. Time to get off the fence: the need for definitive international guidance on statistical analysis of ecotoxicity data. Integr. Environ. Assess. Manag., 8: 242 - 245.

VAN DER HOEVEN N, NOPPERT F, LEOPOLD A, 1997. How to measure no effect: part I: towards a new measure of chronic toxicity in ecotoxicology. Introduction and workshop results. Environmetrics, 8: 241 - 248.

VERHULST P F, 1838. Notice sur la loi que la population poursuit dans son accroissement. Corresp. Math. Phys., 10: 113 - 121.

WANG Y, ZHANG L, MENG F, et al., 2014. Improvement on species sensitivity distribution methods for deriving site-specific water quality criteria. Environ. Sci. Pollut. Res., 22: 5271 - 5282.

WARNE MSTJ, VAN DAM R, 2008. NOEC and LOEC data should no longer be generated or used. Australas. J. Ecotox., 14: 1 - 5.

WARNE MSTJ, BATLEY G E, BRAGA O, et al., 2013. Revisions to the derivation of the Australian and New Zealand guidelines for toxicants in fresh and marine waters. Environ. Sci. Pollut. Res., 21: 51 - 60.

WARNE MSTJ, BATLEY G E, VAN DAM R A, et al., 2014. Revised method for deriving Australian and New Zealand water quality guideline values for toxicants // Prepared for the council of australian government's standing council on environment and water. Sydney, Australia.

WEBB J M, SMUCKER B J, BAILER A J, 2014. Selecting the best design for nonstandard toxicology experiments. Environ. Toxicol. Chem., 33: 2399 - 2406.

WHEELER J P, GRIST E P M, LEUNG K M Y, et al., 2002a. Species sensitivity distributions: data and model choice. Mar. Pollut. Bull., 45: 192 - 202.

WHEELER J R, LEUNG K M Y, MORRITT D, et al., 2002b. Freshwater to saltwater toxicity extrapolation using species sensitivity distributions. Environ. Toxicol. Chem., 21: 2459 - 2467.

WICKHAM H, 2009. Ggplot2: elegant graphics for data analysis. Springer, New York, NY, USA.

WICKHAM H, 2014. Tidy data. J. Stat. Softw., 59: 1 - 23.

Wikipedia, 2015. Big data. https://en.wikipedia.org/wiki/Big_data.

Wikipedia, 2016. Determinant. https://en.wikipedia.org/wiki/Determinant.

ZAJDLIK B A, 2006. Potential statistical models for describing species sensitivity distributions // Prepared for canadian council of ministers for the environment CCME project #382-2006. http://www.ccme.ca/assets/pdf/pn_1415_e.pdf.

ZAJDLIK B A, 2015. A statistical evaluation of the safety factor and species sensitivity distribution approach to deriving environmental quality guidelines. Integr. Environ. Assess. Manag. http://dx.doi.org/10.1002/ieam.1694.

ZHANG J, BAILER A J, ORIS J T, 2012. Bayesian approach to estimating reproductive inhibition potency in aquatic toxicity testing. Environ. Toxicol. Chem., 31: 916-927.

3 化学物质吸收和效应的动态模型

T. Jager[①]

---

① 荷兰德 DEBtox 研究所。

3.1 引　　言

　　模型是对部分现实世界的简化表达。在自然科学中，模型通常以数学模式呈现，以便能够定量地评估其性能。构建模型的目的不在于模型本身，而是获得一种方法，以深入理解引起事物发展变化的潜在机制。如果没有模型，就不可能解释多因素对我们所关注系统的集成效应（通常是非线性的）。此外，模型对预测未经验证的状态至关重要。在生态毒理学中，模型预测有益于实验的优化设计，尤其是对基于科学的风险评价，比如评估某种新化学物质进入环境后的影响或评价不同应对策略的有效性。在环境化学研究中，已证明了模型对达成监管目的具有助力，因为归趋模型几乎是现在所有环境风险评价框架中必不可少的一部分。归趋模型集成了化学物质迁移和转化的定量知识。根据化学物质的化学特性和排放情况，可用归趋模型解析化学物质在环境中的归趋，以及预测其环境浓度（通常随时空变化）。通过这种方式，归趋模型可以对新兴化学污染物（排放之前）和新状况（如某一特定地点的石油泄漏）进行风险评价。虽然模型是对现实的简化，并且总会存在"误差"，但它是我们追求从机制上理解到有效地管理我们周围世界过程中不可缺少的一部分。

　　生态毒理学中最常用的模型是生物个体随时间变化吸收化学物质的模型［毒物代谢动力学（toxicokinetic，TK）模型将在第3.3节中介绍］。通常使用假设检验和剂量-反应曲线（见第2章）来阐释化学物质对生物个体的毒性效应。虽然这些方法被视为粗略的模型，但其目的并不是解释依据基本原理观察到的效果，而是在给定的数据组中检验显著效应或插值。这些描述性方法无法完成建模的两个最重要目标（理解和预测），因此本章不予讨论。研究时间变化对个体毒性效应影响机制的模型［毒物效应动力学（toxicodynamic，TD）模型，将在第3.4节中介绍］越来受到人们关注，但其在生态毒理学中的应用不如TK模型普遍。此外，生态毒理学模型在低层次生物组织（分子和细胞水平）和高层次对象（如种群、食物链和生态系统）中的应用也越来越广泛。但本章不详细讨论这些模型。基于多种原因，个体是生物组织的核心层次，但是对于生态毒理学家而言，最重要的原因是双重的：较低层次组织的影响在生态上几乎没有生态意义，除非它们影响个体的生活史特征（即生长、繁殖、生存），以及基于有毒物质胁迫导致的个体生活史特征的变化最终影响更高级别生物。学习生态毒理学模型时，个体层次是一个很好的切入点，本章将围绕此层次展开。

　　对于海洋生态毒理学，通常也使用与淡水和陆地生物相同的模型。建模原则是相同的，物种间的相似性比差异性更重要。但与淡水和陆地生物相比，海洋生物的一些生物学特征更加显著。大多数海洋无脊椎动物（及大多数鳍刺类鱼）的生命周期包括幼虫阶段，该阶段生物的形态、摄食习惯和生活方式与成虫阶段截然不同（Pechenik，1999）。许多海洋无脊椎动物生命周期的另一个典型特征是具有应对食物资源的季节性变化和为周期性产卵提供能量的能量存储阶段（Giese，1959；Lee et al.，2006）。在尝试对毒物吸收和效应建模时，需要仔细考虑生物生命周期中形态、生态和构成的主要变化。

本章首先介绍了模型构建的基本原则、模型有效性的评价和数据的比对。随后在个体层次上讨论了生态毒理学领域中最重要的模型（较高层次组织将在第 3.7 节中做简要介绍）。本章的重点是概念，而不是数学运算的细节。因为牢固地掌握概念对于生物学家理解和解释模型研究极其重要，也是深入学习数学运算和进行编程前必不可少的。此外，在设计实验时，理解这些概念有利于建模研究。因此本章的运算量限制至最小，主要在于阐述一般原则。本章不是对生态毒理学领域现有各种模型（或模型类型）的综述，而是着重于使用几个简单的例子来说明模型所采用的方法、假设及用途。关于本章的更多信息请见网站（http://www.debtox.info/marecotx.html），其中包含更多详细内容及相关软件（如用于实例研究的 Matlab 原始文件）。

3.2 建模基本原则

在我们开始学习典型的（海洋）生态毒理学模型之前，了解一些建模基本原则是非常重要的。本节首先介绍这些原则，并将在下一节（毒物动力学）采用一个简单的毒代动力学模型来进行验证。

3.2.1 系统和状态

在建模术语中，"系统"通常表示一系列相互作用部分所形成的整体，其边界将它与外部世界分隔。在生物学中，系统可以是单独的生物个体，也可以是生物体内的器官、细胞或种群中的个体，甚至整个生态系统。在本章中，单个生物体是我们要研究的系统，因为重点是个体对化学物质的吸收，以及化学物质对个体的效应。

系统的状态由其状态变量（系统的一个相关属性）决定，此变量通常随时间变化而变化。例如，体内浓度是 TK 模型中一个状态变量。为了预测系统的发展，我们需要知道状态变量的当前值，而非历史值。根据给定模型目的的需要，这些变量至少应充分捕获系统的当前状态。因此，选择合适的状态变量是模型设计的关键步骤。对于同一系统，不同研究目的很可能导致不同的状态变量集。

一般来说，状态变量随时间的变化取决于状态的当前值。因此，动力学模型通常用常微分（ordinary differential equations, ODE）的形式表示方程，其中导数（状态的变化）是状态本身值的函数方程。

3.2.2 假设的作用

建模不是始于数学运算，而是始于科学问题的确定和研究目标的制定。基于对拟建模型的系统观察（如文献查阅），提出一组简化假设。为了在科研中起作用，这些假设应是对所假定系统行为基础机制的简化。显然，生物系统是复杂的，以有效方式简化复杂生物系统是一项艰巨的智力任务。

因此，模型源自对（生物学）现实假设的简化；实用准确的模型遵循清晰且一致的假设。然而模型通常仅以方程的形式呈现，由读者来重建已有或隐含的基本假设。这种做法妨碍了模型（及其结果）在科学界和环境管理中的可接受性。

3.2.3 模型的复杂性和普适的必要性

模型是复杂系统的简化。然而复杂系统并不一定需要复杂的模型。模型的复杂性应当与研究目的和可用于参数化模型的信息紧密相关，和系统本身的复杂程度联系不大。复杂的模型测试困难（错误易被忽略），并且需要大量的信息进行参数化。更重要的是，从复杂的模型中我们只能得到少量信息。因此，建模基本的策略应当是尽可能从简单的开始，只有在必要的情况下才增加更多细节。

确保模型的通用性是驱动模型设计和模型复杂性发展的另一因素。在一系列环境条件 C 下，化学物质 A 对生物 B 的效应模型可以包含 A、B 和 C 的诸多细节，因此很容易变得非常复杂。但针对 A、B 和 C 的各种组合，每次都从零开始开发新模型是对时间和资源的浪费，并且会导致模型（及其结果）无法进行比较。因此，关注不同物种和不同化学物质的共性，而不是关注它们的特性，会有更多的收获。

模型设计的一个重要指导原则是从空间和时间两方面选择合适的尺度。将不同的尺度组合到一个模型中是低效的，且必然会产生问题。例如，分子水平对应的时间尺度为毫秒，空间尺度为纳米；多细胞动物生命史对应的时间尺度为几天至几年，空间尺度为毫米至米，若使用分子尺度的模型去阐释化学物质对多细胞动物生活史的效应，将不会获得什么有用的结果。

3.2.4 机理模型和描述性模型

机理模型和描述模型之间的区别比较复杂。生态毒理学中的剂量－反应曲线和统计检验（详见第 2 章）显然是描述性的模型。这些方法仅被用来描述数据，很少涉及潜在机制。此外，这些方法也不允许在实验的条件之外进行合理的推断。因此描述性方法（如 ECx & NOEC，见第 2 章）产生的总结性统计结果对科学和风险评价的作用是有限的（Jager，2011）。机理模型应当能够为模式提供一个解释，并应当（至少在原则上）为超出测试条件的外推法提供一个平台（例如，从定量到随时间变化的暴露，从随机量到限制食物供应）。然而，如果我们研究得足够深入，所有的机理模型都应包括描述性的元素，因为外推法的作用在实践中有局限性。

3.2.5 房室模型和基础构件

房室模型是用来简化复杂生物体的一种重要方法。单独的房室通常被定义为有明确的边界,可以假定房室是同质性的或是均匀混合的,若干个房室组成一个完整的模型。这种模型在模拟化学物质的环境归宿(MacLeod et al., 2010)和毒代动力学方面有着悠久的历史(Barron et al., 1990)。毒代动力学的例子将在第3.3节呈现,我们把生物体(更准确地说是内部的集中)作为单独的房室。房室可以有明确的物理边界(例如单个生物体或器官);但也可以代表同一物理边界内的不同部分(如化合物及其代谢物可以存在于同一生物体内两个不同房室)。

模型可以由单个房室或以某种方式连接的多个房室组成。对于互通的房室,必须区分广度量和强度量。如果将一个系统分成两个房室,它的广度量(如质量或化学物质的摩尔数)也会被分成两个;而它的强度量(如系统包含的化合物浓度)将保持不变(假设系统是均质的)。当广度量从一个房室转移到另一个房室时,我们需要考虑守恒定律:比如当1 mg的化合物从房室A进入房室B时,房室B中化合物的质量应当增加1 mg。这听起来可能微不足道,但却是人们常常会忽视的。在生态毒理学中,相对于化合物的总量,我们更关注化合物在生物体内的浓度。然而对于浓度等强度量,守恒定律并不适用。比如当化合物从房室A转移到房室B后,房室A浓度减少1 mg/L,房室B的浓度并不一定增加1 mg/L(只适用于A和B的体积相等时)。因此,在质量的基础上开始建立模型,然后转向浓度(按房室的适当体积或质量划分质量)是一个明智的策略。

有大量的数学方程函数可供选择,用于描述影响状态变量的过程。然而,大多数生态毒理学模型是由基础构件组成的,如表3.1所示。借用化学术语来说,它们被称为具有一定"顺序"的"动力学"(表3.1中方程右边A的次幂)。这些简单的动力学方程是构成更复杂模型的基础构件,并可以通过ODE中的加减进行组合(将下一节展示)。在本章背景下,术语"kinetics"和"dynamics"是同义词。然而在传统药理学中,术语"toxicokinetics"和"toxicodynamics"则是不同过程(将在第3.3和3.4节中进行解释)。

表3.1 生态毒理学模型中4种最常用的动力学类型

动力学	释义	ODE方程
零级动力学	反应速率与反应物浓度无关	$\frac{d}{dt}A = k$
一级动力学	反应速率与反应物浓度的一次方成正比	$\frac{d}{dt}A = kA$
二级动力学 (质量作用定律)	反应速率与反应物浓度的乘积成正比	$\frac{d}{dt}A = kA^2$
双曲线动力学 [米凯利斯-门坦 (Michaelis-Menten)动力学]	反应速率与反应物浓度呈双曲线函数关系	$\frac{d}{dt}A = \frac{kA}{K+A}$

注:举例方程中,A是状态变量,k和K是与状态变量无关的参数。注意,每个方程中,k的释义和单位都不相同。

3.2.6 数学的转化和模型测试

在模型的应用过程中，首先要将相关假设转化为数学运算，再将数学运算转化为计算机代码。这是两个技术性步骤，这里不进行详细介绍。但是在深入进行数学研究或使用软件编程前，仍可以对模型进行测试。一般地，如果给定了明确假设，那么从模型假设开始测试是有益的。如果假设不切实际，就不需进入数学运算和编码流程。违反物理定律或与观测结果不一致的假设不太可能产生有用的模型。相反，如果模型产生了不合理的结果，在确保方程转换和编码无误之后，就有必要返回基本假设进行检查。

量纲分析是快速而有效地检验模型方程正确性的方法。我们只能加减相同量纲（和单位）的量；如果对量进行乘法或除法，则也应除以其量纲（和单位）。等式的两边应当具有相同的量纲（和单位），超越函数（如指数和对数）的参数应该是无量纲的。不能进行量纲分析的模型无法表征实际机理。

许多情况下，通过研究微分方程可以更好地理解模型行为。微分方程将状态的变化转化为状态变量和模型参数（和附加力）。如果状态变化为零，表明系统处于平衡状态，进而分析平衡是否稳定（或系统是否会在一个小扰动后回到平衡状态）。

3.2.7 基于数据的模型验证

生态毒理学中，常用实验或监测数据验证模型。这一步骤可在校准过程（拟合模型参数与观测值）或验证过程（将模型预测与实际观测相比较）中完成。首先，区分质反应（离散的）和量反应（连续的）很重要（见第2章）。在一个质反应中，只有反应存在或不存在两种状态，并取决于每个个体（如个体是死还是活）。因此，LC_{50}表示预计受试群体会半数死亡的毒物浓度（在指定的暴露时间后）。相反，量反应表明的是每个个体确切的反应程度（如雌性个体的产卵数）。因此应用于种群繁殖的EC_{50}代表了雌性平均产卵量减少50%时的毒物预测浓度。不同类型的反应需要不同的模型和统计程序来进行数据验证。

在模型校正中，我们需要判断模型与观测值之间的偏差，以便在不同拟合值之间进行筛选，从而找到与数据"最佳拟合"的参数。因为实测值包含误差，所以偏差是不可避免的。模型总是"错误的"，因为它是对现实的简化。因此，我们还需要为这些偏差或残差建立一个模型，这属于统计学的范畴。最常用的统计模型是假设标准差为独立正态分布的随机样本值，其均值为零且方差为常数。这组假设是最小二乘法分析的基础。然而许多实际情况与这些假设并不相符。比如，质反应（如存活）是离散反应。因此，假设残差服从正态分布是存在问题的（尤其当个体样本量较小时）。又比如，生物体体重是一种量反应，其残差服从正态分布较合理。但若方差不变则是有问题的（大体重值往往表现出更多变化）。此外，如果对同一个体进行长期的测量，则违背了独立性原则。这种违反在何种程度上影响数据集的解释并不总是很清楚。总之，最好明确地提及为统计模型所做的假设，就像描述过程的动态模型一样。

当我们选择了残差的统计模型时，可以将模型与数据拟合，并计算参数估计的置信区间。只有当过程模型和统计模型都有意义时，置信区间才有意义（即置信区间为"真"，模型才为"真"）。实际上，没有一个模型是"真的"，这意味着所有置信区间都是近似的。宽的置信区间可能意味着：参数不会（实质上）影响模型输出，或数据没有（或不充分）提供关于这个参数的信息，或者该参数与另一个参数密切相关（即固定另一个参数可能导致第一个参数的区间变窄）。一个窄置信区间意味着这些问题都不存在。几种可以构造不同置信区间的方法已在第 2 章进行介绍。

建模者经常面对他们的模型是否经过验证的问题。但经过验证的模型意味着什么？模型验证是将模型输出值（系统的）与现实世界的实际观察值进行比较，从而判断输出值在定性和定量上与现实是否相符。因此，已验证的模型是已进行比较过的模型，但这也无法说明结果。验证模型是否"有效"，并没有严格的区分，这在很大程度上取决于使用该模型的目的。

3.3 毒物代谢动力学

毒物代谢动力学（TK）是对化学物质在生物体内的吸收、分布、转化和排出过程的全面描述。TK 模型试图通过与暴露水平和暴露时间的函数关系来解释/描述化学物质在生物体内部的浓度。这种建模形式在生态毒理学中已经很成熟，很大程度上受环境化学中使用的模型（如归趋建模）的影响。TK 模型与归趋模型有着天然联系，因为 TK 模型也研究迁移和分布过程。尽管研究对象从环境房室变为生物体模型，但所涉及的大多数过程在概念上是相似的。例如，扩散过程在两种模型中都起着重要的作用，而生物体内的生物转化（从模型角度考虑）相当于在一个环境房室中的生物降解。

TK 模型通常由一个或多个房室组成，房室与环境、房室与房室间可以进行化学物质交换。需要考虑的转运机制包括扩散（如通过皮肤或鳃吸收和清除可溶解的化学物质）、对流（如化学物质随食物吸收进入肠道）和主动运输（如利用运输酶主动排除化学物质）。此外，化学物质可能通过生物转化产生与母体化合物性质不同的代谢物（但这一过程可能需要多个房室，详见第 4 章）。

3.3.1 基于一级动力学的单室模型

最简单实用的 TK 模型是具有一级动力学的单室模型。下文将通过该模型来描述 TK 模型，并具体介绍之前章节中提到的抽象建模原则。这是本章唯一详细讨论的模型。

假设生物体可以被视为单一的、均匀混合的房室，选择与外界水环境有物质交换的生物个体作为研究系统，选择化学物质的体内浓度为该系统唯一的状态变量。生物体在现实中显然更加复杂，但上述简化在实践中往往有奇效。此外，需要假设生物体积恒定（如没生长或萎缩）。对于化学物质的交换，最简单有效的假设是对环境中化学物质的吸收通量与其外部浓度成正比，化学物质从生物体到环境的清除通量与生物体内部浓度成正比（一级动力学，见表3.1和图3.1）。另一个常见的简化是假设外部环境是无限大且均匀的，因此生物体的吸收和清除对外部浓度没有影响。对于海洋生物来说，这一假设是合理的。但对于实验测试，其适用性需要仔细考虑。或者外部介质应有自己的房室，选择的系统是包含动物在内的测试容器，且至少有两个状态变量。

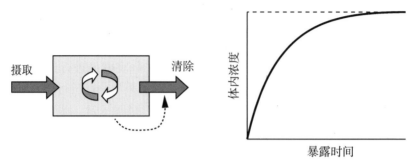

图3.1 基于一级动力学的单室模型

简单的单室模型假设可以转化为体内浓度 C_i 的微分方程：

$$\frac{dC_i}{dt} = k_u C_w - k_e C_i, \quad C_i(0) = 0 \tag{3.1}$$

其中，C_w 是外部浓度（即系统驱动力），吸收速率常数 k_u 和清除速率常数 k_e 是模型参数。内部浓度的变化（导数 dC_i/dt）取决于体内浓度的当前值（C_i）。该 ODE 是一个零阶项（$k_u C_w$）和一个一阶项（$k_e C_i$）的组合，如表3.1所示。微分方程本身只表明体内浓度是如何随时间变化的，而无法展示具体时间点的体内实际浓度。因此需要指定某一时间点状态变量的值，如 $t=0$（另外一个随之的假设是在 $t=0$ 时，$C_i=0$）。

从式（3.1）可以推导出系统的部分特征。初始时，$C_i=0$，故 $k_u C_w > k_e C_i$，因此导数是正值（如体内浓度会开始增加）。接着还应当考虑系统是否存在平衡状态。平衡时，状态变量 C_i 保持不变，$dC_i/dt=0$，由式（3.1）可轻易得出：

$$C_i = \frac{k_u}{k_e} C_w \tag{3.2}$$

比值 k_u/k_e 也称为生物浓缩系数，它代表平衡时内外浓度之比。通过 dC_i/dt 的正负，可以判断平衡是否稳定：C_i 小于平衡值时，导数为正（C_i 将随时间推移而增加）；C_i 大于平衡值时，导数为负（C_i 将减少）。因此，平衡态的小扰动将产生一个导数，使系统返回平衡状态。在化学中，平衡（如热力学平衡）和稳态（外界持续影响下，状态变量不变）之间通常存在区别。而在数学中，两者不存在区别。虽然某些情况下，"稳态"可能更为合适，但本章选用"平衡"一词。

不使用两个速率常数，我们可以用生物浓度系数（K，L/kg）重写式（3.1）的 ODE：

$$\frac{dC_i}{dt} = k_e(KC_w - C_i), \quad \text{其中} \ C_i(0) = 0 \tag{3.3}$$

该公式在数学范畴里完全等同于式（3.1）。两种公式的选择有很大的自主性。如果化学物质交换被认为是由被动扩散驱动的，则式（3.3）更合适（即化学物质交换与浓度差成正比）。如果吸收和清除通量为单独的过程（如通过主动过程介导），式（3.1）更合适。

我们接下来需要进行维度分析，以确定是否存在明显的一致性问题。我们可以通过检查 ODE 中所有元素的单位来完成上述分析。假设 C_i 的单位是 mol/kg，如果我们以天（d）作为时间单位，则 dC_i/dt 的单位是 mol/kg·d^{-1}。为了使式（3.1）中等号两边单位相同，k_e 的单位应当是 d^{-1}。如果 C_w 的单位是 mol/L，则 k_u 的单位为 L/kg·d^{-1}。显然，吸收速率常数与清除速率常数的性质不同。有人可能会说 1 L 的水重 1 kg，因此 k_u 的单位可以简化为 d^{-1}。但 1 L 的水与 1 L 的生物体不同，这两种性质不能用除法抵消。即使 C_i 的单位是 mol/L，k_u 的单位也相应为 $L_水/L_{生物体}·d^{-1}$。

一旦确定了模型的参数值，并指定了系统的驱动力（外部浓度可能是时间变量），就可以预测系统如何从一种状态变化为另一种状态。当然通常需要在软件中将模型实现才行。如果外部浓度（和其他参数）恒定，式（3.1）可解析求解为：

$$C_i(t) = \frac{k_u}{k_e} C_w (1 - e^{-k_e t}) \tag{3.4}$$

至于上述方程是如何通过式（3.1）计算得出，不在本章的讨论范围之内。更多的背景知识可以从有关数学建模的教科书中查阅（如 Doucet & Sloep，1993）。但通过计算式（3.4）中导数 dC_i/dt，以及检验式（3.1）中的等式是否成立，则很容易检查方程推导是否正确。

根据式（3.4）可以得到一些结果。整个系统的动力学由括号中的部分决定，即主要由 k_e 决定。因此清除速率常数决定了达到平衡状态某个百分比所需的时间。通过括号前面的部分可发现吸收速率常数（k_u）只（和 k_e、C_w 一起）决定最终体内浓度。这一发现强调了两个速率常数的性质不同（可从其单位论证），以及 k_e 对系统的动力学的重要性［这也是 k_e 作为式（3.3）唯一速率常数的原因］。

可以很容易地将式（3.4）编写为程序（包括电子表格）进行数据分析和拟合。但只能对最简单的模型进行 ODE 的数值求解。在实际中为进行 ODE 的数值求解时，必须使用包含求解器的专业软件。

3.3.2 简单单室模型扩展

在许多情况下需要对式（3.1）的简单模型进行扩展，这可能会涉及更多房室或不同类型的房室动力学。当不能将生物体视为一个混合均匀的整体，不能通过单室模型解释所观察到的动力学现象时，就需要将其扩展为多房室模型。我们可为生物体准备多个房室，形成比较抽象的"中心"和"外围"房室，或者表示生物体中的不同组织（见 Barron et al.，1990）。我们可能还需要考虑为各种不同的化学池（如惰性和活性池，母体化合物和代谢物池）准备额外房室。最精细的多室模型采用"基于生理的"TK 模型（"physiologically based" TK models，PBTK）形式，用不同房室代表特定器官或器官组，以血液流动将它们进行连接（Nichols et al.，1990）。目前人们只为少数动物群体（哺乳动物、鸟类和鱼类）构建了此种模型。增加模型的复杂性并不总能提高模型的预测能力。将更多的生理学和毒理学背景知识引入模型是很有吸引力的，但只有在必须明晰系统动态时含更多房室才是一个很好的策略。如果相对于体内分布过程,化学物质与环境交换缓慢,整个生物体就像一个混合均匀的单室。此时,使用 PBTK 模型只会增加更多参数,而不会提高模型的性能。

建立多室模型时,最重要的是确保质量平衡:化学物质从 A 室到 B 室,A 室减少的分子数必须等于 B 室增加的分子数(见第 3.2.5 节)。因此通常建议初始以质量为参数构建模型(即以各房室中化学物质质量为状态变量),再在后期重新计算浓度。

许多海洋生物会储存能量,目的是在不利于生存的季节周期变化的条件下,为生存和/或周期性产卵提供能量。这种储存物的积累和使用对 TK 有影响,因为它影响生物体型大小和身体成分(以及对化学物质的亲合力)。这种储存是否需要房室取决于物质与生物体其他部分的交换率;如果交换很快(相对于与环境的交换),整个生物体仍可视为一个房室(尽管参数会随储存量变化而变化)。Van Haren et al.(1994)以贻贝为例介绍了这种更为复杂的单室模型。产卵时,一部分储存物质与配子一起脱离母体,带走一部分母体中的化学物质(母源性传递)。因此,产卵可成为清除疏水化合物的重要途径。图 3.2 展示了包含这一过程的模型概念。

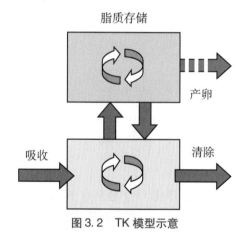

图 3.2 TK 模型示意

图中包括脂质的储存和周期性繁殖造成的损耗过程,箭头代表化学物质转移。

其他扩展还包括一级动力学的偏差。例如,使用主动转运来吸收或清除化学物质的情况(如由转运酶介导)。当化学物质浓度较小时,转运速率与浓度成正比:每个分子都会找到一个自由酶,所以绝对转运速率(分子单位时间)受化学物质可用性的限制(假设酶的作用时间为定值)。在化学物质为高浓度时,转运酶的数量可能无法满足毒物分子的供应量,则运输速率达到饱和:由于酶的数量限制,化学物质浓度增加 2 倍,只能增加不到 2 倍的转运速率。双曲线关系(Michaelise-Menten 动力学)可能更适合这种情况(表 3.1)。

另一种类型的扩展考虑了房室大小的变化。显然动物体型可能会发生变化,并影响毒代动力学特征。例如,体型增大会降低体内化学物质的浓度(生长稀释作用);而体型减小则会对化学物质浓度起浓缩作用。体型变化通常也会影响吸收和清除速率常数的值,因为这些参数取决于动物表面积和体积之比(Gobas et al.,1986)。如果在生命周期中,生物形状没有发生改变,表面积将随体积的 2/3 次方而变化。即表面积和体积的比值将随体积的 1/3 次方而减小,因此其比值与生物体的长度成正比。这可通过假设一个立方体形状的生物体进行证明。假设 a 是立方体边长,则体积为 a^3(a 为边长),表面积为 $6a^2$,表面积/体积 $=6/a$,与边长成反比。只要生物体不随生长而发生形态改变,这一规则同样适用于其他形状,但比例常数取决于形状和所用测量的长度方式。

综上所述,在所模拟的单室模型中,生长稀释和表面积和体积之比的变化将导致式 (3.3) 单室模型的一般性扩展(Kooijman & Bedaux,1996;Jager & Zimmer,2012):

$$\frac{dC_i}{dt} = k_e \frac{L_m}{L}(KC_w - C_i) - \frac{C_i}{V}\frac{dV}{dt}, \quad 其中 C_i(0) = 0 \quad (3.5)$$

方程右边最后一项为体积(V)变化而引起的生长稀释,由化学物质的质量和微分标准规则导出。如果没有增长,$dV/dt = 0$,则该项不存在。如果生长呈指数增长,则相对生长速率(dV/dt 除以 V)是恒定的,生长稀释和清除过程的效果相似(也可以用速率常数表示,见第 4 章生物富集)。式(3.5)中,L_m/L 为清除速率修正因子。随着体长 L 的增加,有效 k_e(即 $k_e L_m/L$)减小,k_e 的值可以理解为最大体型 L_m 时的清除速率。

3.3.3 实例研究

本节采用地中海贻贝(*Mytilus galloprovincialis*)对药物四氢西泮的积累和净化过程来分析具有一级动力学的单室 TK 模型(Gomez et al.,2012)。实验过程中先将贻贝暴露 7 天,然后置于清水中净化 7 天。图 3.3 是两种暴露浓度下,组织残留量基于式(3.1)的同时拟合(参数估算值见表 3.2)。通过改变暴露浓度,获得不同的吸收速率常数 k_u(低暴露产生较低的值),以此优化拟合。根据分析目的和检测值的置信度,可以决定同时拟合或分别拟合两组数据集(正如原作者所做)。

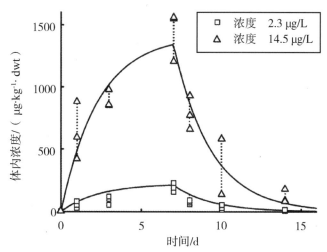

图 3.3 两种暴露浓度下,四氢西泮在地中海贻贝体内组织残留量的拟合
前 7 天为暴露阶段,随后为清水中的净化阶段。

表 3.2　根据图 3.3 拟合的参数估算值（基于 95% 概率的置信区间）

参数	估算值（95% 置信区间）
吸收速率常数（k_u）	47（38～56）L·kg^{-1}·dwt·d^{-1}
清除速率常数（k_e）	0.48（0.39～0.59）d^{-1}

统计模型中，拟合的前提是假设模型和数据间存在独立的正态分布偏差（具有常数方差）。组织残留量的检测是破坏性的过程，所以假设观测具有独立性是合理的。体内浓度是连续变量，所以偏差服从正态分布也合理。但一个恒定的方差具有不确定性。浓度测定的方差可能会随着均值增加而增加。为了符合这一点，可以使用随平均值变化的方差（如增加残差权重）或数学变换（如对数或平方根）。值得强调的是，数学变换也会改变分布的形态；假设残差的对数服从正态分布，则等同于残差服从对数正态分布。

3.4　毒物效应动力学概述

TK 模型可以单独使用（如研究食物链积累），也可与其他模型配对来解释和预测毒性效应。毒性效应通常不是由外部浓度直接引起的，而是在化学物质被生物体吸收并转运到靶部位后才产生。毒效动力学（TD）主要研究化学物质进入机体后与靶部位的相互作用，及其对机体产生毒性效应的过程。TK 和 TD 模型的组合称为 TK-TD 模型。与 TK 相反，TD 模型在生态毒理学中的应用较少。传统上，直接将毒物外部浓度与其在一个特定时间点对生活史性状的影响相关联（如图 3.4 中虚线箭头所示），如使用剂量 - 反应曲线。也可使用类似于已应用于 TK 的环境化学方法对 TD 进行简化。在 TKTD 模型中，将 TK 模型与 TD 模型相连，展示了毒性效应的动态特征（图 3.4）。TKTD 模型比描述性的剂量 - 反应模型具有优势（Jager et al., 2006; Ashauer & Escher, 2010）。最重要的是，这能完善我们对导致毒性效应潜在机制的认识，并对此机制进行定量研究。这对于理解并最终预测物种敏感性差异、化学物质毒性差异，以及化学混合物之间、化学物质和环境胁迫之间的相互作用都是至关重要的一步。TKTD 模型明确地包含了毒性的时间层面，这意味着模型可以解释和预测化学物质毒性的时间变化，以及暴露后毒性的时间效应（如一场漏油事故）。TD 模型是否实用，既取决于其所提出的问题，又取决于终点的性质。比如，研究行为终点的模型可能就与研究生殖毒性效应的模型不同。

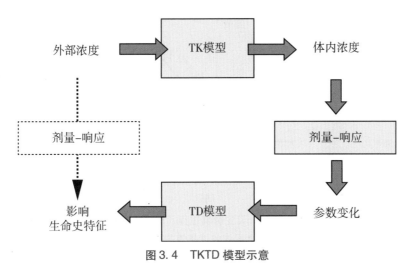

图 3.4 TKTD 模型示意

TKTD 模型没有将外部浓度与毒性效应直接联系起来（虚线箭头），而是将 TK 和 TD 模型"串联"，来研究外部浓度（也可能随时间变化）和随时间变化对生命史特征毒性效应之间的因果关系。

3.4.1 TK 和 TD 模型的连接

我们假设某个化学物质在生物靶部位的内部浓度，该浓度可引起能被观察到的毒性效应。但具体浓度值是多少？具体部位又在哪里？最简单的切入点是假设母体化合物的全身浓度是连接到 TD 模型的最佳指标。即使这一假设不成立，但只要靶部位的"真实"剂量指标与母体化合物的全身浓度或多或少成比例，则仍可给观察到的效应提供一个合理的解释。使用更精确的 TK 模型可能会将效应与特定组织中化合物或特定代谢物的浓度联系起来。

一旦在 TK 模型中建立了最有效的剂量指标，下一个问题就是如何将其与 TD 模型建立联系。可假设剂量指标值（如体内浓度）的增加影响 TD 模型中的一个或多个过程（参数），从而导致可在终点观察到的效应（图 3.4）。例如，体内浓度可能与生物的摄食速率有关，能量摄入的减少可以转化为在暴露时间内对生长和繁殖的影响（详见第 3.6 节）。

剂量指标和受影响的过程之间应该建立什么样的关系？目前几乎没有理论和经验可供参考。线性阈值关系（图 3.5）在实践应用中效果良好（Kooijman，1996；Jager et al.，2011），具有明显优势。水生生物并非生活在纯水中，其周围介质中含有的大量化学物质都可能具有毒性。因此每次毒性测验本质上是一个混合毒性测试。为了避免测试"自然背景"中的所有化学物质，可以假设它们的浓度低于各自的毒性阈值，忽略它们对毒性效应的贡献。阈值以上的浓度与受影响过程之间的线性关系具有直观的意义：每个化学物质分子的增加对整体效应的贡献相同。至少对于较小的效应，这是一个合理的假设；但对于较大的效应，分子和靶部位间的相互作用可能会变得更复杂（例如，随着剂量的增加而导致饱和效应，见图 3.5 中的虚线）。

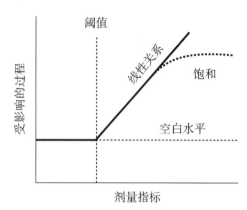

图 3.5　剂量指标（如体内浓度）与受影响过程（即 TD 模型中控制生理过程的参数）之间的线性阈值关系
虚线曲线代表高剂量时潜在的饱和效应。

3.4.2　缺乏机体残留数据时 TK 模型的使用

在大多数情况下，我们常模拟化学物质在一段时间内对生物的影响，但未检测机体残留。此时仍可以应用 TKTD 模型；根据毒性效应随时间推移的发展，可以推导出达到平衡所需时间的信息（和清除速率常数 k_e）。但由于缺乏体内浓度绝对值数据，因此不可能仅从效应数据估算出 k_u［式（3.1）］或 K［式（3.3）］。可以将 K 设为任意值，但更好的方法是对方程进行缩放，将式（3.3）的两边除以 K，可产生一个新的机体残留状态变量，即经缩放的体内浓度 C_i^*：

$$\frac{dC_i^*}{dt} = k_e(C_w - C_i^*) \tag{3.6}$$

上式中，$C_i^* = \frac{C_i}{K}$ 且 $C_i^*(0) = 0$ 新状态与实际状态（但未知）的体内浓度成正比，并且与外部浓度的量纲一致（平衡状态下，$C_i^* = C_w$）。如果 C_i^* 用作图 3.5 中的剂量指标，则阈值也将具有外部浓度的量纲。C_i^* 作为时间的函数可随后被连接到 TD 模块。缩放后的 TK 模型现在只有一个参数（k_e），k_e 可以根据随时间推移的效应数据估计出来。

3.4.3　基于损伤房室或受体动力学的扩展

目前为止，我们已经假设体内某些化学物质浓度是造成毒性效应的原因。但这也可能是体内的化学物质导致某种形式的损伤，且该损伤以一定的速率被修复。这种损伤可能是造成毒性效应的直接原因，而接触毒物是最终原因。损伤可作为一个具有一级动力学的附加房室纳入 TKTD 模型（Lee et al.，2002；Jager et al.，2011）。若速率常数不同，损伤房室可以表现出不同于体内浓度的时间进程。这有助于解释为什么体内浓度的动力学并不总是与毒性的动态模式相匹配（Ashauer et al.，2010）。

在 TKTD 模型中，损害通常被视为抽象的概念，无法被测量。一种更具体的扩展是使用显式受体动力学。假设化学物质被吸收后就与受体结合，形成的复合物导致了毒性效应（Jager & Koijman，2005）。化学物质与受体的结合取决于该化学物质与自由受体的接触频率，因而符合二级动力学（见表3.1）。已结合受体的释放速率取决于已结合受体的数量（一级动力学）。这种受体模型的行为不同于一级损伤扩展，事实上它将在较高浓度下达到饱和（在某些时刻，自由受体的数量过少，从而限制了化学物质的结合率）。

一般来说，只有能测量体内残留物浓度，确定内部浓度的动力学特征时，这些扩展才可用。若仅有随着时间推移的效应数据，则很难区分毒物动力学和损害/受体动力学。

3.5 对生存的影响

3.5.1 动物为什么会死亡？

任何关于终点生存的 TD 模型都应该解决这样一个问题：为什么处于化学胁迫下的动物会死亡，或者为什么它们不会在给定的浓度下同时死亡？对该现象最普遍的解释是个体耐受假说。这一假说的基本假设是，当个体的体内浓度超过阈值时，个体会立即死亡；而群体中不同个体阈值不同，并且服从某种概率分布（通常是对数正态分布或对数-逻辑斯蒂分布）。当毒性试验中有 50% 的个体在一定的暴露时间后死亡，剩下的存活者则被认为对测试毒物更耐受。这种死亡过程的观点与临界机体残留概念（Critical Body Residue，CBR；McCarty & Mackay，1993）联系紧密。CBR 认为机体的有害结局（如死亡）与体内浓度超过临界有关。在试验群体中，无论暴露浓度或暴露时间如何，一定的体内浓度将导致相同的致死率。因此，只有当体内浓度随时间增加而增加时，毒性效应才呈现时间依赖性。

另一种假说是"随机死亡"，它假设所有个体均相同，但死亡是随机过程（Bedaux & Kooijman，1994）。体内毒物浓度增加了个体死亡的可能性。在毒性实验中，50% 的死亡个体并不比存活个体对毒物更敏感，只是运气不好。死亡不太可能是随机的，但其可能涉及非常多的过程和变量，因此随机性可以提供一个很好的描述（类似于掷骰子不是固有的随机过程，只是随机性可以很好地描述掷骰子）。该解释与个体耐受相反，浓度超过阈值不会立即死亡，但致死率会增加。因此实际内部浓度与致死率之间没有固定关系。

尽管参数值不同，但两种假说都倾向于更好地描述数据模式，难以在两者之间做出选择。这两种机制在实际应用中都可能发挥作用。因此研究者提出了混合模型（Jager et al.，2011）。本节只详细讨论随机死亡模型，这已足够论证生存终点 TKTD 模型的构建与应用。个体耐受假说及其一些不切实际的特性，可以参考其他资料（Kooijman，

1996；Newman & McCloskey，2000）。

3.5.2 随机死亡模型

随机死亡模型的核心是"风险率"，这是一种统计原理，用来描述系统或产物随时间推移的随机"失效"。假设刚开始时生物体是活着的，风险率与很短时间间隔的乘积，即为该时间间隔内的死亡率。随着时间的推移，固定的风险率会导致存活率呈指数下降。当处在化学物质的胁迫下时，风险率通常不是固定的，而是随着内部浓度（靶部位）的增加而增加。

不同于式（3.6）的缩放 TK 模型，我们可以定义缩放的内部浓度（C_i^*）和致死率（h）之间存在线性阈值关系（图 3.5）：

$$h(t) = k_k(0, C_i^*(t) - c_0) + h_b \tag{3.7}$$

其中，c_0 是阈值浓度，k_k 是杀灭率（即高于阈值的缩放内部浓度和风险率之间的比例）。h_b 包括背景死亡率，如果毒性试验中的死亡可以简单地被视为是随机的（如处理或其他意外），则 h_b 为恒定值。独立死因可能会增加风险率。在本节，风险率在生物体个体中被视为一个生理过程。

对随时间推移的危险率进行积分，可以得到某一段时间内随时间推移的存活率 S：

$$S(t) = \exp\left(-\int_0^t h(\tau)\mathrm{d}\tau\right) \tag{3.8}$$

完整的模型将外部浓度与随时间变化的生存率联系起来，由 3 个方程组成，即式（3.6）、式（3.7）和式（3.8），只有 4 个参数可以从数据中被估计出来。这表明有意义的 TKTD 模型可以非常简单，且不需要太多参数。考虑到日常存活率数据中存在的诸多细节，拟合更多的参数不一定有益，除非获得更多信息源（如个体残留量、其他物种或化学品的信息）。同时对多个数据集进行拟合是一种非常有效的方法，但要假设每个数据集中的某些参数相同，而另一些参数在一个数据集中是唯一的（Jager & Kooijman，2005）。通过保持每个数据集较低的模型参数数量，则可以使用具有更多参数的、更复杂的模型。

为了使 TKTD 模型与数据进行匹配，我们需要考虑偏差（残差）的性质。残差的正态独立分布并非首选：生存数据只能假设为离散值（每个动物非死即活），存活率为 0~1（正态分布的末端延伸到正无穷和负无穷）。幸运的是，一个更好的选择是多项式分布（Bedaux & Kooijman，1994；Jager et al.，2011）。这是将二项式分布扩展，可处理具有 2 个以上离散结果的随机样本。对拟合生存模型统计数据的深入研究超出了本章范围，具体内容可参考其他文献（Bedaux & Kooijman，1994；Jager et al.，2011）。

3.5.3 案例分析

本节以北方海洋中桡足生物飞马哲水蚤（*Calanus finmarchicus*）暴露于汞的研究为例（Øverjordet et al.，2014），来验证上述定义的随机死亡模型。实验采用第 5 阶段，即成体前最后一个发育阶段的桡足幼体。数据包括 4 天内的 7 个暴露浓度（每组 30 只）

和一个对照组（60只）幼体的存活率。将上一节概述的TKTD模型与此数据集进行拟合，结果见图3.6，参数估计结果见表3.3。采用多项式分布建立判断模型与数据偏差的统计模型（Jager et al.，2011）。多项分布很好地适用于随时间推移的离散随机事件。将得到的似然函数最大化，以寻找最佳拟合参数值。此外，似然函数可用于构造稳健的置信区间（Meeker & Escobar，1995）。

尽管拟合不完美，但该模型合理地解释了存活率的时间变化模式。由于该模型很简单，我们很容易从基础假设出发，通过修改并扩展模型而提供一个更好的拟合结果。但我们需要了解数据的本质，并从观察到的试验群体死亡频率中来估计存活率。死亡过程的随机性，可能导致重复实验产生明显不同的结果。因此，对模型扩展应当慎重，只有当测试包含很大的样本量时，模型扩展才最有意义。无论如何，在该数据集上尝试另一种模型（个体耐受）也是有意义的；这个阶段桡足类的脂质含量是可变的，可以转化为个体间敏感度差异（Hansen et al.，2011）。

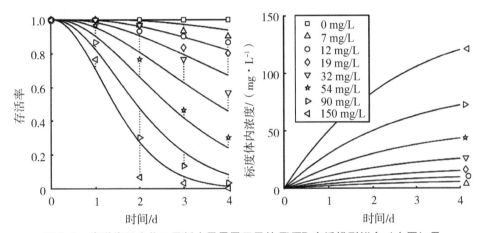

图3.6 海洋桡足生物飞马哲水蚤暴露于汞的TKTD存活模型拟合（左图）及由存活率拟合得出的预测的缩放体内浓度（右图）

估计参数见表3.3。

表3.3 根据图3.3拟合的参数估计值（基于95%似然函数的置信区间）

参数	估计值（95% CI）
清除速率常数（k_e）	0.41（0.034～0.92）d^{-1}
影响阈值（c_0）	2.9（0.35～4.9）$mg \cdot L^{-1}$
杀灭率比例（k_k）	0.015（0.0087～0.11）$L \cdot mg^{-1} \cdot d^{-1}$
本底风险率（h_b）	0（0～0.030）day^{-1}

汞的毒代动力学源于随时间变化的存活率模式（实验中未检测到机体残留）。只使用了一个TK参数，即消除率（k_e），它决定了达到平衡所需的时间。由于采用了缩放TK模型［式（3.6）］，缩放的内部浓度与外部浓度量纲一致；且在平衡状态下，缩放的内部浓度值等于外部浓度值（图3.6的右图）。出于同样的原因，阈值（c_0）和杀灭率（k_k）参数也与外部浓度的单位一致（图3.5）。

3.6 对亚致死终点的影响

3.6.1 亚致死效应的 TD 模型

对于亚致死效应，如影响生长和繁殖，不能使用与生存研究相同的 TD 模型。生存问题可以简单地将死亡描述为随机过程，并将体内浓度与风险率联系起来。显然生长和繁殖不能被视为随机事件，因为它们是分级反应，需要观察每个个体的反应程度（见第 3.2.7 节）。因此，我们需要一种包含生物体生长和生殖机制的 TD 模型。从能量收支的角度能为此类模型提供有用的切入点。动物必须遵守质量和能量守恒定律，当由于化学物质暴露而导致生殖能力降低时，与对照组相比，暴露组用于生殖消耗的能量更少。这些能量去哪里了？可能一开始这些能量就没有被吸收（如摄食率或从食物中同化能量的效率降低），或者应激生物以其他方式利用了这些能量（如调节化学应激的负面效应）。当在构成能量汇的终点，观察到如生长和生殖效应时，跟随能量的流动是一种自然有效的方法。当然对于某些终点，如行为的改变和发育畸形，能量收支可能无法成为一个有用的起点。

能量收支模型明确了动物通过摄食如何从环境中获取能量和基础物质，以及被如何用来为资源需求过程提供能量（图 3.7）。这些模型的核心是质量和能量守恒，必须逐一记录并计算所有的质量和能量变化过程。此外，模型必须符合热力学第二定律，即质量或能量不可能 100% 转化，每个转化步骤中都不可避免地会产生损失。

在简单的能量收支试验中，可以尝试测量一些通量（通常为摄食、排泄和呼吸）来创造一个封闭的质量/能量平衡。吸收和消耗能量的差异则被称为"生长指示器"（Widdows & Johnson，1988），此处"生长"既指身体生长，也指生殖。该方法的局限性在于，它严重依赖于摄食、排泄和呼吸的测量（毒性试验中的常规测试通常不包含这些通量）；同时它是描述性的（需要在每个时间点测量通量）。为在 TD 模型中应用能量收支理论，需要一种动态且更机制化的方法。

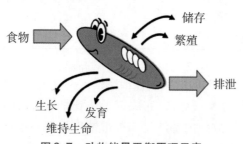

图 3.7 动物能量平衡原理示意

食物被吸收后，如图所示，食物中的部分能量被同化后用于资源需求过程。

3.6.2 动态能量收支

动态能量收支（dynamic energy budget，DEB）理论是目前最全面、最可靠的能量收支模型框架（Nisbet et al.，2000；Kooijman，2001）。关于 DEB 理论的概念及其在生态毒理学中的应用可参考其他资料（Jager，2015b）。基于 DEB 理论已经衍生了一系列实用的模型，通常被称为"DEBtox"（Kooijman & Bedaux，1996；Billoir et al.，2008；Jager & Zimmer，2012）。这些模型使用的是易于理解的复合参数（如最大生殖率），是对直接联系到能量过程（如摄食率或维持生命的特定消耗）的主要基础参数的组合。DEB 模型使用相对简单，参数易于解释，但复合参数的使用缺乏透明度、易导致错误（Billoir et al.，2008；Jager & Zimmer，2012）。此外，由于缺少显式质量平衡，模型扩展（扩展到呼吸或脂质储存等其他终点）受到阻碍。具有显式质量平衡的简单 DEB 模型的详细介绍请参见其他资料（Jager，2015a）。

能量收支模型代表对生物体进行了极简化，不可避免地丢失了许多生物学"细节"。简化的程度可以与 TK 模型相媲美。这意味着有可能根据生态毒理学中（相对）有限的数据集基础，对该模型进行参数化并应用。与 DEB 理论不同，没必要为每个物种从头开始构建一个全新的模型。所有的生物体都是通过进化而联系在一起，相似的生物体具有相似的代谢组织结构。从新陈代谢角度来说，动物形成一个非常相似的群体，因为它们都是通过以其他生物为食来获取资源。然而，对许多的物种和应用程序，我们需要考虑修改标准的 DEB 模型。例如，海洋生物的幼体发育阶段很常见。许多生物幼体与成体在摄食和代谢等方面有很大不同。例如，多数生物在幼体时期代谢率较低，还有些则在生命周期中会发生某种形式的加速（Kooijman，2014）。海洋生物需要考虑的另一个变化是，许多海洋生物会季节性地建立储存缓冲，为度过食物短缺时期和/或为产卵活动提供能量。尽管这种储存能量的建立和使用规则具有物种特异性，有待进一步的详细研究。在 DEB 理论中，这种储存可被视为"生殖缓冲"（Saraiva et al.，2012；Jager & Ravagnan，2015）。修正模型时，应始终牢记：组织代谢的规律遵从自然选择，受进化影响。因此需要对相关物种进行类似的修正。

使用 DEB 模型作为 TD 模型的优势在于，可以在一个完整的框架之下解释生命周期中所有终点的毒性效应。生长和生殖过程是紧密相连的，孤立地看待这些终点没有意义。生殖一般开始于一个恒定的体型，而体型影响摄食率，从而影响生殖率。因此，对生长的毒理作用也会间接地影响生殖。这些特质之间以及与其他特质（如呼吸和摄食）间的相互作用，都是能量收支中不可缺少的部分。另一个优势是，可以使用 DEB 模型对未经测试的状态进行有根据地推断，如食物限制、温度和暴露时间的变化，以及联合毒性。对于哪种能量收支模型最合适（针对哪个问题），学者们可能有不同看法。但一个 TD 模型若从机制上研究能量需求特征，如生长和生殖等的效应时，某种形式的能量平衡是必须的。

3.6.3 TK 与能量收支的连接

在生态毒理学的应用中可以将能量收支模型作为 TD 模型,并将其与选择的 TK 模型连接。在研究如何将 TK 与 TD 连接之前,可以逆向思考一下:动物的生命史特征(可受毒物影响)是如何影响毒代动力学的? 本节主要讨论亚致死效应,该效应下生物会生长,因此可以应用式 (3.5) 来扩展研究。显然有更多通过生命史特征来改变 TK 的方法(尤其是生殖和储存)。具体内容可以参考其他资料 (Van Haren et al., 1994)。

TK 模型提供了剂量指标(如缩放的体内浓度)。可以将剂量变化与代谢过程联系起来,假设其与阈值呈线性关系(图 3.5)。代谢过程则是能量收支的一个主要模型参数。但哪个参数会受到影响? 对不同参数的影响都会对动物整个生命周期的生长和生殖模式(以及其他能量的过程)产生特定的影响。这种模式称为代谢或生理活动模式 (Alda Alvarez et al., 2006)。例如,对同化过程的毒性效应,将导致生物体的能量摄入减少、生长减缓、最终体型变小、生殖延迟和生殖率降低 (Jager, 2015b)。相反地,对配子的毒性效应会导致生殖率降低,但对生长或生殖行为的启动无影响。因此,可以通过分析化学物质对生物体生命史的效应,推断出最可能受影响的参数。

3.6.4 TKTD 能量收支模型的数据拟合

拟合亚致死效应模型时,需要为模型和实测值之间的偏差(残差)选择合适的统计模型。本章不深入讨论统计方面的内容,仅给出一些一般性的结论。与存活模型相比,亚致死效应终点的生长和繁殖是分等级的:可以测量个体的反应程度(而不仅仅是生或死的区别)。因此,可以背离作为最小二乘分析基础的一系列假设:常数方差的独立正态分布。我们在第 3.2.7 节已经介绍过,如果长期监测一组动物,独立性则被破坏,常数方差也可能不成立。此外,如果以配子数量来衡量生殖,它属于离散变量,当观测到的后代数量较少时,数据将不完全服从正态分布。对数据进行更恰当的统计分析需要深入研究模型和实测偏差的本质。在测定体型和生殖行为时,通常测量误差(在进行最小二乘分析时通常会考虑的问题)很小或不存在。然而试验中的个体都存在差异,个体间的能量收支参数不同,对化学物质的反应也不同。模型是对生物复杂性的简化,因此被认为是"错误的"(增加了模型和数据之间的偏差)。

实际上,提出更好的替代方案相当复杂,所以通常忽略这些问题,并使用最小二乘法优化(例如,使用转换或加权来适应与实测平均值的方差变化)。只要模型充分符合实际,统计处理的细节不太可能对最佳拟合参数产生实质性的影响。但置信区间将受到更大的影响,因此这些细节确实需要非常仔细地加以解析。

3.6.5 个案研究

本节以壬基酚对海洋多毛纲海蠕虫(*Capitella teleta*)生长和生殖的影响为例,研究基于 DEB 的 TKTD 模型的应用。实例中使用了 DEBtox 模型(Jager & Zimmer,

2012), 关于方程的详细信息可参见其他资料 (包括推导过程)。TK 模型应用了缩放版的式 (3.5) (包含生长的效应), TD 模型是一个简单的能量收支。所选数据来自 (Hansen et al., 1999), 并由 Jager & Selck 应用 DEB 理论进行了完整的分析 (Jager & Selck, 2011)。

本节所选数据与许多数据集一样, 不能进行简单分析。原因有二: 其一, 第一次暴露处理始终比对照组表现更好 ("毒物兴奋效应"); 其二, 从观测结果来看, 初始生长一定比预期的要慢 ("加速度")。"毒物兴奋效应" 是用来描述 (并非解释) 随着暴露剂量增加, 效应方向发生逆转。此处可观察到在最低暴露剂量时, 生长和生殖增加, 表明生物体有更大的质量和能量消耗 (假设配子大小没有明显差异)。由于生物体需遵循守恒定律, 增加的消耗必须通过增加摄食或提高同化效率来实现。当然在生活史上总是存在种内变异, 所以这种 "刺激" 可能只是偶然的。但如果刺激是真实的, 基本上只有两种解释: 壬基酚增加食物的可获得性或质量 (如通过刺激细菌生长), 或者它如同药物一样可以抵消对照组中未知的应激。关于毒物兴奋效应反应更详细的讨论参见其他资料 (Jager & Selck, 2011), 本节不做讨论。本节忽视毒物兴奋效应反应, 并假设最低剂量代表了 "真实的" 控制行为。

另一个模型偏差包括明显的 "加速度", 因为生长曲线的初始部分与标准 DEBtox 模型中 von Bertalanffy 生长曲线不匹配 (见图 3.8, 左下角)。这一特征较为常见, 尤其是对于具有幼体阶段的生物而言 (Kooijman, 2014)。实验最初释放可自由游动的 *C. teleta* 幼体。一段时间后幼体栖息于沉积物中, 并蜕变为成体。在缺乏生长曲线初始部分信息的情况下, 任何模型的适应都只不过是猜测而已。所以我们简单地引入了在生长和摄入毒物之前的 "滞后时间"。

有了以上补充假设, 可以对数据进行拟合 (图 3.8, 参数见表 3.4)。一共需要从数据中估算出 8 个参数。因为有 6 条曲线 (每条曲线 11 个点), 所以根据有效数据可说明参数总数是合理的。但统计模型的匹配出现困难。测量同一组个体随时间变化的生长和生殖情况, 违背了独立原则。此外, 这两个终点不是独立的, 因为生殖行为取决于体型, 且在同一只动物上测量。身体体积是连续变量, 而生殖是离散变量 (尽管与大量的产卵数量相比, 这是一个小问题)。另一个需要考虑的问题是生殖以速率 (产卵数/天) 衡量, 测量的是某一时间间隔内的产卵数量。解决该问题的一种方法是计算产卵的累积数量 (图 3.8)。这对于计算和解释来说很简单, 但是通过增加观测值之间的依赖关系, 增加了统计模型的问题。这一问题没有简单的解决办法。较实用的方法是假设经过平方根转化后, 常数方差的残差是独立的正态分布 (减少大观测值的权重)。

最能描述这些数据的变化规律是在相同因素下生长成本和生殖成本的增加。体型数据清楚地表明生长成本受到影响: 这种行为模式使动物的生长成本更高, 但不影响最终体型。对生长的影响也间接地影响生殖, 延迟生殖行为的启动, 并需要更长时间达到最大繁殖率 (因为体型小的生物吃得少, 所以生殖的后代也更少)。但为了在这些数据集中充分地掌握毒物对生长的影响, 还需要研究对生殖的额外影响。

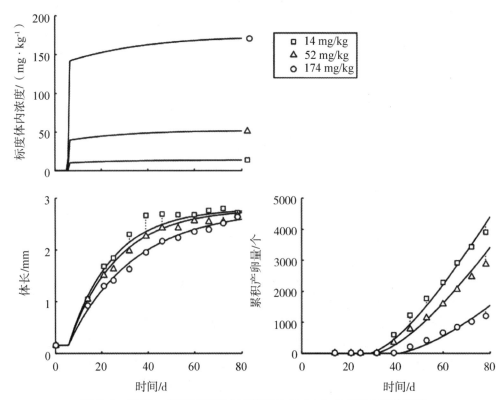

图 3.8 DEBtox 模型同时拟合壬基酚对 *C. teleta* 生长和生殖的影响

上部分的图为预测的缩放内浓度随时间的变化。下部分的图分别为三种暴露浓度下生长和生殖随时间的变化。所有化学物质浓度以 mg/kg 干沉积物为单位。

表 3.4 图 3.8 中同时拟合的参数估计（基于 95%似然函数的置信区间）

参数	估计值（95% CI）
初始体长	0.16（n.e.）mm
成体体长	2.1（2.0～2.1）mm
最大体长	2.8（2.7～2.9）mm
最大繁殖率	120（110～137）个卵细胞/天
能量投资率	10（n.e.）[−]
von Bertalanffy 速率常数	0.053（0.045～0.062）d^{-1}
生长开始的滞后时间	5.6（3.9～7.2）d
清除率常数	0.44（0.081～∞）d^{-1}
效应阈值	14（14～21）mg·kg^{-1}
耐受浓度	240（160～310）mg·kg^{-1}

注：n.e. 表示参数为非估计值，而是设置为一个固定值。

图 3.8 分别为缩放的内浓度和相应生长及生殖的拟合。从效应数据中估计的唯一 TK 参数是清除率（k_e）。由于式（3.5）对应的 TK 模型考虑了体型随时间增加的影响（稀释内浓度，改变动物表面积和体积的比值），所以内浓度累积模式与图 3.1 中的一般模式不同。与生存的示例一样，使用缩放的 TK 模型。这就是为什么在暴露一段时间后，内浓度接近外浓度（缩放内浓度的维度与外浓度维度相同，见第 3.4.2 节）。清除率的置信区间可扩展到无穷大，这表明瞬时平衡也提供了良好的数据拟合。

那么从这个实例中可以学到什么呢？首先，本试验为此物种和试验条件提供了更多新的基础资料：低浓度壬基酚如何明显增加食物的可获得性，以及生物在早期生命阶段发生了什么？这些问题需要专门进行实验研究。其次，本试验为我们提供了一种壬基酚的生理作用模式，这种模式与本数据集的所有信息完全一致。为了检验这一结果，下一步将是使用校准模型对其他特征（如摄食和呼吸速率）或不同暴露状况（如食物限制或脉冲暴露）进行预测。然后再通过实验验证这些预测，得出新的/附加的假设，进一步改进模型，又产生新的预测。这种模型和实验交替验证是理解某种化学物质对特定物种作用的最佳方法，但在实际中却很少得到应用。

3.7 种群及以上水平

环境风险评价所关注的不是单个动物的健康，而是种群、群落和生态系统的长期健康。如果对个体无效应（在所有现实的环境条件下），那么对生态系统也无效应，但反过来也如此吗？如果个体繁殖减少 $x\%$，那么更高层次的群体会如何变化？这些更高层次无法进行系统的实验验证，因此模型是解决这些问题的基本工具。种群和食物链模型研究在生态学方面有着悠久的历史，已有许多经典教材（可访问参考网页 http://www.debtox.info/marecotx.html）。本章重点是在个体层次的模型，关于更高层次只会简单地提及。

个体的生活史特征（如摄食、生存、生长和生殖）最终决定了种群动态。因此生态毒理学的种群建模是通过能够描述个体整个生命周期内的生活史特征的模型来实现，形成一个由化学暴露、食物可利用性和温度等组成的函数。通常某种形式的能量收支模型可以完成这一任务。本节将讨论生态毒理学中 3 种最常见的种群模型。与应用其他模型相同，应当依据研究目标和可用数据来选择最合适的种群模型。

3.7.1 基于个体的模型

从概念上讲，由个体外推到种群的最简单方法是使用基于个体的模型。在这类模型中，种群中的所有个体都被明确地建模，种群动态是由个体行为决定的。这种方法很容易与个体的能量预算模型相结合（Martin et al., 2012），但也容易因计算密集而变得复杂（尤其是研究过程包括了多个相互作用的物种）。生态毒理学中使用基于个体模型方法的深入讨论可参见其他资料（Topping et al.，2009）。

3.7.2 矩阵模型

矩阵模型在生态学和生态毒理学中有着悠久的历史。在这类模型中，根据年龄或阶段（或体型）将种群划分为有限数量的离散类型。能量收支模型也可以与矩阵种群模型相结合（Klanjscek et al.，2006）。然而，由于简单的矩阵模型只包含一个生物体的状态变量（年龄、大小或阶段），很快将变得不足以捕捉个体的化学应激（如不能简单地将包括毒性动力学作为时间和体型的函数来考虑），所以需要寻找捷径。生态毒理学中使用矩阵模型的深入讨论可参见其他资料（Charles et al.，2009）。

3.7.3 内禀增长率

作为研究种群动态的最后一种方法，经典的 Euler-Lotka 方程可以用来计算种群内禀增长率。在一个恒定环境中，所有种群最终都会呈指数增长，该增长率可以作为化学暴露和食物可利用性的函数来计算。模型的输入是由生物体整个生命周期的存活率和生殖率构成，作为时间的连续或离散函数。输出由本章前文讨论的 TKTD 模拟提供，使内禀增长率直接与 TKTD 模型链接起来（参见 Jager et al.，2006）。实际上，环境条件不会长时间保持不变，所以计算结果所反映的生态真实性有限。然而，内禀增长率综合考虑了生长、生殖和生存效应，是一种易于解释种群适应性的衡量方法（一种种群增长的内在能力）。因此，内禀增长率可以作为环境风险评估中一个有价值的统计数据（Forbes & Calow，1999）。

3.8 未来的可能性

最后一节展望了 TKTD 模型的未来。本章的参考网页（http：// www. debtox. info/ marecotox. html）提供了追寻这些模型发展论文的链接，以期为未来的研究提供支撑。

3.8.1 学科间更密切的合作

建模通常被视为与实验截然不同的方法。然而，这两种方法是相互促进的；生态毒理学的科学进步需要促进两种方法更密切的结合。例如，在实验的设计阶段让建模人员参与。大多数毒性试验都使用标准化的试验方案，但 TKTD 模型对实验设计有不同的要求，这意味着需要建模人员有更多的观察；另一方面，也可以减少一些约束。比如，只要可以建立符合实际的浓度-时间模式，就不需要保持暴露浓度恒定。此外，更紧密的合作需要在第一轮模型分析后，评估后续实验的可行性；当然模型分析不可避免地会产生新的（更基本的）的问题。机理模型和实验相互交替是深入理解化学物质效应的最佳途径。

我们同样还需要促进模型和统计之间更密切的协作。统计学家非常依赖描述性模型，而建模者通常使用的统计数据不能很好地匹配当前的问题。例如，模型和观测之间的偏差通常被视为随机噪声或测量误差；然而事实上大多数偏差是由个体间实际差异造成的。在 TKTD 模型中，个体差异转化为模型参数的差异。但由于这些模型参数与生理过程有关，不可能在种群中独立变化。因此，需要更多的机理模型来处理模型和观测之间的偏差（见第 2 章）。

TKTD 建模者和分子生物学家之间也需要更密切的合作（如生物标志物；见第 5.2 节）。过去十年，生态毒理学中的分子研究取得了爆发式进展。但如何将分子水平与个体生活史特征联系起来的问题尚未得到解决。将分子水平与 TKTD 模型参数相结合，是一种比直接将分子与生活史特征联系起来的更有前途的方法。例如，将分子水平上的变化与对生殖的影响联系起来，未必能产生有效的模型。这是因为有许多间接因素影响生殖（如维持生存的消耗增加，或摄食率的下降）。然而，将分子水平与能量收支的某些组成部分（如同化和维持）联系起来可能会更有效。

3.8.2 未来研究课题

TKTD 模型的校准需要更多生物体和化学效应的信息，而不是拟合一个简单的剂量-反应曲线。这些信息可以通过更多的实验获得，但也可以通过其他的方式获得。TKTD 模型参数具有一定的生理意义（不同于剂量-反应曲线），因此有可能根据统计关系或第一性原理来预测参数。需要系统地研究不同物种和不同化学物质之间模型参数的差异。

机体残留与毒性效应只有在极少情况下会同时被确定；机体残留的消除率在大多数情况下与模型的 TD 部分所需的消除率不符（如需了解毒性效应随时间变化的模式）。整体浓度的变化不一定与靶部位浓度保持一致。可能是因为生物体不能被视为一个混合均匀的房室，或者需要考虑代谢产物和额外的损伤阶段（或者这些因素的结合）。无论如何，根据效应数据估算的 k_e（表 3.3 和 3.4）不应解释为全身清除率。梳理出导致机体残留量和效应动力学之间差异的原因，有助于前文所述的预测关系的发展。

TK 的实验通常尽量避免复杂因素，如生长、生殖和毒性效应。而在 TKTD 建模中，我们对动物生长和生殖的 TK 模型特别感兴趣，尤其是在产生毒性的浓度范围内。根据式（3.5），引入了一种用于生长中生物体的模型扩展，当然也可能包括生殖（如转移到产卵）的影响。如果毒性影响生长和生殖，那么也会影响 TK。但毒性可能对 TK 有额外的影响，如通过影响动物的活动和行为进而影响 TK。在更真实的条件下，需要测试 TK 模型性能。

TK & TKTD 模型中，配子和胚胎受到的关注相对较少。化学物质可能由母体转移到卵细胞中，卵细胞也可能与环境交换化学物质。在胚胎发育过程中，卵的总质量会减少（因为胚胎会消耗能量来维持自身和发育），从而影响 TK。此外，胚胎也会如幼体和成体一样受到毒物胁迫的影响；胚胎也具有能量收支，如果毒物浓度过高，胚胎会死亡。对胚胎发育的有害效应可能对种群动态产生较大影响。因此，对胚胎阶段的研究给予更多关注非常重要。

生物体从不会孤立地暴露在单一压力源中，通常多重压力源并存。这不仅与化学物质的混合有关，还与化学物质胁迫和环境因素（如食物可利用性和温度）的结合有关。气候变化和海洋酸化对海洋环境的影响特别明显（见第 10 章，全球变化）。这种压力源会影响海洋生物的生活史，并可能与化学物质胁迫相互作用。由于 TKTD 模型是基于机制而非描述，所以非常适合用来解释和预测多重压力源的效应。但当前有必要对 TKTD 模型的发展潜力和局限性进行更系统的研究。

TKTD 模型仍然需要找到进入风险评价框架的方法（见 9.2 节）。机理模型已在暴露评估（如归趋模型）和生物累积（如 TK 模型）方面起到重要作用。相比之下，效应评价仍严重依赖于描述性方法。这种情况表明了 TKTD 模型在生态毒理学中尚未得到广泛应用。但即使是简单的 TKTD 模型，也比目前的描述性方法的优势明显。例如，TKTD 模型能够对所有可用数据（所有时间点和终点）进行集成分析，并且能够对未经测试的情况（如多重应激和暴露浓度波动）做出有根据的预测。模型合适与否以及它们在监管环境下的应用还需要进一步的研究工作来验证。

参考文献

ALDAÁLVAREZ O, JAGER T, MARCO REDONDO E, et al., 2006. Physiological modes of action of toxic chemicals in the nematode *Acrobeloides nanus*. Environ. Toxicol. Chem., 25: 3230 – 3237.

ASHAUER R, ESCHER B I, 2010. Advantages of toxicokinetic and toxicodynamic modelling in aquatic ecotoxicology and risk assessment. J. Environ. Monit., 12: 2056 – 2061.

ASHAUER R, HINTERMEISTER A, CARAVATTI I, et al., 2010. Toxicokinetic and toxicodynamic modeling explains carry-over toxicity from exposure to diazinon by slow organism recovery. Environ. Sci. Technol., 44: 3963 – 3971.

BARRON M G, STEHLY G R, HAYTON W L, 1990. Pharmacokinetic modeling in aquatic animals I. Models and concepts. Aquat. Toxicol., 18: 61 – 86.

BEDAUX J J M, KOOIJMAN S A L M, 1994. Statistical analysis of bioassays based on hazard modelling. Environ. Ecol. Stat., 1: 303 – 314.

BILLOIR E, DELIGNETTE-MULLER M L, PÉRY A R R, et al., 2008. Statistical cautions when estimating DEBtox parameters. J. Theor. Biol., 254: 55 – 64.

CHARLES S, BILLOIR E, LOPES C, et al., 2009. Matrix population models as relevant modeling tools in ecotoxicology // DEVILLERS J (Eds.). Ecotoxicology modelling. Springer, New York, USA: 261 – 298.

DOUCET P, SLOEP P B, 1993. Mathematical modelling in the life sciences. Ellis Horwood Limited, Chichester, UK.

FORBES V E, CALOW P, 1999. Is the per capita rate of increase a good measure of population-level effects in ecotoxicology? Environ. Toxicol. Chem., 18: 1544 – 1556.

GIESE A C, 1959. Comparative physiology: annual reproductive cycles of marine invertebrates. Annu. Rev. Physiol., 21: 547-576.

GOBAS F A P C, OPPERHUIZEN A, HUTZINGER O, 1986. Bioconcentration of hydrophobic chemicals in fish: relationship with membrane permeation. Environ. Toxicol. Chem., 5: 637-646.

GOMEZ E, BACHELOT M, BOILLOT C, et al., 2012. Bioconcentration of two pharmaceuticals (benzodiazepines) and two personal care products (UV filters) in marine mussels (*Mytilus galloprovincialis*) under controlled laboratory conditions. Environ. Sci. Pollut. Res., 19: 2561-2569.

HANSEN B H, ALTIN D, RØRVIK S F, et al., 2011. Comparative study on acute effects of water accommodated fractions of an artificially weathered crude oil on *Calanus finmarchicus* and *Calanus glacialis* (Crustacea: Copepoda). Sci. Total Environ., 409: 704-709.

HANSEN F T, FORBES V E, FORBES T L, 1999. Effects of 4-n-nonylphenol on life-history traits and population dynamics of a polychaete. Ecol. Appl., 9 (2): 482-495.

JAGER T, 2011. Some good reasons to ban ECx and related concepts in ecotoxicology. Environ. Sci. Technol., 45: 8180-8181.

JAGER T, 2015a. DEBkiss: a simple framework for animal energy budgets. Version 1.4. Leanpub, 12 August 2015. https://leanpub.com/debkiss_book.

JAGER T, 2015b. Making sense of chemical stress: applications of dynamic energy budget theory in ecotoxicology and stress ecology. Version 1.2. Leanpub, 11 August 2015. https://leanpub.com/debtox_book.

JAGER T, ALBERT C, PREUSS T G, et al., 2011. General unified threshold model of survival: a toxicokinetic-toxicodynamic framework for ecotoxicology. Environ. Sci. Technol., 45 (7): 2529-2540.

JAGER T, HEUGENS E H W, KOOIJMAN S A L M, 2006. Making sense of ecotoxicological test results: towards application of process-based models. Ecotoxicology, 15: 305-314.

JAGER T, KOOIJMAN S A L M, 2005. Modeling receptor kinetics in the analysis of survival data for organophosphorus pesticides. Environ. Sci. Technol., 39: 8307-8314.

JAGER T, RAVAGNAN E, 2015. Parameterising a generic model for the dynamic energy budget of Antarctic krill *Euphausia superba*. Mar. Ecol. Prog. Ser., 519: 115-128.

JAGER T, SELCK H, 2011. Interpreting toxicity data in a DEB framework: a case study for nonylphenol in the marine polychaete *Capitella teleta*. J. Sea Res., 66: 456-462.

JAGER T, ZIMMER E I, 2012. Simplified Dynamic Energy Budget model for analysing ecotoxicity data. Ecol. Modell., 225: 74-81.

KLANJSCEK T, CASWELL H, NEUBERT M G, et al., 2006. Integrating dynamic energy budgets into matrix population models. Ecol. Modell., 196: 407-420.

KOOIJMAN S A L M, 1996. An alternative for NOEC exists, but the standard model has to be abandoned first. Oikos, 75: 310-316.

KOOIJMAN S A L M, 2001. Quantitative aspects of metabolic organization: a discussion of concepts. Phil. Trans. R. Soc. B. , 356: 331 - 349.

KOOIJMAN S A L M, 2014. Metabolic acceleration in animal ontogeny: an evolutionary perspective. J. Sea Res. , 94: 128 - 137.

KOOIJMAN S A L M, BEDAUX J J M, 1996. Analysis of toxicity tests on *Daphnia* survival and reproduction. Water Res. , 30: 1711 - 1723.

LEE J H, LANDRUM P F, KOH C H, 2002. Prediction of time-dependent PAH toxicity in *Hyalella azteca* using a damage assessment model. Environ. Sci. Technol. , 36: 3131 - 3138.

LEE R F, HAGEN W, KATTNER G, 2006. Lipid storage in marine zooplankton. Mar. Ecol. Prog. Ser. , 307: 273 - 306.

MACLEOD M, SCHERINGER M, MCKONE T E, et al. , 2010. The state of multimedia mass-balance modeling in environmental science and decision-making. Environ. Sci. Technol. , 44: 8360 - 8364.

MARTIN B T, ZIMMER E I, GRIMM V, et al. , 2012. Dynamic Energy Budget theory meets individual-based modelling: a generic and accessible implementation. Methods Ecol. Evol. , 3: 445 - 449.

MCCARTY L S, MACKAY D, 1993. Enhancing ecotoxicological modeling and assessment: body residues and modes of toxic action. Environ. Sci. Technol. , 27: 1719 - 1728.

MEEKER W Q, ESCOBAR L A, 1995. Teaching about approximate confidence regions based on maximum likelihood estimation. Am. Stat. , 49: 48 - 53.

NEWMAN M C, MCCLOSKEY J T, 2000. The individual tolerance concept is not the sole explanation for the probit dose-effect model. Environ. Toxicol. Chem. , 19: 520 - 526.

NICHOLS J W, MCKIM J M, ANDERSEN M E, et al. , 1990. A physiological based toxicokinetic model for the uptake and disposition of waterborne organic chemicals in fish. Toxicol. Appl. Pharmacol, 106: 433 - 447.

NISBET R M, MULLER E B, LIKA K, et al. , 2000. From molecules to ecosystems through dynamic energy budget models. J. Anim. Ecol. , 69: 913 - 926.

ØVERJORDET I B, ALTIN D, BERG T, et al. , 2014. Acute and sublethal response to mercury in Arctic and boreal calanoid copepods. Aquat. Toxicol. , 155: 160 - 165.

PECHENIK J A, 1999. On the advantages and disadvantages of larval stages in benthic marine invertebrate life cycles. Mar. Ecol. Prog. Ser. , 177: 269 - 297.

SARAIVA S, VAN DER MEER J, KOOIJMAN S A L M, et al. , 2012. Validation of a dynamic energy budget (DEB) model for the blue mussel *Mytilus edulis*. Mar. Ecol. Prog. Ser. , 463: 141 - 158.

TOPPING C J, DALKVIST T, FORBES V E, et al. , 2009. The potential for the use of agent-based models in ecotoxicology // DEVILLERS J (Eds.). Ecotoxicology Modelling. Springer, New York, USA: 205 - 235.

VAN HAREN R J F, SCHEPERS H E, KOOIJMAN S A L M, 1994. Dynamic Energy Budgets affect kinetics of xenobiotics in the marine mussel *Mytilus edulis*. Chemosphere, 29: 163 – 189.

WIDDOWS J, JOHNSON D, 1988. Physiological energetics of *Mytilus edulis*: scope for growth. Mar. Ecol. Prog. Ser., 46: 113 – 121.

4 生物富集和生物监测

W. ‑X. Wang[①]

[①] 香港科技大学。

生物富集通常被定义为水生生物从周围环境介质中吸收污染物，导致体内污染物浓度增加的现象。因此，浓度是生物富集研究的核心部分，必须要了解浓度的重要性。许多研究已经确定了世界不同地区水生生物中不同污染物的浓度，文献中也包含了大量的相关数据。

水生生物有不同的污染物吸收来源，如水（经水源性吸收）和/或食物颗粒（经食源性吸收）。生物富集主要关注生物对污染物的吸收和消除过程。另一个重要的概念是生物可利用率，它被定义为潜在可吸收或实际从环境中吸收的污染物的比例。虽然这两个概念紧密相关，但本质上是不同的。

生物富集可检测生物体内污染物浓度的变化，而生物可利用率则描述了环境中潜在的可用于生物富集的污染物的百分比。因此，许多关于生物可利用率的研究都关注环境中污染物的化学性质。环境中污染物化学性质的生物学改变，以及此变化对生物可利用率的影响同样值得关注。最后，生物富集还涉及生理生化作用对污染物吸收和消除的调控。

生态毒理学研究需要同时考虑生物富集和生物可利用率。如果不考虑生物可利用率，就不可能研究生物富集，反之亦然。两者均可以应用于生物监测。传统的生态毒理学主要包括三个框架，即环境迁移、生物蓄积和污染物与生物相互作用的毒性（Wang，2011a）。环境风险评价以这些框架为基础，并为环境污染物的管理提供依据。

生物富集是环境污染物和暴露生物体之间的直接联系。生物体发生生物富集后，可表现出毒性效应。生物富集直接将环境化学/过程和生物生理/生物化学联系起来，可被视为环境化学和生物学之间的一个交叉点。

在过去的几十年里，已有大量关于水生生物的污染物生物富集研究。所有这些研究不可能被全部总结在本书的一个章节中，读者可以参考一些关于这个主题的早期研究（Spacie et al.，1995）。本章主要讨论生物富集的基本原理和用于量化生物蓄积的方法、生物富集评价中涉及的一些重要因素及其在生物监测中的应用。本章主要关注重金属，对有机污染物仅作简要讨论。

4.1 生物富集的通则

简单地说，如果把生物个体当作一个单独的房室（或盒子）来处理，且不进一步考虑任何内部转运，则生物体（盒子）中污染物的浓度（生物富集）是由流入和流出之间的平衡决定的，如图4.1所示。

图4.1提供了污染物在生物体中生物富集的基本图解。图示虽然较简单，但有助于理解生物富集是一个动态的过程，它同时受到流入系统（流入）和流出系统（流出）的影响。图4.1表明研究污染物的流入和流出对了解生物蓄积是至关重要的，仅单独对流入或流出进行研究，不足以充分认识生物富集的重要性和过程。

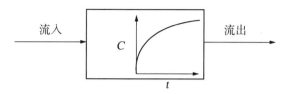

图 4.1　生物体内生物富集的简单图解
C 为生物体中的净富集浓度，t 为暴露时间。

生物富集是流入（吸收）和流出的净结果。当流入大于流出时，生物体中的污染物浓度会随着暴露时间的增加而增加。当流入小于流出时，污染物出现净损失，导致生物体内污染物浓度随着暴露时间的增加而降低。在稳态条件下，流入和流出相等，污染物浓度保持不变。稳态条件在生物富集模型中是一个重要的假设。

流入被定义为从不同环境基质（水、食物）中吸收污染物；流出则被定义为因代谢过程而导致生物体中污染物损失（如排泄、蜕皮、生殖）。流出的定义比较广泛，其他的术语（如净化、消除或损失）也常被用来描述流出。不同的术语之间存在细微差别，例如：净化通常被定义为从污染环境转移到清洁环境后生物体内污染物的损失；消除通常是指一些代谢控制的过程。生长是另一个重要的术语，其可以起到稀释生物中污染物浓度的作用。

对于大多数水生动物而言，经水源性和食源性吸收的污染物均会导致生物富集。动力学过程可用来量化吸收。

4.1.1　吸收

吸收被定义为从溶解相的吸收。在某些情况下，内化被用来描述污染物的吸收。从溶解相的污染物吸收涉及初始吸着和内化过程，但任何吸收测量都应明确去除动物表皮的初始吸着，因为这是外部，而非内部。对于某些金属，螯合物（如乙二胺四乙酸，EDTA，一种极易与金属离子形成螯合物的结晶酸）被定义为弱交换性组分，可用于洗去表面吸附的金属。对于双壳类滤食性动物，可通过吸收效率来量化吸收，公式如下：

$$Influx = k_u \times C_w = \alpha \times FR \times C_w \quad (4.1)$$

式中，k_u 为溶解相来源的吸收速率常数，L/g·h^{-1}；C_w 为溶解相的污染浓度，mg/L；FR 为动物的过滤（或清除）速率，L/g·h^{-1}；α 为吸收效率，%，被视为一级速率常数。严格来说，如果吸收不受扩散限制，则吸收效率与 C_w 和 FR 无关。

4.1.2　同化作用

相对于从溶解相的吸收，同化是指从食物源的吸收。当水生动物摄取食物时，消化在消化系统（如胃）中立即发生。未消化的物质以粪便的形式排泄出来，而其余的成分则通过肠壁被吸收。随着进一步的新陈代谢，这些成分中的某些部分通过消除、呼吸或排泄从体内排出；剩下的部分最终融入组织中，这一过程称为同化（图 4.2）。以下方程式对此进行了总结：

$$IR = AR + Feces \tag{4.2}$$
$$AR = AE + Excretion + Respiration \tag{4.3}$$

其中，IR 是总摄食（喂养）率，AR 是吸收效率，AE 是同化效率。同化效率（AE）= 同化率÷总摄食量。

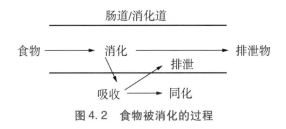

图 4.2　食物被消化的过程

4.1.3　生物浓缩系数

生物浓缩系数（bioconcentration factor，BCF）被定义为污染物在生物体内浓度与溶解在水中浓度的比值，可通过以下公式计算：

$$BCF = C/C_w \tag{4.4}$$

式中，C 是平衡条件下生物体内污染物的浓度（mg/kg），C_w 是水中污染物的浓度（mg/L）。计算或测量 BCF 的一个关键假设是污染物和生物之间的平衡。对于生长速度快的小型生物，平衡是很容易达到的。但是对于鱼这样的大型生物，在它们的整个生命史中平衡也很难达到。

对不同水生生物种群中不同污染物（金属和有机物）的 BCF 有许多测量方法，但很多具有潜在问题。如果使用野外种群进行测量，那么生物富集包括水和食物来源的吸收。因此，需要用另一个重要参数来补充 BCF，即生物富集系数（BAF），其计算方法是污染物在生物体内总生物富集量除以环境浓度（溶解相和颗粒相）。如果在实验室进行测量，则应明确达到生物与水体之间的平衡。然而在许多实验室研究中，若平衡状态未知，可以使用浓缩系数代替 BCF 来量化生物富集。

BCF 是环境风险评价中的一个重要概念，它提供了水生生物吸收污染物能力的定量信息。它常被用作持久性、可生物富集和有毒物质的首要筛选参数之一。但 BCF 不是一个常数（或暗示因子）。相反，BCF 是取决于不同环境和生物条件的一个变量。BCF 与水中污染物浓度（金属）或辛醇水分配系数（K_{ow}）（有机物）成反比（McGeer et al.，2003）。与 BCF 相似，BAF 不是一个常数，主要取决于环境浓度（DeForest et al.，2007）。

毒物代谢动力学描述了污染物在生物中的动力学过程，包括吸收、同化、储存、隔离、转运和消除。因此与生物蓄积相比，毒物代谢动力学的定义更为广泛。它还研究污染物被吸收和积累后的内部过程。毒物代谢动力学是生态毒理学中的一个重要研究领域，尤其对于有机污染物，因为它们涉及在生物体内的重要生物转化。

4.2 生物富集模型

许多模型被用来模拟污染物的生物富集。总体上可将它们可以分为两种类型：平衡分配模型（EqP）和动力学模型。

4.2.1 平衡分配模型（EqP）

平衡分配模型（equilibrium partitioning model，EqP）相对简单，假设水是富集的唯一来源（图4.3）。因此，BCF 可用于量化生物对污染物的生物富集。

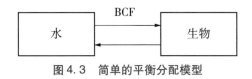

图4.3 简单的平衡分配模型

由于有机污染物生物富集与其化学物质（如 K_{ow}）有关，因此 EqP 被广泛用于有机污染物的研究，BCF 则可用于预测有机污染物的生物富集。对于金属而言，这个模型可用于预测细菌和浮游植物等小型生物中的生物富集，因为这些生物中可出现平衡。但这种方法对不存在平衡的大型生物来说并不可靠。表4.1 总结了海洋生物中一些典型的 BCF 值。不同金属或生物之间的 BCF 值差异很大，这与细胞体积或金属的化学性质有关（Fisher，1986）。BCF 在不同的环境中存在很大的差异，不能用一个 BCF 来代表不同生物体对每种化学物质（金属或有机化合物）的富集。BCF 只作为生物中不同污染物潜在生物浓缩的初始筛选值。

EqP 的另一个缺点是，它只考虑生物体暴露于水体中污染物的情况，而忽略了生物体可能接触食物来源的污染物。由于营养传递在海洋生物污染物富集中起着重要的作用（Wang，2002；Wang & Rainbow，2008），BCF 一般不能单独用于模拟生物体内污染物的生物富集。然而对于只从溶解相中富集污染物的生物（如浮游植物和水生植物）而言，BCF 仍然是一种很实用的模拟生物富集的方法，特别是用于初筛。

表4.1 不同海洋生物中9种金属的生物浓缩系数（L/kg）

生物	Ag	Cd	Cs	Cr	Hg	Ni	Se	Pb	Zn
浮游植物	5×10^4	10^3	20	5×10^3	10^5	3×10^3	3×10^4	10^5	10^4
大型植物	5×10^3	2×10^4	50	6×10^3	2×10^4	2×10^3	10^3	10^3	2×10^3
浮游动物	2×10^4	6×10^4	40	10^3	4×10^3	10^3	6×10^3	10^3	10^5

续表 4.1

生物	Ag	Cd	Cs	Cr	Hg	Ni	Se	Pb	Zn
软体动物	6×10^4	8×10^4	60	2×10^3	2×10^3	2×10^3	9×10^3	5×10^4	8×10^4
甲壳动物	2×10^5	8×10^4	50	10^2	10^4	10^3	10^4	9×10^4	3×10^5
鱼类	10^4	5×10^3	100	2×10^2	3×10^4	10^3	10^4	2×10^2	10^3

注：在本章中，"金属"一词包括类金属和非金属，如砷（As）和硒（Se）。

4.2.2 动力学模型

动力学模型不受平衡因素的限制，可用于模拟污染物生物富集作用随时间的动力学变化。现有几种已开发的动力学模型，从最简单的一室模型到复杂的多室模型（图 4.4）。

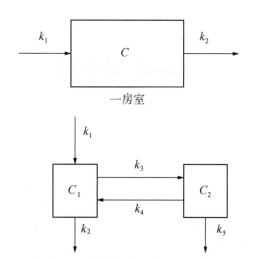

图 4.4 具有不同速率常数的一房室和二房室模型示意
C 为浓度，k 为速率常数。

简单的一室动力学模型可以通过以下方程式表示：

$$\mathrm{d}C/\mathrm{d}t = k_u \times C_w - k_e \times C \tag{4.5}$$

式中，k_u 为从水中的吸收速率常数，L/g/h；C_w 为水中污染物浓度，mg/L；L/g·L^{-1}；k_e 为排出速率常数，C 为富集浓度，mg/g。

整合后，污染物的蓄积浓度随时间 t 的变化可由以下方程表示：

$$C_t = k_u \times C_w/k_e \times [1 - e^{-k_e t}] \tag{4.6}$$

在稳态条件下，$\mathrm{d}C/\mathrm{d}t = 0$，稳态浓度（$C_{ss}$）可计算为：

$$C_{ss} = k_u \times C_w/k_e \tag{4.7}$$

利用已知的 k_u、k_e 和 C_w，可以预测稳态条件下污染物的浓度。由于 $BCF = C_{ss}/C_w$，因此也可以使用两个动力学参数来计算 BCF：

$$BCF = k_u/k_e \qquad (4.8)$$

上式直接量化了 BCF，尤其是对于大型生物而言。简单测定 k_u 和 k_e 即可准确预测 BCF。鉴于 k_u 和 k_e 都是动力学参数，表明 EqP 和动力学模型之间存在内在联系。对于许多水生动物来说，溶解相和食物来源的污染物都导致了污染物的富集。简单的动力学模型可以包含食物的暴露来源（图4.5）：

$$dC/dt = [k_u \times C + k_f \times C_f] - (k_e \times C) \qquad (4.9)$$

式中，k_f 是食物来源的吸收速率常数，C_f 是食物中的污染物浓度。同样地，该模型可用于预测特定时间 t 的富集浓度：

$$C_t = [k_u \times C_w + k_f \times C_f]/k_e \times [1 - e^{-k_e t}] \qquad (4.10)$$

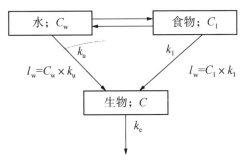

图4.5 同时考虑水和食物来源的吸收，并使用不同的动力学参数来量化生物富集

在稳态条件下，$dC/dt = 0$，可以通过以下公式计算富集浓度（C_{ss}）：

$$C_{ss} = [k_u \times C_w + k_f \times C_f]/k_e \qquad (4.11)$$

动力学模型是20世纪70年代在放射生态学被广泛使用的一个相对简单的概念。Thomann（1981）将生物能量学概念引入方程中，进一步发展了这种模型：

$$k_u = \alpha \times FR \qquad (4.12)$$

$$k_f = AE \times IR \qquad (4.13)$$

式中，α 表示从水中吸收污染物的效率，FR 为生物的过滤率（如过滤或净化的水量），AE 为饮食同化效率，IR 为生物摄食率。有时将动力学方程式称为基于生物能量学的模型或生物动力/生物动力学模型。C_{ss} 可通过以下方程计算：

$$C_{ss} = [\alpha \times FR \times C_w + AE \times IR \times C_f]/k_e \qquad (4.14)$$

上述模型忽略了生物的生长，而生长对于表现出高生长速率的生物（如细菌和浮游植物等小型生物）来说是一个重要的参数。因此，更完整的动力学模型还应考虑生物的生长速率常数（g）：

$$C_{ss} = [\alpha \times FR \times C_w + AE \times IR \times C_f]/(k_e + g) \qquad (4.15)$$

上述模型将生物视为一个单室，因而没有进一步研究生物内部的动力学过程（如运输、再分配、储存和隔离）。有了这个基本方程式，就可以进一步考虑一些特殊的情况。例如，海水是海洋浮游植物的唯一吸收源，其 g 远高于 k_e。因此浮游植物中污染物的浓度可以计算为：

$$C_{ss} = (k_u \times C_w)/g = I_w/g \qquad (4.16)$$

式中，I_w 是指从水源的摄入。该式在研究浮游植物对污染物的吸收方面有着重要的应用，可以直接使用已知的 C_{ss} 和 g 进行计算。

如果营养转移是生物富集的唯一来源，则式（4.15）也可以简化为：
$$C_{ss} = (k_f \times C_f)/k_e \tag{4.17}$$

营养转移因子（TTF）或生物放大因子（BMF）的计算如下：
$$TTF = BMF = C_{ss}/C_f = k_f/k_e \tag{4.18}$$

TTF 和 BMF 的概念与 BCF 非常相似。量化 TTF 或测定污染物是否有可能在食物链中被生物放大（例如，一种物质浓度的增加是否仅通过 3 个或更多营养等级的食物摄取）的简单方法就是比较两个动力学参数 k_f 和 k_e。TTF 大于 1，且 k_f 大于 k_e。

生物动力学模型提供了生物体内污染物生物富集的完整过程。在模型中确定的生物或化学参数是准确的，但它们的测量需要更专业的方法（如放射性示踪剂或清洁技术）。我们需要对这些动力学参数有更全面的了解，因为这将最终影响到对它们的测量。

4.3　动力学参数

动力学模型在研究海洋生物对污染物（特别是金属）的生物富集中起着重要作用（Wang & Rainbow，2008）。模型的建立需要很多确定的参数。这些参数的准确测量是生态毒理学家所面临的挑战之一。例如，C_w 和 C_f 需要专门的分析方法，尤其是水体中低 C_w 的测量。FR、IR 和 g 是生理生态学研究的主题（如复杂生态条件下海洋生物的摄食和生长过程）。AE、k_u 和 α 是生态毒理学研究的对象。过去的几十年里，在改进这些参数的量化方法方面取得了重大进展。如今，研究人员需要仔细考虑针对不同目标生物或污染物的测量方法和注意事项。随着方法学的发展，这些参数在不同环境和生物条件下的变化变得更为重要。

4.3.1　溶解吸收速率常数 k_u

表 4.2（Wang，2016）总结了在不同海洋生物类群中某些金属的 k_u 值，此表表明：①与较大的生物相比，相对较小的生物往往具有较大的 k_u 值；②B 类金属的 k_u 值比 A 类金属高。SeO_3^{2-} 和 CrO_4^{2-} 的 k_u 值最低，可能因为它们是通过阴离子通道转运。

与 k_u 相比，一级动力学参数 α（吸收效率）很少被了解和研究；α 可通过 k_u/FR 计算，而在大多数 k_u 动力学测定中，FR 通常不能被同时量化。严格地说，不同生物间生物可利用率的比较应该基于 α 而不是 k_u，因为不同海洋生物物种的 FR 存在较大的差异。

表4.2 海洋动物金属水溶性吸收速率常数（L/g·d^{-1}）

		Ag	Cd	Zn	Se	Hg（Ⅱ）	CH$_3$Hg
桡足类	锥形宽水蚤（*Temora sp.*）	8.45～12.84	0.626～0.796	2.388～3.993	0.017～0.035		
双壳类	紫贻贝（*Mytilus edulis*）	1.794	0.365	1.044	0.035		
	翡翠贻贝（*Perna viridis*）	0.638～8.212	0.206	0.637	0.019	10.51	99.6
	菲律宾帘蛤（*Ruditapes philippinarum*）	2.62	0.064	0.234			
	近江牡蛎（*Crassostrea rivularis*）	0.719	2.05	0.06			
	团聚牡蛎（*Saccostrea glomerata*）	0.534	1.206	0.064	2.604	3.445	
	白樱蛤（*Macoma balthica*）	0.032	0.091				
	华贵栉孔扇贝（*Chlamys nobilis*）	0.455	0.677				
多毛类	沙蚕（*Nereis succinea*）	1.853	0.028	0.359	0.006	1.27	2.58
腹足类	疣荔枝螺（*Thais clavigera*）	0.03	0.069		0.079	0.108	
	九孔螺（*Haliotis diversicolor*）	1.78	0.056			0.32	
星虫动物门	裸体方格星虫（*Sipuncula nudus*）	0.0018	0.035				
鱼类	黑鲷（*Acanthopagrus schlegeli*）	0.002	0.0055				
	黄斑胡椒鲷（*Plectorhinchus gibbosus*）		0.195	4.515			
	笛鲷（*Lutjanus*）	0.005	0.01	0.0008			
	紫红笛鲷（*argentimaculatus*）						
	金头鲷（*Sparus auratus*）	0.005	0.004				

大量研究已明确了不同化学和生物因素对k_u的影响。例如，Veltman et al.（2008）证明了在17种水生生物中，代表金属/生物配体结合亲和力的共价指数与10种金属的k_u值之间有显著的相关性（Veltman et al.，2008）。共价指数可以通过$x_m^2 r$计算，其中x_m是鲍林电负性值，r是离子半径。此外，金属的α也与共价指数显著相关。这些关系表明促进膜转运可能是吸收这些金属的主要机制。

吸收速率k_u也与生物体的生物学特性密切相关，如动物的体型大小。相对体型较小的动物的k_u值比体型较大的动物高（Zhang & Wang，2007a；Wang & Dei，1999）。吸收是否取决于生物体的生长速度是生态毒理学中一个有趣的问题。根据式（4.16）可知，如果生长速率与吸收速率（或k_u）无关，那么生物的生长速度越快就可能导致富集的污染物浓度越低（如k_u保持不变）。相反，如果吸收与g有关，生物富集将取决于这两个参数的相对变化幅度。如何回答这个问题是解释浮游植物金属富集反馈机制的基础（Sunda & Huntsman，1998）。例如，金属可能会抑制浮游植物的生长，从而导致浮游生物的g值减少，进而导致细胞中金属浓度的增加。这被认为是对浮游植物的一种正反馈机制。相反，必需金属的限制（如 Cu 或 Zn 的可利用率降低）可能会抑制浮游植物的生长，从而导致细胞内金属浓度的增加。这种反馈缓解了细胞内必需金属的限制作用。以往针对浮游植物对金属吸收依赖性的研究表明，细胞对金属的吸收随着浮游植物的生长而增加（Miao & Wang，2004；Wang et al.，2005）。因此随着细胞的生长，细胞中富集金属的浓度将取决于这两个参数的相对变化程度（吸收 VS 生长）。

当然环境因素可显著影响污染物的 k_u。在这些环境因子中，盐度、温度、溶解性有机质（DOM）以及如 H^+、Ca^{2+}、Mg^{2+}、溶解氧（DO）等其他竞争离子是最受关注的。这些影响大多是由于污染物形态变化以及动物的生理生化过程导致。其中，盐度可能是研究得最多的影响 k_u 的环境因子（Wang et al.，1996；Wang & Dei，1999）。

盐度除了对物种形成有直接影响，还会引起生物的生理变化。例如，Zhang & Wang （2007b） 使黑鲷（*Acanthopagrus schlegeli*）适应不同盐度，并对其吸收的 Cd 和 Zn 进行定量分析。随着盐度从 35 psu 降低到 0 psu，黑鲷对 Cd 的吸收增加了 31 倍，对 Zn 的吸收增加了 16 倍。摄入量的增加与无水离子浓度之间存在显著相关性（鳃吸收的 Cd 与其浓度比值为 1∶1），这表明游离 Cd^{2+} 的变化完全可以解释盐度对此鱼吸收 Cd 的影响。在不同盐度条件下，金属通过离子通道（钙通道）的运输是不同的。在高盐度条件下，钙离子通道不参与黑鲷对 Cd 和 Zn 的吸收，主要起促进作用。在低盐度下（如淡水），钙离子通道积极参与鱼对 Cd 和 Zn 的吸收（跨细胞吸收）。这项研究说明在不同盐度下，金属化学和鱼类生理在影响鱼类对金属吸收方面的复杂性。

在实验室用不同盐度驯化后，Blackmore & Wang（2003a）比较了香港水域两个不同盐度位点的翡翠贻贝（*Perna viridis*）对金属的吸收。从高盐度地点采集的贻贝金属富集的速度比从低盐度地点采集的贻贝（已经过中高实验盐度驯化，>17 psu）快了 1.2~2.2 倍。这种差异不能用鳃表面积（两个贻贝种群鳃表面积相似）和过滤速率来解释。相反，高盐度种群的显著透水性平均约比低盐度种群高 1.6 倍，这可能是两个群体对金属吸收产生差异的部分原因。这项研究还表明，盐度对金属吸收的影响取决于金属的生物地球化学特性以及一系列的生理响应。

DOM 在吸收金属过程中的作用对于不同的生物来说，具有差异性，可能是由于 DOM 的数量和质量的不同。一种观点认为，DOM 可以与水中的金属络合，从而有效降低金属的生物可利用率而减少其被吸收。然而也有证据表明 DOM – 金属络合物可直接被如底栖滤食性贝类等生物利用。这种 DOM – 金属络合物的协同运输已在双壳类动物中得到证实（Roditi et al.，2000；Pan & Wang，2004）。显然，在研究不同因素对海洋生物吸收金属的影响时，应考虑动物的功能解剖学特征。这可能是未来生物富集研究中需考虑的最重要因素之一。

4.3.2 同化效率

同化效率（assimilation efficiency，AE）的概念在生物富集研究中已得到了很好发展。许多研究已经量化了海洋动物对污染物（主要为金属）的摄食同化。这主要归功于放射性示踪技术的应用。表 4.3（Wang，2011b）总结了不同海洋动物中不同金属的可用 AEs。与 k_u 相似，AEs 受到生物、化学和环境因子的影响（Wang & Fisher，1999；Wang & Rainbow，2008）。

膳食金属富集的一个热门领域是捕食者同化过程中对猎物不同细胞组分中重金属的调控。在早期的研究中，将浮游植物中金属的分布分为两部分：细胞壁/膜和细胞质。研究发现，与胞质部分结合的金属对于桡足类和双壳类等海洋食草动物来说，表现出更高的生物可利用率（Reinfelder & Fisher，1991；Wang & Fisher，1996）。Wallace & Luoma（2003）的研究主要集中于猎物体内金属隔离对捕食者金属生物可利用率的调控。他们提出了营养用金属（trophically available metal，TAM）的概念，解释了以草虾（*Palaemon macrodatylus*）为食的双壳类的 Cd 和 Zn 营养有效性。此概念已在一些海洋肉食性动物如腹足类和鱼类中得到验证（Cheung & Wang，2005；Zhang & Wang，2006；Rainbow et al.，2007）。

表4.3　不同海洋动物对金属的摄食同化效率（%）

物种		Ag	Cd	Zn	Se	Hg（Ⅱ）	CH₃Hg
桡足类	刺尾纺锤水蚤（*Acartia* sp.）	66	9	38			
	锥形宽水蚤（*Temora* sp.）	8～19	33～53	52～64	50～59		
双壳类	紫贻贝（*Mytilus edulis*）	4～34	28～34	32～45	56～72		
	翡翠贻贝（*Perna viridis*）	13～32	11～25	21～32	59		
	菲律宾帘蛤（*Ruditapes philippinarum*）	30～52	38～55	33～59		41～70	
	近江牡蛎（*Crassostrea rivularis*）	58～75	68～80	56～74			
	团聚牡蛎（*Saccostrea glomerata*）	52～67	60～65	52～68			
	白樱蛤（*Macoma balthica*）	88	50	74			
	华贵栉孔扇贝（*Chlamys nobilis*）	94	83				
多毛类	沙蚕（*Nereis succinea*）	12～27	5～44	24～57	36～60	20	70
腹足类	疣荔枝螺（*Thais clavigera*）	75	80		70	95	
	九孔螺（*Haliotis diversicolor*）	58～83	33～59			65～78	
星虫动物门	裸体方格星虫（*Sipuncula nudus*）	6～30	5～15				
鱼类	花身鯻（*Terapon jarbua*）	3～9	2～52	13～26	23～43	90	
	黄斑胡椒鲷（*Plectorhinchus gibbosus*）		20	80			
	笛鲷（*Lutjanus*）	20	40	65			
	紫红笛鲷（*argentimaculatus*）						
	金头鲷（*Sparus auratus*）	45	18				

Guo et al.（2013）研究了被捕食者中亚细胞金属分布和积存量如何影响金属向海鱼花身鯻的转移。收集不同被污染历史的近江牡蛎（*Crassostrea hongkongensis*）并将其分离为3个亚细胞组分：富金属颗粒、细胞碎片和 TAM（细胞器组分、热变性蛋白和类金属硫蛋白）。这些纯化的组分的金属浓度区间较宽。以相当于鱼体重3%的日摄食率喂food 7 d。用营养转移因子（trophic transfer factor，TTF）对 Cd、Cu 和 Zn 的生物蓄积进行定量分析发现，3种亚细胞组分对于鱼都具有生物可利用性。金属之间存在一定程度的差异，TTFs 呈现金属吸收顺序为：细胞碎片 > TAM > 富金属颗粒，说明金属在

被捕食者体内的亚细胞分布影响金属生物可利用率。然而在这项研究中，TTF 与食物中的金属浓度之间，尤其是 Cd 和 Zn 的浓度，存在显著的负相关，表明 TTF 对被捕食者体内金属浓度的高度依赖性。在这种情况下，亚细胞的金属分布对营养转移的重要性可能不如被捕食者体内总金属浓度。

虽然被捕食者体内富集金属的物化形态是影响金属 AE 和营养转移的重要因素（Luoma & Rainbow, 2008; Rainbow et al., 2011），但生理条件也必然会影响同化作用（Wang & Rainbow, 2005）。消化过程，如肠道通过时间、细胞外和细胞内消化的分配以及各种结合配体的诱导，都是需要考虑的重要因素。

4.3.3 外排

外排是海洋生物富集金属的一个重要决定因素。放射性示踪技术的应用为海洋生物中污染物外排的量化研究提供了可能（表 4.4）。但是外排测量必须小心进行，特别需注意放射性标记的持续时间（使体内组织中的放射性同位素达到充分平衡）。图 4.6 说明了利用放射性标记测量海洋动物金属外排的结果。

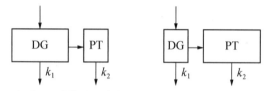

（A）以消化过程为主　（B）以生理周转过程为主

图 4.6　消化（快速）室（DG）和生理周转（慢速）室（PT）中金属损失的图示

与流入或吸收相比，外排对生物富集的作用尚未得到充分的认识。主要是由于海洋动物的外排行为相对保守，以及实际确定外排所需的时间较长。在测量外排量时，将生物暴露于放射性示踪剂，然后进行长时间的净化作用（例如，双壳类或鱼类等较大的动物，净化时间为数周至数月）。因此与溶解吸收或摄食同化的测定相比，k_e 的测量显得很繁冗乏味。对于外排的一个普遍共识是金属之间的差异小于生物之间的差异。例如，不同金属在海洋双壳类中的排出速率常数范围为 $0.01 \sim 0.03 \, d^{-1}$；而对于如桡足类的小型生物，其具有更高的单位体重代谢速率，k_e 值更高。其中一些动物的周转率甚至达到每天 1 次 [表 4.4（Wang, 2011b）]。

表4.4　不同海洋生物中金属的典型流出速率常数（d^{-1}）

		Ag	Cd	Zn	Se	Hg（Ⅱ）	CH$_3$Hg
桡足类							
	刺尾纺锤水蚤（*Acartia sp.*）			0.59	0.62	0.89	
	锥形宽水蚤（*Temora sp.*）		0.173～0.294	0.108～0.297	0.079～0.108	0.155	
双壳类							
	紫贻贝（*Mytilus edulis*）		0.034	0.011	0.02	0.026	
	翡翠贻贝（*Perna viridis*）		0.032～0.087	0.02	0.029		
	菲律宾帘蛤（*Ruditapes philippinarum*）			0.01	0.023		
	近江牡蛎（*Crassostrea rivularis*）		0.014	0.014	0.034		
	团聚牡蛎（*Saccostrea glomerata*）		0.004	0.003	0.013		
	白樱蛤（*Macoma balthica*）		0.018	0.012			
	华贵栉孔扇贝（*Chlamys nobilis*）		0.005～0.009	0.012～0.023			
多毛类							
	沙蚕（*Nereis succinea*）					0.027	0.014
腹足类							
	九孔螺（*Haliotis diversicolor*）	0.003		0.011		0.011	
鱼类							
	黑鲷（*Acanthopagrus schlegeli*）		0.089	0.016	0.043		
	黄斑胡椒鲷（*Plectorhinchus gibbosus*）					0.028～0.055	0.010～0.013
	紫红笛鲷（*Lutjanus argentimaculatus*）		0.025～0.047	0.015	0.027～0.031		
	金头鲷（*Sparus auratus*）		0.016	0.006			

外排可导致金属的超富集（Luoma & Rainbow，2005）。牡蛎是众所周知的富集高浓度 Cu 和 Zn 的海洋生物；藤壶则是典型的 Zn 超富集生物。这些高浓度金属主要是由极少量的金属外排引起的。例如，纹藤壶（*Balanus amphitrite*）Zn 的 k_e 值可低至 0.001 d^{-1}（表4.4），牡蛎中 Cu 和 Zn 的外排量也很低。k_e 背后的生理或生化机制鲜为人知。例如，金属在生物体内是如何周转，以及它们的周转是否与它们潜在结合配体的周转相结合（如蛋白质）。

图4.7提供了一个金属外排中可能涉及的亚细胞系统的概念框架。尽管目前还缺乏内部金属池中金属分布的动力学变化及这些变化如何影响整体外排的相关知识。但这显

然是未来重要的研究领域。

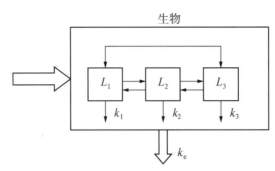

图4.7 假设（L_1，L_2，L_3）亚细胞配体与金属结合，并可能从不同房室流出 k_e是生物体的总流出量。

Ng & Wang（2005）通过将消除过程中的金属分为不溶性组分（IF）、热敏感蛋白和类金属硫蛋白（MTLP），研究了翡翠贻贝（*P. viridis*）体内 Cd、Ag 和 Zn 的亚细胞分布动力学。在外排过程中，可溶组分中的金属介导了净化作用；而不可溶组分中的金属充当了最终的存储池。Ag 和 Cd 的高外排率与 MTLP 的高分配和 IF 的低分配有关。

Pan & Wang（2008a）研究了可富集高浓度 Cd 和 Zn 的栉孔扇贝（*Chlamys nobilis*）中金属亚细胞分布的动力学。Cd 在细胞器和 MTLP 中均以无毒形式储存，这是扇贝中 Cd 外排率低的原因。在外排过程中，Cd 在细胞器（如溶酶体）中的百分比不断增加，但在 MTLP 组分中的分布则保持稳定，这表明 MTLP 在金属解毒过程中可能只是作为一个汇。生物体内 Cd 可能最终缓慢地在溶酶体中被清除，或者移至并沉积在 MTLP 中储存，导致 MTLP 中 Cd 的百分比最高。相比之下，Zn 在扇贝体内被消除的速度高于 Cd；Zn 在各亚细胞间的再分配速度远大于 Cd，表明具有某种相对有效的调控机制。

外排可受环境条件影响，如温度、组织中污染物的浓度、暴露途径和食物条件（Wang & Fisher，1998）或内部隔离。例如，一旦翡翠贻贝（*P. viridis*）组织中的 Ag 与 S 发生络合，基本上不会发生外排作用（Shi et al.，2003）。Buchwalter et al.（2008）也发现 21 种水生昆虫体内 Cd 的外排作用与 Cd 在 MTLP 中的分配直接相关。Poteat et al.（2013）探索了物种特异性对物种间环境敏感性差异的影响。Cd 和 Zn 的外排作用在水生昆虫科的不同物种中具有很强的协变性（小蜉科和纹石蛾科）。发生类群（节肢动物、软体动物、环节动物和脊索动物，共 77 种）在外排方面表现出明显的变异性，表明某些类群清除金属的能力与其他类群相比，受到更大的限制。

4.4 生物富集模型的应用

Luoma & Rainbow（2008）及 Wang & Rainbow（2008）综述了生物富集概念模型，为生物富集的研究提供了一个总体框架。该模型不仅可用于模拟和预测生物富集，还用于检测各种暴露途径，并有助于理解生物富集过程。众所周知，海洋鱼类中 Hg 浓度随着鱼体型的增大而增加，这与大多数其他金属形成了强烈的对比（如金属浓度随体型增大而减少）。这种反生长稀释的现象让生态毒理学家感到困惑，他们推测这是由于鱼类饮食需求变化（如从食草性向食肉性的转变）和大型鱼体中 Hg 的消除速度降低导致的。Dang & Wang (2012) 通过量化野外采集的黑鲷（*A. schlegeli*）幼鱼 Hg 浓度的尺寸依赖性，解决了这个有趣的问题。由于动力学测量的尺寸限制，仅对幼鱼（体长为 3.8~12 cm）进行测量。Hg 的测量结果表明，在一定范围内，总汞（THg）和甲基汞（MeHg）的浓度与鱼体质量相关，其动力学系数为 0.19~0.33。表 4.5 总结了鱼体中 Hg（Ⅱ）和 MeHg 不同生物动力学参数的体型尺寸依赖性。用 k_u、k_e 和 g 来表示 Hg 的生物动力学参数与鱼体大小的负相关关系，Hg（Ⅱ）的 AE 随鱼体的增大而增加，而 MeHg 在不同大小的鱼体中具有非常相似的 AE。

表 4.5　黑鲷（*Acanthopagrus schlegeli*）中汞（Ⅱ）和甲基汞的异速生长关系

参数	Hg（Ⅱ）	MeHg
k_u（L/g·d^{-1}）	$0.24W^{-0.68}$	$0.36W^{-0.54}$
AE（%）	$25.6W^{0.260}$	80–100
k_e（d^{-1}）	$0.050W^{-0.36}$	$0.0062W^{-0.40}$
g（d^{-1}）	$0.0077W^{-0.42}$	
含量（ng/g）	$5.91W^{0.33}$	$21.5W^{0.19}$

数据来源于 Dang, F., Wang, W. -X., 2012. Why mercury concentration increases with fish size? Biokinetic explanation. Environ. Pollut. 163, 192–198.

根据这些确定的关系，Dang & Wang (2012) 将 Hg 的富集标度指数建模：MeHg 为 0.21、THg 为 0.21~0.25，这个结果非常接近独立的野外测量结果（MeHg 为 0.33，THg 为 0.19）。因此异速生物动力学参数合理地解释了在自然环境观察到的幼鱼个体大小依赖的 Hg 富集模式。敏感性分析进一步表明，随着鱼体增大，g 和 k_e 的降低有效地增加了 Hg 的浓度，这是这种关系的关键驱动因素。

Dang & Wang (2012) 的研究表明，较慢的生长加上较低的 Hg 外排率会增加 MeHg 和 THg 的浓度，并产生了尺寸依赖异速正相关。为了管理鱼中的 Hg 污染，应探索提高鱼类 g 和 k_e 的因素。有趣的是，由于养殖鱼类在养殖系统中的快速生长以及人工饲料中的低 Hg 含量，养殖鱼类的 Hg 浓度较低；在富营养化水体系统中，鱼类生长速度比贫营养水域快，体内的 Hg 浓度也较低。Chen & Folt (2005) 分析了美国东北部 38 个湖泊中鱼类的 Hg 积存量和浮游生物密度，发现浮游动物、草食性及肉食性鱼类中 Hg 浓度与浮游动物的密度呈负相关。浮游动物的密度可以解释湖泊中 40% 以上肉食性鱼类体内的 Hg 浓度变化。高质量食物消费的快速增长也会显著减少淡水食物网中 MeHg 的富集和营养转移（Karimi et al., 2007）。

Ward et al. (2010) 通过测量 15 个溪流栖息地大西洋鲑鱼（生长范围跨越一个数量级）中 7 种金属（As、Cd、Cs、Hg、Pb、Se、Zn）的浓度，研究生长对鱼体中微量元素浓度的影响。快速生长的鲑鱼体内的金属浓度低于缓慢生长的，这表明较大生物量中的金属稀释作用导致快速生长的鱼体内金属浓度降低。

生物动力学模型的另一个应用是预测海洋动物中仅涉及营养转移（饮食或食物摄入）的不同污染物 TTF。假设食物是海洋动物污染物富集的唯一来源，则 TTF 的计算公式为：

$$TTF = BMF = C/C_f = AE \times IR/k_e \tag{4.19}$$

显然，污染物潜在的生物放大作用由 3 个动力学参数决定，即 AE、IR 和 k_e。通过对这些生物动力学参数的大量测量，Wang（2002）对不同海洋动物的 TTF 进行预测。基于这一预测，污染物和捕食者在沿着食物链的潜在生物放大作用上存在较大的差异。众所周知，MeHg 呈生物放大性，而其他金属如 Cs、Se 和 Zn（对于小鱼来说）在海洋鱼类中的 TTF 值可能大于 1。由于各种原因，许多底栖无脊椎动物潜在的 TTF 值可能大于 1，如双壳类和肉食性的腹足类动物。双壳类动物有很高的 IR 及膳食金属 AE，而肉食性腹足动物具有很高的 AE，但它们的 IR 通常很低。对于大多数金属来说，鱼类 AE 和 IR 相对较低，其潜在的 TTF 值最低。生态毒理学家在研究污染物的食物链转移过程中，需要认识到金属生物学的多样性。在做出有关污染物的生物放大作用结论时，这种复杂性常常被忽略。

严格地说，TTF 不是一个常数，而是由环境条件和动物生理学高度控制的 3 个动力学参数（AE、IR 和 k_e）变化决定。在某些情况下，海洋动物体内的金属浓度与食物链中的金属浓度无关。滤食性动物牡蛎超富集 Cu 和 Zn，扇贝超富集 Cd，藤壶则超富集 Zn。在采集自中国污染严重河口的牡蛎中发现，Cu 和 Zn 的浓度最高，约占组织干重的 2.5%（Wang et al.，2011；Wang & Wang，2014）。Wang et al.（2011）和 Tan et al.（2015）对如此高金属浓度的机制进行了研究。

与金属污染相比，海洋动物中有机污染物的 TTF 记录相对较少；对海洋动物中有机污染物（如 DDTs、PCBs、PAH）的生物动力学过程的研究有限。表 4.6 总结了这些污染物的部分典型值。

表 4.6　海洋动物中典型有机污染物的生物动力学

		$k_u [L \cdot (g^{-1} \cdot h^{-1})]$	AE/%	k_e/d^{-1}	TTF	参考文献
苯并(A)芘	红纺锤水蚤 (Acartia erythraea)	1.2	2～25	0.82～1.66	<0.1	Wang & Wang (2006)
	紫红笛鲷 (Lutjanus argentimaculatus)	0.157	30～50			
DDT	红纺锤水蚤 (Acartia erythraea)	1.2	10～30	0.01～0.05	1～9	Wang & Wang (2005)
	紫红笛鲷 (Lutjanus argentimaculatus)	0.3	70～100	0.002	4～45	
二噁英	刺尾纺锤水蚤 (Acartia spinicauda)	0.36	30～60	0.02～0.30	2～22	Zhang et al. (2011)
	黑鲷 (Acanthopagrus schlegeli)	0.08	30～70			

4.5 生物监测

基于生物富集的生物监测经典例子是 20 世纪 70 年代首次提出的贻贝观察计划（Phillips, 1976, 1977; Goldberg et al., 1978）。由于海水中许多污染物的浓度普遍较低，难以被精确测量，因此利用生物监测物来反映环境污染物的浓度无疑是一种很有吸引力的方法。所监测生物中污染物的浓度比环境介质高几个数量级（表 4.1），可以合理且确定地进行测量。使用生物监测的最重要原因是这些所测得的污染物浓度表明了污染物在环境中的生物可利用率。因此，与简单测量水（或沉积物）中的浓度相比，生物监测更具有生物学意义。

但随着分析技术的显著进步，如今可以准确地量化环境中极低的污染物浓度，并在监测项目中使用这些数据。凝胶扩散是目前广泛用于测量水中不稳定金属浓度的技术（Zhang & Davison 1995, 2000）；半透膜装置则被用于测量有机污染物的浓度（Preset et al., 1992; Huckins et al., 1993）。然而这些化学测量值和生物群中生物可利用率之间的关系仍不确定，尤其在脉冲或长期暴露的情况下。生物监测指示物提供污染物浓度生物可利用率的时间集成测量，从而实现了"从苹果到苹果"（指对两个东西的各个方面作一一对应的比较，译者注）的时空比较。

4.5.1 指示生物的选择

实用性和科学性决定了生物监测指示生物的选择。最实际的考虑是物种的分布，以便进行时空上的长期监测和比较。指示生物常见的选择为定栖的无脊椎动物，如贻贝、牡蛎和藤壶。这些物种已被广泛应用于各国和国际监测项目中，大量相关文献已经发表。在一个国家的范围内，单一物种的可用性可能是有限的，需要具有广泛可用性的不同指示生物。在美国的国家现状和趋势计划（National Status and Trends Program, NS & T）中，东部和西部海岸使用贻贝，而南部海岸使用牡蛎（O'Connor & Lausenstein, 2006）。

大多数的生物监测项目都是在如盐度等环境条件相对稳定的沿海水域实施。在过渡性水域（如河口）进行的监测项目，需要选择能够承受大范围盐度变化的广盐性生物。河口牡蛎，如香港牡蛎（*C. hongkongensis*）的耐受盐度范围为 5~25 psu，已被用于金属污染的监测。

沉积物的污染监测主要利用以如多毛类等的泥食性无脊椎动物进行（Bryan & Langston, 1992）。然而沉积物中金属的生物可利用率也是一个复杂的问题，广泛依赖于沉积物的地球化学和环境过程，如沉积物中金属的不同"形态"、向孔隙水迁移或生物活动（可能通过如挖洞等方式影响污染物的分布，也可能通过如躲避等行为影响污染物的暴露）。早期大多数的研究都试图了解受污染沉积物的生物可利用率和毒性，目前利用泥食性动物进行生物监测的理论已经得到了较好的发展。在最近的一项研究中，Fan et al.（2014）考察了中国北部锦州湾泥食性多毛类 Neanthes japonica 的金属富集和沉积物中金属浓度及地球化学组分（Cd、Cu、Pb、Zn、Ni）之间的关系。他们的研究表明，金属在多毛类中的积累受 Fe 和 Mn 含量的影响较大，受有机质的影响较小。将沉积物中的金属浓度归一化为锰含量，大大提高了对多毛类中金属富集预测的准确性。沉积物中地球化学组分金属的可交换态、有机质和 Fe/Mn 氧化物对调控沉积物金属对多毛类的生物可利用率具有重要意义。在另一项研究中，Tan et al.（2013）测量了在中国厦门潮间带采集的弓形革囊星虫（Phascolosoma arcuatum）和沉积物中的金属浓度。沉积物中 Cr、Ni、Cu、Zn、Cd 和 Pb 的浓度与虫体组织及体腔液的对应金属浓度存在显著相关性。这项研究表明，测量体腔液中的金属浓度可以作为快速评估海洋沉积物中金属生物可利用率的方法。

4.6 生物监测原理及注意事项

理论上指示生物的污染物生物富集应与被监测环境中生物可利用污染物的浓度成正比。动物体内污染物浓度的测量结果表明了环境中生物可利用污染物的浓度。结合动力学模型的应用，C_{ss} 与环境中总浓度（C_t）的关系可以通过以下方程式描述（Wang et al.，1996）：

$$C_{ss} = [k_u + AE \times IR \times K_d]/[k_e \times (1 + TSS \times K_d)] \times C_t \quad (4.20)$$

$$BAF = C_{ss}/C_t = [k_u + AE \times IR \times K_d]/[k_e \times (1 + TSS \times K_d)] \quad (4.21)$$

式中，K_d 为食物中金属的分配系数，TSS 为总悬浮颗粒物载荷。若只考虑 C_w，式（4.21）可进一步简化为：

$$C_{ss}/C_w = [k_u + AE \times IR \times K_d]/k_e \quad (4.22)$$

因此，不同站点和时间点的比较要基于假设 BAF 为恒定值（式 4.21）。若 BAF 在不同的站点或不同时间点存在变化，则应谨慎处理数据。为确保 BAF 的稳定性或同质性，式（4.21）中的生物动力学参数需要在不同站点和/或不同时间段间进行比较。这构成了生物监测项目基本原则的基础。Blackmore & Wang（2003b）比较了全球贻贝中金属的生物动力学，并验证了这一假设。贻贝（Mytilus spp.）样本分别来自纽约（紫贻贝 Mytilus edulis，美国）、阿拉斯加（油黑壳莱蛤 Mytilus trossulus，美国）、普利茅斯（紫贻贝 Mytilus edulis，英国）和大林（紫贻贝 Mytilus galloprovincialis，中国）。翡翠贻贝（P. viridis）则采集自中国香港的不同地区。结果显示，在不同的贻贝种群中，Cd 和 Zn 的生物动力学（AE、k_u 和 FR）具有相似性，表明这些动力学参数不受到贻贝体内组织金属浓度的显著影响。因此，理论上不同站点之间的 BAF 应该是一致的，可以对全球范围内的贻贝进行比较。

然而环境条件可能会显著影响污染物的生物动力学，并最终导致 BAF 的改变。例如，预暴露于金属的生物体可能会改变金属的生物动力学（Wang & Rainbow，2005）。

另一个需要考虑的重要因素是指示生物的生长速率。如前文所述，生长受到如温度和食物资源环境条件的影响，也可能导致生物富集的改变。在许多无脊椎动物的生物富集过程中，生长通常被忽略，因为它远低于污染物的外排量。Pan & Wang（2008b）利用扇贝 *C. nobilis* 进行了一项迁移实验。该实验研究不同海洋环境中，扇贝在相似的环境金属浓度下如何富集不同浓度的金属。研究表明，直接测量水相和食物相中的金属浓度不能提供对海洋环境中金属可生物利用率的全面了解。相反，当研究海洋双壳类中金属生物富集时，必须同时考虑食物的可获得性和水文等环境条件。在生物动力学模型中，生长速率不仅作为一个生长稀释项，而且还与如摄食率等其他可影响动物体内金属浓度的生理参数相关。总之，如何阐释生物生理学潜在地改变生物监测数据仍是一个热门的研究领域。

鉴于生理和生物化学对污染物生物动力学和 BAF 的潜在影响，Liu 和 Wang（2015）提出，将金属浓度与常量营养元素和常量元素（如 K、Na、Ca 和 Mg）浓度联系起来的概念。从 8 个国家/地区（美国南部和北部、欧洲、澳大利亚、亚洲）的 21 个地点采集贻贝（*M. edulis* & *P. viridis*），分析了贻贝软组织中常量营养元素（C、N、P、S）、主要阳离子（Na、Mg、K、Ca）和痕量元素（Al、V、Mn、Fe、Co、Ni、Cu、Zn、As、Se、Mo、Cd、Ba、Pb）的含量。不同地区贻贝组织中 Na 和 Mg 存在显著差异，表明不同站点的盐度（这些主要离子和其他主要离子的组织浓度是海水盐度的代用指标）存在差异（Liu & Wang，2015）。盐度除影响生物生理外，还能通过影响金属形态而影响金属在贻贝（和其他海洋动物）体内的生物可利用率。在这项研究中，*P. viridis* 组织中的大多数痕量阳离子（如 Al、Mn、Fe、Co、Cu、Zn、Cd 和 Ba）与主要阳离子呈负相关，这与先前大多研究发现的 Cu、Zn 和 Cd 的组织浓度与盐度呈负相关结果相似（Wright，1995；Lee et al.，1998；Blackmore & Wang，2003a）。相反，贻贝中主要阳离子和痕量含氧阴离子（如 As、Se 和 Mo）之间存在正相关关系。总体上，微量元素中 12%～84% 的变化与主要阳离子有关；因此，当贻贝（和其他海洋动物）被用作指示生物时，必须考虑盐度这一因素。

Liu 和 Wang（2015）发现，贻贝中微量元素与常量营养元素浓度间呈显著相关性，痕量元素中 14% 到 69% 的变化与常量营养元素有关。常量营养元素浓度与动物生长及生殖密切相关，相关分析表明这些生物过程对某些痕量元素的生物富集有很大的影响。这些相关性在解释贻贝生物监测的数据时具有重要意义。例如，同时测量组织中的 N 和 P 是有用的，因为 N 和 P 是动物生长和生殖的重要指标，会对某些痕量元素的生物富集产生强烈影响。这在很多营养丰富（如富营养化）的沿海和河口水域尤为重要。

常量营养元素对海洋浮游食物链中的金属吸收和营养迁移有显著影响（Wang & Dei，2001；Wang et al.，2001）。常量营养元素、主要阳离子和痕量元素之间的耦合关系是今后研究的重点，尤其是在环境变化较大的区域。机理研究也将通过实验测试来阐释环境变量/应力对金属生物富集和生物可利用率的影响。

4.7 展　　望

生物富集的研究已走过了漫长的道路。如今模型为生物富集研究提供了重要的途径。我们对波动环境条件下的生物富集的研究仍面临着重大挑战，如金属在不规则模式下的释放。认识水生生物中金属生物学的复杂性和多样性非常重要，如果不能重视这一点就无法进一步推动这一领域的发展。但在这方面，生物富集的研究还存在许多问题。此外，模型在生物监测中的应用仍然是生态毒理学者们面临的一个巨大挑战。

参考文献

BLACKMORE G, WANG W-X, 2003a. Inter-population differences in Cd, Cr, Se, and Zn accumulation by the green mussel Perna viridis acclimated at different salinities. Aquat. Toxicol., 62: 205-218.

BLACKMORE G, WANG W-X, 2003b. Variations of metal accumulation in marine mussels at different local and global scales. Environ. Toxicol. Chem., 22: 388-395.

BRYAN G W, LANGSTON W J, 1992. Bioavailability, accumulation and effects of heavy metals in sediments with special reference to United Kingdom estuaries: a review. Environ. Pollut., 76: 89-131.

BUCHWALTER D B, CAIN D J, MARTIN C A, et al., 2008. Aquatic insect ecophysiological traits reveal phylogenetically based differences in dissolved cadmium susceptibility. Proc. Natl. Acad. Sci. USA, 105: 8321-8326.

CHEN C Y, FOLT C L, 2005. High plankton densities reduce mercury biomagnification. Environ. Sci. Technol., 39: 115-121.

CHEUNG M, WANG W-X, 2005. Influence of subcellular metal compartmentalization in different prey on the transfer of metals to a predatory gastropod. Mar. Ecol. Prog. Ser., 286: 155-166.

DANG F, WANG W-X, 2012. Why mercury concentration increases with fish size? Biokinetic explanation. Environ. Pollut., 163: 192-198.

DEFOREST D K, BRIX K V, ADAMS W J, 2007. Assessing metal bioaccumulation in aquatic environments: the inverse relationship between bioaccumulation factors, trophic transfer factors and exposure concentration. Aquat. Toxicol., 84: 236-246.

FAN W, XU Z, WANG W-X, 2014. Metal pollution in a contaminated bay: relationship between metal geochemical fractionation in sediments and accumulation in a polychaete. Environ. Pollut., 191: 50-57.

FISHER N S, 1986. On the reactivity of meals for marine phytoplankton. Limnol. Oceanogr., 31: 443-449.

GOLDBERG E D, BOWEN V T, FARRINGTON J W, et al., 1978. Mussel watch. Environ. Conserv., 5: 101-125.

GUO F, YAO J, WANG W-X, 2013. Bioavailability of purified subcellular metals to a marine fish. Environ. Toxicol. Chem., 32: 2109-2116.

HUCKINS J N, MANUWEERA G K, PETTY J D, et al., 1993. Lipid containing semipermeable membrane devices for monitoring organic contaminants in water. Environ. Sci. Technol., 27: 2489-2496.

IAEA (International Atomic Energy Agency), 2004. Sediment distribution coefficients and concentration factors for biota in the marine environment. Technical Report Series, No. 422. Vienna, Austria.

KARIMI R, CHEN C Y, PICKHARDT P C, et al., 2007. Stoichiometric controls of mercury dilution by growth. Proc. Natl. Acad. Sci. USA, 104 (18): 7477-7482.

LEE B G, WALLACE W G, LUOMA S N, 1998. Uptake and loss kinetics of Cd, Cr and Zn in the bivalves *Potamocorbula amurensis* and *Macoma balthica*: effects of size and salinity. Mar. Ecol. Prog. Ser., 175: 177-189.

LIU F J, WANG W-X, 2015. Linking trace element variations with macronutrients and major cations in marine mussels *Mytilus edulis* and *Perna viridis*. Environ. Toxicol. Chem., 34 (9): 2041-2050.

LUOMA S N, RAINBOW P S, 2005. Why is metal bioaccumulation so variable? Biodynamics as a unifying concept. Environ. Sci. Technol., 39: 1921-1931.

LUOMA S N, RAINBOW P S, 2008. Metal contamination in aquatic environments: science and lateral management. Cambridge University Press, London, UK.

MCGEER J C, BRIX K V, SKEAF J M, et al., 2003. Inverse relationship between bioconcentration factor and exposure concentration for metals: implications for hazard assessment of metals in the aquatic environment. Environ. Toxicol. Chem., 22: 1017-1037.

MIAO A J, WANG W-X, 2004. Relationships between cell specific growth rate and uptake rate of cadmium and zinc by a coastal diatom. Mar. Ecol. Prog. Ser., 275: 103-113.

NG T Y T, WANG W-X, 2005. Dynamics of metal subcellular distribution and its relationship with metal uptake in marine mussels. Environ. Toxicol. Chem., 24: 2365-2372.

O'CONNOR T P, LAUSENSTEIN G G, 2006. Trends in chemical concentrations in mussels and oysters collected along the US coast: update to 2003. Mar. Environ. Res., 62 (4): 261-285.

PAN J-F, WANG W-X, 2004. Differential uptake of particulate and dissolved organic carbon by the marine mussel *Perna viridis*. Limnol. Oceanogr., 49: 1980-1991.

PAN K, WANG W-X, 2008a. The subcellular fate of cadmium and zinc in the scallop *Chlamys nobilis* during waterborne and dietary metal exposure. Aquat. Toxicol., 90: 253-260.

PAN K, WANG W-X, 2008b. Validation of biokinetic model of metals in the scallop *Chlamys nobilis* in complex field environments. Environ. Sci. Technol., 42: 6285-6290.

PHILLIPS D J H, 1976. Common mussel *Mytilus edulis* as an indicator of pollution by zinc, cadmium, lead and copper. I. Effects of environmental variables on uptake of metals. Mar. Biol., 38: 59 - 69.

PHILLIPS D J H, 1977. Use of biological indicator organisms to monitor trace metal pollution in marine and estuarine environments: a review. Environ. Pollut., 13 (4): 281 - 317.

POTEAT M D, GARLAND T, FISHER N S, et al., 2013. Evolutionary patterns in trace metal (Cd and Zn) efflux capacity in aquatic organisms. Environ. Sci. Technol., 47: 7989 - 7995.

PRESET H F, JARMAN W M, BURNS S A, et al., 1992. Passive water sampling via semipermeable-membrane devices (SMPDs) in concert with bivalves in the Sacramento San Joaquin River Delta. Chemosphere, 25: 1811 - 1823.

RAINBOW P S, AMIARD J - C, AMIARD - TRIQUET C, et al., 2007. Trophic transfer of trace metals: subcellular compartmentalization in bivalve prey, assimilation by a gastropod predator and in vitro digestion simulations. Mar. Ecol. Prog. Ser., 348: 125 - 138.

RAINBOW P S, LUOMA S N, WANG W-X, 2011. Trophically available metals: a variable feast. Environ. Pollut., 159: 2347 - 2349.

REINFELDER J R, FISHER N S, 1991. The assimilation of elements ingested by marine copepods. Science, 251: 794 - 796.

RODITI H A, FISHER N S, SANUDO-WILHELMY S A, 2000. Uptake of dissolved organic carbon and trace elements by zebra mussels. Nature, 407: 78 - 80.

SHI D, BLACKMORE G, WANG W-X, 2003. Effects of aqueous and dietary preexposure and resulting body burden on the biokinetics of silver in the green mussels, *Perna viridis*. Environ. Sci. Technol., 37: 936 - 943.

SPACIE A, MCCARTY L S, RAND G M, 1995. Bioaccumulation and bioavailability in multiphase systems // RAND G M (Eds.). Fundamental of aquatic toxicology: effects, environmental fates, and risk assessments. CRC Press: 493 - 521.

SUNDA W G, HUNTSMAN S A, 1998. Processes regulating cellular metal accumulation and physiological effects: phytoplankton as model systems. Sci. Total Environ., 219: 165 - 181.

TAN Q G, KE C, WANG W-X, 2013. Rapid assessments of metal bioavailability in marine sediments using coelomic fluid of sipunculan worms. Environ. Sci. Technol., 47: 7499 - 7505.

TAN Q G, WANG Y, WANG W-X, 2015. Synchrotron study of the speciation of Cu and Zn in two colored oyster species. Environ. Sci. Technol., 49: 6919 - 6925.

THOMANN R V, 1981. Equilibrium model of fate of microcontaminants in diverse aquatic food chains. Can. J. Fish. Aquat. Sci., 38 (3): 280 - 296. https://doi.org/10.1139/f81 - 040.

VELTMAN K, HUIJBREGTS MA J, VAN KOLCK M, et al., 2008. Metal bioaccumulation in aquatic species: quantification of uptake and elimination rate constants using physicochemical properties of metals and physiological characteristics of species. Environ. Sci. Technol., 42 (3): 852 - 858. https://pubs.acs.org/doi/10.1021/es071331f.

WALLACE W G, LUOMA S N, 2003. Subcellular compartmentalization of Cd and Zn in two bivalves. Ⅱ. Significance of trophically available metal (TAM). Mar. Ecol. Prog. Ser., 257: 125 – 137.

WANG L, WANG W-X, 2014. Depuration of metals by the green-colored oysters (*Crassostrea sikamea*). Environ. Toxicol. Chem., 33: 2379 – 2385.

WANG W X, DEI R C H, XU Y, 2001. Cadmium uptake and trophic transfer in coastal plankton under contrasting nitrogen regimes. Mar. Ecol. Prog. Ser., 211: 293 – 298.

WANG W-X, DEI R C H, HONG H, 2005. Seasonal study on the Cd, Se, and Zn uptake by natural coastal phytoplankton assemblage. Environ. Toxicol. Chem., 24: 161 – 169.

WANG W-X, DEI R C H, 1999. Factors affecting trace element uptake in the black mussel *Septifer virgatus*. Mar. Ecol. Prog. Ser., 186: 161 – 172.

WANG W-X, DEI R C H, 2001. Effects of major nutrient additions on metal uptake in phytoplankton. Environ. Pollut., 111: 233 – 240.

WANG W-X, FISHER N S, LUOMA S N, 1996. Kinetic determinations of trace element bioaccumulation in the mussel, *Mytilus edulis*. Mar. Ecol. Prog. Ser., 140: 91 – 113.

WANG W-X, FISHER N S, 1996. Assimilation of trace elements and carbon by the mussel *Mytilus edulis*: effects of food composition. Limnol. Oceanogr., 41: 197 – 207.

WANG W-X, FISHER N S, 1998. Accumulation of trace elements in a marine copepod. Limnol. Oceanogr., 43: 273 – 283.

WANG W-X, FISHER N S, 1999. Assimilation efficiencies of chemical contaminants in aquatic invertebrates: a synthesis. Environ. Toxicol. Chem., 18: 2034 – 2045.

WANG W-X, RAINBOW P S, 2005. Influence of metal exposure history on trace metal uptake and accumulation by marine invertebrates. Ecotoxicol. Environ. Saf., 61: 145 – 159.

WANG W-X, RAINBOW P S, 2008. Comparative approaches to understand metal bioaccumulation in aquatic animals. Comp. Biochem. Physiol. C Toxicol. Pharmacol., 148: 315 – 323.

WANG W-X, YANG Y, GUO X, et al., 2011. Copper and zinc contamination in oysters: subcellular distribution and detoxification. Environ. Toxicol. Chem., 30: 1767 – 1774.

WANG W-X, 2002. Interactions of trace metals and different marine food chains. Mar. Ecol. Prog. Ser., 243: 295 – 309.

WANG W-X, 2011a. Incorporating exposure into aquatic toxicological studies: an imperative. Aquat. Toxicol., 105 (3 – 4 Suppl.): 9 – 15.

WANG W-X, 2011b. Trace metal ecotoxicology and biogeochemistry. Science Press, Beijing: 322.

WANG X H, WANG W-X, 2005. Uptake, absorption efficiency, and elimination of DDT by marine phytoplankton, copepods, and fish. Environ. Pollut., 136: 453 – 464.

WANG X H, WANG W-X, 2006. Bioaccumulation and transfer of benzo (a) pyrene in a simplified marine food chain. Mar. Ecol. Prog. Ser., 312: 101 – 111.

WARD D M, NISLOW K H, CHEN C Y, et al., 2010. Reduced trace element concentrations in fast-growing juvenile Atlantic salmon in natural streams. Environ. Sci. Technol., 44: 3245 – 3251.

WRIGHT D A, 1995. Trace metal and major ion interactions in aquatic animals. Mar. Pollut. Bull., 31: 8 – 18.

ZHANG H, DAVISON W, 1995. Performance characteristics of diffusion gradients in thin-films for the in situ measurement of trace metals in aqueous solution. Anal. Chem., 67 (19): 3391 – 3400.

ZHANG H, DAVISON W, 2000. Direct in situ measurements of labile inorganic and organically bound metal species in synthetic solutions and natural waters using diffusive gradients in thin films. Anal. Chem., 72: 4447 – 4457.

ZHANG L, WANG W-X, 2006. Significance of subcellular metal distribution in prey in influencing the trophic transfer of metals in a marine fish. Limnol. Oceanogr., 51: 2008 – 2017.

ZHANG L, WANG W-X, 2007a. Size dependence of the potential for metal biomagnification in early life stages of marine fish. Environ. Toxicol. Chem., 26: 787 – 794.

ZHANG L, WANG W-X, 2007b. Waterborne cadmium and zinc uptake in the euryhaline teleost acclimated at different salinities. Aquat. Toxicol., 84: 171 – 181.

ZHANG Q, YANG L, WANG W-X, 2011. Bioaccumulation and trophic transfer of dioxins in marine copepods and fish. Environ. Pollut., 159: 3390 – 3397.

5 生物标记物和效应

M. Hampel[①], **J. Blasco**[②], **M. L. Martín Díaz**[①]

---

① 西班牙加迪斯大学。
② 西班牙安达卢西亚海洋科学研究所(CSIC)。

5.1 引　　言

生态系统的衰退与人类活动带来的压力息息相关，其中环境污染物的潜在有害效应已成为人们日益关注的问题（United Nations，2013）。欧盟（EU）市场每年至少生产3万种产量超过1吨的化合物，并最终排放至环境（European Commission，2006）。因此我们亟需采取以下措施：①评价新兴和现有化学物质对生态系统和人类健康的潜在有害效应；②监测污染区域及敏感生态系统的环境状况，以确保地球环境的完整性。

本章主要描述污染物的暴露效应如何体现在对生物生理或细胞功能的损害上，以及如何通过生物标记物来识别这些变化。在生态毒理学中，生物标记物通常指将特定环境暴露与健康结局联系起来的某种生物状态或状况的可测量指标。生物标记物在了解个体和种群水平暴露于环境化学物质与有害效应发展之间的关系方面起着重要作用。而在过去的几十年里，科学家们更关注分子、细胞或更高层次水平对个体的效应。直到近年来，鉴定和验证新生物标记物方面的研究取得了巨大进步。通过将行为和生殖参数纳入环境风险评价（ERA）过程，这些生物标记物可用于基于种群的污染诱导效应的研究。

自20世纪80年代初以来，生物标记物在环境毒理学研究和监测中得到了广泛的应用，并在灵敏性和特异性方面得到了迅速发展。除化学品胁迫后的效应评价外，生物标记还广泛应用于环境监测［如地中海污染监测和研究项目（Mediterranean Pollution Monitoring and Research Programme，MEDPOL）和奥斯陆巴黎保护东北大西洋海洋环境公约（OSPAR）］，以确定哨点生态系统的生物状况。随着近年来高通量分子技术的发展，产生大量的基因和蛋白表达数据，传统ERA方法已越来越不实用。污染物-特异性的生物标记物的开发和ERA过程合理化的提高，打开了前景广阔的新大门。

本章首先综述重要的生物污染标记物及其在海洋环境中的应用；然后介绍可作为发现新特异性生物标记物基本工具的不同高通量和组学分子技术，以及这些技术能够应对未来出现的污染挑战；最后探讨该领域的未来发展和趋势。

5.2　生物标记物

"生物标记物"一词被定义为与环境化合物暴露或毒性效应相关的生物反应的变化——从分子到细胞、从生理到行为（van der Oost et al.，2005）。生物标记物响应已被应用于生物监测项目（Solé et al.，2009），并作为对生物体的有毒化学效应早期预警的初步筛选工具（Cajaraville et al.，2000；Martín-Díaz et al.，2004）。同时，它们也被用于历史遗留和新兴污染物的环境风险评价（Blasco & DelValls，2008）。生物标记物和特异性生物标记物的描述见表 5.1 和 5.2。

表 5.1 不同生物标记物的描述、作用机制、应用范例及其在海洋环境监测应用的不足

生物标记物组	效应	应用	缺点/不足
生物转化酶	生物转化反应的目的是将外源化学物转化为更易于排出的水溶性形式	环境污染物生物可利用率的测定	① 缺乏标准化方法； ② 诱导酶催化活性结果不一定与基因表达的上调相关； ③ 暴露于污染物后的酶活性抑制； ④ 不能对所有海洋物种的亚型基因进行识别； ⑤ 低生态相关性
免疫毒性	生物体暴露于外源化学物而引起的免疫功能障碍	外源物质或富营养化导致免疫抵抗力损害的测定	① 主要研究对象为哺乳动物；基于此效应可寻找新的生物标记物，可能是一个有前景的领域； ② 不特定于某一组污染物
抗氧化酶	污染物引起的酶活性变化反应	氧化还原状态变化和活性氧还原的测定	① 不限于污染物； ② 对中度污染难以解释； ③ 解释时应考虑涉及抗氧化反应和其他活性氧清除剂的整个酶系统
氧化损伤的生化指标	氧化剂攻击脂质（特别是多不饱和脂肪酸）引起的脂质过氧化	外源化学物暴露引起效应的测定	① 不特定于某一组污染物； ② 反应需要长时间暴露才可观察到； ③ 似乎高度受生物体状况和季节性的影响
金属硫蛋白	暴露于某些金属（如 Cd）后的诱导作用	金属暴露和解毒机制的测定	① 涉及不同的方法和单位，难以比较； ② 水平的变化可能与其他环境变量有关； ③ 营养或生殖状况呈现出季节性变化； ④ 生物监测研究中缺乏关于亚型的信息
δ-氨基乙酰丙酸脱水酶（δ-ALAD）	与铅暴露相关的活性降低	铅暴露的鉴定	① 抑制与健康状况间无明确关系； ② 反应可能具物种特异性； ③ 可能对其他金属有反应（—SH 基团）

续表 5.1

生物标记物组	效应	应用	缺点/不足
生殖改变	暴露于外源化学物中而导致生殖能力下降	外源性雌激素污染物对内分泌干扰的测定	① 在测量生物标记物之前,应了解指示生物的生殖周期; ② 分析前应了解指示生物性别
神经毒性	抑制与神经功能有关的酶	外源性药物引起神经毒性风险的测定	还需要更多的研究来确定酶抑制过程中其他可能的生理变化
遗传毒性	生物群体遗传物质的变化	遗传物质序列的检测和定量可作为生物体暴露于遗传毒性化合物的暴露和效应生物标记物	① 缺乏标准化方法; ② 不特定于某一组污染物

表 5.2 海洋环境监测项目中最常用生物标记物的分析方法概述

	类型	作用机制和方法	文献来源
细胞色素 450	暴露	CYP450 亚型催化活性的诱导 分光光度法	Gagnè & Blaise 1999; Gagnè et al., 2007
谷胱甘肽转移酶活性(GST)	暴露	酶活性增加 分光光度法	McFarland et al., 1999
抗氧化酶	暴露/效应	酶活性变化 分光光度法	Regoli et al., 2012
金属硫蛋白	暴露/解毒	蛋白质水平改变 分光光度法 微分脉冲极谱法 银饱和分析法 MT 基因表达	Viarengo et al., 1997; Olafson & Olsson 1991; Scheuhammer & Cherian 1991; Romero et al., 2008; Russo et al., 2003
α-氨基乙酰丙酸脱水酶 (α-ALA-D)	暴露	酶活性	Berlin & Schaler 1974; Company et al., 2011
吞噬作用 溶酶体膜稳定性(LMS) NO 环氧合酶(COX)活性	暴露/效应	吞噬活性降低 染料中性红保留时间减少 亚硝酸盐含量增加 酶活性增加	Blaise et al., 2002; Lowe & Pipe 1994; Martínez-Gómez et al., 2008; Verdo et al., 1995; Fulimoto et al., 2005
VTG 水平	暴露/效应	VTG 水平的增加或降低; a. 酶联免疫测定法; b. 碱不稳定磷酸盐水平	Pateraki & Stratakis 1997; Gagnè et al., 2003

续表 5.2

	类型	作用机制和方法	文献来源
DNA 链断裂	暴露/效应	DNA 链断裂增加	Gagnè et al., 1995; Olive [AU1] 1988
脂质过氧化	暴露/效应	脂质过氧化产物增加	Janero 1990; Wills 1987
乙酰胆碱酯酶活性	暴露	酶活性增加	Ellman et al., 1961; Guilhermino et al., 1996

效应生物标记物：机体中可测出的生理、生化、行为或其他改变的指标，反映与不同靶剂量的外源化学物或其代谢物有关联的对健康有害效应的信息。暴露生物标记物：反映机体组织、体液或排泄物中的外源化学物质或其代谢产物或其与内源性物质相互作用产物数量的指标。解毒生物标记物：一种涉及毒性效应减少机制的测量参数。本表摘录自 Nikimann, M., 2014. An Introduction to Aquatic Toxicology. Academia Press, Oxford, 240 p.

5.2.1 生物转化酶

外源化学物通常具有亲脂性，难以被排出体外。然而这些化学物质中的大多数在体内以一种称为"生物转化"的方式进行代谢转化反应。这些生化反应由酶介导，使化学物质转化为极性更强、更易于排泄的代谢物。若有毒物质不排出体外，则会积累到有害水平，因此这种生物转化反应的结果之一是促进动物体内有毒化学物质的清除（Buhler & Williams, 1988）。生物转化酶水平和活性的变化是最敏感的生物标记物之一。这些酶的活性可因外源化学物暴露而被诱导或抑制（Bucheli & Fent, 1995）。酶的诱导是指酶数量或活性的增加，或两者兼有。酶的诱导主要与环境中存在的具生物可利用性外源亲脂性化学物有关。生物和非生物因素会干扰外源物质的生物可利用性。一旦暴露的生物体对这些外源化学物具有生物可利用性并吸收至体内，则会发生一系列生物转化反应，目的是将这些外源化学物转化为更易于排泄的水溶性形态。这些过程将避免具生物可利用性外源化学物的生物富集以及可能对生物造成的有害效应。参与外源化学物生物转化的两类主要酶为：阶段Ⅰ酶、阶段Ⅱ酶。

5.2.1.1 阶段Ⅰ酶

阶段Ⅰ代谢：脱去或加上反应性官能团，包括氧化、还原或水解（Goeptar et al., 1995）。这些反应引入或暴露了母体化合物上的官能团（—OH、—NH$_2$、—SH），形成非活性的代谢物，但在某些情况下也可形成活性代谢物。

后者研究最多的是细胞色素 P450 依赖的混合功能氧化酶（MFO）系统。它们被称为 MFOs 和单加氧酶。因为底物的代谢会消耗一个分子氧，并产生一个被氧化的底物，另一个氧分子作为副产品出现在水中。

细胞色素 P450 单加氧酶（CYP）是一类古老且广泛分布的蛋白质超家族。最新公布的记录给出了超过 750 个序列，属于超过 107 个不同家族（Nelson, 1998）。近几年来，更多的 CYP 基因被描述。可在定期更新的数据中查找到连续注释的总数（参见

Nelson 博士网页，http：//drnelson.utmem.edu;homepage.html）。P450 蛋白存在于多种生物中，包括细菌、植物、真菌和动物。大多数氧化阶段 I 反应是由属于微粒体单加氧酶（MO）的细胞色素 P450 亚型酶催化。

在海洋环境中，对细胞色素 P450 超家族研究得最多的是 CYP1A1，主要由二噁英、多环芳烃（PAHs）和多氯联苯（PCBs）诱发。目前已生产出多种鱼类的抗 CYP1A1 抗血清，通过免疫印迹和酶联免疫吸附试验（ELISA）可成功地检测酶蛋白水平的变化（Goksøyr & Larsen 1991；Goksøyr & Förlin 1992）。其他海洋生物（尤其是无脊椎动物）中也存在着可受外源化学物影响的 CYP 蛋白，但上述具有前景的方法尚未被广泛应用，（Snyder，2000）。

CYP 谱的变化通常被用作暴露于环境污染物的指标（Andersson & Lars 1992；Stegeman & Lech 1991）。针对相同的目标可用不同的方法（达到不同的反应）测定这些 CYP 谱，即测定污染区域外源化学物的生物转化水平和生物可利用率：①CYP 相关基因 mRNA 的表达水平；②采用酶联免疫吸附试验（ELISA）、免疫印迹法（Western blotting）或组织化学技术，用单抗或多克隆抗体检测 CYP 蛋白水平（Bucheli & Fent 1995）；③CYP 同工酶的催化活性。CYP1A 可用于环境风险评价过程的各个步骤，如痕量有机污染物影响和暴露的量化、生物体和生态系统健康的环境监测、微弱早期毒性效应的识别、调控作用的触发、特定化合物暴露的识别、毒理学筛选以及外源化学物的毒性机制研究等（Stegeman et al.，1992）。乙氧基异吩噁唑－O－去乙基酶（ethoxyresorufin O-deethylase，EROD）似乎是测定鱼类 cyt P450 系统诱导反应中最灵敏的催化探针（Goksøyr & Förlin，1992）。依据反应产物——间苯二酚荧光强度的变化来测定 ERODE 活性（Burke & Mayer，1974）。已有综述较全面地汇总和评估了关于鱼类 EROD 活性作为化学物暴露生物标记物的应用、局限性以及规程设计等科学信息（Whyte et al.，2000）。一般来说，CYP1A 蛋白水平与 EROD 活性之间存在较大的相关性（van der Oost et al.，1996）。大量野外实验研究表明，在被污染的环境中，鱼类肝脏 CYP1A 蛋白水平和活性都有显著升高。鱼肝脏 EROD 活性是一种非常敏感的生物标记物，因而可能在环境风险评价过程中具有重要的应用价值。尽管某些化学物质可能会抑制 EROD 的诱导或活性，但这种干扰通常不被认为是 EROD 作为生物标记的缺点（Whyte et al.，2000）。结合 CYP1A 蛋白和 mRNA 的表达水平，CYP1A 催化活性的诱导不仅可用于暴露程度的评价，也可作为许多痕量有机污染物潜在有害效应的预警信号。对 CYP1A 诱导毒性机制的研究表明，EROD 活性不仅可以指示化学暴露，还可以预测不同生物组织水平上的效应（Whyt et al.，2000）。然而在解释这些参数的响应时，必须考虑某些可能影响酶活性的混淆变量。海洋环境中 CYP1A1 被研究得最多，其次则为鱼类中的 CYP3A。目前主要测定鱼体内 CYP1A & CYP3A-like 蛋白酶活性，缺乏对 CYP2B-like 蛋白酶的测定。甲壳动物 Aristeus antennatus 缺乏 CYP1A 和 CYP3A-like 的生物转化活性，可表现出 CYP2 家族相关酶活性。鱼类和甲壳类之间的代谢差异与 PCB 的生物富集特征一致。A. antennatus 体内显著积累了 CYP1A 诱导同系物，但其代谢的同系物被称为哺乳动物 CYP2B 诱导物。还需要进一步的研究来确定这类动物类群中同工酶的存在，并证实这些结果可外推到其他甲壳类动物。现有的 209 种 PCB 同系物大部分都是由 CYP2B-like 酶代谢的，这一事实支持了甲壳类动物富集的 PCBs 水平通常低于鱼类的观点。此外，鱼类代谢 PCBs 的能力更强，也表明鱼类 CYPs 对 PCBs 有更大的潜在反应能力（Koenig et al.，2012）。经废弃物（Martín-Díaz et al.，2007，2008）、石油泄漏（Morales Caselles et al.，2008a，b）、不同药物产品（Aguirre-Martínez et al.，2016；Maranho et al.，2015a-c）和污水处理厂污水（Maranho et al.，2012）等暴露后，甲壳类动物滨蟹 *Carcinus maenas* 和菲律宾蛤仔 *Ruditapes philippinarum* 体内由 CYP1A1、1A2、1B1 等代表的 EROD 催化活性被显著诱导。甲壳类和双壳类，如滨蟹 *C. maenas* 和菲律宾蛤蜊 *R. philippinarum* 暴露于海洋环境中不同药物和污水处理厂废水后，其体内已检测出 CYP3A-like 异构体活性的同工酶催化活性，被称为二苯基荧光素（dibenzylfluorescein dealkylase，DBF）活性（Aguirre-Martínez et al.，2016；Maranho et al.，2012，2015a-c）。双壳类经不同化学处理后，不同的 CYP 基因转录水平有所提高。Zanette et al.（2010）发现紫壳菜蛤 *Mytilus edulis* 经 AhR 受体激动剂处理后，出现不同的 CYP 转录模式。贻贝经 β−萘黄酮（25 mg/g）处理后，鳃的 CYP3-like-1 和 CYP3-like-2 基因表达上调；经 3，3'，4，4'，5−五氯联苯（PCB−126）（2 mg/g）处理后，消化盲囊中 CYP3-like-2 基因转录率增加，表明在贻贝紫壳菜蛤 *M. edulis* 中 CYP3-like-2 的基因激活机制各不相同。菲律宾蛤蜊 *R. philippinarum* 的 CYP 基因转录对布洛芬暴露表现出不同的响应（Milan et al.，2013）。暴露于 1 mg/L 布洛芬 3 d 和 5 d 后，CYP3A65 和 CYP2U1 基因表达上调，而其他编码 CYP 亚家族 4（CYP4）蛋白的基因表达下调。CYP356A1 基因是在太平洋牡蛎（*Crassostrea gigas*）样本经污水暴露后发现的，它是属于 CYP 亚家族的一个基因（Rodrigues-Silva et al.，2015）。

5.2.1.2 阶段Ⅱ酶

阶段Ⅱ酶和辅酶因子参与外源性母体化合物或其代谢物与内源性配体的结合作用。结合作用是将大的（通常是极性的）化学基团或化合物（如糖和氨基酸）共价结合到外源化合物和药物中的加成反应（Lech & Vodicnik，1985）。大部分阶段Ⅱ酶催化这些加成反应，从而通过增加更多极性基团来促进化学物质的排出。谷胱甘肽 S−转移酶（GST）催化谷胱甘肽与含亲电中心的外源化合物结合。细胞色素 P450 系统与芳香族化合物作用产生的氧化物可以通过 GST 与谷胱甘肽结合。处理活性亲电体对生物而言非常重要。因为它们可以与控制细胞生长的大分子发生反应，如 DNA、RNA 和蛋白质。大多数化学致癌物是亲电体（Miller & Miller，1979）。GST 在具有毒性、诱变性和致癌性特性的强亲电体的解毒过程中起着重要的作用（Miller & Miller，1979；Lee，1988）。

GST 属于阶段Ⅱ解毒酶的超家族（Boutet et al.，2004），是已知所有生命体的必需酶。实际上 GSTs 是细胞防御外源化学物的多功能同工酶，保护着生物体（Blanchette et al.，2007）。基于 N 端氨基酸序列、底物特异性、抗体交叉反应性和对抑制剂的敏感性（Kim et al.，2009），GSTs 亚科可进一步分为至少 14 类，即 alpha（α）、beta（β）、delta（δ）、epsilon（ε）、zeta（ζ）、theta（θ）、kappa（κ）、lambda（λ）、mu（μ）、pi（π）、sigma（σ）、tau（τ）、phi（φ）和 omega（Ω）（Navaneethaiyer et al.，2012）。每个 GST 包含一个 G 位点（谷胱甘肽底物结合位点）和一个 H 位点（疏水亚基团结合位点）（Mannervik & Danielson 1988）。不同种群的 G 位点均位于 N 端区域，而 H 位点具有高度多样性，其显著的序列和拓扑变异特征说明了 GSTs 酶活性的可变性（Navaneethaiyer et al.，

2012；Liu et al.，2015）。

GST 酶主要分布于肝、肝胰腺、消化腺组织的胞质部分（Sijm & Opperhuizen，1989）。大多数研究使用人工底物 1 - 氯 - 2，4 - 二硝基苯（CDNB）来测定总 GST 活性。除 q 类酶外，所有 GST 亚型都可与 CDNB 结合（George 1994；Van der Aar et al.，1996）。许多外源化合物的毒性可以通过诱导 GSTs 来调节。如通过 CDNB 结合作用来测定诱导剂对几种鱼类总肝 GST 活性的影响（George，1994）。甲壳类和双壳类动物暴露于含有药物和个人护理产品的经初级处理城市污水后，GST 活性水平显著提高。

尽管金属不是这些酶的天然底物，但暴露于几种金属元素中后，水生无脊椎动物体内 GST 可被诱导（Moreira et al.，2006）。贻贝暴露于 Hg 后，其体内 GSTs 活性升高（Verlecar et al.，2007）；螃蟹中砷、镉、铬、铜、铁、汞、锰及铅和蛤蚌中镉、铅、锌、铜、铬、汞及砷的存在均可显著诱导 GSTs 活性（Martín-Díaz et al.，2007）；多毛类经 Cd、Pb、Zn、Cu、Cr、Hg 和 As 污染的沉积物暴露后也表现出相同的结果（Moreira et al.，2006）。此外，一些研究发现鱼类暴露于 PAHs、PCBs、OCPs 和 PCDDs 等污染物后，肝 GSTs 活性有所增加（van der Oost et al.，2003）；然而鱼体内总 GST 活性作为 ERA 的生物标记物似乎并不适合，因为同一物种暴露于同一外源化学物时，会观察到 GST 活性的增加或减少。总而言之，更多关于这个参数的研究对于主要解毒过程的了解至关重要，可能有助于明晰对污染物有更敏感和选择性反应的特异性同工酶。

5.2.2 氧化应激参数

5.2.2.1 抗氧化酶活性

分子氧（O_2）通过 ADP 氧化磷酸化的氧化耦合能量传递来为真核生物提供能量。这个过程由线粒体电子传递链完成，四个电子 O_2 被还原生成水。各种内源性过程的部分还原反应会导致少量高活性氧（ROS，也称为氧自由基）的持续产生，包括超氧阴离子自由基和过氧化氢（图 5.1）。这些过程会导致 DNA、蛋白质、碳水化合物和脂质等出现结构性损伤。由 ROS 引起的氧化损伤称为"氧化应激"。外源化学物可以增加氧自由基种类的产生，包括与过渡金属（如 Fe、Cu、Ni 和 Co）和有机自由基的氧化还原反应以及外源化学物的氧化还原反应循环（Livingstone et al.，2000）。实际上 ROS 的产生可作为许多污染物具有毒性的证据。

为避免氧自由基的负面效应（如氧化还原平衡和细胞内游离钙的变化、酶钝化、脂质过氧化作用等），可以利用以下几种化合物的作用，包括低分子量化合物（谷胱甘肽，维生素 A、C 和 E，尿酸，等等）和抗氧化酶［如超氧化物歧化酶（superoxide oxidoreductase，SOD）］、过氧化氢酶，以及谷胱甘肽过氧化物酶和谷胱甘肽还原酶。它们都表现出清除 ROS 和保护机体免受氧化应激的能力。

SOD（EC 1.15.1.1）可以将超氧自由基（O_2^-）还原为过氧化氢。一般情况下，超氧自由基无法穿过生物膜，它应该在产生的同一区域被还原，因此可以发现超氧自由基多于 SOD。Cu、Zn-SOD 存在于胞浆中，也存在于过氧化物酶体、细胞核和过氧化物酶体中。Mn-SOD 是一种线粒体酶，虽然它存在于过氧化物酶体膜中（Dhanunsi et al.，

1993；Singh 1996）。另一种胞外 SOD 形式已在结缔组织和原核细胞中被发现（Fridovich，1995）。

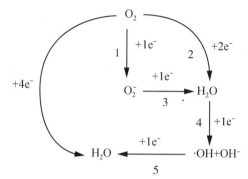

图 5.1 参与氧化还原代谢的反应

1—超氧阴离子自由基的产生；2—通过还原分子氧（2e）形成过氧化氢；3—通过还原超氧化物（1e）形成过氧化氢；4—通过还原过氧化氢（1e）形成羟基自由基；5—还原（1e）羟基自由基转化为水。

GPX（谷胱甘肽过氧化物酶，hydrogen peroxide oxidoreductase；EC 1.11.1.9）是最重要的解毒氢过氧化物的过氧化物酶。GPX 催化 H_2O_2 和有机氢过氧化物的谷胱甘肽依赖还原反应。GPX 有几种形式，如依赖硒型和非依赖硒型。在超微结构水平，细胞内 GPX 位于细胞核和线粒体中，同时在细胞质基质、溶酶体和过氧化物酶体基质中也均被发现（Orbea et al.，2000）。

过氧化氢酶（过氧化氢氧化还原酶，hydrogen peroxide oxidoreductase；EC 1.11.1.6）的主要功能是将过氧化氢分解为水和氧气。该酶位于过氧化物酶体中，并被用作该细胞区室的标记物（Fahimi & Cajaraville，1995）。但在哺乳动物和无脊椎动物细胞的其他位置也发现了该酶的存在。

Orbea et al.（2000）对软体动物肝胰腺、甲壳类消化腺中 SODs 和 GPX 进行免疫定位，发现与 Cu、Zn-SOD 和 Mn-SOD 的位置相似；它们存在于无脊椎动物的消化管、贻贝和螃蟹的消化细胞。蟹类结缔组织中也发现被染色的 Cu、Zn-SOD 和 GPX；并发现贻贝消化腺、蟹肝胰腺和鱼肝脏的过氧化物酶体中也有过氧化氢酶存在。

除此以外，其他依赖于谷胱甘肽的酶活动也可能属于抗氧化系统。Regoni et al.（1997）在南极扇贝 *Adamussium colbecki*、海扇 *Pecten japonicus* 和紫贻贝 *Mytilu galloprovincialis* 中发现了抗氧化酶活性，包括乙二醛酶 I（EC 4.4.1.5，以 GSH 为辅助因子，将细胞氧化过程中形成的有毒的 α-酮醛转化为中间产物硫醇酯）、乙二醛酶 II（EC 3.1.2.6，将乙二醛酶 I 催化的反应产物水解成相应的 D-羟基酸并再生 GSH）、GSTs（EC 2.5.1.18，参与 GSH 与外源性化合物亲电中心的结合作用）和谷胱甘肽还原酶（EC 1.6.4.2，将氧化型 GSSG 转化为还原型 GSH）。上述物种表现出相似的抗氧化系统，但扇贝在较低温度时，其抗氧化活性更有效。

痕量金属可以通过参与哈伯-韦斯（Haber-Weiss）和芬顿（Fenton）反应生成 ROS。

虽然以 Fe（Ⅲ）/Fe（Ⅱ）为例，但其他金属能以不同的效率催化反应（如 Cu（Ⅰ）、Cr（Ⅲ）（Ⅳ）和（Ⅵ））（Regoli，2012）。

有机污染物（如 PAHs、PCBs、二噁英和二噁英类化合物）可通过细胞色素 P450（详见本章第 5.2.1 节）催化数种氧化反应来增加 ROS 的生成，从而产生更多的亲水化合物。这些代谢物可以通过结合反应完成进一步转化，以便易于排出和解毒。在某些情况下，代谢物被醌类化合物、醇类和过渡金属螯合物等激活后可以有助于氧化作用（Livingstone，2001）。

文献中经常出现利用抗氧化酶活性作为海洋生物暴露于残留和新型污染物的应激生物标记物（Chandurvelan et al.，2013；Hariharan et al.，2014；Franzelliti et al.，2015；Katsumiti et al.，2015；Macías-Mayorga et al.，2015；Volland et al.，2015）。但在野外研究中应谨慎使用，因为生物多态性（如营养状况）（Gonzalez-Fernández et al.，2015）会影响生物标记物的反应。Campillo et al.（2013）在马尔梅诺尔潟湖（Mar Menor lagoon）进行了笼中实验。发现过氧化氢酶（CAT）活性的下降与污染物水平的增加有关。在从意大利海岸采集的紫贻贝 *M. galloprovincialis* 和须鲷 *Mullus barbatus*（Lionetto et al.，2003）中发现，虽然选择物种之间的反应具高度特异性，但可以利用抗氧化酶（CAT 和 GPX）监测污染物引起的暴露效应。Liu & Wang（2016）将两种牡蛎移至受金属污染的河口中，发现这种特异性物种对抗氧化酶的变化敏感。酶活性对有毒化学物质的响应呈钟形曲线趋势，最初因酶合成的激活而升高，后因直接毒性效应的分解代谢而降低。Viarengo et al.（2007）指出酶的测定应与其他生物标记物同时进行，以发现生理变化的意义。

综上所述，抗氧化酶作为污染物的生物标记具有潜在的应用价值，同时它们已经作为生物标记物应用于河口和沿海地区的监测项目中。但抗氧化酶对污染的响应缺乏一致性。与对照站点相比，受污染站点生物体中的这些酶活性水平可表现出上升趋势。但受不同污染物输入的站点之间，这些指标也会出现不显著或非一致变化。事实上，使用某些酶活性作为应激生物标记物涉及到更好地了解其季节性和自然变化规律的必要性（Viarengo & Canesi 1991；Blasco et al.，1993；Solé et al.，1995）。因此，在野外研究中使用这些生物标记物之前，确定哨兵物种的酶活性正常值范围是至关重要的。

5.2.2.2　脂质过氧化反应

细胞毒性 ROS，也称为活性氧中间体（ROIs）、氧自由基（Di Giulio et al.，1989），是已知的能产生氧化毒性的物质。ROS 包括超氧阴离子自由基（O_2^-）、过氧化氢（H_2O_2）和羟基自由基（OH^+），是能够与关键细胞大分子发生反应的强氧化剂，可能会导致酶钝化、脂质过氧化（LPO）、DNA 损伤和最终的细胞死亡（Winston & Di Giulio，1991）。

ROS 的形成发生在许多生物过程中，它可能通过自然产生或暴露于如外源化学物等不同环境应激剂中而诱导产生。它们是在有氧代谢的数种细胞途径中自然产生的，包括氧化磷酸化、线粒体和微粒体中的电子传递链、经氧化还原酶催化产生的 ROS 作为中间产物或最终产物，甚至如主动吞噬作用的免疫反应（Halliwell & Gutteridge，1999）。在环境生态毒理学中，目前的研究热点主要集中在一些外源化学物通过氧化还原循环过程增加细胞内 ROS 生成的能力上。

如前文所述，ROS 可以与关键的细胞大分子发生反应而导致酶钝化、LPO、DNA 损伤和细胞死亡（Winston & Di Giulio，1991）。目前脂质过氧化（LPO）被认为是细胞结构氧化损伤和导致细胞死亡毒性过程中的主要分子机制（Repetto et al.，2012）。

LPO 是一个已知发生在植物和动物中的复杂过程。在此过程中，如自由基或非自由基的氧化剂会攻击含碳碳双键的脂质，特别是多不饱和脂肪酸（PUFAs），涉及碳链脱氢反应并伴随氧原子的插入，生成脂质过氧化氢自由基和氢过氧化物（Yin et al.，2011）。LPO 已经被用作生物监测研究中脊椎和无脊椎海洋生物的生物标记物。

LPO 是一种自由基介导的链式反应，一旦启动就会导致多不饱和脂类的氧化变质。最常见的目标是生物膜成分。当氧化剂以脂质为目标时，它们可以启动脂质过氧化过程，通过一个链式反应产生多种分解分子和降解产物，如醛、丙酮和丙二醛（MDA）。De zwart et al.（1997）阐述了通过检测醛、丙酮和丙二醛（MDA）等降解产物来证明 LPO 的发展趋势。

脂质过氧化也可以通过检测共轭二烯烃、乙烷和戊烷气体、异丙烷和 4-羟基壬烯醛（4-HNE）等来进行评估。另一种方法是通过蛋白质和 DNA 的修饰来监测氧化损伤。但这些物质有时可通过自由基以外的途径生成。MDA 是目前最常见的细胞和组织氧化损伤指标。MDA 由不饱和脂肪断裂产生，被广泛用作脂质过氧化的指标。20 世纪 60 年代以来，已经开发了几种评价 MDA 的方法，包括采用分光光度法或荧光检测的定量方法、高效液相色谱（HPLC）、气相色谱法和免疫学技术。MDA 是通过某些初级和次级脂质过氧化产物分解而形成的一类低分子量终产物。在低 pH 和高温条件下，MDA 易与 β-硫代巴比妥酸（2-thiobarbituric acid，TBA）发生亲核加成反应，生成红色荧光的 MDA：TBA（1：2）加合物。基于上述现象，加上灵敏量化 MDA（作为游离醛或其 TBA 衍生物）方法的可用性和操作简易性，导致 MDA 测定的普遍使用，特别是"TBA 特测"，以检测和定量广泛样品类型中的脂质过氧化（Janero，1990）。

源于人类活动的污染物如重金属、PAHs、PAHs、农药残留和纤维素工业等会诱导抗氧化和脂质过氧化防御（van der Oost et al.，2003；McDonagh & Sheehan 2008）。痕量金属和有机外源化学物是典型促氧化作用的环境污染物（Regoli et al.，2014）。暴露于 Cd 和 Cu 的鱼类（如鲈鱼，Romeo et al.，2000）和暴露于受金属污染的海洋沉积物中的双壳类（如菲律宾蛤仔 R. philippinarum，Ramos Gomez et al.，2011）各组织中的 LPO 含量均有所增加。

其他的典型促氧化化学品，包括芳香族外源化学物 [如 PAHs、PCBs、卤代烃、二噁英（TCDD）和二噁英类化学物质]，通过诱导细胞色素 P450 途径增加细胞内 ROS 的生成。经苯并（a）芘暴露后，在紫贻贝 M. galloprovincialis 组织中发现脂质过氧化含量增加（Maria & Bebianno，2011）。

然而在海洋环境中，关于暴露于外源化学物后 LPO 水平上升的研究较少。这种反应需长时间暴露才能观察到，并似乎受生物体状态和季节性的高度影响。

5.2.3 金属硫蛋白

虽然金属硫蛋白（MTs）和第一个含锌金属酶是同一时间被发现的，但 MTs 缺乏科学的证据证明，因此被遗忘了很长一段时间。MTs 被认为是一类特殊的应激蛋白（van der Oost et al.，2005）。MTs 的两个特性引起了科学界的关注：①高半胱氨酸残基含量；②高金属含量（每摩尔含 7 g 原子量金属）（Kojima et al.，1976）。MTs 其他特性是体积小、在体内对某些金属（必需的和非必需）具诱导性，当然其他化合物（如糖皮质激素、黄体酮、雌激素、乙醇和干扰素，Vallee，1991）也对某些金属具有诱导性。MTs 的一个相关特性是可以调控必需金属（Cu 和 Zn）及其在 Cd 或 Hg 解毒过程中的作用（Roesijadi & Robinson，1994）。这些蛋白最初从马的肾脏中分离而出，但在动物界（无脊椎动物和脊椎动物）广泛存在。

根据金属硫蛋白的命名法，MT 蛋白分为三类（Fowler et al.，1987）：第Ⅰ类 MTs 的半胱氨酸位置与哺乳动物中 MTs 定位密切相关。一些软体动物和甲壳动物中的 MTs 属于这一类，如贻贝（Mackay et al.，1990）、牡蛎（Roesijadi et al.，1989）、螃蟹（Lerch et al.，1982）及龙虾（Brouwer et al.，1989）。第Ⅱ类 MTs 与哺乳动物 MTs 缺乏密切的相似性。第Ⅲ类 MTs 由非蛋白质 MTs 组成，也被称为植物螯合肽。Binz & Kagi（1999）在考虑系统发育特征的基础上提出了一种新的分类方法，将动物、植物、原核生物和真菌界中Ⅰ类和Ⅱ类的 MTs 分为 15 个科。

MTs 不是单一的蛋白质，目前已经分离出几种亚型，不同的生物学特性可能与不同的亚型相关。在不同的环境因素和动物群体中，MTs 可能发挥不同的生理作用，在解毒过程中被依赖程度也有所不同。目前已经鉴定出多种亚型，与哺乳动物相比，多态性在无脊椎动物中似乎特别重要（Amiard et al.，2006）。

MTs 在细胞过程中有两个主要的作用：一个是影响 Zn 和 Cu 稳态，另一个是与非必需金属结合以减少毒性效应，它们可以作为抗氧化应激的保护剂。Viarengo et al.（1999）研究了 MTs 对贻贝的保护作用；基于转录和翻译水平对 MT 的诱导进行评价，发现金属含量增加的信息在细胞水平上可以通过被金属激活的转录因子（在结合特定金属后启动表达）传递到 MT 基因（Roesijadi，1994）。Zn 置换是 MT 诱导的基础假设，已在美洲牡蛎（*Crassostrea virginica*）中得到了验证（Roesijadi，1996）。

目前已在大约 50 种不同种类的水生无脊椎动物中鉴定出 MTs，其中多数是软体动物或甲壳类动物。因此，这些物种已成为评价海洋环境污染的重要研究对象，MTs 则被视为软体动物和鱼类金属暴露的潜在生物标记物（Langston et al.，1998）。MTs 主要在细胞质中生成，但在实验室或野外暴露于必需和非必需金属离子中后，也可在细胞核和溶酶体中检测到。MT 诱导已被证明存在于受污染种群或实验室暴露于 B 类金属（如 Ag、Cd、Cu、Hg 和 Zn）的生物体中。MT 在鳃、消化腺和肾脏中参与金属隔离更为明显，反映这些组织对金属的吸收、储存和排泄的重要性（Bebianno et al.，1993，1994）。不同物种和组织之间 MT 的诱导程度存在差异。另外，由于工业活动、农业径流和城市污水的排放，水生生物受到污染物（如金属）的混合影响。相较于单一金属暴露，金属混合暴露下蛤蜊（*Ruditapes decussatus*）体内 MT 的诱导更高。这被认为是金属间的相互作用导致,其取决于不同金属对 MTs 的亲和力差异(Serafim & Bebianno, 2010)。

MT 水平受环境(盐度、季节、潮间带位置等)和生物(性成熟度、体重等)因素的影响(Hamza-Chaffai et al., 1999; Mouneyrac et al., 1998)。因此,在野外研究中需区分 MTs 浓度受金属污染影响的变化和自然影响的变化。Trombini et al. (2010) 发现,牡蛎(*Crassostrea angulata*)体内金属硫蛋白模式与金属污染并不完全相关。因此,在分析 MT 浓度的季节变化时,应考虑其他环境变量。虽然 MT 的诱导受到其他环境胁迫因子的影响,而不只是受金属赋存影响,但它主要用作金属暴露的通用标记物(如 MEDPOL 和 OSPAR 项目)。金属与 MTs 的结合信息可以提供诱导介质的额外信息。Amiard et al. (2006) 指出,如果在精心设计的采样程序中合理使用,MTs 确实可以在作为生物标记物方面发挥作用。当然还应慎重选择生物体、组织和分析方法。Oaten et al. (2015) 回顾了取样和样品处理方案对 MT 结果的影响时表明,样品储存、组织类型选择和样品预处理会影响 MT 结果。由此他们对 MT 在生物监测方法中的广泛应用提出了质疑。

由于 MTs 独特的一级结构和相对较低的分子质量,对其进行检测和定量分析并不简单。目前的方法并没有提供一个绝对值。研究人员提出了几种 MT 定量方法(Cosson & Amiard 2000; Dabrio et al., 2002),并进行了改进(El Hourch et al., 2003, 2004)。对不同技术的比较分析表明,微分脉冲极谱(DPP)和金属饱和度测定法(Onosaka & Cherian 1982)、DPP 和分光光度法(Romeo et al., 1997)、反相高效液相色谱法(RP-HPLC-FL)和 DPP 等方法之间显示良好的相关性(Romero et al., 2008)。但是采用不同的技术检测 MTs 浓度时,通常表示结果的单位不同。Adam et al. (2010) 综述了 MT 分析方法,指出可以利用光谱仪器联用方法来揭示 MTs 与其他生物活性化合物的新转运机制和相互作用。

5.2.4　δ-氨基乙酰丙酸脱水酶

δ-氨基乙酰丙酸脱水酶(ALA-D),也称胆色原素(PBG)合成酶(PBGs, EC 4.2.1.24)或氨基酮戊酸脱水酶(δ-ALA-D),可以催化两个氨基乙酰丙酸(ALA)生成一个 PBG 分子。PBG 是血红蛋白的前体。这种酶广泛分布于自然界中,需要 Zn^{2+} 辅因子才能活化; ALA-D 结构中高含量的硫醇基团可使其与其他金属结合,导致其活性和 PBG 生成水平被抑制(Finelli, 1977)。与 Zn^{2+} 相比,Pb^{2+} 与 ALA-D 具有更高的亲和力,可以改变酶的性质,是合成血红蛋白(Hb)的关键步骤。其他金属(如 Hg、Cu 和 Ag)对 δ-ALA-D 的调节作用可能由于金属导致的中毒而不具特异性(Rocha et al., 2012)。尚未在鱼类中发现 ALA-D 的抑制与贫血的关系。地中海西班牙沿岸须鲷(*M. barbatus*)体内酶活性的降低与 Pb 含量有关(Fernández et al., 2015)。但尚未有报道指出 Pb 含量对其他鱼类(如蟾鱼)肾、肝和血细胞中的酶活性有显著影响(Campana et al., 2003)。Pb 含量对酶活性的影响与须鲷(*M. barbatus*)的性别、性腺状况和体长之间没有显著的相关性。但其他研究者发现鱼体大小与 ALA-D 之间具有相关性。如大西洋鳕鱼(*Gadus morhua*)中鱼体大小与 ALA-D 之间的负相关(Tudor, 1984)和正相关（Hylland et al.，2009）与肝脏中 Zn 水平有关，且相关性随着体型增大而增加。条纹鲮脂鲤（*Prochilodus lineatus*）血液样品的 ALA-D 含量与 Pb 水平呈负相关，但未发现有显著性别差异。研究者通过研究 Zn 和二硫苏糖醇对酶活性的单一和联合作用发现，在 Zn 为 15 μmol 时可以再活化酶活性，而且血液样本中的再活化指数与血液浓度之间存在显著的静态关系（$p < 0.001$）（Lombardi et al.，2010）。事实上，这些研究者认为使用再活化指数可能是一个更可靠、更敏感的 Pb 暴露生物标记物。

目前在不同软体动物双壳类中，已有 Pb 对 δ-ALA-D 活性抑制作用的相关报道，如鸡帘蛤（*Chamelea gallina*）（Kalman et al.，2008）、河蚬（*Corbicula fluminea*）（Company et al.，2008）、紫贻贝（*M. galloprovincialis*）（Company et al.，2011）。此外，有研究表明 5-氨基乙酰丙酸或 δ-ALA-D 是影响氧化还原状态的促氧化剂（Rocha et al.，2012）。δ-ALA-D 已被纳入环境监测的多生物标记物方法框架。Cravo et al.（2009）在葡萄牙南海岸不同地点采集的紫贻贝（*M. galloprovincialis*）中发现该酶呈现季节性变化：夏季的含量低于冬季。这一模式与航运量增加导致的 Pb 水平增加有关。因为一些船舶发动机可能使用含铅汽油。但是相对于 Pb 水平，Cravo et al.（2013）没有发现 δ-ALA-D 的季节性变化和抑制作用。研究人员认为过低的 Pb 水平是引起抑制的原因。

虽然 Hylland（2004）发表了适用于测定鱼类血液 δ-ALA-D 的方法，并得以在 OSPAR 监测框架内实施，但目前测量该酶活性的方法是基于测定血液中 δ-ALA-D 活性的欧洲标准方法（Berlin & Schaller，1974）。

5.2.5 免疫学参数

大量的环境化学物质可能会损害免疫系统的组成部分。Vos et al.（1989）的综述表明，抗体和细胞介导的免疫都可能被某些污染物抑制。虽然大多数关于这一系统的研究是在哺乳动物上进行的，但这一研究方向可能是一个在寻找新生物标记物方面有前途的领域（Wester et al.，1994）。免疫毒理学是一门研究生物体暴露于外源化学物引起免疫功能障碍的学科。免疫功能障碍的形式可能是免疫抑制，也可能是过敏、自身免疫或任何数量的炎症性疾病或病理表现。免疫系统在宿主抵抗疾病和维持机体正常稳定中起着至关重要的作用。因此识别免疫毒性风险对于人类、动物和野生动物健康的保护具有重要意义（van der Oost et al.，2003）。免疫系统的重大变化迅速表现为生物体的显著发病率甚至是死亡率，但在它们发生之前，免疫系统某些组成部分会发生细微的变化，这些变化可作为免疫毒性的早期指标或生物标记物（Fournier et al.，2000；Brousseau et al.，1997；Dean & Murray，1990）。由于这些效应通常发生在低于诱导急性毒性效应的化合物暴露浓度，因此该研究领域引起研究人员极大的兴趣（Brousseau et al.，1997；Koller & Exon，1985）。免疫系统在细胞和体液免疫水平上都体现的特征包括：吞噬作用、自然杀伤（NK）细胞毒活性、环氧合酶（COX）活性、NO 和溶菌酶的分泌。侵入性抗原在细胞内被降解之前的吞噬作用是生物体细胞防御的组成部分。环境应激源被认为可通过改变血细胞的吞噬活性来降低机体免疫能力，吞噬细胞存在于所有现存的生物物种之中，因为其活性在整个进化过程中都是一种保守的功能。在进化程度更高的物种中,这一功能对于更复杂的免疫反应(如体液和细胞介导的免疫)的发展至关重要(Fournier et al.,2000)。对于大多数受关注的物种,可以使用非侵入性技术从外周血或循环液中采集吞噬细胞。这些细胞的主要活动是吞噬外来颗粒或产生抗微生物分子。这些功能可以在多种物种中使用标准化方法进行评估(Brousseau et al.,1999)。多种鱼类血细胞吞噬活性的抑制与高水平的PAHs有关(Weeks et al.,1989;1990a,b;Seeley & Weeks-Perkins,1991)。Pipe(1992)证实氯化烃和痕量金属原位暴露紫贻贝(*M. galloprovincialis*)会导致其血细胞总数、各类血细胞数量、吞噬功能等变化,以及影响活性氧的产生和释放。这些反应都与污染物负荷有关。此外,酚类化合物和低剂量的某些金属暴露会导致吞噬作用的兴奋效应(Pipe et al.,1995)。

贻贝血细胞对中性红的吸收可通过胞饮作用或细胞膜的被动扩散实现。中性红是一种阳离子染料,可在细胞的溶酶体中积累。对其吸收的改变可反映血淋巴细胞质膜的损伤、溶酶体室体积的变化和溶酶体膜的损伤。环境污染物的免疫毒性效应可以通过检测中性红保留时间(NRTT)来进行评价。可以较容易地从不同生物体中收集血淋巴细胞,并通过NRRT方法分析溶酶体膜稳定性(LMS)(Lowe & Pipe,1994)。监测研究在伊比利亚半岛不同地区(Martínez-Gomez et al.,2008)和地中海野生贻贝种群中验证了NRTT法的适用性(Gorbi et al.,2008;Dagnino et al.,2007;Viarengo et al.,2007)。LMS是一种分析新兴污染物(如药物)免疫毒性的灵敏方法。研究人员在菲律宾蛤仔(*R. philippinarum*)、滨蟹(*C. maenas*)(Aguirre-Martínez et al.,2013a,b)和紫壳菜蛤(*M. edulis*)(Martín-Díaz et al.,2009)物种中验证了这一方法的可行性。

溶菌酶的产生主要是由细菌诱导(Chu & Peyre 1989;Hong et al.,2006;Li et al.,2008)。尽管如此,雌激素化合物、三氯生和降血压药物可能是通过与细胞膜的相互作用而激活贻贝 *Mytilus* sp. 的溶菌酶活性(Canesi et al.,2007a,b,c)。例如,一些PCBs同系物可以诱导溶菌酶的自发释放,并抑制了对大肠杆菌(*Escherichia coli*)的杀菌性(Canesi et al.,2003)。据Oliver et al.(2001)报道,野生美洲牡蛎(*C. virginica*)中溶菌酶活性与Cu和PCBs含量呈正相关,而与Cd和Hg含量(Dondero et al.,2006)呈负相关,表明城市废水影响环境的复杂性。血浆溶菌酶活性的测定方法主要参考Lee & Yang(2002)。

一氧化氮的含量是通过测量血浆中亚硝酸盐浓度来估计。由于NO很容易与氧反应产生亚硝酸盐(NO^{2-})和硝酸盐(NO^{3-}),因此通过添加硝酸还原酶来测定血浆中亚硝酸盐的总浓度(Ver-don et al.,1995)。NO的产生是为了在吞噬过程中协助吞噬体破坏被吞噬的细菌(Gourdon et al.,2001)。在贻贝 *Mytilus* sp. 中发现17β-雌二醇可以诱导吞噬作用和NO分泌(Canesi et al.,2007c)。

NO合成与炎症反应密切相关,并参与诱导COX活性的表达(Grisham et al.,1999)。菲律宾蛤仔(*R. philippinarum*)(Maranho et al.,2015b)和海洋多毛类沙蚕(*Hediste diversicolor*)(Maranho et al.,2015c)经城市污水暴露后,其性腺组织中的COX活性被诱导。用花生四烯酸氧化2,7-二氯荧光素的方法可以测定血细胞和其他生物组织如性腺中COX的活性(Fuji-moto et al.,2002)。

由于免疫系统的基本生理作用,任何一种对海洋双壳类动物免疫抗性的损伤都可能被解释为ERA的重要信号。但对海洋双壳类动物免疫抗性的研究还有待进一步深入。

5.2.6 遗传毒性参数

在世界各地受污染地区的海洋沉积物中已检测到 1000 多种致癌化合物，其中许多化学物质富集在受影响水域的各种生物体内。如今越来越多的证据表明，海洋生物（特别是生活在受污染水域的鱼类）患病的风险不断增加。在鱼类和贝类种群中已发现了不同类型的肿瘤（Bolognesi et al.，1996）。生物体暴露在遗传毒性化学物质中可能会引发一系列变化（Shugart et al.，1992）：DNA 结构的改变、DNA 损伤过程和突变基因的后续表达。遗传生态毒理学可以定义为研究污染物诱导的生物遗传物质的变化。对这一系列事件的检测和定量可作为生物暴露于环境遗传毒性物质的暴露和效应生物标记物（van der Oost et al.，2003）。在分子和细胞水平上有不同的终点效应，如基因突变、染色体改变、DNA 损伤和修复的诱导。

如今，已有不同的遗传毒性参数被用作海洋环境中遗传毒性化学物质暴露的生物标记物。研究对象主要集中在鱼类和贝类。对 DNA 加合物的研究是一个确定化学和物理因素如何导致遗传疾病和损伤频率机制的重要方法。更普遍的方法包括检测由有毒化学物质（或其代谢物）或者是在结构损伤过程中直接产生的 DNA 链断裂（Shugart et al.，1992）。此外，具有不可逆作用的染色体突变也被用作海洋污染的生物标记物。

软体动物暴露于污染物（PAHs、氯化烃、PCBs、重金属）与组织病理学病变和肿瘤之间的因果关系已在野外调查和实验室研究中得到明确证明（Yevich et al.，1987；Auffret，1988；Gardner et al.，1991，1992）。目前已将哺乳动物化学致癌理论（Barrett，1992）扩展到了对鱼类更广泛的研究中。底栖鱼类肝肿瘤的发生与沉积物中 PAHs 的含量呈正相关（Landahl et al.，1990；Myers et al.，1994）。在实验室中，PAHs（包括苯并[a]吡喃，B[a]P）（Hendricks et al.，1998；Fong et al.，1993）暴露诱发了生物体癌变，并激活所诱导肿瘤中 K-ras 癌基因（Wirgin et al.，1989；MacMahon et al.，1990；Fong et al.，1993）。由于 PAHs 是环境中的主要污染物，基于模型化合物 B[a]P 的 PAH 诱导 DNA 损伤在脊椎动物中已被广泛研究（Colapietro et al.，1993；Chen et al.，1996；De Vries et al.，1997）。

5.2.6.1 DNA 加合物

在分子遗传学中，DNA 加合物是一段与（致癌）化学物质共价结合的 DNA 片段。这个过程可能是一个癌细胞或癌变的开始。在科学实验中，DNA 加合物通常被用作暴露生物标记物（La et al.，1996）。除用作遗传毒物暴露和效应标记物外，DNA 加合物可提供有关化学物质的生物效应和潜在风险的信息。有学者建议任何形成 DNA 加合物的化学物质，即使在非常低的水平，都应被认为具有致癌和致突变的潜力（Maccubbin，1994）。DNA 加合物的形成主要是用来研究 PAHs 暴露的生态毒理学方法。DNA 加合物的研究不仅可以通过观察化学物质如 PAHs 与 DNA 的结合，还可以通过与活性因子的结合来观察。负责有效清除生物可降解物质（如 PAH）的代谢过程（阶段Ⅰ和阶段Ⅱ）也能将环境致癌物激活成 DNA-反应形式（Dunn，1991）。阶段Ⅰ或阶段Ⅱ的诱导越有效就可能会导致越高的致癌风险。当然普遍认可的是海洋生物肝脏 PAH-DNA 水平反映致癌或致突变的 PAHs 暴露程度。虽然 DNA 加合物主要在肝、肝胰腺和消化腺中测量，但目前大多的研究集中在鳃组织。提取 DNA 后（Helbock et al.，1998），利用 Genevois et al.（1998）描述的 ^{32}P 后标记技术来测定 DNA 加合物。贻贝（Solé et al.，1996；Akcha et al.，2000）、鱼类（Ericson et al.，1996）和甲壳类动物（James et al.，1992）经 PAHs 暴露后，其组织内可观察到大量的 DNA 加合物形成。

5.2.6.2　二级 DNA 修饰

DNA 损伤可能是生物体化学损伤中最重要的后果之一。因为这不仅可能会影响生物个体，还可能影响到整个种群。DNA 损伤是 DNA 化学结构的改变，如 DNA 链断裂、DNA 骨架上的碱基缺失或碱基的化学变化（如 8-OhdG）。自然发生的 DNA 损伤可由代谢或水解过程引起。新陈代谢会释放破坏 DNA 的化合物，包括 ROS、活性氮自由基、活性羰基类物质、脂质过氧化物和烷基化剂等，而水解则会裂解 DNA 中的化学键。DNA 损伤可以反映近期海洋环境的污染状况。DNA 断裂可以通过不同的机制修复（Turner 和 Parry，1989）。不同物种和个体之间的 DNA 修复能力差异很大，这些差异与物种的最长寿命有关。此外，修复 DNA 所用时间也因 DNA 损伤剂的不同而不同。DNA 损伤与突变都是 DNA 的错误类型，但二者存在明显差异。DNA 损伤是 DNA 中一种异常的化学结构，而突变是标准碱基对序列的变化。碱性展开法和沉淀法采用荧光测定来检测受损 DNA 是以往两种常用方法（Brunk，1979）。基于 DNA 沉淀的沉淀法相对更简单和快速（Olive et al.，1986，1988）。彗星试验（单细胞凝胶电泳）可快速筛选潜在的 DNA 损伤（Frenzilli et al.，2009；Picado et al.，2007）。彗星试验是一种测定真核细胞中脱氧核糖核酸（DNA）链断裂的简单方法。将细胞包埋于琼脂糖凝胶中，铺于载玻片上，用去污剂和高盐溶液溶解，最后形成与核基质连接的超螺旋 DNA 环的类核。用荧光显微镜观察，高 pH 下的电泳结果呈类似彗星的结构；彗星尾部相对于头部的荧光强度反映了 DNA 断裂的数量（Nandhakumar et al.，2011）。鱼类如纹首鮨（*Serranus scriba*）、双壳类如紫贻贝（*M. galloprovincialis*）、蟹类（*Maja crispata*）、海参（*Holothuria tubulosa*）（Bihari & Fafan，2004）和宽吻海豚（Lee et al.，2013）中观察到的 DNA 损伤已被应用于环境风险评价过程中的环境污染物评价。

5.2.6.3　不可逆的遗传毒性事件

DNA 损伤和突变有不同的生物学后果。虽然对于大多数的 DNA 损伤可以进行 DNA 修复，但这种修复并非 100% 有效。未修复的 DNA 损伤会在非复制型细胞（如成年哺乳动物大脑或肌肉细胞）中富集，并导致细胞衰老（Bolognesi et al.，1990）。错误经常发生在复制型细胞 DNA 模板链对过去损伤的复制或 DNA 损伤的修复过程中。这些错误会导致突变或表观遗传学变异（Brunnetti et al.，1988）。这两种类型的改变都可以复制并传递给后续的细胞世代，且可以改变基因功能或基因表达调控，从而导致癌症的发生。在细胞的生命周期里，微核率是常用于指示遗传损伤积累的指标。暴露于致突变化合物的动物在停止暴露后可能会出现微核率的增加（Mayone et al.，1990）。基于上述阐述，该实验适用于监测早于取样时间所引起的遗传毒性损害。

有某种 DNA 损伤的细胞往往具有微核（MN）的特征。可由辐射、有害化学物质和发生在整个基因组随机突变等所造成的损害引起。微核是从新分裂的子细胞中出芽产生的一种小体,可以包含整个染色体或部分染色单体。微核的增加通常表明 DNA 损伤或突变增加,其特性是存在于癌细胞或暴露于危险因素增加的细胞中。DNA 双链断裂产生一个单独的线性片段是微核形成的另一种机制。微核在癌症中常被忽视。在显微镜下可观察到微核,并且通常紧邻其他较大的核。这种方法是否可用于预测未来的癌症风险,目前还在调查和研究中。但与染色体畸变相比,它们更容易分析。

在众多监测系统中,微核测试是用于确定污染水域和复杂混合物暴露下生物体遗传变化的最可靠技术之一。近年来,通过对许多水生生物的实验,该试验方法得到了改进。MN 试验是一种能快速检测染色体损伤的方法,因为它的几个优点有助于确定染色体是否丢失或断裂:①在检测染色体损伤方面比其他试验结果更客观;②容易学习;③细胞分裂中期难以观察染色单体和染色体损伤,但此方法不需要对染色体进行计数;④快速的制备阶段;⑤使得每次实验中检测成千上万的细胞成为可能,而不是局限于数百个细胞(OECD,2004)。

MN 试验在研究污染物的致染色体断裂和遗传效应方面取得了较好的结果。鱼类和双壳类是作为评价水生环境健康的主要指示物种,如双壳类的紫贻贝(*M. galloprovincialis*)、太平洋牡蛎(*C. gigas*)、贝类(*Chamelea galina*),以及欧洲舌齿鲈(*Dicentrarchus labrax*)(Hooftman & Raat 1982;Manna et al.,1985;Metcalfe,1988;Rodriguez-Ariza et al.,1992;Al-Sabti et al.,1994;Arslan et al.,2010;Tsarpalias & Dailianis 2012)。

5.2.7 生殖和内分泌参数

暴露于外源化学物的生物体生殖可能发生改变。这一问题的严重性已促使诸如经济合作与发展组织(OECD)和欧洲联盟(EU)等国际机构发起大规模研究项目,并制定新的指南和条例。内分泌干扰物(EDCs)有合成和天然 2 种来源。合成来源包括塑料、洗涤剂、药物(如口服避孕药)、化妆品、阻燃剂、除草剂和杀虫剂等。它们通过工业和污水排放、主动使用和径流进入环境。天然来源包括人类和动物的激素、污水和畜牧业径流中发现的植物性和真菌性雌激素、有意或无意地作为食物和饲料的配料(Goksøyr 2006)。

环境污染物暴露可能会损害生物的激素调节(Spies et al.,1990)。野生生物生殖能力下降被认为是人类排放持久性污染物最具破坏性的影响之一。一些在环境中广泛分布的外源化学物具有内分泌活性,可能通过影响生殖从而威胁易感物种的生存(Colborn et al.,1993;Peterson et al.,1993;White et al.,1994)。因此,生殖率有限且处于高营养级水平的动物可能最脆弱。

这些外源化学物的作用机制可分为以下几类:①激动/拮抗效应("拟激素");②自然激素产生、运输、代谢或分泌的干扰;③激素受体产生和/或功能的干扰(Rotchell 和 Ostrander,2003;Goksøyr et al.,2003)。广泛的潜在生物标记物可以应用于水生环境内分泌干扰物的研究。对于海洋生物来说,上述潜在生物标记物包括配子形态学改变、卵母细胞的成熟、排卵和产卵、卵细胞数量和活力、卵黄发生、下丘脑和垂体激素、配子类固醇及肝脏分解代谢。

　　海洋生物化学物暴露的阶段和持续时间将在很大程度上决定对卵巢影响的差异。对于长期暴露（通常从生殖周期的早期开始，直至卵母细胞成熟时），通常使用性腺指数（GSI）、卵母细胞分期、组织学检查、产/排卵数和活力来衡量污染物的毒性效应。短期暴露，或体外实验，更适合于研究特定的作用机制，如卵黄生成作用、类固醇生成及垂体活动。长期污染物暴露几乎必然会导致 GSI 减少、较小的发育不全卵母细胞和较少的大成熟卵母细胞及闭锁卵泡数量的增加。此外，其也会导致卵母细胞卵黄颗粒减少（Sukumar & Karpagaganapathy，1992）、卵壁破裂、卵黄囊泡（Kulshrestha & Arora，1984），以及核仁受损、细胞质发生较大变化（Kirubagaran & Joy，1988；Murugesan & Haniffa，1992）。Cd 和 Hg 暴露会导致卵泡皮质广泛空泡化、卵泡膜坏死及卵泡细胞肥厚，而 Hg 对卵黄蛋白原转入卵母细胞的影响很小，但 Cd 似乎抑制了营养物质的转移（Victor et al.，1986）。

　　由于卵黄蛋白原转入发育中的卵母细胞是卵巢发育生长的关键过程，因此海洋生物暴露于污染物后 GSI 较低可能是由于卵黄蛋白原转入受到抑制所致。一个更直接的测量方法是检测血浆、肝脏和卵巢卵黄原蛋白浓度。

　　EDCs 被定义为能够改变生物内分泌系统功能，并因此对完整生物体或其子代或（亚）种群造成有害健康效应的化学物质（世界卫生组织/国际化学品安全规划署，WHO/IPCS，2002）。EDCs 作为外源雌激素，可与 ERs 结合并诱导卵黄蛋白原（Vtg）表达，因此 Vtg 可作为鱼类中拟雌激素的特异性生物标记物（Goksøyr et al.，2003；WHO/IPCS，2002）。

　　卵黄蛋白前体分子——卵黄蛋白原在雄鱼或幼鱼体内的诱导是一种众所周知的外源雌激素污染物效应。在实验室和野外研究中已被广泛用作生物标记物（Matthiessen & Sumpter 1998；WHO/IPCS 2002；Arukwe & Goksøyr 2003；Goksøyr et al.，2003；Ortiz-Zarragoitia & Cajaraville，2005）。Vtg 水平也被确定为海洋无脊椎动物 EDCs 暴露的生物标记物。雄性无脊椎动物暴露于雌激素后未检测到 Vtg 的诱导。因此，对雌性海洋无脊椎动物正常 Vtg 水平异常的测定一直是研究的重点。

　　在卵巢雌二醇的刺激下肝脏会合成卵黄蛋白和卵黄蛋白原，它们在卵巢发育中起着重要的作用。卵黄蛋白原是一种来源于肝脏脂质的糖磷脂蛋白。胆固醇也是所有类固醇激素的前体，因此污染物对肝脏脂质合成的影响可能会影响卵巢的发育（Kime et al.，1995）。肝脏、卵巢和血浆中脂质含量的变化可能反映了污染物抑制肝脂质进入血浆以及与卵巢发育中的卵母细胞结合（Lal & Singh 1987；Singh，1992）。由于脂质动员在卵黄形成阶段是最强的，这一阶段污染物可能对脂质分布的影响最大。伴随着卵黄蛋白原转入卵巢，发育中卵母细胞营养输入的改变可能与 GSI、排卵数以及产卵和孵化成功率的下降有关。

　　Vtg 在脊椎动物/无脊椎动物的血液/血淋巴中可以被测量。因此测定作为外源性雌激素污染生物标记物的 Vtg 水平干扰可能成为一种非侵入性方法。这在海洋鱼类和甲壳类动物中可行，因为可以通过观察外部身体特征来判断生物的性别；但对于双壳类动物则不可行，因为它们个体性别的判断必须经显微镜观察性腺组织涂片后完成。在生物指示物种的肝胰腺、消化腺、肝脏和性腺组织中也可以测定 Vtg 浓度。Vtg 浓度可能因生物体所处生殖阶段的不同而不同。因此，重要的前提是验证生物是否处于生殖期，以及何时处于繁殖期，以确定卵巢成熟阶段。如果将属于不同卵巢成熟期或生殖期个体之间的 Vtg 水平进行比较，可能会得到混淆的结果。此外，鉴于 Vtg 水平在所关注生物组织中可能会随时间推移而变化，建议进行一项长期的研究。需要注意的是，Vtg 从肝胰腺、肝脏、消化腺通过循环液（血液和血淋巴）运输到卵巢。

目前已经开发出几种测定 Vtg 的方法：基于特定抗体的免疫技术，如放射性免疫测定法、酶联免疫吸附测定（ELISAs）、蛋白质印迹法和免疫组化等；分子工具，如 RNA 保护分析、利用 RNA 印迹技术或聚合酶链反应（PCR）的转录分析；基于蛋白质组学技术的蛋白质表达研究（Denslow et al.，1999；Arukwe & Goksøyr，2003；Marin & Matozzo，2004）。这些方法中，通过碱性磷酸酶（ALP）法测定磷蛋白的方法已广泛应用于鱼类和软体动物双壳类等不同的水生生物（Kramer，1998；Blaise，1999；Verslycke，2002；Marin & Matozzo，2004）。鱼类中 ALP 水平已被证实与 Vtg 水平有关（经特异性免疫技术和基因表达工具测定）（Versonnen et al.，2004；Robinson et al.，2004）。但目前还没有海洋软体动物如紫壳菜蛤（*M. edulis*）和紫贻贝（*M. galloprovincialis*）中 Vtg 样蛋白的特异性抗体。因此，ALP 方法有可能成为一种简单有效的生物标记物，用于检测贻贝和其他软体动物哨兵物种的内分泌干扰作用。对不同双壳类软体动物的季节性研究表明，雌性体内的 ALP 水平变化与配子发生周期具有相同的趋势（Blaise et al.，1999，2002；Ortiz-Zarragoitia，2005）。雌性紫贻贝（*M. galloprovincialis*）ALP 水平在活跃的配子发生期开始上升，在性腺成熟时达到最高（Porte et al.，2006）。根据 Gagné et al.（2003）的测定方法，可以测定每个动物性腺匀浆中的 ALP 水平。

5.2.8 神经毒性参数

胆碱酯酶类（ChE）被认为是研究神经功能的酶（Payne et al.，1996）。乙酰胆碱酯酶（AChE）在中枢和外周神经系统的大量胆碱能通路中通过快速水解神经递质乙酰胆碱，而参与脉冲传递的终止。

AChE 属于 ChEs 家族，是一种专门分解胆碱脂的羧酸酯水解酶。ChEs 家族包括水解神经递质乙酰胆碱的 AchE 或以丁酰胆碱为底物的丁酰胆碱酯酶（BChE）（Lionetto et al.，2013）。ChEs 对中枢和外周神经系统的正常功能维持至关重要。它有两种类型，第一类对 AChE 具有高亲和性；第二类对丁酰胆 BChE 具有高亲和性，也称为非特异性酯酶或伪胆碱酯酶（Walker & Thompson，1991；Sturm et al.，2000）。与 BchE 相比，关于 AChE 及其作用方式、抑制和激活的信息较多。AChE 通过激活突触后膜上的乙酰胆碱受体，将乙酰胆碱水解成胆碱和乙酸盐。AChE 活动可终止突触传递，阻止神经末梢的连续神经放电。因此，AChE 对中枢和外周神经系统的正常功能维持至关重要。然而目前关于 BChE 的生理功能尚不清楚（Daniels，2007）。

AChE 抑制剂可抑制 ChE 酶分解乙酰胆碱，增加神经递质活动水平和持续时间。基于 AChE 抑制剂的作用模式可将其分为两种：不可逆和可逆类。竞争性或非竞争性可逆抑制剂大多用于治疗，其毒性效应则往往与不可逆的 AChE 活性调节剂有关（Lionetto et al.，2013）。各种抑制剂诱导的酶失活导致乙酰胆碱积累、烟碱受体和毒蕈碱受体的过度刺激和神经传递的中断（Colovic et al.，2013）。近年来，关于外源性化学物质对人类和动物体内 AChE 抑制作用的报道越来越多（Goldstein，1992；Lionetto et al.，2004；Jebali et al.，2006；Vioque-Fernández et al.，2007）。AChE 抑制作为 ERA 生物标记物的应用是研究的重点。

有机磷和氨基甲酸酯农药是乙酰胆碱酯酶催化活性的特异性抑制剂，其诱导的生理失活机制已被广泛研究。自有机氯农药被淘汰后，它们已经成为当今使用最广泛的农药。它们通过磷酸化或去氨基化作用与酯基位点结合而使酶失活。有机磷化合物被认为是 AChE 的功能性不可逆抑制剂，因为将酶从抑制中释放所需时间可能超过合成新 AChE 所需时间。此外，氨甲酸酯类有一个较快的脱氨基反应步骤，因此可在有限时间内大量恢复酶活性（Lionetto et al.，2013）。AChE 抑制剂已被广泛用作农药污染的生物标记物，包括作为具有神经毒性效应的有机磷酸酯类和氨基甲酸酯类。

虽然许多研究已成功地将 AChE 抑制剂作为鱼类的生物标记物，但关于混合毒物对海洋无脊椎动物 AChE 活性影响的研究却很少。例如在 Bocquené et al.（1990）的研究中，测定了鱼类（*Solea solea*）、无脊椎动物（*Palaemon serratus*、*Crangon crangon*、*M. edulis*、*C. gigas*）和多毛动物（*Nereis* sp.）中的 AChE 活性。

以 AChE 活性作为 ERA 的生物标记物来评价氨基甲酸酯类农药的研究较多；此外，使用 AChE 活性评价海洋环境中金属、PAHs、清洁剂和复杂污染物混合物成分毒性效应的研究报道也越来越多（Pérez et al.，2004；Lionetto，2004；Jebali et al.，2006）。

纳米材料在不同领域的应用越来越广泛，不同种类的纳米颗粒，包括金属、氧化物和碳纳米管（SiO_2、TiO_2、Al_2O_3、Al、Cu、导电铜涂层、多壁碳纳米管、纳米金、单壁碳纳米管）在沙蚕、软体动物双壳类（Buffet et al.，2014）和鱼类（Ferreira et al.，2016）中显示出对 AChE 的高度亲和性（Wang et al.，2009a，b）。

参考 Ellman et al.（1961）的方法，AChE 活性可以通过分光光度法进行测量。该方法的原理是以乙酰硫代胆碱用作底物，测量其水解后生成硫代胆碱的速率。该分光光度法可用于测定组织提取物、匀浆液、细胞悬浮液等的乙酰胆碱酯酶活性。通过硫代胆碱与 5，5′-二硫代双（2-硝基苯甲酸）（DTNB）反应时所产生的黄色产物增量来测量酶活性。每毫克蛋白每分钟催化产生 1 nmol TNB 为 1 个酶活单位。

许多研究已经成功地使用 AChE 抑制剂作为农药暴露的生物标记物；但该参数应用于 REA 项目前还需更多的研究。如需要进一步的研究来更好地解释 AChE 抑制与死亡率相关性的物种特异性，以及研究与 AChE 抑制相关的其他生理干扰（Fulton & Key，2001）。这种生物标记物的实用性在于它可以提供一种综合的测量方法，来衡量环境中存在的生物可利用污染物所带来的整体神经毒性风险。

5.3 高通量筛选技术或"组学"

自 20 世纪 50 年代早期首次发现和描述 DNA 及其功能以来（Watson & Crick，1953），人类在理解细胞过程背后的分子机制方面取得了巨大的进步。分子生物学尤其关注生物学信息的流动及其在基因和蛋白质水平上的影响。虽然现在对于 DNA 发现者的争议仍存在，但 DNA 双螺旋结构的发现清楚地表明，基因是具有特定核苷酸序列的功能性 DNA 分子，细胞一定有一种利用基因来合成蛋白质的方法。基因转录和翻译后来被发现是基因表达和随后蛋白质合成及更高层次的驱动因素。随着医学研究对基因调控和细胞功能认识的日益加深，毒理学家也开始关注污染物暴露引发分子效应的评价。这一趋势大大促进了评价污染物分子效应的科学研究，以及分子和细胞方法在生态毒理学中的应用。对于这项相对较新领域中产生的相关术语完整列表，读者可参考《经济合作与发展（OECD）组织工作定义汇编》（2012）。

毒理学中，分析和理解由污染物引起的分子效应越来越受到重视，其原因是多方面的。

（1）传统的污染物 ERA 建立在如死亡和组织病理学等极端效应的基础上，这些效应是在 20 世纪开始测量的，当时不加考虑地释放了大量污染物，导致了大规模的死亡和其他瞬时的惊人反应。然而现今环境中污染物暴露浓度相对较低，但具有持久性或持续增加。虽然慢性低剂量效应可能不会引起明显的毒性，但可能通过生物的健康和生理（如行为）发生相当微妙的变化，从而在种群水平和生物多样性方面造成有害生态结局（Relyea & Hoverman，2006）。这些可以使用分子生物学工具检测和监测。

（2）分子和细胞工具在野外监测中也很实用，可用于诊断实际污染对野生动物的影响，并可用于检测亚致死、慢性效应及其与生态系统水平变化的关系（Relyea & Hoverman，2006）。由于在 DNA 和蛋白质水平上的效应被认为是对污染的首要反应，分子和细胞工具在（早期）检测这种细微的毒物效应方面尤为重要，生物标记物在环境评价中的效用突出了这一点。

（3）了解某些污染物引发有毒事件背后的分子和细胞水平过程有助于确定化学物质风险试验的优先次序以及进行物种间推断。欧盟市场上每年生产 30000 多种产量超过 1 t 的化学物质，并最终以各种形式排放到环境当中，传统的描述性、逐个化合物评估的方法从经济和时效角度来看已经到了极限。因此效应评价需要一种新的方法，能够从机理上理解引起的毒性，并能够阐明参与毒性反应的分子通路和网络，从而阐明物质的作用方式（mode of action，MoA）。MoA 是指生物体暴露于某种物质后在细胞水平上发生功能或解剖学变化。因此，分子和细胞方法有助于克服上述局限性，取代目前依赖于动物表型反应的检测方法。通过探索与污染物相互作用后而以某种方式改变的主要分子过程或通路，可以确定病理的具体起源，并可以根据其 MoA 对化学物质进行分类（Diamond et al.，2011；Segner，2011）。根据 Ankley et al.（2010）提出的有害结局路径（adverse outcome pathway，AOP）概念，组织的低水平效应（如某些基因或蛋白质的表达量减少或降低）应该联系组织更高水平效应［通过使用如高通量筛选技术、毒理基因组学分析（揭示分子效应）］，结合组织病理学和/或行为评价（称为生理锚定），以鉴别化学损伤的生物体或种群有效性及其表征的分子来源。因此，在与风险评价相关的组织生物层面上，AOP 在概念上是将一个直接的分子起始事件（如污染物与特定生物分子间的相互作用）与有害结局（如组织病理、死亡等）联系起来（图 5.2）。具有特定 MoAs 的化学物质也可能针对特定物种的特定生态受体或特定生命阶段，从而导致标准生态毒理学风险评价无法预测的毒性效应。对此，分子和细胞方法可以明确哪些生态受体群、哪个生命阶段,或哪些功能和特征可能会因特定化学物质而处于危险之中(Hutchinson et al.,2006;Yadetie et al.,2012;Segner et al.,2013;Brown et al.,2014)。总之,在明显不确定的情况下,基于机制的评价方法支持基于证据的风险评价,提高了我们诊断和预测化学污染有害效应影响的能力。

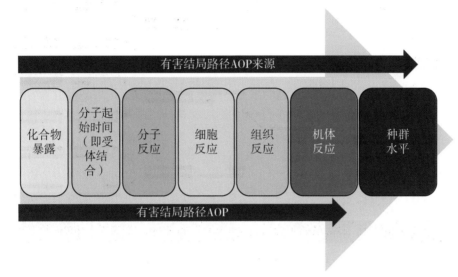

图 5.2　有害结局途径(AOP)

AOP 是从分子启动体内结局(群落或种群水平)事件的顺序进程。一般来说,它指的是一系列更广泛的途径,包括:①从化学物质与生物目标物相互作用(如 DNA 结合、蛋白质氧化)的分子起始事件(MIEs)开始;②继续一系列连续的生物活动(如基因激活、组织发育的改变);③最终导致人类或生态风险评价目标的有害效应(如死亡、生殖损伤、癌症、灭绝)(OECD,2011)。

分子和细胞方法在生态毒理学中日益广泛的应用与过去几十年相关方法学的巨大技术进步有关。1983 年 PCR 技术的发展(Mullis et al.,1986)使特定特征编码的 DNA 片段的大量复制成为可能(图 5.3)。这项技术基于通过调节温度实现 DNA 双链的变性和杂交,在生物学和医学领域引发了一场革命。它使穆利斯(Mullis)获得了 1993 年诺贝尔化学奖;并在 20 世纪 90 年代随着逆转录聚合酶链反应(RT-PCR)的发现而得到进一步的发展。传统 PCR 和 RT-PCR 的区别在于 RT-PCR 是在 PCR 过程中实时监测目标 DNA 分子的扩增,而不是像传统 PCR 那样在 PCR 结束时监测。RT-PCR 可分为定量 RT-PCR(quantitative real-time PCR)和半定量 RT-PCR(semi-quantitative RT-PCR),即针对一定数量的 DNA 分子(半定量 RT-PCR)或定性(定量 RT-PCR)分析。关于 RT-PCR 程序和数据规范化步骤的进一步信息,读者可参考 Gause & Adamovicz(1994)、Bustin(2002)& Bustin et al.(2005)。在 Pfaffl(2001)、Paffl et al.(2002)& Pfaffl(2004)等文章中广泛讨论了量化和扩增效率方面的问题。

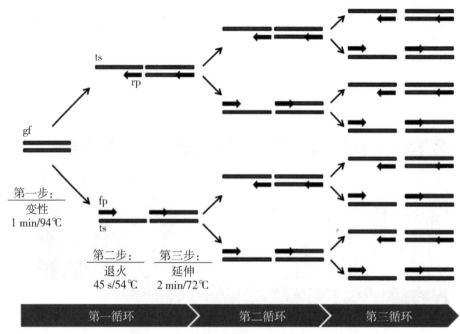

图 5.3 聚合酶链反应（PCR）示意

第一步，将目标双链基因片段（gf）在高温下物理反应解旋为两个单模板链（ts）（变性）；第二步，降低温度，使特定正向引物（fp）和反向引物（rp）与模板结合（退火）；第三步，根据模板链的互补性，聚合酶将连续的核苷酸结合到复制的 DNA 链中。连续的 PCR 循环以指数方式放大目标序列区域或基因片段。PCR 结果的选择性是由于在特定的热循环条件下，使用了与 DNA 区域互补的引物进行扩增。给定的温度是定向的。

毒理学者可利用 RT-PCR 测定不同条件下（环境或实验）的个体或种群中单个基因或 DNA 片段的差异表达，如以受污染暴露和未受暴露的生物体作为比较。在毒理学中，转录分析的进展已使人们认识到基因表达的改变可能是一种早期、快速和灵敏的应激反应检测手段。但这项技术是基于针对特定转录的预期效应的有根据猜测，并不能提供大量基因相互作用的信息，也不适用于检测意外或未知的过程和通路。因此，PCR 技术并不适用于寻找新的生物标记物，除形成基于最新高通量技术的概念之外，由于其极高的灵敏性，在海洋生态毒理学中可以用来验证遗传诱导或某些生物标记物酶的沉默或其他特征。Seo et al.（2004）研究了潮间带桡足类日本虎斑猛水蚤（*Tigriopus japonicas*）在不同的氧化应激源（盐度、重金属、H_2O_2）作用下谷胱甘肽还原酶的表达，这是 RT-PCR 技术在海洋（生态）毒理学中应用的例子。Won et al.（2013）以暴露于原油的海洋多毛类双齿围沙蚕（*perinereis nuntia*）为对象，测定了 3 个新的细胞色素 P450（CYP）和抗氧化基因的表达，报道了在原油中的双齿围沙蚕对这种化学胁迫的反应良好。另一个例子就是 Rhee et al.（2013）应用 RT-PCR 技术在海洋环境中的研究。这些研究者通过 γ 射线照射海洋鱼类斑纹隐小鳉（*Kryptolebias marmoratus*）来研究抗氧化酶的共表达以及 p53、DNA 修复、热休克蛋白基因的表达。Du et al.（2013）在研究太平洋牡蛎 *C. gigas* 基因表达时提出将管家基因作为内参的方法。这些只是这一实用技术在海洋生态毒理学中广泛应用的一些例子。

在过去的十年中，已开发了几种不同的技术来同时分析多种目标基因及感兴趣的特征，其中包括基因表达系列分析（SAGE）。SAGE 是一种以对应于转录片段的小标签形式在样品中产生信使 RNA 群体快照的技术（Velculescu et al.，1995），作为一种 cDNA 序列短片段的标签或表达序列标签，可以通过互补来识别基因转录（Adams et al.，1991）。此后人们通过改进原始技术，开发出了不同的变种，如 Long-SAGE（Saha et al.，2002）、RL-SAGE（Gowda et al.，2004）& SuperSAGE（Matsumura et al.，2005），以捕获更长的标签，从而可以更可靠地识别基因源。但是直到高通量或组学技术（DNA 微阵列和 RNAseq）发展起来，人们才有可能在一定条件下同时分析数千个基因的表达。

"组学"表明对"某种整体"的研究。它在生物学中用于非常大规模的数据收集和分析，可以分为三个主要类别：基因组学/转录组学、蛋白质组学和代谢物组学/代谢组学（Gehlenborg et al.，2010）。组学分析是同时测定成千上万个基因（转录组学）、蛋白质（蛋白质组学）或代谢物（代谢物组学/代谢组学）的活性或表达，以创建一个细胞功能的全局图。通过测量这些表达水平，组学可以研究化合物的详细作用模式（Dulin et al.，2012）。此外，DNA 测序技术的进步促进了对包括人类在内（Venter et al.，2001；Lander et al.，2001）的许多基因组序列的了解，再加上我们使用 DNA 微阵列和其他技术来量化 mRNA 表达变化的能力，更复杂的技术使我们能够确定参与毒性事件的特定基因启动或沉默（转录组学）、蛋白质丰度的增加或减少（蛋白质组学）或代谢物改变（代谢物组学）（表 5.3）。

表 5.3 生态毒理学中最常用的组学技术、分析目标和预期结果的概要

技术	目标	相关信息
转录组学	所有的转录产物（mRNA）	基因启动或沉默
蛋白组学	所有蛋白质	遗传信息被翻译成蛋白质
代谢组学	细胞代谢产物	蛋白质在细胞中的作用

5.3.1 转录组学

一般而言，转录组学是环境毒理学中最常用的组学技术，主要平台为 cDNA 微阵列或寡核苷酸阵列（Hartung & McBride，2011）。DNA 微阵列是附着在类似于显微镜载玻片的固体表面上的微小 DNA 点的集合。每个 DNA 点包含一个特定的 DNA 序列，称为探针（或报告基因或寡核苷酸）。Pollack et al.（1999）& Carter（2007）对 cDNA 微阵列的基本原理和应用做了详细的介绍。微阵列技术要求以样本 mRNA 分子为目标，转录成与芯片上的探针相结合 cDNA。微阵列技术的一个主要缺点是，探针可能会受到阵列上的目标特征（数量和性质）的限制，以及在意想不到的特征上"失去"有趣反应的可能性。这一问题正在被新的有前景的基因表达检测技术所克服，如二代测序（NGS）和转录组测序技术（RNA-seq）。RNA-seq 利用 NGS 的功能来揭示特定时刻基因

组中 RNA 存在和数量的快照（Chu & Corey, 2012）。RNA-seq 可同时产生成千上万的序列。RNA-seq 很可能在将来取代基于微阵列的转录组学，那时这种技术的成本将变得越来越低。尽管 RNA-seq 仍然是一项正在积极发展的技术，但是它与现有技术相比有几个关键优势。与基于杂交的方法如 cDNA 微阵列相比，cDNA 微阵列研究中可以获得的信息仅限于与存在的基因组序列相对应的表面探针，而 RNA-seq 可以通过测序将样本中的所有 RNA 样本分析出来，因此完全独立于任何预先设计的平台。这使得对基因组序列尚未确定的非模式生物相关研究特别有用（Wang et al., 2009a, b）。当然微列阵技术在基因表达的定向鉴定方面仍有用武之地，使其成为监管监督的理想工具。

5.3.2 蛋白质组学

与转录组学类似，质谱（MS）领域的发展使蛋白质（蛋白质组学）和代谢物（代谢组学）在不同条件下的定量和鉴定成为可能，并使我们能够深入了解生物对污染物暴露应激的机制。蛋白质组学是一组技术，可用于尝试分离、量化和鉴定样品中的数百种蛋白质，同时分析细胞和组织范围内的蛋白质表达。因此，基因组学主要研究基因表达，而蛋白质组学是分析基因的蛋白产物。传统上确定单个蛋白质的表达水平是通过结合到一个特定抗体（＝识别）的蛋白质印迹技术（Towbin et al., 1979）。在目前的蛋白质组学技术中，通过有（双向凝胶电泳 2D-PAGE；二维差异凝胶电泳 2D-DIGE）或无（同位素标记相对和绝对定量 iTRAQ）凝胶电泳分离的 MS 来识别和量化蛋白质。自 20 世纪 60 年代末以来，二维聚丙烯酰胺凝胶电泳（2D-PAGE）和质谱技术已分别被用于蛋白质分离和鉴定，以及比较对照和患病或化学暴露的生物样品之间的蛋白质表达（Martyniuk et al., 2012）。蛋白质组学最首要的创新是将 MS 与 2D-PAGE 联合，作为鉴定差异表达蛋白质的一种方法（Lilley et al., 2001；Beranova-Giorgianni 2003；Herbert et al., 2001）。利用该技术，样品中的蛋白质首先通过等电聚焦（IEF，一维）根据其电荷分离，然后根据其分子量分离。一维分离是允许通过单个正电荷或负电荷来分辨不同的蛋白质，使得该方法除定量分析蛋白质表达变化外，可用于区分蛋白质异构体、翻译后修饰的蛋白质。差异凝胶电泳（DIGE）方法提高了蛋白质组学方法的标准，允许通过对样品差异荧光染色，可以在同一凝胶中将一组对照组蛋白从治疗组或疾病组中分离出来的蛋白共分离（Tonge et al., 2001）。最近基于不同标记物蛋白质片段 MS 使用不依赖凝胶的技术。在此背景下，通过结合稳定同位素标记来定量分析蛋白质已成为现代蛋白质组学研究的核心技术。通过 iTRAQ（用于相对和绝对定量的等量异位标签）定量是毒理学中监测受干扰生物系统中蛋白质丰度相对变化的技术手段之一（Ross et al., 2004）。iTRAQ 方法可以在一个 MS 实验中比较多达 8 种不同的样品，并避免了凝胶中蛋白质分离的低灵敏度。氧化还原蛋白质组学是蛋白质组学中一门发展迅速的学科。氧化还原蛋白质组学特别关注在氧化还原活性条件下蛋白质组的修饰，这种修饰常出现在受污染情况下。有许多化学物质可以触发细胞内 ROS 形成，引发炎症过程。氧化还原蛋白质组由可逆和不可逆的共价修饰组成。这些共价修饰将氧化还原代谢与生物结构和功能联系起来。这些修饰作用于蛋白质折叠和成熟、催化活性、信号传导和大分子间相互作用等分子水平，从而在分泌和细胞形态的调控等宏观水平上发挥作用。氧化还原蛋白质组与氧化还原活性化学物质的相互作用是生命周期中大分子结构、调控和信号传导的中心环节，并在对摄食和环境挑战的耐受性和适应性方面具有核心作用（McDonagh & Sheehan, 2006；Sheehan, 2006）。

5.3.3 代谢物组学

代谢组由细胞、组织、器官、生物液体或整个生物体内的代谢物或小分子等构成（Miller, 2007）。代谢物组学旨在通过识别和量化生物系统（细胞、组织、器官和生物体）中的所有小分子来测量生命系统对生物刺激的全局动态代谢反应。最常用的代谢物组学技术有液相色谱 - 质谱（LC-MS）、气相色谱 - 质谱（GC-MS）和核磁共振（NMR）（Nicholson & Lindon, 2008）。虽然代谢物组学被认为是对其他组学技术的补充，但它实际上可能为其他组学方法所具有的许多缺陷提供了解决方案（Griffin & Bollard, 2004；Bilello, 2005；van Rav-enzwaay et al., 2007）。基因表达和蛋白产量变化的测量受到各种体内平衡控制和反馈机制的影响。这可能导致代谢物组学比其他组学技术能更敏感地反映外部压力（Ankley et al., 2006；van Ravenzwaay et al., 2007）。虽然已经开发了检测基因组、转录组和蛋白质组图谱变化的方法，但根据这些数据进行有意义的解释所需的信息有时并不容易获取（Ankley et al., 2006）。另外，大多数代谢物的结构和功能特征相当明显，而且可能比基因和蛋白质的数量要少（van Ravenzwaay et al., 2007）。

类似地，还有其他的、现有的或新兴的组学技术，如脂质组学［描述一个细胞、组织或生物体内完整的脂质概况，是"代谢组学"的一个子集，还包括其他三大类生物分子：蛋白质/氨基酸、糖和核酸（Wenk 2005）］，以及金属组学［描述游离金属离子在每个细胞隔间中的分布（Banci & Bertini, 2013）］和金属代谢组学［描述金属药物的代谢物在样品中的分布（Ge et al., 2016）］等。

5.3.4 海洋基因组资源和实例

目前可获得的有关污染物暴露的分子效应（生态）毒理基因组学数据大多局限于管制试验的物种或淡水物种（Handy et al., 2008a, b；Blaise et al., 2008；Federici et al., 2007；Warheit et al., 2007；Lovern & Klaper, 2006）。缺乏经常用于毒性试验和风险评价的大多数海洋鱼类和无脊椎动物物种的已有遗传信息，是目前可用的不同基因组技术更广泛应用的一个主要限制。由于非模式生物中很难识别转录本和蛋白质，许多环境组学研究依赖于通过物种同源搜索来识别不同表达的特征（Hampel et al., 2015；Silvestre et al., 2006；Romero-Ruiz et al., 2006；Vioque-Fernández et al., 2009）。这种方法已成功地应用于 Cu 或盐度对贻贝胁迫的研究（Shepard et al., 2000），以及 Cu、As 和三丁基锡等浓度梯度暴露鸡帘蛤 *C. gallina* 的研究（Rodríguez-Ortega et al., 2003）。目前，海洋非模式物种的基因组测序正在开展中，以下为常用于经典毒性试验的海洋生物的部分或完整公开基因组：红鳍东方鲀（*Fugu rubripes*）、底鳉（*Fundulus heteroclitus*）、太平洋牡蛎（*C. gigas*）、地中海贻贝（*M. galloprovincialis*）、菲律宾蛤仔

(*R. philippinarum*)、欧洲比目鱼（*Platichthys flesus*）、海鲷（*Sparus aurata*）、欧洲舌齿鲈（*Dicentrarchus labrax*）以及其他一些溯河鱼类，如大西洋鲑鱼（*Salmo salar*）和三刺鱼（*Gasterosteus aculeatus*）等。

尽管用于海洋生物的基因组资源较少，但人们正在努力提高其可用性和应用性。转录组学方法也已用于环境监测工作。例如，Milan et al.（2011）对菲律宾蛤仔（*R. philippinarum*）的转录组进行了测序，并开发了一种微阵列，后来将其应用于威尼斯潟湖的环境监测（Milan et al.，2015）。同样，通过焦磷酸测序获得的紫贻贝（*M. galloprovincialis*）的转录组数据，提供了广泛的基因组信息，并对不同组织、线粒体和相关微生物的表达产生了新的观察结果，还促进所需生物寡核苷酸芯片的生产（Craft et al.，2010）。

Hampel et al.（2015）通过将大西洋鲑鱼（*Salmo salar*）暴露于环境相当浓度的抗癫痫药物卡马西平（Carbamazepine）中，观察到其大脑转录组表达和不同代谢途径的表达发生了改变，包括细胞凋亡、血红素生物合成、铁离子和无机阳离子的动态平衡等的负调控。

相似地，Knigge et al.（2004）分析了来自挪威卡姆岛受污染海洋栖息地周边的紫壳菜蛤（*M. edulis*）的蛋白质组，在两个不同的污染点（重金属 VS 多环芳烃）发现了蛋白/肽的差异表达，表现出与原发地点相关的特异性或一般诱导。通过结合一些蛋白质标记的建树算法，研究者能够以90%的准确率从这些站点正确地对样品进行分类。

5.4 展　　望

组学技术在各种科学领域中不断发展。这些技术相对较新，但发展迅速，因此产生的大量数据对科学家来说是另一个新的挑战。除处理这些海量信息的物理存储场所外，这些海量数据的实际使用和汇总输出也让科学家们倍感困难。由高通量技术获得的组学数据为揭示环境污染物（众多中的）暴露所引发的机制提供了巨大的潜力。然而生成的大量数据往往没有得到充分利用。"组学"技术的最新进展为生物研究创造了前所未有的机遇，但目前的软件和数据库资源极度分散（Henry et al.，2014），迫切需要对生物信息学资源进行组织整理（Cannata et al.，2005）。此外，转录组学、蛋白质组学、代谢物组学和其他生命科学领域的新兴技术正在产生越来越多复杂的数据和信息，这些数据和信息无法用传统方式解释。生物信息学是一个快速发展的领域，需要解析大量的数据，如今与这些新兴技术相关的变化使得计算机科学的作用显得更为关键。为了解决与新兴和未来生命科学挑战相关的日益复杂的问题，生物信息学、计算生物学研究者和开发者需要探索、开发和应用新的计算概念、方法、工具和系统。此外，工具的细节和访问常常随着版本的改变而改变（Dellavalle et al.，2003），这使得研究团队保持最新的研究越来越具有挑战性（例如 van der Ven et al.，2005，2006）。这些研究的设计和性能在相对较多的不同模式下的生物暴露时间和环境、使用的剂量和发育阶段方面差异显著。上述情况导致这些数据难以解释和统一,以及得出共同的结论。因此,应鼓励采取主动行动,最终将各种各样的数据整合到一个单一的生物功能模型中,形成一个"模拟空间"。一个"组学"模拟包含每个基因对每个关键参数的估计以及在每个层次上的相互作用规则,以提供相互作用网络,并像计算机生物学实验室一样能对生物功能进行估计(Evans, 2000)。

参考文献

ADAM V, FABRIK I, KIZEK R, et al., 2010. Vertebrate metallothioneins as target molecules for analytical techniques. TrAC-Trends Anal. Chem., 29 (5): 409 - 418.

ADAMS M D, KELLEY J M, GOCAYNE J D, et al., 1991. Complementary DNA sequencing: expressed sequence tags and human genome project. Science, 252 (5013): 1651 - 1656.

AEBI H, 1984. Catalase in vitro // SIES H, KAPLAN N, COLOWICK N (Eds.). Methods in enzymology, 105: 121 - 126.

AGUIRRE-MARTINEZ G V, BURATTI S, FABBRI E, et al., 2013a. Using lysosomal membrane stability of haemocytes in *Ruditapes philippinarum* as a biomarker of cellular stress to assess contamination by caffeine, ibuprofen, carbamazepine and novobiocin. J. Environ. Sci. China, 25: 1408 - 1418.

AGUIRRE-MARTINEZ G V, BURATTI S, FABBRI E, et al., 2013b. Stability of lysosomal membrane in *Carcinus maenas* acts as a biomarker of exposure to pharmaceuticals. Environ. Monit. Assess., 185: 3783 - 3793.

AGUIRRE-MARTINEZ G V, DELVALLS T A, MARTIN-DIAZ M L, 2016. General stress, detoxification pathways, neurotoxicity and genotoxicity evaluated in *Ruditapes philippinarum* exposed to human pharmaceuticals. Ecotoxicol. Environ. Saf., 124: 18 - 31.

AKCHA F, IZUEL C, VENIER P, et al., 2000. Enzymatic biomarker measurement and study of DNA adduct formation in B [a] P contaminated mussels, *Mytilus galloprovincialis*. Aquat. Toxicol., 49: 269 - 287.

AL-SABTI K, FRANKO M, ANDRIJANIC B, et al., 1994. Chromium induced micronuclei in fish. J. Appl. Toxicol., 14: 333 - 336.

AMIARD J C, AMIARD - TRIQUET C, BARKA S, et al., 2006. Metallothioneins in aquatic invertebrates: their role in metal detoxification and their use as biomarkers. Aquat. Toxicol., 76 (2): 160 - 202.

ANDERSSON T, LARS F, 1992. Regulation of the cytochrome P450 enzyme system in fish. Aquat. Toxicol., 24: 1 - 20.

ANKLEY G T, DASTON G P, DEGITZ S J, et al., 2006. Toxicogenomics in regulatory ecotoxicology. Environ. Sci. Technol., 40 (13): 4055 - 4065.

ANKLEY G T, BENNETT R S, ERICKSON R J, et al., 2010. Adverse outcome pathways: a conceptual framework to support ecotoxicology research and risk assessment. Environ. Toxicol. Chem., 29 (3): 730 - 741.

ARSLAN O C, PARLAK H, KATALAY S, et al., 2010. Detecting micronuclei frequency monitoring pollution of Izmir Bay (Western Turkey). Environ. Monit. Assess., 165: 55 –66.

ARUKWE A, GOKSØYR A, 2003. Eggshell and egg yolk proteins in fish: hepatic proteins for the next generation: oogenetic population, and evolutionary implications of endocrine disruption. Comp. Hepatol., 2: 4.

AUFFRET M, 1988. Histopathological changes related to chemical contamination in *Mytilus edulis* from field and experimental conditions. Mar. Ecol. Prog. Ser., 46: 101 – 107.

BANCI L, BERTINI I, 2013. Metallomics and the cell: some definitions and general comments // BANCI L (Eds.). Metallomics and the Cell. Metal ions in life sciences. vol. 12. chapter1. Springer: 1 – 13.

BARRETT J C, 1992. Mechanisms of action of known human carcinogens // VAINIO H, MAGEE P N, MCGREGOR D B, et al (Eds.). Mechanisms of carcinogenesis in risk assessment. International Agency for Cancer Research, Lyon, France: 115 – 134.

BEBIANNO M J, NOTT J A, LANGSTON W J, 1993. Cadmium metabolism in the clam *Ruditapes decussata*: the role of metallothioneins. Aquat. Toxicol., 27: 315 – 334.

BEBIANNO M J, SERAFIM M A, RITA M F, 1994. Involvement of metallothionein in cadmium accumulation and elimination in the clam *Ruditapes decussata*. Bull. Environ. Contam. Toxicol., 53: 726 – 732.

BERANOVA-GIORGIANNI S, 2003. Proteome analysis by two-dimensional gel electrophoresis and mass spectrometry: strengths and limitations. TrAC-Trends Anal. Chem., 22 (5): 273 –281.

BERLIN A, SCHALLER K N, 1974. European standardized method for determination of δ-ALAD activity in blood. Z. Klin. Chem. Klin. Biochem., 12 (8): 389 – 390.

BIHARI N, FAFANDEL M, 2004. Interspecies differences in DNA single strand breaks caused by benzo (a) pyrene and marine environment. Mutat. Res., 552: 209 – 217.

BILELLO J A, 2005. The agony and ecstasy of "OMIC" technologies in drug development. Curr. Mol. Med., 5 (1): 39 –52.

BINZ P A, KAGI J H R, 1999. Metallothionein: molecular evolution classification // KLAASSEN C (Eds.). Metallothionein IV. Birkhäuser Verlag, Basel: 7 – 13.

BLAISE C, GAGNÉ F, PELLERIN J, et al., 1999. Measurement of a vitellogenin-like protein in the hemolymph of *Mya arenaria* (Saguenay Fjord, Canada): a potential biomarker for endocrine disruption. Environ. Toxicol., 14 (4): 455 –465.

BLAISE C, TROTTIER S, GAGNÉ F, et al., 2002. Immunocompetence of bivalve hemocytes as evaluated by a miniaturized phagocytosis assay. Environ. Toxicol., 17: 160 – 169.

BLAISE C, GAGNÉ F, FÉRARD J F, et al., 2008. Ecotoxicity of selected nano-materials to aquatic organisms. Environ. Toxicol, 23: 591 –598.

BLANCHETTE B, FENG X, SINGH B R, 2007. Marine glutathione S-transferases. Mar. Biotechnol., 9: 513 –542.

BLASCO J, DELVALLS T A, 2008. Impact of emergent contaminants in the environment: environmental risk assessment // BARCELÓ D, PETROVIC M (Eds.). Emerging

Contaminants from Industrial and Municipal Waste. The handbook of environmental chemistry, vol. 5. S/1. Springer-Verlag, Heiderberg: 169 – 188.

BLASCO J, PUPPO J, SARASQUETE M C, 1993. Acid and alkaline phosphatase activities in the clam *Ruditapes philippinarum*. Mar. Biol., 115 (1): 113 – 118.

BOCQUENÉ G, GALGANI F, TRUQUET P, 1990. Characterization and assay conditions for use of AChE activity from several marine species in pollution monitoring. Mar. Environ. Res., 30: 75 – 89.

BOLOGNESI C, RABBONI R, ROGGIERI I, 1996. Genotoxicity biomarkers in *M. galloprovincialis* as indicators of marine pollutants. Comp. Biochem. Phys. C, 113 (2): 319 – 323.

BOLOGNESI C, 1990. Carcinogenic and mutagenic effects of pollutants in marine organisms: a review // GRANDJEAN E (Eds). Carcinogenic, mutagenic, and teratogenic marine pollutants: impact on human health and the environment. Portfolio Publishing Company, The Woodland, TX, USA: 67 – 83.

BOUTET I, TANGUY A, MORAGA D, 2004. Response of the Pacific oyster *Crassostrea gigas* to hydrocarbon contamination under experimental conditions. Gene, 329: 147 – 157.

BROUSSEAU P, DUNIER M, DE GUISE S, et al., 1997. Marqueurs immunologiques // LAGADIC L, et al (Eds.). Biomarqueurs en ecotoxicologie: aspects fondamentaux. Masson, Paris: 287 – 315.

BROUSSEAU P, PAYETTE Y, BLAKLEY B R, et al., 1999. Manual of immunological methods. CRC Press, Boston, USA.

BROUWER M, WINGE D R, GRAY W R, 1989. Structural and functional diversity of copper-metallothioneins from the American lobster *Homarus americanus*. J. Inorg. Biochem., 35: 289 – 303.

BROWN A R, GUNNARSSON L, KRISTIANSON E, et al., 2014. Assessing variation in the potential susceptibility of fish to pharmaceuticals, considering evolutionary differences in their physiology and ecology. Philos. Trans. R. Soc. B, 369 (1656): 20130576.

BRUNETTI R, MAJONE F, GOLA I, et al., 1988. The micronucleus test: examples of application to marine ecology. Mar. Ecol. Prog. Ser., 44: 65 – 68.

BRUNK C F, 1979. Assay for nanogram quantities of DNA in cellular homogenates. Anal. Biochem., 92: 497 – 500.

BUCHELI T D, FENT K, 1995. Induction of cytochrome P450 as a biomarker for environmental contamination in aquatic ecosystems. Crit. Rev. Environ. Sci. Technol., 25: 201 – 268.

BUFFET P E, ZALOUK-VERGNOUX A, CHÂTEL A, et al., 2014. A marine mesocosm study on the environmental fate of silver nanoparticles and toxicity effects on two endobenthic species: the ragworm *Hediste diversicolor* and the bivalve mollusc *Scrobicularia plana*. Sci. Tot. Environ., 470 – 471: 1151 – 1159.

BUHLER D R, WILLIAMS D E, 1988. The role of biotransformation in the toxicity of chemicals. Aquat. Toxicol., 11: 19 – 28.

BURKE M D, MAYER R T, 1974. Ethoxyresorufin: direct fluorometric assay of a microsomal O-dealkylation which is preferentially inducible by 3 methylcholanthrene. Drug Metab. Dispos., 2: 583 – 588.

BUSTIN S A, BENES V, NOLAN T, et al., 2005. Quantitative real-time RT-PCR: a perspective. J. Mol. Endocrinol., 34 (3): 597 – 601.

BUSTIN S A, 2002. Quantification of mRNA using real-time reverse transcription PCR (RT-PCR): trends and problems. J. Mol. Endocrinol., 29: 23 – 39.

CAJARAVILLE M P, BEBIANNO M J, BLASCO J, et al., 2000. The use of biomarkers to assess the impact of pollution in coastal environments of the Iberian Peninsula: a practical approach. Sci. Total Environ., 247 (2 – 3): 295 – 311.

CAMPANA O, SARASQUETE C, BLASCO J, 2003. Effect of lead on ALA-D activity, metallothionein levels, and lipid peroxidation in blood, kidney, and liver of the toadfish *Halobatrachus didactylus*. Ecotoxicol. Environ. Saf., 55 (1): 116 – 125.

CAMPILLO J A, ALBENTOSA M, VALDÉS N J, et al., 2013. Impact assessment of agricultural inputs into a Mediterranean coastal lagoon (Mar Menor, SE Spain) on transplanted clams (*Ruditapes decussatus*) by biochemical and physiological response. Aquat. Toxicol., 142 – 143: 365 – 379.

CANESI L, CIACCI C, BETTI M, et al., 2003. Effects of PCB congeners on the immune function of Mytilus hemocytes: alterations of tyrosine kinase-mediated cell signaling. Aquat. Toxicol., 63: 293 – 306.

CANESI L, CIACCI C, LORUSSO L C, et al., 2007a. Effects of triclosan on *Mytilus galloprovincialis* hemocyte function and digestive gland enzyme activities: possible modes of action on non target organisms. Comp. Biochem. Physiol. C Toxicol. Pharmacol., 145: 464 – 472.

CANESI L, LORUSSO L C, CIACCI C, et al., 2007b. Effects of blood lipid lowering pharmaceuticals (bezafibrate and gemfibrozil) on immune and digestive gland functions of the bivalve mollusc, *Mytilus galloprovincialis*. Chemosphere, 69: 994 – 1002.

CANESI L, LORUSSO L C, CIACCI C, et al., 2007c. Immunomodulation of *Mytilus hemocytes* by individual estrogenic chemicals and environmentally relevant mixtures of estrogens: in vitro and in vivo studies. Aquat. Toxicol., 81: 36 – 44.

CANNATA N, MERELLI E, ALTMAN R B, 2005. Time to organize the bioinformatics resourceome. PLoS Comput. Biol., 1: 76 – 81.

CARTER N P, 2007. Methods and strategies for analyzing copy number variation using DNA microarrays. Nat. Genet., 39 (7 Suppl.): 16 – 21.

CHANDURVELAN R, MARSDEN I D, GAW S, et al., 2013. Biochemical biomarker responses of green-lipped mussel, *Perna canaliculus*, to acute and subchronic waterborne cadmium toxicity. Aquat. Toxicol., 140 – 141: 303 – 313.

CHEN L, DEVANESAN P D, HIGGINBOTHAM S, et al., 1996. Expanded analysis of benzo [a] pyrene-DNA adducts formed in vitro and in mouse skin: their significance in tumor initiation. Chem. Res. Toxicol., 9: 897 – 903.

CHU Y, COREY D R, 2012. RNA sequencing: platform selection, experimental design, and data interpretation. Nucl. Acid Ther., 22 (4): 271 – 274.

CHU F E, PEYRE J F L, 1989. Effect of environmental factors and parasitism on hemolymph lysozyme and protein in American oysters (*Crassostrea virginica*). J. Invertebr. Pathol., 54: 224 – 232.

COLAPIETRO A M, GOODELL A L, SMART R. C, 1993. Characterization of benzo [a] pyrene-initiated mouse skin papillomas for Ha-ras mutations and protein kinase C levels. Carcinogenesis, 14: 2289 – 2295.

COLBORN T, SAAL F V S, SOTO A M, 1993. Development effects of endocrine-disrupting chemicals in wildlife and humans. Environ. Health Perspect., 101: 378 – 384.

COLOVIC M B, KRSTIC D Z, LAZAREVIC-PASTI T D, et al., 2013. Acetylcholinesterase inhibitors: pharmacology and toxicology. Curr. Neuropharmacol., 11: 315 – 335.

COMPANY R, SERAFIM A, LOPES B, et al., 2008. Using biochemical and isotope geochemistry to understand the environmental and public health implications of lead pollution in the lower Guadiana River, Iberia: a freshwater bivalve study. Sci. Total Environ., 405 (1 – 3): 109 – 119.

COMPANY R, SERAFIM A, LOPES B, et al., 2011. Source and impact of lead contamination on δ-aminolevulinic acid dehydratase activity in several marine bivalve species along the Gulf of Cadiz. Aquat. Toxicol., 101 (1): 146 – 154.

COSSON R P, AMIARD J C, 2000. Use of metallothionein as biomarkers of exposure to metals // LAGADIC L, CAQUET T, AMIARD J – C, et al (Eds.). Use of biomarkers for environmental quality assessment. Science Publishers, Inc., Enfield, NH: 79 – 111.

CRAFT J A, GILBERT J A, TEMPERTON B, et al., 2010. Pyrosequencing of *Mytilus galloprovincialis* cDNAs: tissue-specific expression patterns. PLoS One, 5, (1): 8875.

CRAVO A, LOPES B, SERAFIM A, et al., 2009. A multibiomarker approach in *Mytilus galloprovincialis* to assess environmental quality. J. Environ. Monit., 11 (9): 1673 – 1686.

CRAVO A, LOPES B, SERAFIM A, et al., 2013. Spatial and seasonal biomarker responses in the clam *Ruditapes decussatus*. Biomarkers, 18 (1): 30 – 43.

DABRIO M, RODRIGUEZ A R, BORDIN G, et al., 2002. Recent developments in quantification methods for metallothionein. J. Inorg. Biochem., 88: 123 – 134.

DAGNINO A, ALLEN J I, MOORE M N, et al., 2007. Development of an expert system for the integration of biomarker responses in mussels into an animal health index. Biomarkers, 12: 155 – 172.

DANIELS D, 2007. Functions of red cell surface proteins. Vox Sang., 93: 331 – 340.

DE VRIES A, DOLLÉ M E T, BROEKHOF J L M, et al., 1997. Induction of DNA adducts and mutations in spleen, liver and lung of XPA-deficient/lacZ transgenic mice after oral treatment with benzo [a] pyrene: correlation with tumor development. Carcinogenesis, 18 (12): 2327 – 2332.

DE ZWART L L, VENHORST J, GROOT M, et al., 1997. Simultaneous determination of eight lipid peroxidation degradation products in urine of rats treated with carbon tetrachloride using gas chromatography with electron capture detection. J. Chromatogr. B, 694: 227 - 287.

DEAN J H, MURRAY M J, 1990. Toxic responses of the immune system // KLAASSEN C D, AMDUR M O, DOULL J (Eds.). Toxicology: the basic science of poisons. vol. 4. McMillan, New York: 282 - 333.

DELLAVALLE R P, HESTER E J, HEILIG L F, et al., 2003. Information science. Going, going, gone: lost Internet references. Science, 302 (5646): 787 - 788.

DENSLOW N D, CHOW M C, KROLL K J, et al., 1999. Vitellogenin as a biomarker of exposure for estrogen or estrogen mimics. Ecotoxicology, 8: 385 - 398.

DHAUNSI G S, SINGH I, HANEVLD C D, 1993. Peroxisomal participation in cellular responses to the oxidative stress of endotoxin. Mol. Cell. Biochem., 126: 25 - 35.

DI GIULIO R T, WASHBURN P C, WENNING R J, et al., 1989. Biochemical responses in aquatic animals: a review of determinants of oxidative stress. Environ. Toxicol. Chem., 8: 1103 - 1123.

DIAMOND J M, LATIMER H A, MUNKITTRICK K R, et al., 2011. Prioritizing contaminants of emerging concern for ecological screening assessments. Environ. Toxicol. Chem., 30: 2385 - 2394.

DONDERO F, PIACENTINI L, MARSANO F, et al., 2006. Gene transcription profiling in pollutant exposed mussels (*Mytilus* spp.) using a new low - density oligonucleotide microarray. Gene, 376: 24 - 36.

DU Y, ZHANG L, XU F, et al., 2013. Validation of housekeeping genes as internal controls for studying gene expression during Pacific oyster (*Crassostrea gigas*) development by quantitative real-time PCR. Fish Shellfish Immunol., 34: 939 - 945.

DULIN D, LIPFERT J, MOOLMAN M C, et al., 2012. Studying genomic processes at the single-molecule level: introducing the tools and applications. Nat. Rev. Genet., 14 (1): 9 - 22.

DUNN B P, 1991. Carcinogen adducts as an indicator for the public health risks of consuming carcinogenexposed fish and shellfish. Environ. Health Perspect., 90: 111 - 116.

EL HOURCH M, DUDOIT A, AMIARD J C, 2003. Optimization of new voltammetric method for the determination of metallothionein. Electrochim. Acta, 48: 4083 - 4088.

EL HOURCH M, DUDOIT A, AMIARD J C, 2004. An optimization procedure for the determination of metallothionein by square wave cathodic stripping voltammetry: application to marine worms. Anal. Bioanal. Chem., 378 (3): 776 - 781.

ELLMAN G L, COURTNEY K D, ANDRES JR V, et al 1961. A new and rapid colorimetric determination of acetylcholinesterase activity. Biochem. Pharmacol., 7: 88 - 95.

ERICSON G, AKERMAN G, LIEWENBORG B, et al., 1996. Comparison of DNA damage in the early life stages of cod, *Gadus morhua*, originating from the Barents Sea and Baltic Sea. Mar. Environ. Res., 42: 119 - 123.

EuropeanCommission, 2006. Regulation (EC) No 1907/2006 of the European parliament and of the council of 18 December 2006 concerning the registration, evaluation, authorisation and restriction of chemicals (REACH). Establishing a European Chemicals Agency.

FAHIMI H D, CAJARAVILLE M P, 1995. Induction of peroxisome proliferation by some environmental pollutants and chemicals // CAJARAVILLE M P (Eds.). Cell biology in environmental toxicology. Servicio Editorial de UPV/EHU, Leioa: 221 – 255.

FEDERICI G, SHAW B J, HANDY R D, 2007. Toxicity of titanium dioxide nanoparticles to rainbow trout, (*Oncorhynchus mykiss*): gill injury, oxidative stress, and other physiological effects. Aquat. Toxicol, 84 (4): 415 – 430.

FERNÁNDEZ B, MARTÍNEZ-GÓMEZ C, BENEDICTO J, 2015. Deltaaminolevulinic acid dehydratase activity (ALA-D) in red mullet (*Mullus barbatus*) from Mediterranean waters as biomarker of lead exposure. Ecotoxicol. Environ. Saf, 115: 209 – 216.

FERREIRA P, FONTE E, ELISA SOARES M, et al., 2016. Effects of multi-stressors on juveniles of the marine fish *Pomatoschistus microps*: gold nanoparticles, microplastics and temperature. Aquat. Toxicol., 170: 89 – 103.

FINELLI V, 1977. Lead, zinc and δ-aminolevulinate dehydratase // LEE S, PEIRANO B (Eds.). Biochemical effects of environmental pollutants. Ann Arbor Science Publishers, Ann Arbor, Michigan, US: 351 – 364.

FONG A T, DASHWOOD R H, CHENG R, et al., 1993. Carcinogenicity, metabolism and Ki-ras proto-oncogene activation by 7, 12 – dimethylbenz [a] anthracene in rainbow trout embryos. Carcinogenesis, 14 (4): 629 – 635.

FOURNIER M, CYR D, BLAKLEY B, et al., 2000. Phagocytosis as a biomarker of immunotoxicity in wildlife species exposed to environmental xenobiotics. Am. Zool., 40: 412 – 420.

FOWLER B A, HILDEBRAND C E, KOJIMA Y, et al., 1987. Nomenclature of metallothioneins // KAGI J H R, KOJIMA Y (Eds.). Metallothionein Ⅱ. Birhäuser Verlag, Basel: 19 – 22.

FRANZELLITTI S, BURATTI S, DU B, et al., 2015. A multibiomarker approach to explore interactive effects of propranolol and fluoxetine in marine mussels. Environ. Pollut., 205: 60 – 69.

FRENZILLI G, NIGRO M, LYONS B P, 2009. The comet assay for the evaluation of genotoxic impact in aquatic environments. Mutat. Res., 681: 80 – 92.

FRIDOVICH I, 1995. Superoxide radical and superoxide dismutases. Ann. Rev. Biochem., 64: 97 – 112.

FUJIMOTO Y, SAKUMA S, INOUE T, et al., 2002. The endocrine disruptor nonylphenol preferentially blocks cyclooxygenase-1. Life Sci., 70: 2209 – 2214.

FUJIMOTO Y, USA K, SAKUMA S, 2005. Effects of endocrine disruptors on the formation of prostaglandin and arachidonoyl-CoA formed from arachidonic acid in rabbit kidney medulla microsomes. Prost. Leuko. Essent. Fatty Acids, 73: 447 – 452.

FULTON M H, KEY P B, 2001. Acetylcholinesterase inhibition in estuarine fish and invertebrates as an indicator of organophosphorus insecticide exposure and effects. Environ. Toxicol. Chem., 20: 37-45.

GAGNÉ F, BLAISE C, 1999. Toxicological effects of municipal wastewaters to rainbow trout hepatocytes. Bull. Environ. Contam. Toxicol., 63: 503-510.

GAGNÉ F, BLAISE C, PELLERIN J, et al., 2003. Sex alteration in soft-shell clams (*Mya arenaria*) in an intertidal zone of the St. Lawrence River (Québec, Canada). Comp. Biochem. Phys. C, 134: 189-198.

GAGNÉ F, BLAISÉ C, ANDRE C, et al., 2007. Neuroendocrine disruption and health effects in *Elliptio complanata*mussels exposed to aeration lagoons for waste-water treatment. Chemosphere, 68: 731-743.

GARDNER G R, YEVICH P P, HARSHBARGER J C, et al., 1991. Carcinogenicity of Black Rock Harbor sediment to the eastern oyster and trophic transfer of Black Rock Harbor carcinogens from the blue mussel to the winter flounder. Environ. Health Perspect, 90: 53-66.

GARDNER G R, PRUELL R J, MALCOLM A R, 1992. Chemical induction of tumors in oysters by a mixture of aromatic and chlorinated hydrocarbons, amines and metals. Mar. Environ. Res., 34: 59-63.

GAUSE W C, ADAMOVICZ J, 1994. The use of the PCR to quantitate gene expression. PCR Methods Appl., 3 (6 Suppl.): 123-135.

GE R, SUN X, HE Q Y, 2016. Overview of the metallometabolomic methodology for metal-based drug metabolism. Curr. Drug Metab., 12 (3): 287-299.

GEHLENBORG N, O'DONOGHUE S, BALIGA N S, et al., 2010. Visualization of omics data for systems biology. Nat. Methods, 7 (3): 56-68.

GENEVOIS C, PFOHL-LESZKOWICZ A, BOILLOT K, et al., 1998. Implication of cytochrome P450 1A isoforms and the Ah receptor in the genotoxicity of coal tar fume condensate and bitumen fumes condensates. Environ. Toxicol. Pharmacol., 5: 283-294.

GEORGE S G, 1994. Enzymology and molecular biology of phase II xenobiotic-conjugating enzymes in fish//MALINS D C, OSTRANDER G K (Eds.). Aquatic toxicology molecular, biochemical and cellular perspectives. Lewis Publishers, CRC Press: 37.

GOEPTAR A R, SCHEERENS H, VERMEULEN N P E, 1995. Oxygen reductase and substrate reductase activity of cytochrome P450. Crit. Rev. Toxicol., 25: 25-65.

GOKSØYR A, FÖRLIN L, 1992. The cytochrome P450 system in fish, aquatic toxicology and environmental monitoring. Aquat. Toxicol., 22: 287-312.

GOKSØYR A, LARSEN E, 1991. The cytochrome P450 system of Atlantic salmon (*Salmo salar*): I. Basal properties and induction of P450 1A1 in liver of immature and mature fish. Fish Physiol. Biochem., 9: 339-349.

GOKSØYR A, ARUKWE A, LASRSSON J, et al., 2003. Molecular/cellular processes and the impact on reproduction//LAWRENCE J A, HEMINGWAY K L (Eds.). Effects of pollution on fish: molecular effects and population responses. Oxford: Blackwell Science: 179-220.

GOKSØYR A, 2006. Endocrine disruptors in the marine environment: mechanisms of toxicity and their influence on reproductive processes in fish. J. Toxicol. Environ. Health A, 69: 175 – 184.

GOLDSTEIN W, 1992. Neurologic concepts of lead poisoning in children. Pediatr. Ann., 21: 384 – 388.

GONZÁLEZ-FERNÁNDEZ C, ALBENTOSA M, CAMPILLO J A, et al., 2015. Influence of mussel biological variability on pollution biomarkers. Environ. Res., 137: 14 – 31.

GORBI S, VIRNO LAMBERTI C, NOTTI A, et al., 2008. An ecotoxicological protocol with caged mussels, Mytilus galloprovincialis, for monitoring the impact of an offshore platform in the Adriatic Sea. Mar. Environ. Res., 65: 34 – 49.

GOURDON I, GUERIN M C, TORREILLES J, et al., 2001. Nitric oxide generation by hemocytes of the mussel Mytilus galloprovincialis. Nitric Oxide Biol. Chem., 5: 1 – 6.

GOWDA M, JANTASURIYARAT C, DEAN R A, et al., 2004. Robust-LongSAGE (RL-SAGE): a substantially improved LongSAGE method for gene discovery and transcriptome analysis. Plant Physiol., 134 (3): 890 – 897.

GRIFFIN J L, BOLLARD M E, 2004. Metabonomics: its potential as a tool in toxicology for safety assessment and data integration. Curr. Drug Metab., 5 (5): 389 – 398.

GRISHAM M B, JOURD'HEUIL D, WINK D A, 1999. Nitric oxide. I. Physiological chemistry of nitric oxide and its metabolites: implications in inflammation. Am. J. Physiol., G276: 315 – 321.

GUILHERMINO L, LOPES M C, CARVALHO A P, et al., 1996. Inhibition of acetylcholinesterase activity as effect criterion in acute tests with juvenile Daphnia magna. Chemosphere, 32 (4): 727 – 738.

HALLIWELL B, GUTTERIDGE, 1999. Free radicals in biology and medicine. 3rd ed. Oxford University Press, Oxford, UK.

HAMPEL M, ALONSO E, APARICIO I, et al., 2015. Hepatic proteome analysis of Atlantic salmon (Salmo salar) after exposure to environmental concentrations of human pharmaceuticals. Mol. Cell. Prot, 14 (2): 371 – 381.

HAMZA-CHAFFAI A, AMIARD J C, COSSON R P, 1999. Relationship between metallothioneins and metals in a natural population of the clam Ruditapes decussatus from Sfax coast: a non-linear model using Box – Cox transformation. Comp. Biochem. Physiol. C, 123 (2): 153 – 163.

HANDY R D, HENRY T B, SCOWN T M, et al., 2008a. Manufactured nanoparticles: their upake and effects on fish: a mechanistic analysis. Ecotoxicology, 17: 396 – 409.

HANDY R D, VON DER KAMMER F, LEAD J R, et al., 2008b. The ecotoxicology and chemistry of manufactured nanoparticles. Ecotoxicology, 17: 287 – 314.

HARIHARAN G, PURVAJA R, RAMESH R, 2014. Toxic effects of lead on biochemical and histological alterations in green mussel (Perna viridis) induced by environmentally relevant concentrations. J. Toxicol. Environ. Health A, 77 (5): 246 – 260.

HARTUNG T, MCBRIDE M, 2011. Food for thought... On mapping the human toxome.

ALTEX, 28 (2): 83-93.

HELBOCK H J, BECKMAN K B, SHIGENAGA M K, et al., 1998. DNA oxidation matters: the HPLC-electrochemical detection assay of 8 - oxo-deoxy-guanosine and 8 - oxo-guanine. Proc. Natl. Acad. Sci. USA, 95: 288-293.

HENDRIKS A J, PIETERS H, DE BOER J, 1998. Accumulation of metals, polycyclic (halogenated) hydrocarbons, and biocides in zebra mussel and eel from the Rhine and Meuse rivers. Environ. Toxicol. Chem., 17: 1885-1898.

HENRY V J, BANDROWSKI A E, PEPIN A-S, et al., 2014. OMICtools: an informative directory for multi-omic data analysis. Database. http://dx.doi.org/10.1093/database/bau069. Article ID: bau069.

HERBERT B R, HARRY J L, PACKER N H, et al., 2001. What place for polyacrylamide in proteomics? Trends Biotechnol., 19 (10 Suppl.): 3-9.

HONG X T, XIANG L X, SHAO J Z, 2006. The immunostimulating effect of bacterial genomic DNA on the innate immune responses of bivalve mussel, *Hyriopsis cumingii*. Fish Shellfish Immunol, 21: 357-364.

HOOFTMAN R N, RAAT W K, 1982. Induction of nuclear anomalies (micronuclei) in the peripheral blood erythrocytes of the eastern mudminnow *Umbra pygmaea* by ethyl methanesulphonate. Mutat. Res., 104: 147-152.

HUTCHINSON T H, ANKLEY G T, SEGNER H, et al., 2006. Screening and testing for endocrine disruption in fish: biomarkers as "signposts," not "traffic lights," in risk assessment. Environ. Health Perspect., 114 (1 Suppl.): 106-114.

HYLLAND K, RUUS A, GRUNG M, et al., 2009. Relationships between physiology, tissue contaminants, and biomarker responses in Atlantic cod (*Gadus morhua* L.). J. Toxicol. Environ. Health, 72A: 226-233.

HYLLAND K, 2004. Biological effects of contaminants: quantification of δ-aminolevulinic Acid Dehydratase (ALA-D) activity in fish blood. vol. 34. ICES Tech. Mar. Environ. Sci.: 9.

JAMES M O, ALTMAN A H, LI C L J, et al., 1992. Dose-and time-dependent formation of benzo [a] pyrene metabolite DNA adducts in the spiny lobster, *Panulirus argus*. Mar. Environ. Res, 34: 299-302.

JANERO D R, 1990. Malondialdehyde and thiobarbituric acidreactivity as diagnostic indices of lipid peroxidation and peroxidative tissue injury. Free Radic. Biol. Med., 9: 515-540.

JEBALI J, BANNI M, GUERBEJ H, et al., 2006. Effects of malathion and cadmium on acetylcholinesterase activity and metallothionein levels in the fish *Seriola dumerili*. Fish Physiol. Biochem., 32: 93-98.

KALMAN J, RIBA I, BLASCO J, et al., 2008. Is δ-aminolevulinic acid dehydratase activity in bivalves from southwest Iberian Peninsula a good biomarker of lead exposure? Mar. Environ. Res., 68 (1): 38-40.

KATSUMITI A, GILLILAND D, AROSTEGUI I, et al., 2015. Mechanisms of toxicity of

Ag nanoparticles in comparison to bulk and ionic Ag on mussel hemocytes and gill cells. PLoS One, 10 (6).

KIM M, AHN I Y, CHEON J, et al., 2009. Molecular cloning and thermal stress-induced expression of a pi-class glutathione S-transferase (GST) in the Antarctic bivalve *Laternula elliptica*. Comp. Biochem. Physiol. A, 152: 207 – 213.

KIME D E, 1995. The effects of pollution on reproduction in fish. Rev. Fish Biol. Fish., 5: 52 – 96.

KIRUBAGARAN R, JOY K P, 1988. Inhibition of testicular 3l3 – hydroxy-As-steroid dehydrogenase (3l3 – HSD) activity in catfish *Clarias batrachus* (L.) by mercurials. Ind. J. Exp. Biol., 26: 907 – 908.

KNIGGE T, MONSINJON T, ANDERSEN O K, 2004. Surface-enhanced laser desorption/ionization-time of flightmass spectrometry approach to biomarker discovery in blue mussels (*Mytilus edulis*) exposed to polyaromatic hydrocarbons and heavy metals under field conditions. Proteomics, 4: 2722 – 2727.

KOENIG S, FERNÁNDEZ P, SOLÉ M, 2012. Differences in cytochrome P450 enzyme activities between fish and crustacea: relationship with the bioaccumulation patterns of polychlorobiphenyls (PCBs). Aquat. Toxicol., 108: 11 – 17.

KOJIMA Y, BERGER C, VALLEE B L, et al., 1976. Aminoacid sequence of equine renal metallothionein-IB. Proc. Natl. Acad. Sci. USA, 73 (10): 3413 – 3417.

KOLLER L D, EXON J H, 1985. The rat as a model for immunotoxicity assessment // DEAN J H, LUSTER M I, MUNSON A E, et al (Eds.). Immunotoxicology and immunopharmacology. Raven Press, New York: 99 – 111.

KRAMER V J, MILES – RICHARDSON S, PIERENS S L, et al., 1998. Reproductive impairment and induction of alkaline-labile phosphate, a biomarker of estrogen exposure, in fathead minnows (*Pimephales promelas*) exposed to waterborne 17β-estradiol. Aquat. Toxicol., 40: 335 – 360.

KULSHRESTHA S K, ARORA N, 1984. Impairments induced by sublethal doses of two pesticides in the ovaries of a fresh water teleost *Channa striatus* Bloch. Toxicol. Lett., 20: 93 – 98.

LA ROCCA C, CONTI L, CREBELLI R, et al., 1996. PAH content and mutagenicity of marine sediments from the Venice lagoon. Ecotoxicol. Environ. Saf., 33: 236 – 245.

LAL B, SINGH T P, 1987. The effect of malathion and BHC on the lipid metabolism in relation to reproduction in the tropical teleost, *Clarias batrachus*. Environ. Pollut., 48: 37 – 47.

LANDAHL J T, MCCAIN B B, MYERS M S, et al., 1990. Consistent associations between hepatic lesions in English sole (*Parophrys vetulus*) and polycyclic aromatic hydrocarbons in sediment. Environ. Health Perspect., 89: 195 – 203.

LANDER E S, LINTON L M, BIRREN B (International Human Genome Sequencing Consortium), et al., 2001. Initial sequencing and analysis of the human genome. Nature, 409

(6822): 860-921. https://doi.org/10.1038/35057062.

LANGSTON W J, BEBIANNO M J, BURT G R, 1998. Metal handling strategies in molluscs//BATLEY G E, LANGSTON W J, BEBIANNO M J (Eds.). Metal metabolism in aquatic environments. Chapman and Hall, London: 219-283.

LAWRENCE R A, BURK R F, 1976. Glutathione peroxidase activity in selenium-deficient rat liver. Biochem. Biophys. Res. Comm., 71: 952-958.

LECH J J, VODICNIK M J, 1985. Biotransformation//RAND G M, PETROCELLI S R (Eds.). Fundamentals of aquatic toxicology: methods and applications. Hemisphere Publishing Corporation, New York, USA: 526-557.

LEE R F, 1988. Gluthatione-S-transferase in marine invertebrates from Langsundford. Mar. Ecol. Progr. Ser, 46: 33-36.

LEE Y C, YANG D, 2002. Determination of lysozyme activities in a microplate format. Anal. Biochem., 310: 223-224.

LEE R F, BULSKI K, ADAMS J D, et al., 2013. DNA strand breaks (comet assay) in blood lymphocytes from wild bottlenose dolphins. Mar. Pollut. Bull., 77: 355-360.

LERCH K, AMMER D, OLAFSON R W, 1982. Crab metallothionein: primary structures of metallothionein 1 and 2. J. Biol. Chem., 257: 2420-2426.

LI H, PARISI M G, TOUBIANA M, et al., 2008. Lysozyme gene expression and hemocyte behaviour in the Mediterranean mussel, *Mytilus galloprovincialis*, after injection of various bacteria or temperature stresses. Fish Shellfish Immunol., 25: 143-152.

LILLEY K S, RAZZAQ A, DUPREE P, 2001. Two-dimensional gel electrophoresis: recent advances in sample preparation, detection and quantitation. Curr. Opin. Chem. Biol., 6: 46-50.

LIONETTO M G, CARICATO R, GIORDANO M E, et al., 2003. Integrated use of biomarkers (acetylcholinesterase and antioxidant enzymes activities) in *Mytilus galloprovincialis* and *Mullus barbatus* in an Italian coastal marine area. Mar. Pollut. Bull., 46 (3): 324-330.

LIONETTO M G, CARICATO R, GIORDANO M E, et al., 2004. Biomarker application for the study of chemical contamination risk on marine organisms in the Taranto marine coastal area. Chem. Ecol., 20: 333-343.

LIONETTO M G, CARICATO R, CALISI A, et al., 2013. Acetylcholinesterase as a biomarker in environmental and occupational medicine: new insights and future perspectives. Biomed. Res. Int., 2013: 321213.

LIU X, WANG W X, 2016. Time changes in biomarker responses in two species of oyster transplanted into a metal contaminated estuary. Sci. Total Environ., 544: 281-290.

LIU H, HEA J, ZHAO R, et al., 2015. A novel biomarker for marine environmental pollution of pi-class glutathione S-transferase from *Mytilus coruscus*. Ecotoxicol. Environ. Saf., 118: 47-54.

LIVINGSTONE D R, CHIPMAN J K, LOWE D M, et al., 2000. Development of biomarkers to detect the effects of organic pollution on aquatic invertebrates: recent molecular, genotoxic, cellular and immunological studies on the common mussel (*Mytilus edulis* L.) and other mytilids. Int. J. Environ. Pollut., 13 (1): 56-91.

LIVINGSTONE D R, 2001. Contaminated-stimulated reactive oxygen species production and oxidative damage in aquatic organisms. Mar. Pollut. Bull., 42: 656-666.

LOMBARDI P E, PERI S I, VERRENGIA GUERRERO N R, 2010. ALA-D and ALA-D re-activated as biomarkers of lead contamination in the fish *Prochilodus lineatus*. Ecotoxicol. Environ. Saf., 73: 1704-1711.

LOVERN S B, KLAPER R, 2006. Daphnia magna mortality when exposed to titanium dioxide and fullerene nanoparticles. Environ. Toxicol. Chem, 25 (4): 1132-1137.

LOWE D M, PIPE R K, 1994. Contaminant induced lysosomal membrane damage in marine mussel digestive cells: an in vitro study. Aquat. Toxicol., 30: 357-365.

MACCUBBIN A E, 1994. DNA adduct analysis in fish: laboratory and field studies // MALINS D C, OSTRANDER G K (Eds.). Aquatic toxicology: molecular, biochemical and cellular perspectives. Lewis Publishers, CRC Press: 267-294.

MACÍAS-MAYORGA D, LAIZ I, MORENO-GARRIDO I, et al., 2015. Is oxidative stress related to cadmium accumulation in the mollusc Crassostrea angulata? Aquat. Toxicol, 161: 231-241.

MACKAY E A, DUNBAR B, DAVIDSON I, et al., 1990. Polymorphism of cadmium-induced mussel metallothionein. Experentia, 46 (A36).

MACMAHON G, HUBER L J, MOORE M J, et al., 1990. Mutations in c-Ki-ras oncogenes in diseased livers of winter flounder from Boston harbor. Proc. Natl. Acad. Sci. USA, 87: 841-845.

MANNA G K, BANERJEE G, GUPTA S, 1985. Micronucleus test in the peripheral erythrocytes of the exotic fish. Nucleus, 23: 176-179.

MANNERVIK B, DANIELSON U H, 1988. Glutathione transferases structure and catalytic activity. CRC Crit. Rev. Biochem., 23: 283-337.

MARANHO L A, SEABRA PEREIRA C D, BRASIL CHOUERI R, et al., 2012. The application of biochemical responses to assess environmental quality of tropical estuaries: field surveys. J. Environ. Monit, 14: 2608-2615.

MARANHO L A, MOREIRA L B, BAENA-NOGUERAS R M, et al., 2015a. A candidate short-term toxicity test using *Ampelisca brevicornis* to assess sublethal responses to pharmaceuticals bound to marine sediments. Arch. Environ. Contam. Toxicol., 68: 237-258.

MARANHO L A, ANDRÉ C, DELVALLS T A, et al., 2015b. In situ evaluation of wastewater discharges and the bioavailability of contaminants to marine biota. Sci. Total Environ., 15: 876-887.

MARANHO L A, ANDRÉ C, DELVALLS T A, et al., 2015c. Toxicological evaluation of sediment samples spiked with human pharmaceuticals products: energy status and neuroendocrine effects in marine polychaetes *Hediste diversicolor*. Ecotoxicol. Environ. Saf., 118: 27-36.

MARIA V L, BEBIANNO M J, 2011. Antioxidant and lipid peroxidation responses in Mytilus galloprovincialis exposed to mixtures of benzo (a) pyrene and copper. Comp. Biochem. Physiol. C, 154 (1): 56-63.

MARIN M G, MATOZZO V, 2004. Vitellogenin induction as a biomarker of exposure to estrogenic compounds in aquatic environments. Mar. Pollut. Bull., 48: 835-839.

MARTÍN-DÍAZ M L, BLASCO J, SALES D, et al., 2004. Biomarkers as tools to assess sediment quality: laboratory and field surveys. TrAC-Trends Anal. Chem., 23 (10-11): 807-818.

MARTÍN-DÍAZ M L, BLASCO J, SALES D, et al., 2007. Biomarkers study for sediment quality assessment in Spanish ports using the crab Carcinus maenas and the clam Ruditapes philippinarum. Arch. Environ. Contam. Toxicol., 53: 66-76.

MARTÍN-DÍAZ M L, BLASCO J, SALES D, et al., 2008. Field validation of a battery of biomarkers to assess sediment quality in Spanish ports. Environ. Pollut., 151: 631-640.

MARTÍN-DÍAZ M L, FRANZELLITTI S, BURATTI S, et al., 2009. Effects of environmental concentrations of the antiepileptic drug carbamazepine on biomarkers and cAMP-mediated cell signaling in the mussel Mytilus galloprovincialis. Aquat. Toxicol., 94: 177-185.

MARTÍNEZ-GÓMEZ C, BENEDICTO J, CAMPILLO J A, et al., 2008. Application and evaluation of the neutral red retention (NRR) assay for lysosomal stability in mussel populations along the Iberian Mediterranean coast. J. Environ. Monit., 10: 490-499.

MARTYNIUK C J, ALVAREZ S, DENSLOW N D, 2012. DIGE and iTRAQ as biomarker discovery tools in aquatic toxicology. Ecotoxicol. Environ. Saf., 76: 3-10.

MATSUMURA H, ITO A, SAITOH H, et al., 2005. SuperSAGE. Cell. Microbiol., 7 (1): 11-18.

MATTHIESSEN P, GIBBS P E, 1998. Critical appraisal of the evidence for tributyltin-mediated endocrine disruption in mollusks. Environ. Toxicol. Chem., 17: 37-43.

MAYONE F, BRUNETTI R, FUMAGALLI O, et al., 1990. Induction of micronuclei by mitomycin C and colchicine in the marine mussel Mytilus galloprovincialis. Mutat. Res., 244 (2): 147-151.

MCCORD J M, FRIDOVICH I, 1976. Superoxide dismutase: an enzymatic function for erythrocuprein (hemiocuprein). J. Biol. Chem., 244: 6049-6055.

MCDONAGH B, SHEEHAN D, 2006. Redox proteomics in the blue mussel Mytilus edulis: carbonylation is not a prerequisite for ubiquitination in acute free radicalmediated oxidative stress. Aquat. Toxicol., 79 (4): 325-333.

MCDONAGH B, SHEEHAN D, 2008. Effects of oxidative stress on protein thiols and disulphides in Mytilus edulis revealed by proteomics: actin and protein disulphide isomerase are redox targets. Mar. Environ. Res., 66: 193-195.

MCFARLAND V A, INOUYE S L, LUTZ C H, et al., 1999. Biomarkers of oxidative stress and genotoxicity in livers of field collected brown bullhead, Ameiurus nebulosus. Arch. Environ. Contam. Toxicol., 37: 236-241.

METCALFE C D, 1988. Induction of micronuclei and nuclear abnormalities in the erythrocytes of mudminnows (*Umbra limi*) and brown bullheads (Ictalurusnebulosus). Bull. Environ. Contam. Toxicol., 40: 489-495.

MILAN M, COPPE A, REINHARDT R, et al., 2011. Transcriptome sequencing and microarray development for the Manila clam, *Ruditapes philippinarum*: genomic tools for environmental monitoring. BMC Genomics, 12: 234.

MILAN M, FERRARESSO S, CIOFI C, et al., 2013. Exploring the effects of seasonality and chemical pollution on the hepatopancreas transcriptome of the Manila clam. Mol. Ecol, 22 (8): 2157-2172.

MILAN M, PAULETTO M, BOFFO L, et al., 2015. Transcriptomic resources for environmental risk assessment: a case study in the Venice lagoon. Environ. Pollut., 197: 90-98.

MILLER J A, MILLER E C, 1979. Perspectives on the metabolism of chemical carcinogens//EMMELOT P, KRIEK E (Eds.). Environmental carcinogenes. Elsevier, Amsterdam: 25-100.

MILLER M G, 2007. Environmental metabolomics: a SWOT analysis (strengths, weaknesses, opportunities, and threats). J. Proteome Res., 6 (2): 540-545.

MORALES-CASELLES C, MARTÍN-DÍAZ M L, RIBA I, et al., 2008a. The role of biomarkers to assess oil-contaminated sediment quality using toxicity tests with clams and crabs. Environ. Toxicol. Chem., 27: 1309-1316.

MORALES-CASELLES C, MARTÍN-DÍAZ M L, RIBA I, et al., 2008b. Sublethal responses in caged organisms exposed to sediments affected by oil spills. Chemosphere, 72: 819-825.

MOREIRA S M, MOREIRA-SANTOS M, GUILHERMINO L, et al., 2006. An in situ post exposure feeding assay with *Carcinus maenas* for estuarine sediment-overlying water toxicity evaluations. Environ. Pollut., 139: 318-329.

MOUNEYRAC C, AMIARD J C, AMIARD-TRIQUET C, 1998. Effects of natural factors (salinity and body weight) on cadmium, copper, zinc and metallothionein-like protein levels in resident populations of oysters *Crassostrea gigas* from a polluted estuary. Mar. Ecol. Progr. Ser., 162: 125-135.

MULLIS K F, FALOONA F, SCHARF S, et al., 1986. Specific enzymatic amplification of DNA in vitro: the polymerase chain reaction. Cold Spring Harb. Symp. Quant. Biol., 51: 263-273.

MURUGESAN A G, HANIFFA M A, 1992. Histopathological and histochemical changes in the oocytes of the air-breathing fish *Heteropneus tesfossilis* (Bloch) exposed to textile-mill effluent. Bull. Environ. Contam. Toxicol., 48: 929-936.

MYERS M S, STEHR C M, OLSEN O P, et al., 1994. Relationships between toxicopathic hepatic lesions and exposure to chemical contaminants in English sole (*Pleuronectus vetulus*), starry flounder (*Platichthys stellatus*), and white croaker (*Genyonemus lineatus*) from selected marine sites on the Pacific coast, USA. Environ. Health Perspect, 102: 200-215.

NANDHAKUMAR S, PARASURAMAN S, SHANMUGAM M M, et al., 2011. Evaluation of DNA damage using single-cell gel electrophoresis (Comet Assay). J. Pharmacol. Pharmacother., 2: 107 – 111.

NAVANEETHAIYER U, KASTHURI S R, YOUNGDEUK L, et al., 2012. A novel molluscan sigma-like glutathione S-transferase from Manila clam, *Ruditapes philippinarum*: cloning, characterization and transcriptional profiling. Comp. Biochem. Physiol. C, 155: 539 – 550.

NELSON D R, 1998. Metazoan cytochrome P450 evolution. Comp. Biochem. Physiol. C Pharmacol. Toxicol. Endocrinol., 121 (1 – 3): 15 – 22.

NICHOLSON J K, LINDON J C, 2008. Systems biology: metabonomics. Nature, 455 (7216): 1054 – 1056.

NIKIMANN M, 2014. An introduction to aquatic toxicology. Academia Press, Oxford: 240.

OATEN J F P, HUDSON M D, JENSEN A C, et al., 2015. Effect of organism preparation in metallothionein and metal analysis in marine invertebrates for biomonitoring marine pollution. Sci. Total Environ., 518 – 519: 238 – 247.

OECD, 2004. Oecd guideline for the testing of chemicals. Draft proposal for a new guideline 487: in vitro micronucleus test, 487.

OECD, 2011. Report of the workshop on using mechanistic information in forming chemical categories. OECD environment, health and safety publications series on testing and assessment. No. 138. ENV/JM/MONO (2011) 8.

OECD, 2012. Workshop on using mechanistic information in forming chemical categories. Appendix I: collection of working definitions.

OLAFSON E W, OLSSON P E, 1991. Electrochemical detection of metallothionein // RIORDAN J F, VALLE B L (Eds.). Methods in enzymology. vol. 205. Academic Press Inc., San Diego: 205 – 213.

OLIVE P L, HILTON J, DURAND R E, 1986. DNA conformation of Chinese hamster V79 cells and sensitivity to ionizing radiation. Radiat. Res., 107: 115 – 124.

OLIVE P L, CHAN A P S, BRITISH C S C, 1988. Comparison between the DNA precipitation and alkali unwinding assays for detecting DNA strand breaks and cross-links. Can. Res., 48: 6444 – 6449.

OLIVER L M, FISHER W S, WINSTEAD J T, et al., 2001. Relationships between tissue contaminants and defense-related characteristics of oysters (*Crassostrea virginica*) from five Florida bays. Aquat. Toxicol., 55: 203 – 222.

ONOSAKA S, CHERIAN G, 1982. Comparison of metallothionein determination by polarographic and cadmium – saturation methods. Toxicol. Appl. Pharmacol., 63: 270 – 274.

ORBEA A, FAHIMI H D, CAJARAVILLE M P, 2000. Immunolocalization of four antioxidant enzymes in digestive glands of mollusks and crustaceans in fish liver. Histochem. Cell Biol., 114: 393 – 404.

ORTIZ-ZARRAGOITIA M, 2005. Effects of endocrine disruptors on peroxisome proliferation, reproduction and development of model organisms, zebrafish and mussel (Ph. D. thesis). University of the Basque Country: 164.

PATERAKI L E, STRATAKIS E, 1997. Characterization of vitellogenin and vitellin from land crab *Potamon potamios*: identification of a precursor polypeptide in the molecule. J. Exp. Zool., 279: 597 – 608.

PAYNE J F, MATHIEU A, MELVIN W, et al., 1996. Acetylcholinesterase, an old biomarker with a new future? Field trials in association with two urban rivers and a paper mill in Newfoundland. Mar. Pollut. Bull., 32: 225 – 231.

PÉREZ E, BLASCO J, MONTSERRAT S, 2004. Biomarker responses to pollution in two invertebrate species: *Scrobicularia plana* and *Nereis diversicolor* from the Cádiz bay (SW Spain). Mar. Environ. Res., 58: 275 – 279.

PETERSON R E, THEOBALD H M, KIMMEL G L, 1993. Developmental and reproductive toxicity of dioxin and related compounds: cross-species comparisons. Crit. Rev. Toxicol., 23: 283 – 335.

PFAFFL M W, HORGAN G W, DEMPFLE L, 2002. Relative expression software tool (REST) for group-wise comparison and statistical analysis of relative expression results in real-time PCR. Nucl. Acids Res., 30 – 36.

PFAFFL M W, 2001. A new mathematical model for relative quantification in real-time RT-PCR. Nucl. Acids Res., 29 (9): 45 – 53.

PFAFFL M W, 2004. Quantification strategies in real-time PCR // BUSTIN S A (Eds.). A-Z of Quantitative PCR, IUL Biotechnology Series. International University Line, La Jolla, CA: 87 – 120.

PICADO A, BEBIANNO M J, COSTA M H, et al., 2007. Biomarkers: a strategic tool in the assessment of environmental quality of coastal waters. Hydrobiologia, 587: 79 – 87.

PIPE R K, COLES J A, 1995. Environmental contaminants influencing immune function in marine bivalve molluscs. Fish Shellfish Immunol., 5: 581 – 595.

PIPE R K, 1992. Generation of reactive oxygen metabolites by the haemocytes of the mussel *Mytilus edulis*. Dev. Comp. Immunol., 16: 111 – 122.

POLLACK J R, PEROU C M, ALIZADEH A A, et al., 1999. Genome-wide analysis of DNA copy-number changes using cDNA microarrays. Nat. Gen., 23: 41 – 46.

PORTE C, JANER G, LORUSSO L C, et al., 2006. Endocrine disruptors in marine organisms: approaches and perspectives. Comp. Biochem. Phys. C, 143: 303 – 315.

RAMOS – GÓMEZ J, COZ A, VIGURI J R, et al., 2011. Biomarker responsiveness in different tissues of caged *Ruditapes philippinarum* and its use within an integrated sediment quality assessment. Environ. Pollut., 159: 1914 – 1922.

VAN RAVENZWAAY B, CUNHA G C P, LEIBOLD E, et al., 2007. The use of metabolomics for the discovery of new biomarkers of effect. Toxicol. Lett., 172 (1 – 2): 21 – 28.

REGOLI F, GIULIANI M E, 2014. Oxidative pathways of chemical toxicity and oxidative stress biomarkers in marine organisms. Mar. Environ. Res., 93: 106 – 117.

REGOLI F, PRINCIPATO G B, BERTOLI E, et al. , 1997. Biochemical characterization of the antioxidant system in the scallop *Adamussium colbecki*, a sentinel organism for monitoring the Antarctic environment. Polar Biol. , 17: 251 – 258.

REGOLI F, BOCCHETTI R, FILHO D W, 2012. Spectrophotometric assays of antioxidants // ABELE D, VÁZQUEZ-MEDINA J P, ZENTENO-SAVÍN T (Eds.). Oxidative stress in aquatic ecosystems. Wiley-Blackwell, Chichester: 367 – 380.

REGOLI F, 2012. Chemical pollutants and the mechanisms of reactive oxygen species generation in aquatic organisms // ABELE D, VÁZQUEZ-MEDINA J P, ZENTENO-SAVÍN T (Eds.). Oxidative stress in aquatic ecosystems. Wiley-Blackwell, Chichester: 308 – 316.

RELYEA R, HOVERMAN J, 2006. Assessing the ecology in ecotoxicology: a review and synthesis in freshwater systems. Ecol. Lett. , 9: 1157 – 1171.

REPETTO M, SEMPRINE J, BOVERIS A, 2012. Lipid peroxidation: chemical mechanism, biological implications and analytical determination // CATALA A (Eds). Biochemistry, genetics and molecular biology. CC BY 3.0 License: 1 – 30.

RHEE J S, KIM B M, KIM R O, et al. , 2013. Co-expression of antioxidant enzymes with expression of p53, DNA repair, and heat shock protein genes in the gamma ray-irradiated hermaphroditic fish Kryptolebias marmoratus larvae. Aquat. Toxicol, 140 – 141: 58 – 67.

ROBINSON C D, BROWN E, CRAFT J A, et al. , 2004. Effects of prolonged exposure to 4-tert-octylphenol on toxicity and indices of oestrogenic exposure in the sand goby (*Pomatoschistus minutus*, Pallas). Mar. Environ. Res. , 58: 19 – 38.

ROCHA J B T, SARAIVA R A, GARCIA S C, et al. , 2012. Aminolevulinate dehydratase (δ – ALA-D) as marker protein of intoxication with metals and other pro-oxidant situations. Toxicol. Res. , 1 (2), 85 – 102.

RODRIGUES - SILVA C, FLORES - NUNES F, VERNAL J I, et al. , 2015. Expression and immunohistochemical localization of the cytochrome P450 isoform 356A1 (CYP356A1) in oyster *Crassostrea gigas*. Aquat. Toxicol, 159: 267 – 275.

RODRIGUEZ-ARIZA A, ABRIL N, NAVAS J I, et al. , 1992. Metal mutagenicity and biochemical studies on bivalve mollusks from Spanish Coasts. Environ. Mol. Mutagen, 19: 112 – 124.

RODRÍGUEZ-ORTEGA M, GROSVIK B E, RODRÍGUEZ-ARIZA A, et al. , 2003. Change in protein expression profiles in bivalve molluscs (*Chamaelea gallina*) exposed to four model environmental pollutants. Proteomics, 3: 1535 – 1543.

ROESIJADI G, ROBINSON W E, 1994. Metal regulation in aquatic animals: mechanisms of uptake, accumulation and release // MALINS D C, OSTRANDER K G (Eds.). Aquatic toxicology: molecular, biochemical and cellular perspectives. CRC/Lewis Publisher, Boca Raton (FL): 387 – 420.

ROESIJADI G, KIELLAND S, KLERKS P, 1989. Purification and properties of novel molluscan metallothioneins. Arch. Biochem. Biophys. , 273 (2): 403 – 413.

ROESIJADI G, 1994. Behavior of metallothionein-bound metals in a natural population of an estuarine mollusc. Mar. Environ. Res. , 38 (3): 147 – 168.

ROESIJADI G, 1996. Metallothionein and its role in toxic metal regulation. Comp. Biochem. Physiol. C, 113 (2): 117 – 123.

ROMÉO M, COSSON R P, GNASSIA-BARELLI M, et al., 1997. Metallothionein determination in the liver of the sea bass *Dicentrarchus labrax* treated with copper and B (a) P. Mar. Environ. Res., 44 (3): 275 – 284.

ROMRO M, BENNANI N, GNASSIA-BARELLI M, et al., 2000. Cadmium and copper display different responses towards oxidative stress in the kidney of the sea bass. Aquat. Toxicol., 48: 185 – 194.

ROMERO-RUIZ A, CARRASCAL M, ALHAMA J, et al., 2006. Utility of proteomics to assess pollutant response of clams from the Doñana bank of Guadalquivir Estuary (SW Spain). Proteomics, 6 (1 Suppl.): 245 – 255.

ROMERO-RUIZ A, ALHAMA J, BLASCO J, et al., 2008. New metallothionein assay in *Scrobicularia plana*: heating effect and correlation with other biomarkers. Environ. Pollut., 156 (3): 1340 – 1347.

ROSS P L, HUANG Y N, MARCHESE J N, et al., 2004. Multiplexed protein quantitation in *Saccharomyces cerevisiae* using amine-reactive isobaric tagging reagents. Mol. Cell. Proteomics, 3: 1154 – 1169.

ROTCHEL J M, OSTRANDER G K, 2003. Molecular markers of endocrine disruption in aquatic organisms. J. Toxicol. Environ. Health B, 6: 453 – 495.

RUSSO R, BONAVENTURA R, ZITO F, et al., 2003. Stress to cadmium monitored by metallothionein gene induction in *Paracentrotus lividus* embryos. Cell Stress Chaperones, 8 (3): 232 – 241.

SAHA S, SPARKS A B, RAGO C, et al., 2002. Using the transcriptome to annotate the genome. Nat. Biotechnol., 20 (5): 508 – 512.

SCHEUHAMMER A M, CHARIAN M G, 1991. Quantification of metallothionein by silver saturation // RIORDAN J F, VALLE B L (Eds.). Methods in enzymology. vol. 205. Academic Press Inc., San Diego: 78 – 83.

SEELEY K R, WEEKS – PERKINS B A, 1991. Altered phagocytic activity of macrophages in oyster toadfish from a highly polluted subestuary. J. Aquat. Anim. Health, 3: 224 – 227.

SEGNER H, CASANOVA-NAKAYAMA A, KASE R, et al., 2013. Impact of environmental estrogens on fish considering the diversity of estrogen signaling. Gen. Comp. Endocrinol., 191: 190 – 201.

SEGNER H, 2011. Moving beyond a descriptive aquatic toxicology: the value of biological process and trait information. Aquat. Toxicol., 105: 50 – 55.

SEO K H, VALENTIN-BON I E, BRACKETT R E, et al., 2004. Rapid, specific detection of Salmonella Enteritidis in pooled eggs by real-time PCR. J. Food Prot., 67 (5): 864 – 869.

SERAFIM A, BEBIANNO M J, 2010. Effect of a polymetallic mixture on metal accumulation and metallothionein response in the clam *Ruditapes decussatus*. Aquat. Toxicol., 99 (3): 370 – 378.

SHEEHAN D, 2006. Detection of redox – based modification in two-dimensional electrophoresis proteomic separations. Biochem. Biophys. Res. Commun. , 349 (2): 455 – 462.

SHEPARD J L, OLSSON B, TEDENGREN M, et al. , 2000. Protein expression signature identified in *Mytilus edulis* exposed to PCBs, copper and salinity stress. Mar. Environ. Res. , 50: 337 – 340.

SHUGART L R, BICKHAM J, JACKIM G, et al. , 1992. DNA alterations // HUGGETT R J, KIMERLY R A, MEHRLE P M, et al. , (Eds.). Biomarkers: biochemical, physiological and histological markers of anthropogenic stress. Lewis Publishers, Chelsea, MI, USA: 155 – 210.

SHUGART L R, 1990. Biological monitoring: testing for genotoxicity // MCCARTHY J F, SHUGART L R (Eds.). Biomarkers of environmental contamination. Lewis Publishers, Boca Raton, FL, USA: 205 – 216.

SIJM D T H M, WEVER H, OPPERHUIZEN A, 1989. Influence of biotransformation on the accumulation of PCDDs from flyash in fish. Chemosphere, 19: 475 – 480.

SILVESTRE F, DIERICK J F, DUMONT V, et al. , 2006. Differential protein expression profiles in anterior gills of *Eriocheir sinensis* during acclimation to cadmium. Aquat. Toxicol. , 76: 46 – 58.

SINGH R B, SINGH T P, 1992. Modulatory actions of ovine luteinizing hormone-releasing hormone and *Mystus gonadotropin* on y-BHC-induced changes in lipid levels in the freshwater catfish, *Heteropneustes fossilis*. Ecotoxicol. Environ. Saf. , 24: 192 – 202.

SINGH I, 1996. Mammalian peroxisomes: metabolism of oxygen and reactive oxygen species. Ann. N. Y. Acad. Sci. , 804: 612 – 627.

SNYDER M, 2000. Cytochrome P450 enzymes in aquatic invertebrates: recent advances and future directions. Aquat. Toxicol. , 48: 529 – 547.

SOLÉ M, PORTE C, ALBAIGÉS J, 1995. Seasonal variation in the mixed function oxygenase system and antioxidant enzymes of the mussel *Mytilus galloprovincialis*. Environ. Toxicol. Chem. , 14: 157 – 164.

SOLÉ M, PORTE C, BIOSCA X, et al. , 1996. Effects of the Aegean Sea oil spill on biotransformation enzymes, oxidative stress and DNA-adducts in digestive gland of the mussel (*Mytilus edulis* L.). Comp. Biochem. Physiol. C, 113: 257 – 265.

SOLÉ M, KOPECKA-PILARCZYK J, BLASCO J, 2009. Pollution biomarkers in two estuarine invertebrates, *Nereis diversicolor* and *Scrobicularia plana*, from a Marsh ecosystem in SW Spain. Environ. Int. , 35 (3): 523 – 531.

SPIES R B, STEGEMAN J J, RICE D W, et al. , 1990. Sublethal responses of *Platichtus stellatus* to organic contamination in San Francisco Bay with emphasis on reproduction // MCCARTHY J F, SHUGART L R (Eds.). Biomarkers of environmental contamination. Lewis Publishers, Boca Raton, FL, USA: 87 – 122.

STEGEMAN J J, LECH J J, 1991. Cytochrome P450 monooxygenase systems in aquatic species: carcinogen metabolism and biomarkers for carcinogen and pollutant exposure. Environ. Health Perspect., 90: 101 – 109.

STEGEMAN J J, BROUWER M, RICHARD T D G, et al., 1992. Molecular responses to environmental contamination: enzyme and protein systems as indicators of chemical exposure and effect//HUGGETT R J, KIMERLY R A, MEHRLE P M, et al., (Eds.). Biomarkers: biochemical, physiological and histological markers of anthropogenic stress. Lewis Publishers, Chelsea, MI, USA: 235 – 335.

STURM A, WOGRAM J, SEGNER H, et al., 2000. Different sensitivity to organophosphates of acetylcholinesterase and butyrylcholinesterase from three-spined stickleback (*Gasterosteus aculeatus*): application in biomonitoring. Environ. Toxicol. Chem., 19: 1607 – 1615.

SUKUMAR A, KARPAGAGANAPATHY P R, 1992. Pesticideinduced atresia in ovary of a freshwater fish, *Colisa ialia* (Hamilton-Buchanan). Bull. Environ. Contam. Toxicol., 48: 457 – 462.

TONGE R, SHAW J, MIDDLETON B, et al., 2001. Validation and development of fluorescence twodimensional differential gel electrophoresis proteomics technology. Proteomics, 1: 377 – 396.

TOWBIN H, STAEHELIN T, GORDON J, 1979. Electrophoretic transfer of proteins from polyacrylamide gels to nitrocellulose sheets: procedure and some applications. Proc. Natl. Acad. Sci. U. S. A., 76 (9): 4350 – 4354.

TROMBINI C, FABBRI E, BLASCO J, 2010. Temporal variations in metallothionein concentration and subcellular distribution of metals in gills and digestive glands of the oyster *Crassostrea angulata*. Sci. Mar., 74 (1 Suppl.): 143 – 152.

TSARPALIAS V, DAILIANIS S, 2012. Investigation of landfill leachate toxic potency: an integrated approach with the use of stress indices in tissues of mussels. Aquat. Toxicol., 125: 58 – 65.

TUDOR M, 1984. Preliminary evaluation of α-aminolevulinic acid dehydratase in blood of lesser spotted dogfish (*Sciliorhinus canicula* L.) from the middle Adriatic. Institut ZA Oceanografijuiribarstvo-Splitsfr Jugosla Vij AN° 55v. Bil-jeske-Notes. 55.

TURNER J E, PARRY J M, 1989. The induction and repair of DNA damage in the mussel Mytilus edulis. Mar. Environ. Res, 28: 346 – 347.

UnitedNations, 2013. The millennium development goals report 2013.

VALLEE B L, COLEMAN J E, AULD D S, 1991. Zinc fingers, zinc clusters and zinc twists in DNA-binding protein domains. Proc. Natl. Acad. Sci. USA, 88: 999 – 1003.

VAN DER AAR E M, BUIKEMA D, COMMANDEUR J N M, et al., 1996. Enzyme kinetics and substrate selectivities of rat glutathione S-transferase isoenzymes towards a series of new 2 – substituted 1 – chloro-4-nitrobenzenes. Xenobiotica, 26: 143 – 155.

VAN DER OOST R, GOKSØYR A, CELANDER M, et al., 1996. Biomonitoring aquatic pollution with feral eel (*Anguilla anguilla*): Ⅱ. Biomarkers: pollutioninduced biochemical responses. Aquat. Toxicol., 36: 189 – 222.

VAN DER OOST R, BEYER J, VERMEULEN N P E, 2003. Fish bioaccumulation and biomarkers in environmental risk assessment: a review. Environ. Toxicol. Pharmacol., 13: 57 – 149.

VAN DER OOST R, PORTE-VISA C, VAN DEN BRINK N W, 2005. Biomarkers in environmental assessment // DEN BESTEN P J, MUNAWAR M (Eds.). Ecotoxicological testing of marine and freshwater ecosystems: emerging techniques, trends and strategies. Taylor & Francis, Boca Raton, FL: 87 – 152.

VAN DER VEN K, DEWIT M, KEIL D, et al., 2005. Development and application of a brain-specific cDNA microarray for effect evaluation of neuro-active pharmaceuticals in zebrafish (*Danio rerio*). Comp. Biochem. Physiol. B, 141: 408 – 417.

VAN DER VEN K, KEIL D, MOENS L N, et al., 2006. Effects of the antidepressant mianserin in zebrafish: molecular markers of endocrine disruption. Chemosphere, 65: 1836 – 1845.

VELCULESCU V E, ZHANG L, VOGELSTEIN B, et al., 1995. Serial analysis of gene expression. Science, 270 (5235): 484 – 487.

VENTER J C, ADAMS M D, MYERS E W, et al., 2001. The sequence of the human genome. Science, 291 (5507): 1304 – 1351.

VERDON C P, BURTON B A, PRIOR R L, 1995. Sample pretreatment with nitrate reductase and glucose-6 – phosphate dehydrogenase quantitatively reduces nitrate while avoiding interference by NADP + when the Griess reaction is used to assay for nitrite. Anal. Biochem., 224 (2): 502 – 508.

VERLECARA X, JENA K, CHAINY G, 2007. Biochemical markers of oxidative stress in Perna viridis exposed to mercury and temperature. Chem. Biol. Interact., 167: 219 – 226.

VERSLYCKE T, VANDENBERGH G F, VERSONNEN B, et al., 2002. Induction of vitellogenesis in 17 α-ethinylestradiol-exposed rainbow trout (*Oncorhynchus mykiss*): a method comparison. Comp. Biochem. Physiol. Part C, 132: 483 – 492.

VERSONNEN B J, GOEMANS G, BELPAIRE C, et al., 2004. Vitellogenin content in European eel (*Anguilla anguilla*) in Flanders, Belgium. Environ. Pollut., 128: 363 – 371.

VIARENGO A, CANESI L, 1991. Mussels as biological indicators of pollution. Aquaculture, 94: 225 – 243.

VIARENGO A, PONZANO E, DONDERO F, et al., 1997. A simple spectrophotometric method for metallothionein evaluation in marine organisms: an application to Mediterranean and Antarctic molluscs. Mar. Environ. Res., 44 (1): 69 – 84.

VIARENGO A, BURLANDO B, CAVALETTO M, et al., 1999. Role of metallothionein against oxidative stress in the mussel *Mytilus galloprovincialis*. Am. J. Physiol., 277: R1612 – R1619.

VIARENGO A, LOWE D, BOLOGNESI C, et al., 2007. The use of biomarkers in biomonitoring: A 2-tier approach assessing the level of pollutant-induced stress syndrome in sentinel organisms. Comp. Biochem. Physiol. C, 146: 281 – 300.

VICTOR B, MAHALINGAM S, SAROJINI R, 1986. Toxicity of mercury and cadmium on oocyte differentiation and vitellogenesis of the teleost, *Lepidocephalichthys thermalis*

(Bleeker). J. Environ. Biol., 7: 209-214.

VIOQUE-FERNÁNDEZ A, DE ALMEIDA E A, BALLESTEROS J, et al., 2007. Doñana National Park survey using crayfish (*Procambarus clarkii*) as bioindicator: esterase inhibition and pollutant levels. Toxicol. Lett., 168: 260-268.

VIOQUE-FERNÁNDEZ A, ALVES DE ALMEIDA E, LÓPEZ-BAREA J, 2009. Assessment of Doñana National Park contamination in *Procambarus clarkii*: integration of conventional biomarkers and proteomic approaches. Sci. Total Environ., 407: 1784-1797.

VOLLAND M, HAMPEL M, MARTOS-SITCHA J A, et al., 2015. Citrate gold nanoparticle exposure in the marine bivalve *Ruditapes philippinarum*: uptake, elimination and oxidative stress response. Environ. Sci. Pollut. Res., 23: 17414-17424.

VOS J, VAN LOVEREN H, WESTER P, et al., 1989. Toxic effects of environmental chemicals on the immune system. Trends Pharmacol. Sci., 10: 289-292.

WALKER C H, THOMPSON H M, 1991. Phylogenetic distribution of cholinesterases and related esterases // MINEAU P (Eds.). Cholinesterase-inhibiting insecticides, chemicals in agriculture. Elsevier, Amsterdam: 1-17.

WANG Z, ZHAO J, LI F, et al., 2009a. Adsorption and inhibition of acetylcholinesterase by different nanoparticles. Chemosphere, 77: 67-73.

WANG Z, GERSTEIN M, SNYDER M, 2009b. RNA-Seq: a revolutionary tool for transcriptomics. Nat. Rev. Genet., 10 (1): 57-63.

WARHEIT D B, BORM P J A, HENNES C, et al., 2007. Testing strategies to establish the safety of nanomaterials: conclusions of an ECETOC workshop. Inhal. Toxicol, 19: 631-643.

WATSON J D, CRICK F H, 1953. Molecular structure of nucleic acids: a structure for deoxyribose nucleic acid. Nature, 171 (4356): 737-738.

WEEKS B A, WARINNER J E, RICE C D, 1989. Recent advances in the assessment of environmentally-induced immunomodulation // OCEANS'89 proceedings. vol. 2. Institute of Electrical and Electronics Engineers, New York: 408-441.

WEEKS B A, WARINNER J E, MATHEWS E S, et al., 1990a. Effects of toxicants on certain functions of the lymphoreticular system of fish // PERKINS F O, CHENG T C (Eds.). Pathology in marine science. Academic Press, San Diego: 369-374.

WEEKS B A, HUGGETT R J, WARINNER J E, et al., 1990b. Macrophage responses of estuarine fish as bioindicators of toxic contamination // MCCARTHY J F, SHUGART L R (Eds.). Biomarkers of environmental contamination. Lewis Publishers, CRC Press, Boca Raton: 193-220.

WENK M R, 2005. The emerging field of lipidomics. Nat. Rev. Drug Discov., 4 (7): 594-610.

WESTER P W, VETHAAK D, VAN MUISWINKEL W B, 1994. Fish as biomarkers in immunotoxicology. Toxicology, 86: 213-232.

WHITE R, JOBLING S, HOARE S A, et al., 1994. Environmentally persistent alkylphenolic compounds are estrogenic. Endo, 36: 175-182.

WHYTE J J, JUNG R E, SCHMITT C J, et al., 2000. Ethoxyresorufin-O deethylase (EROD) activity in fish as a biomarker of chemical exposure. Crit. Rev. Toxicol., 30: 347-570.

WILLS E D, 1987. Evaluation of lipid peroxidation in lipids and biological membranes // SNELL K, MULLOC B (Eds.). Biochemical toxicology: a practical approach. IRL Press, USA: 127-150.

WINSTON G W, DI GIULIO R T, 1991. Prooxidant and antioxidant mechanisms in aquatic organisms. Aquat. Toxicol., 19: 137-161.

WIRGIN I, CURRIE D, GARTE S J, 1989. Activation of the K-ras oncogene in liver tumors of Hudson River tomcod. Carcinogenesis, 10: 2311-2315.

WON F J, RHEE J S, SHIN K H, et al., 2013. Expression of three novel cytochrome P450 (CYP) and antioxidative genes from the polychaete, *Perinereis nuntia* exposed to water accommodated fraction (WAF) of Iranian crude oil and Benzo [α]. Mar. Environ. Res, 90: 75-84.

World Health Organization/International Programme on ChemicalSafety, 2002. Global assessment of the state-of-the-science of endocrine disruptors. // DAMSTRA T, BARLOW S, BERGMAN A, et al (Eds.). World health organization (WHO/PCS/EDC/02.2). World Health Organization, Geneva, Switzerland.

YADETIE F, BUTCHER S, FORDE H E, et al., 2012. Conservation and divergence of chemical defense system in the tunicate *Oikopleura dioica* revealed by genome wide responses to two xenobiotics. BMC Genomics, 13: 55-63.

YEVICH P P, YEVICH C, PESCH G, 1987. Effects of black rock harbor dredged material on the histopathology of the blue Mussel *Mytilus edulis* and polychaete worm *Nephtys incisa* after laboratory and field exposures. US EPA Technical Report D-87-8, Narragansett, RI.

YIN H, XU L, PORTER N A, 2011. Free radical lipid peroxidation: mechanisms and analysis. Chem. Rev., 111: 5944-5972.

ZANETTE J, GOLDSTONE J V, BAINY A C D, et al., 2010. Identification of CYP genes in *Mytilus* (mussel) and *Crassostrea* (oyster) species: first approach to the full complement of cytochrome P450 genes in bivalves. Mar. Environ. Res., 69: 1-3.

6 海水毒性试验

B. Anderson[①], **B. Phillips**[①]

---

① 美国加州大学戴维斯分校。

6.1 引　　言

已有多种方法被科学家用于水生资源的生态完整性评价，包括对水质、沉积物和组织的化学分析、生物学评价和毒性试验。毒性试验是评价化学物质对水生生态系统影响的重要组成部分，因为它揭示了复杂化合混合物的毒性效应。水生毒性试验将选定的几组生物暴露于规定条件下的试验材料（这里以海水为样本）中，以确定潜在的有害效应。试验在受控的实验室条件下或现场进行。目前已制定了若干标准化毒性试验规程来确定化学物质对水生物种的毒性。海洋毒性试验的详细指导手册可从美国环境保护署（U. S. EPA）、美国材料与试验学会（ASTM）及美国公共卫生协会（APHA）等其他单位获得。这些规程作为毒性试验的应用指南，指导单一化学品、复杂的废水和环境水样的毒性评价。

本章旨在对可用于海洋和河口水质评价的各种标准化水生毒性试验方案进行概述，以及使读者熟悉生态毒理学家用于环境评估的一种工具，但并不会全面综述所有海洋水生试验方法。尽管目前已报道了多种非标准化的毒性试验方法，但本章重点将放在U. S. EPA、ASTM 和其他单位提供的标准化协议，因为这些在规范化应用中最为常用。这些毒性试验大多数是根据美国和国际环境法规制定。在美国，此类法规包括《联邦杀虫剂、灭真菌剂和灭鼠剂法》（FIFRA）、《清洁水法》（CWA）、《海洋保护、研究和保护区法》（MPRSA）和《有毒物质与防治法》（TSCA）（Ward，1995）。本章重点介绍美国大西洋沿岸、墨西哥湾地区及太平洋沿岸有关的物种和相关协议，在加拿大水域使用的方法和在过渡海岸环境中毒性评价的应用，以及几种海水毒性试验在现场评价中的应用实例。此外，本章还讨论南美和欧洲的标准化协议，并简要介绍基于多种热带物种开发的方法。最后总结未来开展和应用海洋生态毒理学毒性试验研究的需求。

6.2 专业术语

毒性试验依照试验持续时间、生命阶段和终点分类。急性、短期试验通常是经48 或96 h 暴露后测定死亡率，以计算半致死浓度（LC_{50}），即暴露试验群体50% 死亡的浓度。慢性毒性试验周期更长，并纳入了亚致死终点。使用物种的生命阶段因规范化应用而异。在美国 EPA 急性毒性试验方案中，启动的生命阶段与短期慢性毒性试验相同（U. S. EPA，2002a）。这些试验包括鱼类和无脊椎动物的早期生命阶段试验（Goodma et al.，1985；ASTM，2008b）。长期慢性试验可以进行 28 d 或更久，旨在通过仔鱼生长和幼鱼发育来检测胚胎发育过程。此外，对无脊椎动物（如糠虾）也会进行全生命周期试验，检测早期发育、生长和生殖的相关指标（ASTM，2008b）。为了响应对废水慢性毒性筛选的立法要求，会对鱼类、无脊椎动物和大型藻类进行简要的慢性试验。

这些试验已被开发成慢性毒性的短期指标，并纳入亚致死终点，如对大型藻类的萌发和增殖、海胆属动物卵受精、海胆属及软体动物胚胎发育以及鱼类和糠虾幼体存活和生长的测定。这些试验的周期从 1 h 到 7 d（U. S. EPA，1995，2002b）。在各种标准化的急性和慢性毒性试验方案中所使用的物种、程序和终点如下所述。

6.3 一般海水毒性试验方法及规程

表 6.1 至表 6.3 总结了海水毒性试验的标准化方法。该类方法包括 ASTM、APHA/AWWA/WPFC、U. S. EPA 和加拿大环境部的标准化方法。

表 6.1 海水毒性试验指南

标准方法，第 8000 部分（APHA，2012）	
8111	生物刺激（藻类生产力）
8112	浮游植物
8113	海洋大型海藻
8310	纤毛虫类原生动物
8410	造礁石珊瑚
8610	软体动物
8710	节肢动物
8810	棘皮动物
8910	鱼类
ASTM （ASTM，2007，2008a，b，2012a-e，2013）	
E-1191-90	糠虾生命周期毒性试验指南
E-729-88a	鱼类、大型无脊椎动物和两栖动物急性毒性试验指南
E-1241-92	鱼类早期生命周期毒性试验指南
E-1440-91	轮虫（臂尾轮虫属 *Brachionus* 以及河口和海洋轮虫）急性毒性试验指南
E-724-89	四种海洋双壳类胚胎开始的静态急性毒性试验指南
E-1463-92	美国西海岸的糠虾静态和通流急性毒性试验的指南
E-1191-88	废水对鱼类、大型无脊椎动物和两栖动物的急性毒性试验指南
E-1498-92	海藻有性生殖试验指南

续表 6.1

E-1218 – 90	微藻类 96 h 静态毒性试验指南
EPA 急性污水试验（U. S. EPA，2002a）	
	拟糠虾 *Americamysis* (*Mysidopsis*) *bahia* (*Holmesimysis costata* = 西海岸替代种) 急性毒性试验
	安芬拟银汉鱼 *Menidia* sp. (*Atherinops affinis* = 西海岸替代种) 急性毒性试验
	杂色鳉 *Cyprinodon variegatus* 急性毒性试验
EPA 慢性污水试验（U. S. EPA，1995，2002b）	
	杂色鳉（*Cyprinodon variegatus*）仔鱼存活和生长试验方法
	杂色鳉（*Cyprinodon variegatus*）胚胎 – 仔鱼存活率及致畸试验方法
	安芬拟银汉鱼（*Atherinops affinis*）仔鱼存活和生长试验方法
	美洲银汉鱼（*Menidia beryllina*）仔鱼存活和生长试验方法
	拟糠虾（*Americamysis* (*Mysidopsis*) *bahia*）仔鱼、生长和生殖试验方法
	糠虾（*Holmesimysis costata*）存活和生长试验方法[b]
	马粪海胆（*Arabacia punctulata*）受精试验方法
	紫球海胆（*Strongylocentrotus purpuratus*）[b]或沙钱（*Dendraster excentricus*）受精试验方法
	紫球海胆（*Strongylocentrotus purpuratus*）[b]胚胎 – 幼体发育试验方法
	地中海贻贝（*Mytilus galloprovincialis*）[b]或三倍体长牡蛎（*Crassostrea gigas*）胚胎 – 幼体发育试验方法
	红鲍螺（*Haliotis rufescens*）胚胎 – 幼体发育试验方法[b]
	环节藻（*Champia parvula*）有性生殖试验方法
	巨藻（*Macrocystis pyrifera*）配子体萌发和生长试验方法
EPATSCA 试验指南（CODE OF FEDERAL REGULATIONS，1990）	
797.1050	藻类急性毒性试验
797.1075	淡水和海洋藻类急性毒性试验
797.140	鱼类急性毒性试验
797.1440	鱼类急性毒性试验
797.1600	鱼类早期生命阶段毒性试验
797.1800	牡蛎急性毒性试验
797.1930	糠虾急性毒性试验

续表 6.1

797.1950	糠虾慢性毒性试验
797.1970	对虾急性毒性试验
EPA-FIFRA 试验指南（U.S. EPA，1985a-d，1986a-c）	
	SEP[a]：河口和海洋生物的急性毒性试验（河口鱼类 96 h 急性毒性）（EPA-540/9 – 85 – 009，1985）
	SEP：河口和海洋生物急性毒性试验（虾类 96 h 急性毒性）（EPA-540/9 – 85 – 010，1985）
	SEP：河口和海洋生物急性毒性试验（软体动物 96 h 通流外壳沉积研究）（EPA-540/9 – 85 – 011，1985）
	SEP：河口和海洋生物的急性毒性试验（软体动物 48 h 胚胎 – 幼体研究）（EPA-540/9 – 85 – 012，1985）
	SEP：鱼类早期生命阶段（EPA-540/9 – 86 – 138，1986）
	SEP：鱼类生命周期毒性试验（EPA-540/9 – 86 – 137，1986）
	SEP：非靶标植物：1 层和 2 层水生植物的生长和繁殖（EPA-540/9 – 86 – 134，1986）
加拿大环境部（ENVIRONMENT CANADA，2011）	
	海胆属受精试验（海胆和海钱）（2011 年）

[a] SEP = Standard Evaluation Procedure，标准评估程序。
[b] 东太平洋试验物种。

表 6.2　海水毒性试验常用试验种

脊椎动物

杂色鳉 *Cyprinodon variegatus*（大西洋和墨西哥湾沿岸地区）

底鳉 *Fundulus heteroclitus*

长吻底鳉 *Fundulus similis*

银汉鱼科 *Menidia* sp.（大西洋和墨西哥湾沿岸地区）

安芬拟银汉鱼 *Atherinops affinis*（太平洋东部和加拿大地区）

三刺鱼 *Gasterosteus aculeatus*

菱体兔牙鲷 *Lagodon rhomboides*

黄尾平口石首鱼 *Leiostomus xanthurus*

海鲫 *Cymatogaster aggregate*

寡鳞杜父鱼 *Oligocottus maculosus*

眼点副棘鲆 *Citharichthys stigmaeus*

续表 6.2

脊椎动物
大西洋牙鲆 *Paralichthys dentatus*，*P. lethostigma*
星斑川鲽 *Platichthys stellatus*
副眉鲽 *Parophrys vetulus*
大西洋鲱 *Clupea harengus*
塔斯梅尼亚黏鱼 *Parablennius tasmanians*[b]

无脊椎动物
桡足类，克氏纺锤水蚤属 *Acartia clausi*，汤氏纺锤水蚤 *Acartia tonsa*
虾，白对虾 *Penaeus setiferus*，桃红对虾 *P. duorarum*，棕虾 *P. aztecus*
草虾，短刀小长臂虾 *Palaemonetes pugio*，中长臂虾 *P. intermedius*，小长臂虾 *P. vulgaris*
沙虾，褐虾 *Crangon* sp.
虾，齐氏长额虾 *Pandalus jordani*，长额虾 *P. danae*
湾虾，黑尾湾虾 *Crangon nigricauda*
糠虾，Mysid，*Americamysis*（*Mysidopsis*）*bahia*，*M. bigelowi*，*M. intii*，*M. almyra*，*Holmesimysis costata*，*M. juniae*，*Mysidium gracile*[a]
蓝蟹 *Callinectes sapidus*
食草蟹 *Hemigrapsus* sp.，*Pachygrapsus* sp.
普通滨蟹 *Carcinus maenas*
招潮蟹属 *Uca* sp.
牡蛎 *Crassostrea virginica*，*C. gigas*（东太平洋、大西洋以及墨西哥湾沿岸地区）
蛤，硬壳蛤 *Mercenaria mercenaria*，侏儒蛤 *Mulinia lateralis*（大西洋和墨西哥湾沿岸地区）
红鲍螺 *Haliotis rufescens*（东太平洋地区）
贻贝，紫贻贝 *M. edulis*，地中海贻贝 *M. galloprovincialis*，加州贻贝 *M. californianus*（东太平洋及加拿大地区）
海胆，紫海胆 *Strongylocentrotus purpuratus*（东太平洋及加拿大地区），绿海胆 *S. droebachiensis*，*Arabacia punctulata*（大西洋及墨西哥湾沿岸地区），*Lytechinus* sp.，长海胆 *Echinometra lucunter*[a]，红海胆 *Heliocidaris tuberculata*[b]
海钱 *Dendraster excentricus*（东太平洋及加拿大地区）

a 巴西试验物种。
b 澳大利亚温带水域试验物种。
After Ward, G. S., 1995. Saltwater tests. In: Rand, G. M. (Ed.), Fundamentals of Aquatic Toxicology: Effects, Environmental Fate, and Risk Assessment, second ed. Taylor and Francis, Washington, DC.

表 6.3　热带水域海水毒性试验试验物种

无脊椎动物
海胆，白棘三列海胆 Tripneustes gratilla[a]
桡足类，热带纺锤水蚤 Acartia sinjiensis[b]，Glandioferans imparipes[b]
珊瑚，多孔鹿角珊瑚 Acropora millepora[b]，美丽轴孔珊瑚 A. formosa[b]，指状蔷薇珊瑚 Montipora digitata[b]，细柱滨珊瑚 Porites cylindrica[b]，尖枝列孔珊瑚 Seriatopora hystrix[b]
对虾 Prawn，Penaeus monodon[b]

a 夏威夷水域试验物种。
b 澳大利亚热带水域试验物种。

6.3.1　鱼类

在大西洋和墨西哥湾沿岸海域，杂色鳉（Cyprinodon variegates）和银汉鱼属（Menidia sp.）最常作为废水和环境监测的对象。大多数都是使用商业型供应商提供的杂色鳉（C. variegates）及美洲银汉鱼（Menidia beryllina）仔鱼或饲养鱼。目前已经针对这两种物种开发出标准化的急性和慢性毒性试验规程（U.S. EPA，2002a，b）。US. EPA（2002a）中描述了杂色鳉（C. variegates）、大西洋银汉鱼（Menidia menidia）、美洲原银汉鱼（M. beryllina）及潮间美洲原银汉鱼（Menidia peninsulae）急性毒性试验规程。这些物种的急性试验持续时长为 24～96 h，在每个容器内放置 20 条仔鱼（至少 2 组重复）。这些物种为广盐性，可分别在 5‰～32‰（杂色鳉 C. variegates）和 1‰～32‰（美洲原银汉鱼 M. peninsulae）盐度范围内试验。试验可以在静态、半静态或流动条件下的 1 L 或 2 L 试验容器中进行。对这些物种短期慢性毒性试验规程的相关描述可见 US. EPA（2002b）。试验暴露 7 天后评估杂色鳉（C. variegates）或美洲原银汉鱼（M. beryllina）仔鱼生长和存活。杂色鳉 C. variegatus 短期慢性测试是一项为期 7 天的半静态试验，该规程要求试验必须选用刚孵化出的幼体（<24 h 龄），幼体可由商业型供应商运送或室内培养孵化而来。试验开始之前，首先将胚胎在 25 ℃（30‰盐度）条件下孵育 5～6 天。孵化过程中，把先孵化的仔鱼分隔开。大多数在随后 24 h 内孵化的仔鱼可用于试验；若仔鱼数量无法满足实验设计，可用较早期先孵化的仔鱼来补齐。试验在 600 mL 或 1 L 的烧杯中进行，每日换水。每个烧杯中放置 10 条幼鱼，每个实验浓度重复 4 组。每天投喂仔鱼刚孵出的卤虫。7 天后，计算仔鱼存活率以及测定其生长（以干重计）。美洲原银汉鱼（M. beryllina）的慢性试验流程与杂色鳉相似，不同之处在于美洲银汉鱼应选取 7～11 天龄鱼，处于该龄的仔鱼能够食用卤虫无节幼体。

除了使用仔鱼进行试验外，使用杂色鳉（C. variegatus）和拟银汉鱼科（Menidia sp）胚胎的幼体发育试验及早期生命阶段试验也都已被开发。与 7 天仔鱼试验相同，利用杂色鳉（C. variegatus）进行的 9 天胚胎幼体试验从 24 h 龄以下的胚胎开始，室内培养或由商业型供应商提供胚胎来源。该试验中每组 15 个胚胎（共 4 个重复组），在烧杯、结晶皿或组织培养皿中进行暴露实验。通过显微镜观察透明胚胎记录每日胚胎发育情况。在 25 ℃ 环境，胚胎 5～6 天开始孵化，9 天后试验停止。检测终点包括各种形态畸形、游泳行为和孵化成功率。试验中需每天换水，但不需要进行投喂 (U. S. EPA, 2002b)。

Rand 和 Petrocelli (1985) 对杂色鳉 (*C. variegatus*) 早期生命阶段 (28 天) 和全生命周期 (孵化后 120 天) 慢性试验进行了描述，Ward (1995) 对其进行总结。试验开始时每组设置 40～50 个胚胎 (至少有两个重复组)，试验均在通流条件下进行。每日监测胚胎发育。胚胎一旦孵化出仔鱼，每天喂食 3 次以上的卤虫无节幼体 (*Artemia nauplii*)。在早期生命阶段试验中，对鱼苗的存活和发育进行监测，28 天后试验终止，记录其存活率和生长情况 (测量长度和干重)。在全生命周期试验中，鱼苗生长 28～30 天后进行照相测量。然后将鱼苗的密度减至每组 25 条，继续暴露于试验液中至性成熟 (孵化后 3～4 个月)。一旦鱼到达性成熟阶段，开始监测 3 个为期 2 周的产卵期的产卵率。不同产卵期使用不同的产卵组，包括 3 条雌性和 2 条雄性。鱼被置于有筛网的产卵室中。这样的设计是为了使卵被雌鱼排出且经雄鱼授精后穿过大筛孔落到更细的筛网上，以防止成鱼吞食鱼卵。在为期 2 周的产卵期内，每日都要收集并统计鱼卵数量。此外，收集 3 个产卵期中的 1 个产卵期鱼卵，置于孵化室中，以评估胚胎幼体发育和孵化成功率。Ward (1995) 对杂色鳉 (*C. variegatus*) 生命周期试验中可能遇到的困难进行了讨论，包括与载体溶剂有关的细菌生长，以及从成鱼转移到鱼苗的病原体感染等问题。

安芬拟银汉鱼 (*Atherinops affinis*) 是美国西海岸一种常见于海湾和河口的鱼类，同时也是加利福尼亚河口最重要的生态鱼类之一，代表着这些系统最大的鱼生物量。成鱼由一个商业实验室进行饲养，为毒性试验提供胚胎和仔鱼，应用于加利福尼亚州、华盛顿州和加拿大的污水和环境监测。U. S. EPA (1995) 中对安芬拟银汉鱼 7 天仔鱼生长和生存试验进行了描述，其余相关研究也被 Middaugh 和 Anderson (1993)、McNulty et al. (1994)、Anderson et al. (1995) 做了总结。

安芬拟银汉鱼 7 天仔鱼生长和存活试验规程类似于美洲银汉鱼 (*M. beryllina*) 以及其他银汉鱼科物种的试验规程，被设计用于在美国西海岸试验中替代美洲银汉鱼 (*M. beryllina*) 的试验规程。除使用安芬拟银汉鱼 (*A. affinis*) 进行的仔鱼生长和存活规程外，Anderson et al. (1991) 开发了一个 12 天胚胎幼体发育试验。这种检测方法不适用于日常污水检测，但可用于有高危致畸物的污染水域监测。在一项对镍水质量标准的研究中，Hunt et al. (2002) 还使用安芬拟银汉鱼 (*A. affinis*) 进行了 40 天早期生命阶段毒性试验。在 20 ℃ 和 34‰ 通流条件下，试验开始时将 30 个处于早期原肠胚阶段的胚胎暴露于含镍溶液的 10 L 养鱼池。随后检测胚胎发育和第 12 天的仔鱼孵化成功率。再持续暴露 28 天后，在第 40 天对存活的仔鱼进行计数，同时测量其长度和干重。

使用安芬拟银汉鱼 (*A. affinis*) 试验的优点之一是该物种具广盐性。目前已制定了安芬拟银汉鱼和其他银汉鱼科幼鱼的毒性鉴定评估 (TIE) 规程 (U. S. EPA, 1996)。虽然安芬拟银汉鱼表现出与其他银汉鱼科物种类似或更高的敏感性 (Middaugh & Anderson, 1993)，但它们缺乏对其他鱼类和无脊椎动物的敏感性，因此在水质评估中的应用可能会受到限制。Rose et al.（2005）研究提出，与标准规程中建议的仔鱼体重终点指标相比，耳石的生长是安芬拟银汉鱼生长中一个更为敏感的指标。安芬拟银汉鱼幼鱼可在河口环境下作为指示非离子氨毒性的指标（Phillips et al.，2005）。辅助数据表明，安芬拟银汉鱼仔鱼对低溶氧环境尤为敏感（Middaugh D，私人通信）。尽管目前还没有安芬拟银汉鱼的相关研究，但通过其他银汉鱼科物种和杂色鳉（*C. variegatus*）的研究可知，仔鱼对离子失衡很敏感，因而可能会干扰这些物种试验的结果（Pillard et al.，2000）。Anderson et al.（1995）发现，虽然安芬拟银汉鱼的胚胎和仔鱼能够耐受5‰～35‰的盐度，但实验数据表明受渗透胁迫影响，在低盐度（17‰）条件下的仔鱼可能对污染物更敏感。

海洋毒理学研究中常用的另外两个海洋物种分别为底鳉（*Fundulus heteroclitus*；例如，Weis & Weis，1982）和鲱鱼（*Clupea* sp.）。虽然这两物种都被广泛地进行研究，但均未用于国家污染排放消除系统（National Pollution Discharge Elimination System，NPDES）的监测。Dinnel et al.（2005）用太平洋鲱鱼（*Clupea pallasi*）开发了3种不同规程，用于华盛顿州的全流出物毒性测试。一种规程是为期18天的胚胎发育试验，其中包括亚致死终点，如心率、胚胎运动、50%孵化时间、胚胎存活率以及孵化时仔鱼长度。仔鱼试验则纳入10天暴露后的生存和生长终点。在所有试验中，所有的规程都使用野外采集的鲱鱼胚胎或野生捕获鲱鱼中收集的配子，或从这些来源经发育而来的仔鱼。

太平洋鲱鱼的胚胎和仔鱼也被用于评估石油和石油分散剂的毒性效应。这些实验包括实验室标准和生物标记终点（Carls et al.，1999），以及原位暴露（Kocan 1996；Kocan et al.，1996）。其他鱼类胚胎的其他生物标记终点也已被报告（Nacci et al.，1998）。人们愈发认识到鱼类胚胎发育对多核环芳香烃（例如来自石油泄漏）的敏感性。大量研究表明鱼胚胎暴露于碳氢化合物时，蓝囊综合症是一种常见发育反应（Barron et al.，2004；Incardona et al.，2004，2005；Hicken et al.，2011）。Incardona et al.（2005年）发现风化的阿拉斯加北坡（ANS）原油中的三环芳烃（PAH）是影响斑马鱼胚胎发育的主要毒性成分。他们的数据表明菲和二苯并噻吩是风化石油中特别重要的有毒成分。此外，他们还发现心脏形态形成过程中的心功能受损是风化的ANS原油最早和最显著效应。近期一篇文章综述了三环芳烃对从亚热带到亚北极栖息地的多种物种的影响（Scholz & Incardona，2015），结果表明心脏综合征的特征在各物种之间是一致的。除了因石油泄漏而风化的原油外，三环芳烃的来源还包括船只燃油泄漏、木馏油和城市径流。这些也被证明会造成同样的效应。在引起关注的多环芳烃污染环境中，海洋鱼类胚胎发育是一个特别重要的观察终点（图6.1）。

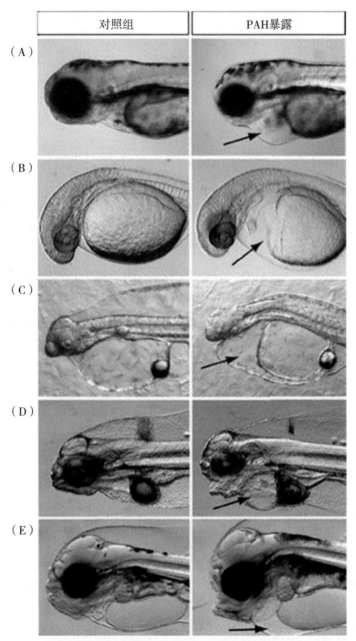

图 6.1 从世界各地不同地质来源的原油中提取的多环芳烃（PAHs）
在多种鱼类中引起一致的心力衰竭损伤表型

所有图像均为受精后不久开始暴露在原油中的孵化期仔鱼（A，C，D，E）或胚胎晚期（B），左栏为对照组，右栏为原油暴露鱼。箭头表示心包积液或水肿，作为循环衰竭的标志。（A）斑马鱼（*Danio rerio*）暴露于路易斯安那（美国）低硫原油；（B）太平洋鲱鱼（*Clupea pallasi*）暴露于阿拉斯加北坡（美国）原油；（C）牙鲆（*Paralichthys olivaceus*）暴露于伊朗重质原油；（D）黄鳍金枪鱼（*Thunnus albacares*）暴露于路易斯安那（美国）低硫原油；（E）大西洋黑线鳕（*Melanogrammus aeglefinus*）暴露于挪威海原油。

6.3.2 软体动物

利用腹足类和双壳类软体动物进行胚胎-幼体发育试验已经在水和沉积物的质量评估中使用了几十年。这些试验也纳入了代表生态和经济上重要的海洋和河口物种关键生命阶段敏感的、亚致死终点，对毒性监测特别有效，因为需要相对较短的暴露时间（≤ 48 h）。用于大西洋和墨西哥湾海岸试验的双壳类物种包括美洲牡蛎（*C. virginica*）、硬壳蛤（*Mercenaria mercenaria*）（ASTM，2012c）和侏儒蛤（*Mulinia lateralis*）（Morrison & Petrocelli，1990）。学者还为太平洋沿岸几种软体动物开发了试验规程，包括红鲍螺（*Haliotis rufescens*，一种在太平洋海岸的海底岩底环境中发现的腹足软体动物，图6.2）、河口贻贝类地中海贻贝（*Mytilus galloprovincialis*）和紫贻贝（*Mytilus edulis*）（U. S. EPA，1995），以及加州壳菜蛤 *Mytilus californianus*（Cherr et al.，1990）。这些物种均在其栖息地中是重要的被捕食者，可以形成一个大的群落为其他生物提供栖息地。加州壳菜蛤（*M. californianus*）和红鲍螺（*H. rufescens*）还被视为人类食物，被用于商业性或个人捕捞。海洋软体动物毒性试验的相关综述见 Hunt & Anderson（1993）。

这些品种的成体可作胚胎-幼体毒性试验的亲鱼。亲鱼可从野生种群中收集，但一般由能培养可靠繁殖个体的商业型供应商提供。通过诱导雄性和雌性亲鱼释放配子（产卵），并结合卵子和精子产生用于试验的胚胎而开始胚胎幼体试验。不同物种的产卵方式有所不同：红鲍螺（*H. rufescens*）在温度较低的曝气海水中产卵，可添加过氧化氢及三羟甲基氨基甲烷试剂；巨牡蛎（*Crassostrea* sp.）和贻贝（*Mytilus* sp.）则须在紫外线照射过的温水中进行处理。一旦产生足够数量的配子，卵子和精子结合，受精发生，就确定了胚胎密度。试验在受精后4 h内开始，即胚胎处于2～8细

图6.2　正常发育48 h后红鲍螺（*Haliotis rufescens*）幼体的显微照片（放大400倍）

胞阶段。美洲牡蛎（*C. virginica*）的试验在 20 ℃ 进行，硬壳蛤（*M. mercenaria*）和侏儒蛤（*M. lateralis*）为 25 ℃，贻贝（*Mytilus* sp.）、长牡蛎（*Crassostrea gigas*）和红鲍螺（*H. rufescens*）则为 15 ℃。

试验通常是在有盖的小玻璃容器中进行，并加入试验溶液（10～200 mL）。试验溶液可由海水样品、盐淡水或河口水样品、海水/盐水对照和参考毒物对照组成。每个试验有 4～5 个重复组。向溶液中加入已知密度的胚胎，这些胚胎在试验过程中会发育成能自主活动的幼体。胚胎的终密度为每毫升 10～25 个，取决于被测试的物种。这些静态的、零换水的试验在 48 h 后通过添加缓冲福尔马林而终止。使用倒置显微镜来计数正常和异常发育的幼体，以测定终点，即正常发育率和存活率（图 6.3）。

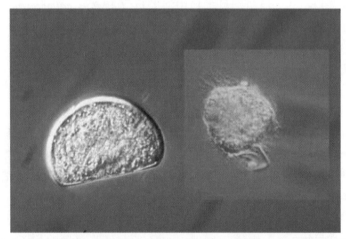

图 6.3 地中海贻贝（*Mytilus galloprovincialis*）幼体在 48 h 时的显微照片
（放大 400 倍）：正常（左），异常（右）

目前针对环境对软体动物胚胎幼体异常发育造成影响的研究较少。以红鲍螺胚胎-幼体发育试验为例，Hunt & Anderson（1989）揭示了贝壳类发育终点的生态后果。例如，暴露于锌中的胚胎未能发育形成正常形状的面盘幼体，被证明不能进一步发育、附着和变态（图 6.4）。Conroy et al.（1996）用锌和漂白硫酸盐厂废水进行了连续脉冲-回复暴露，表明贝壳幼体的异常妨碍了其浮游阶段之后的生存。在评估镍水质标准时，Hunt et al.（2002）表明红鲍螺胚胎-幼体发育和幼体变态试验是已报道的对镍金属最敏感的海水试验之一。胚胎-幼体发育试验的另一个优势是其符合毒性鉴别评估（TIE）规程（U.S. EPA，1996）。在水体毒性评估中使用胚胎幼体试验的一个潜在考虑因素是其对非离子氨毒性的敏感性（Phillips et al.，2005）。

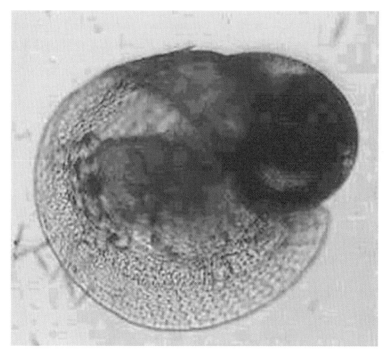

图 6.4 培养 12～14 天后新变态红鲍螺幼体（*Haliotis rufescens*）显微照片（放大 400 倍）

6.3.3 棘皮动物

海胆纲动物（如海胆和沙钱）分布于世界上大部分的海洋栖息地，是以海洋藻类为食的食草动物，同时也作为哺乳动物、鱼类和掠食性无脊椎动物的食物，在生态上具有重要地位。海胆纲动物的许多物种也是人类的食物来源，被用于商业或个人捕捞。自 Kobayashi et al.（1972）引入海洋污染评价规程以来，海胆毒性试验方案是目前最常用的水生毒理学试验。试验有两种，一种是评估胚胎幼体发育，另一种是评估精子暴露于受试溶液后致卵子受精的成功率。这些试验已被学者总结（Dinnel et al., 1989；Nacci et al., 1991；Bay et al., 1993），U. S. EPA（1995）也详细介绍了海胆受精和发育试验方法。

海胆的胚胎幼体试验遵照上述双壳软体动物试验程序进行。即诱导成虫产卵，暴露 72～96 h 后观察发育过程中棘皮动物幼体的发育异常。诱导不同物种产卵的方法也不同：诱导海胆 *Arabacia punctulata* 产卵采用温和的电刺激；而对于紫色球海胆（*Strongylocentrotus purpuratus*）和沙钱（*Dendraster excentricus*），则注射氯化钾溶液，通过渗透压冲击诱导配子释放。一旦获得配子，按照特定精卵比率促使卵子受精，具体比例因物种而异。然后将受精卵接种到试验容器中，步骤同上文所述的软体动物胚胎发育试验步骤。试验容器通常是小容量玻璃瓶，如闪烁计数瓶，以方便我们使用倒置显微镜来观察胚胎的发育。

海胆纲动物幼体发育试验在静态条件下进行。添加缓冲液福尔马林后，对于海胆

(*A. punctulata*)（20 ℃）来说，试验应在 48 h 后终止，而对于紫色球海胆（*Strongylocentrotus purpuratus*）（15 ℃）和海胆（*Heliocidaris tuberculata*）（Woodworth et al., 1999）来说，则在 72～96 h 后终止。通过计算正常和异常发育的幼体来确定终点，即正常发育百分比（图 6.5）。

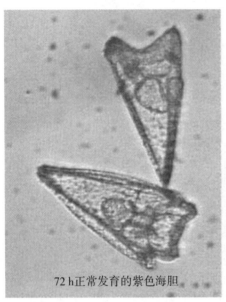
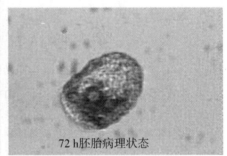

图 6.5　72 h 后紫色海胆（*Strongylocentrotus purpuratus*）幼体的显微照片（放大 400 倍）：正常（左），异常（右）

　　Dinnel et al.（1983）首次采用东太平洋的海胆进行受精试验。试验结果已被证明可作为污水和环境水毒性的一个敏感指标。受精试验是对某些化学物质，特别是金属最敏感的试验之一。并且由于该试验可以快速进行，因此特别适合大批量样本的筛选试验（例如，Bay et al., 1999）。这个特性也使受精测试对研究高挥发性或瞬态化学物质（如氯；Bay et al., 1993）的毒性有用。许多不同的棘皮类动物已经被应用于这个试验。在大西洋和墨西哥湾沿岸，红海胆（*Arabacia punctulata*）为常用物种。虽然紫色海胆（*Strongylocentrotus purpuratus*）是最常用的物种，但西海岸已使用许多其他海胆纲物种作为替代物种，如方斑圆线虫（*Strongylocentrotus franciscanus*）、圆毛线虫（*Strongylocentrotus droebachiensis*）、海钱（*D. excentricus*）、白腹鳉（*Lytechinus pictus*）等。除这些物种以外，还开发了一种用于夏威夷水域的热带海胆（*Tripneustes gratilla*）受精试验。该试验最先由 Nacci & Morrison（1993）实施，由 Vazquez（2003）改良，由 U.S. EPA（2012）细化。在澳大利亚水域，已开发了一种利用温带海胆物种（*H. tuberculata*）的受精试验（Woodworth et al., 1999）。

　　海胆受精试验适用于各种 TIE 规程，一些研究已使用此类试验来确定南加州沿海水域阳离子金属造成环境毒性（如新贝德福德港因 PCBs 导致的沉积物毒性，Ho et al., 1997）和雨水毒性（Bay et al., 1999）的原因。此类试验的另一个积极特性是：海胆纲动物精子对非离子氨浓度的高耐受性。此特性使得此类试验在氨可能掩盖其他污染物毒性的情况下十分有用。

Bay et al.（1993）列出了上述方法存在的局限性。研究者发现使用紫色海胆（*S. purpuratus*）及沙钱（*D. excentricus*）试验时，毒性效应与商业海盐和高盐水有关，并描述了该试验对极端 pH 的敏感性。这些学者还讨论了使用该方法在评价环境毒性时异常高比例假阳性毒性结果的发生。当明显无毒的样品被鉴定为有毒时，就会出现假阳性的结果。

棘皮类动物胚胎试验的另一个特点是：除上述发育终点外，这些试验已被证明可用于测定遗传毒性效应。例如，Anderson et al.（1994）利用紫色海胆（*S. purpuratus*）进行暴露试验后发现，该物种的胚胎发育方案允许纳入直接测量染色体损伤的终点，包括用光学显微镜观察染色胚胎中染色体断裂导致的后期畸变分析。另一些则采用 DNA 解链（如彗星电泳法）方法用于棘皮类动物胚胎（Shugart et al.，1992）。

6.3.4 糠虾

糠虾作为小型类虾甲壳类动物，是海洋和河口鱼类的主要食物。除了生态上的重要性，他们还具有许多特性，因而成为水质监测的理想试验生物。糠虾对各类污染物都敏感，且它们世代时间相对短，适合实验室培养。自最初试验规程被开发以来（Nimmo et al.，1977），大西洋/墨西哥湾沿岸物种拟糠虾（*Americamysis bahia*，前称为 *Mysidopsis bahia*）试验已被用于监管应用数十年。Lussier et al.（1988）介绍了该类物种的培育方法。其生活史和生态描述见 Rodgers et al.（1986）及 U. S. EPA（2002a）。为此物种开发的规程包括急性、短期慢性和长期生命周期试验方案。NPDES 管理监测采用为期 7 天的短期慢性测试（Lussier et al.，1999；U. S. EPA，2002b）。试验选用 7 天龄未成年糠虾幼苗（成体通常由供应商或通过室内培养提供）开始，检测生存、生长和生殖力终点。每个重复试组验容器中放入 5 只糠虾，将其暴露于试验溶液，并每天记录其生存。试验温度为 26±1 ℃，每天投喂两次新孵化的卤虫，每天清洗 1 次容器并更换试验溶液。每天记录存活糠虾数目，并且在 7 天后终止试验（该试验中 4 天的存活率可为 96 h 急性死亡率提供数据）。试验终止后，使用立体显微镜检查存活个体，确定未成熟个体数量、成熟个体性别以及成年雌性输卵管或育卵囊中是否有卵（图 6.6）。有证据表明在测量的 3 个终点中，在实验室鉴定繁殖力终点最为困难。虽然有多种繁殖力定量测定方法，但 U. S. EPA 指出这通常是测试中最敏感的终点（U. S. EPA，2002b）。Lussier et al.（1999）发现，如果在预测试期间糠虾培育温度保持在 26～27 ℃之间，并保持低于每升 10 个的生物密度，则符合繁殖力可接受标准（≥50% 对照组雌性产卵量）的试验比率从 60% 增加到 97%。

Nimmo（1977）等首次描述了拟糠虾（*A. bahia*）的 28 天生命周期试验。试验开始于使用刚孵出的糠虾幼苗（<24 h），在通流条件下发育为成体并繁殖。待试验终止时，评估繁殖成功和存活的 F1 代。新生虾苗在第 10～12 天间发育为性成熟成体，与此同时雌性会发育出卵囊（育幼袋）。卵存于卵囊中，当发育成幼体时，孵化囊会生长并变黑。繁殖的成功是通过隔离孵卵的雌性和计算释放的新生幼体数量来确定。而释放的新生幼体存活率也可通过将它们隔离 96 h 来确定。除非需要 F1 代群体的繁殖数据，否则

试验在 F0 代糠虾暴露 28 天后终止。第一代糠虾的生长也可以通过测定试验终止时的总长度或干重来评估。由于雌性和雄性虾体型存在大小差异，因此需要明确糠虾不同性别的生长。针对糠虾标准化 28 天生命周期试验的详细试验指南和可接受试验标准见 ASTM（2008b）。

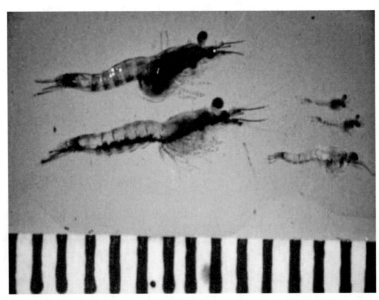

图 6.6　拟糠虾（A. bahia）典型生命阶段和性别的显微照片
左上为成熟的妊娠期雌性糠虾，育幼袋鼓胀；左下为成熟雄性；右上为 2 尾新生糠虾幼苗；右下为稚体。照片由 Kay Ho 提供。

目前其他几种糠虾的标准化试验规程也已开发完成。*Holmesimysis costata* 是一种糠虾类甲壳动物，发现于太平洋沿岸巨型海藻床的表面冠层，是鱼类的重要食物来源。美国加利福尼亚州已为 NPDES 检测开发了一项为期 7 天的短期慢性试验（U. S. EPA, 1995）。试验开始于从野外收集的成年糠虾繁殖而来的 3～4 天龄糠虾新生幼体。每组 5 个重复，每个容器有 5 只糠虾暴露于测试溶液中，并在 48 h 和 96 h 后更新测试溶液。每日观察虾苗生存，以确定 96 h 和 7 天的死亡终点。在 7 天后结束试验时，使用微量天平称重干燥后的糠虾以确定生长终点。其他通常用于糠虾短期试验的物种 *Neomysis mercedis* 是西海岸的另一种糠虾，可以作为低盐度试验的替代物种（ASTM, 2008b）。*N. mercedis* 已被证明在淡水和半咸水环境监测研究中有用。*N. mercedis* 承受的水传导性超出了其他甲壳类动物（如水蚤类）的耐受范围（Finlayson et al., 1991; Hunt et al., 1999）。*N. mercedis* 的最佳试验盐度范围为 1‰～3‰。但其在盐度达到 18‰ 的野外也能生存。此外 Hunt et al.（2002）报告了在对海洋生物的镍水质量标准评估时，可使用太平洋海岸糠虾（*Mysidopsis intii*）进行急性（96 h）和慢性 28 d 的全生命周期暴露试验。试验观察到糠虾较低的生殖成功率。因此，慢性试验中最低观察效应浓度和无观察效应浓度是基于生长和死亡终点。一项为期 7 天的太平洋海岸糠虾（*Mysidopsis intii*）短期慢性毒性试验规程已被开发，但尚未于 NPDES 监测中实施（Harmon & Langdon 1996; Langdon et al., 1996）。此外，用于巴西水域的 *Mysidopsis juniae* & *Mysidium gracile* 短期急性毒性试验也已被开发（Prosperi et al.，1998）。

现今已为拟糠虾 A. bahia（U. S. EPA，1996）& *N. mercedis*（Hunt et al.，1999）开发了 TIE 方法；另外，评估 *H. costata* 对不同 TIE 处理方法的耐受性也成为加利福尼亚州水资源控制委员会的海洋生物试验项目的一部分（Phillips et al.，2003a）。商业型供应商报告了 *H. costata* 在冬季的供应限制，所以利用该物种进行为期 7 天的生长和存活试验可能会受到试验生物可用数量的限制（Hunt J，个人通讯）。此外一些研究人员指出该物种的 7 天试验很难达到可控结果（Phillips et al.，2003a）。糠虾试验可能会受高于或低于特定效应阈值离子浓度的干扰，特别是在某些污水中（如采出水和农业废水；Ho & Caudle，1997）。Pillard et al.（2000）开发了用于预测升高的主要离子浓度（如 K^+、Ca^{2+}、Mg^{2+}、Br^-、SO_4^{2-}、HCO_3^-、$B_4O_7^{2-}$）的毒性和与其缺陷相关效应的模型。糠虾在环境毒性试验中的应用详见 6.3.10 节。

6.3.5 桡足类

因河口浮游生物和海洋桡足类生物适应实验室培养条件，对毒物敏感，具有生态重要性，被广泛用于毒性试验。已开发了哲水蚤（calanoid）及猛水蚤（harpacticoid）的标准化测试方案。Gentile & Sosnowski（1978）描述了利用汤氏纺锤水蚤（*Acartia tonsa*）进行静态急性试验的一般准则。试验将成年桡足类置于装有 100 mL 试验溶液的 250 mL 结晶皿中，暴露于污染物 96 h 后，记录其死亡率。Ward（1979）等对纺锤水蚤（*A. tonsa*）慢性毒性试验进行了描述，随后 Kusk & Petersen（1997）又对其进行了补充。Ward et al.（1979）描述了成年桡足类 30 天多方面的毒性试验结果。第一阶段为桡足类繁殖后，第二阶段为无节幼体在通流条件下暴露 20 天。无节幼体被允许发育到性成熟并繁殖，随后在静态条件下通过计算孵化的无节幼体数量来确定繁殖成功率。Kusk & Petersen（1997）测定了急性（48 h）暴露于三丁基锡和直链烷基苯磺酸盐下的成年汤氏纺锤水蚤（*A. tonsa*）死亡率。这些试验结果与 8 天暴露试验的胚胎-幼体发育测量结果进行比较。幼体发育率以孵化无节幼体与总幼体数之比表示。标准化的急性和慢性试验方案也已被开发用于海水试验，对象为猛水蚤类（*harpacticoid* spp.），如河口物种猛水蚤（*Tigriopus brevicornis*）（Lassus et al.，1984）。一种利用澳大利亚热带纺锤水蚤（*Acartia sinjiensis*）和剑肢水蚤（*Gladioferens imparipes*）评价短期暴露后的固定化规程也已被开发（Evans et al.，1996；Tsvetnenko et al.，1996）。

6.3.6 十足类

许多十足类甲壳动物已被应用于毒性试验，具体见表 6.4。研究表明十足类动物对多种毒物敏感，特别是杀虫剂。这类动物的不同生命阶段，包括幼体、大眼幼体、幼苗、成体，均已被用于进行生态毒理研究。因该类生物较早期的生命阶段不适应实验室试验条件（Gentile et al.，1984），所以早期阶段较少用于试验。后两个阶段更常用于急性暴露（ASTM，2007；APHA，2012）。草虾（*Palaemonetes* sp.）是毒理学研究中最常用的十足类甲壳动物之一。该物种的急性和慢性试验方案也已被开发。可以将带卵的雌虾单独置于玻璃培养皿中，并在其卵释放时收集幼体（Tyler Schroeder，1978a）。25 ℃的静水条件下，将幼虾置于 1 L 玻璃培养皿并将其暴露于试验溶液中。每天更新试验溶液后投喂刚孵出的卤虫无节幼体，在第 96 h 时记录幼虾的死亡率。利用草虾进行的 145 天生命周期试验在其他文献中也有相关描述（Tyler-Schroeder 1978b，1979）。试验开始时每个处理组使用 100 尾幼虾苗（额剑到尾节的长度 < 15 mm），在通流、20‰ 盐度、25 ℃ 条件下进行。每 4 周监测一次幼虾的生长，直到试验终止。在光周期为 8 h 光照和 16 h 黑暗下的条件持续约 3 周至虾性成熟，随后逐渐增加光周期，最终增加至 14 h 光照和 10 h 黑暗，光强度也从 15 W 增加到 100 W。虾达到性成熟后，将产（含）卵雌虾从培养皿分隔开。试验中测定的生殖终点包括产卵雌虾数量、卵的数量和胚胎孵化成功率。因为对草虾进行慢性试验为劳动密集型工作，所以 Ward（1995）认为糠虾短期慢性试验已经在很大程度上取代了草虾的试验。澳大利亚就已经使用热带虾斑节对虾（*Penaeus monodon*）进行短期急性毒性试验（Evans et al.，1996；Tsvetnenko et al.，1996）。

表 6.4　毒性试验十足类甲壳动物常用物种一览表

科	种
无脊椎动物	长臂虾 *Palaemon adspersus*
	巨指长臂虾 *P. macrodactylus*
	草虾 *Palaemonetes pugio*
	普通小长臂虾 *P. vulgaris*
对虾科	棕虾 *Penaeus aztecus*
	桃红对虾 *P. duorarum*
	白对虾 *P. setiferus*
	南美兰对虾 *P. stylirostris*
真虾下目	褐虾 *Crangon* sp.
黄道蟹科	斑纹黄道蟹 *Cancer irroratus*
	首长黄道蟹 *C. magister*
	红黄道蟹 *C. productus*
梭子蟹科	蓝蟹 *Callinectes sapidus*
	岸蟹 *Carcinus maenas*
海螯虾科	美洲螯龙虾 *Homarus americanus*

由于在河口食物网中的生态重要性，草虾是研究相关热点新兴污染物效应的前哨物种。许多相关研究已将草虾纳入生物标记物和生理测量研究。McKenney（1998）综合应用发育、生长和代谢反应评价了生长调节剂烯虫酯对短刀小长臂虾（*Palaemonetes pugio*）的效应。此研究将草虾幼体暴露于烯虫酯中，通过测量蜕皮成功率（有蜕皮壳存在即为蜕皮成功）来观察发育速率。此外，将暴露过程中的幼虾密封于玻璃注射器中，监测其氧气消耗速率、氨排泄率和生长（以干重计）速率。与哈氏泥蟹（*Rhithropan-*

opeus harrisii）的类似研究进行比较，草虾的研究中纳入了生物标记物终点，包括 DNA 损伤（Hook & Lee，2004）和细胞色素 P450（CYP1A）卵黄蛋白诱导的测量（Oberdorster et al.，2000）。

6.3.7 轮虫

轮虫是一种的微型无脊椎动物，其特点是生命周期相对较短，在压力条件下可能会包含一个休眠（胞囊）阶段。这些特征使它们能够经受实验室的毒性试验。Snell & Janssen（1998）描述了河口轮虫——褶皱臂尾轮虫（*Brachionus plicatilis*）的急性毒性试验。试验开始的 24 h 前，于温暖的稀释海水（15‰）中诱导该轮虫的休眠卵孵化，使用小于 2 h 龄的新孵化（幼体）轮虫开始试验。在多孔聚苯乙烯组织培养容器中进行急性试验，每孔 10 个幼体，记录 24 h 暴露后的死亡率。目前还没有针对这一物种的慢性试验方案，但已研究了一些亚致死反应，包括行为终点，如游泳和摄食率。此外，Snell & Janssen（1998）还对酶活性和应激蛋白诱导等生化和分子反应进行了综述。

6.3.8 刺胞动物

刺胞动物（Phylum Cnidaria）包括水螅、水母、海葵和珊瑚。其海水试验已被 Ward（1995）总结（见 Stebbing & Brown 综述，1984）。最常用的试验是利用水螅（集群水螅）和珊瑚进行。一种用于群居海洋水螅类（如 *Laomedea flexuosa*，*Eirene viridula*）的试验方案是评价其群居生长速率和发育中珊瑚虫的形态变化（Karbe et al.，1984；Ward 1995）。刺胞动物群落建立于玻璃板上，暴露至少 2 周。每周投喂 2 次卤虫。在试验开始和喂食时确定群落的大小。测定的终点包括进食珊瑚虫（水螅体）的数量和群落出芽率。Ward（1995）提出，由于水螅的试验和培养需要消耗大量劳力，而且大多数物种在生态或经济上都不像常规试验中使用的其他物种那样重要，因此水螅试验在管理应用中并未得到重视。

对刺胞动物的其他试验还包括针对硬壳珊瑚（珊瑚虫）开发的试验。采用珊瑚的实验室试验描述详见标准化方法（APHA，2012）。这些规程已用于许多物种。Branton（1998）讨论了用珊瑚进行的微尺度试验。鹿角珊瑚（*Acropora cervicornis*）分布广泛、具敏感性，且与印度洋-太平洋物种美丽鹿角珊瑚（*Acropora formosa*）相似（Ward，1995），因此常作为热带大西洋水域的主要试验物种。

依照标准测试规程，在通流条件下暴露 20 个初始大小均一（湿重 10 g）的集落，测量终点包括存活率和几个亚致死反应（如虫黄藻被赶出和珊瑚虫收缩）。群落的生长可以用浮力称重来测定（Branton，1998）。除了使用上述物种进行试验外，还制定了在澳大利亚热带水域使用一些珊瑚物种的规程。这些规程包括使用多孔鹿角珊瑚（*Acropora millepora*）（Negri & Heyward，2000）来评价受精和幼体变态，以及对与美丽鹿角珊瑚（*A. formosa*）、指状蔷薇珊瑚（*Montipora digitata*）、细柱滨珊瑚（*Porites cylindrica*）和尖枝列孔珊瑚（*Seriatopora hystrix*）共生的甲藻叶绿素荧光的测定（Jones et al.，2003）。

6.3.9 海洋藻类毒性试验

Thursby et al.（1993 年）回顾了海洋、河口中微藻与大型藻类在毒性试验中的应用，并指出将藻类试验规程纳入水质评估的主要原因之一是，藻类作为主要的生产者代表了水生食物网的基础。由于在污水和环境水样本中，除草剂、杀菌剂和其他专门用于影响藻类的化学物质盛行，因此藻类试验是水质评价项目中的必要组成部分。Lewis（1995）总结了利用微藻、大型藻类和维管植物的海水和淡水毒性试验。下文将简要概述标准化规程中的常用物种。

作为美国 EPA 为整个污水毒性试验制定的海洋毒性试验方案的一部分，学者评价了红藻（*Champia parvula*）和棕藻（*Laminaria saccharina*）等大型藻类。这项研究还导致了环节藻（*C. parvula*）标准化有性生殖试验的研发（U.S. EPA，2002b）。将性成熟的雄性和雌性海藻分支暴露于试验溶液 2 天。随后在对照海水中恢复 5～7 天，使其繁殖。该试验虽被纳入美国 EPA 海洋毒性试验手册（U.S. EPA，2002b），但未被美国环境保护局颁布用于监管监测，所以并不作为常规方法使用。

美国 EPA 在制定东部和墨西哥湾沿岸试验方法的同时，还制定了一项使用巨藻（*Macrocystis pyrifera*）孢子的 48 h 试验规程，作为美国西海岸的海洋污水监测的一项藻类毒性试验规程。该规程是加利福尼亚常规管理监控中最常用的试验规程之一。该试验规程如同为石莼和岩藻种（下文将简要讨论）量身定做的规程，因此制定的巨藻方法被认为具有代表性。*M. pyrifera* 是一种大型海藻，在东太平洋海岸的近岸地区形成了广阔的巨藻森林。这些森林结构复杂，可为众多物种提供栖息地和食物。该种巨藻生命周期可分为两个阶段：体积较大的孢子形成阶段（孢子体）和体积很小的配子产生阶段（配子体），二者在生命周期中交替进行。从野生巨藻根部收集可以产生孢子的叶子。随后将这些叶子置于实验室凉爽干燥的条件下 24 h 后，将其浸泡于海水中，使孢子释放。将收集到的孢子稀释至特定浓度后，接种到装有 200 mL 试验溶液的 600 mL 容器中。试验溶液可包括海水样品、咸淡水或河口水样，海水（盐水）对照组以及参考毒物对照组。在静水不更新测试液条件下持续 2 天，期间孢子沉降并萌发，发育为配子体。测量 2 个终点：孢子萌发成功率和配子体萌发的芽管长度（Anderson & Hunt 1988；Anderson et al.，1990）。此外，制定的短期慢性（48 h）和长期生殖试验（21 d）被作为加州海洋毒性试验方法开发工作的一部分。考虑到实际情况，仅将短期 48 h 试验应用于常规污水检测，将长期生殖试验用于校准 48 h 试验的相对灵敏度和生态意义。大量研究已证明了对 *M. pyrifera* 孢子 48 h 试验终点的生态意义。这些研究表明抑制了孢子的萌芽和生长的毒物，也抑制了巨藻的生殖［孢子体生产（Anderson et al.，1997），Thursby et al.（1993）& U.S. EPA（1995）对此也做出总结］。目前针对该试验的 TIE 方法已开发完成（U.S. EPA，1996）。

目前已发现一项与 *M. pyrifera* 48 h 试验相关的潜在干扰因素。在试验开始的前一天从野外收集巨藻孢子叶。随后将其运送至实验室，并浸泡于海水中诱导孢子释放。从孢子叶收集到孢子释放之间的时间通常小于 24 h。Gully et al.（1999）发现在以铜作为参考毒物试验中，孢子叶储藏影响了孢子萌发终点的反应。尽管这些研究人员未发现对芽管生长终点的影响，但他们认为对萌发终点的影响可能会增加芽管生长终点的相对灵敏度，从而混淆污水试验结果及对试验结果的解读。

上述规程的另一个局限性是可能不太适合测试河口样本。巨藻是一种生存范围局限于岩石潮下区的近岸物种。Hooten & Carr（1998）提出了一个用于河口生境中藻类毒性研究的替代试验。该试验与 *M. pyrifera* 的试验规程类似，不同点在于使用的是河口藻类裂片石莼（*Ulva fasciata*）的游动孢子。研究人员对沉积物孔隙水试验进行了评估，并提出因裂片石莼孢子对多种毒物相对敏感，且对非离子氨具有耐受性，所以该测试在 NH_3 含量升高作为潜在干扰因素的情况下依然可用。此外，使用墨角藻（*Fucus vesiculosus*）（Brooks et al.，2008）、螺旋墨角藻（*Fcusus spiralis*）（Girling et al.，2015）和绿藻肠浒苔（*Ulva intestinalis*）（Girling et al.，2015）的类似规程也已制定完成。以上物种均已被证明对金属敏感，其中肠浒苔（*U. intestinalis*）对含有金属和除草剂混合物的防污杀菌剂敏感（Girling et al.，2015）。

6.3.10 海洋水体毒性试验在环境监测中的应用

上述规程已广泛应用于污水毒性监测，并在较小范围内用于水环境监测。1995 年，环境毒理学和化学学会（SETAC）在密歇根州佩尔斯顿召开了一个研讨会，以评价目前在污水和水质环境评估中使用全废水毒性测试的方法。出席研讨会的有来自政府、工业界和学术界的专家，他们在毒性试验的应用方面经验丰富。与会者达成共识——按照美国 EPA 方法进行的测试规程在技术上是达标的。尽管与会者的结论肯定了这些试验提供了废水对受纳水体潜在影响的有用信息，但它们在海洋和河口环境水毒性监测方面的应用尚未得到淡水系统中那样全面的评价（Grothe et al.，1996）。研讨会确定了几个需要进一步研究的领域。Schimmel & Thursby（1996）指出，由于各种原因，尚未有研究将海洋或河口受纳水体的环境毒性与其水质或底栖生物群落的影响联系起来。由于复杂的生物和非生物因素可能与河口化学应激物相互作用，所以更难确定这些系统中环境毒性与受纳系统生态影响之间的关系，但也有一些研究已表明河口系统中毒性和底栖生物影响之间存在联系（如 Anderson et al.，2014）。佩尔斯顿研讨会的与会者得出的结论是，标准化测试的水体毒性试验应与生物评价和化学分析结合用于综合决策。

为了污水检测，美国环境保护局（1991）建议至少使用其指导文件中列出的 3 种标准测试方案来筛选污水样品的毒性。若条件许可，测试物种应包括鱼类、无脊椎动物和水生植物各一种，因为不同物种对不同种类毒物的反应可能不同。一套包括代表不同门和类群的测试物种的做法也适用于环境毒性研究（U. S. EPA，1991）。

如上所述，美国联邦和各州关于环境水质监测中水体毒性试验应用的指南建议，在毒性筛查阶段，至少要使用代表包括无脊椎动物、鱼类和植物在内的各种群体中的 3 种物种。随后的测试可以用最为敏感的物种进行。由于不同物种和相同物种的之间对试验方案具有不同的敏感性，因此选择适当方案用于危害评价中的效应特征时，取决于在评价拟定阶段所明确的目标化学物质。例如，相对于糠虾，海胆和软体动物胚胎幼体发育试验以及海胆受精试验对镉都不是特别敏感，因此选用无脊椎动物而不是糠虾试验进行筛查会低估镉带来的生态风险。相反，海胆和软体动物胚胎幼体发育和受精试验是对铜、锌最为敏感的标准化试验方案。在重金属作为主要目标化学物质的情况下，该类测试规程将会更为合适。值得注意的是，经常因为有些胚胎幼体发育的试验具有相似的终点而被归为一组，如地中海贻贝（*M. galloprovincialis*）、紫球海胆（*S. purpuratusus*）、红鲍螺（*H. rufescens*），但这些规程可能对所有有毒物的反应不同。例如，Phillips et al.（2003b）发现了贻贝和海胆胚胎对镉、铜、锌和镍的反应差异巨大。海胆和双壳类动物胚胎之间也存在差异（Gries 1998）。

与其他甲壳纲动物（如端足类）相同，糠虾也对许多抗微生物剂（如叠氮化钠，五氯苯酚）和杀虫剂（尤其是有机氯、有机磷和拟除虫菊酯杀虫剂）敏感（Clark et al.，1989；Cripe 1994；Hunt et al.，1997）。糠虾对其他的有机氯化合物也相对敏感，如PCBs。当这些化合物是主要污染物时，使用拟糠虾（*A. bahia*）或其他替代物种（*N. mercedis*或*H. costata*）进行试验较为合适。*H. costata*，7天生长和生存试验不包括生殖终点。若在特定的风险评价中需要考虑对糠虾或其他甲壳类动物的生殖影响，用墨西哥湾沿岸物种拟糠虾（*A. bahia*）进行试验是一个合适的替代方案。鉴于上述物种的敏感性，糠虾、海胆受精试验以及使用某些物种进行的胚胎幼体发育试验也适用于评价某些有机氯农药（如DDT）和类金属化合物（如TBT）相关的化合物。综上所述，海胆受精试验［紫色球海胆（*S. purpuratus*）或海钱（*D. excentricus*）］对多种有毒物质敏感，尤其适用于筛选高挥发性或瞬态化学物质（如氯、雨水；Bay et al.，1993）。

在某些情况下，并没有足够的数据来确定海洋水体毒性试验方案对特定污染物的相对敏感性。虽然近期研究表明鱼胚胎的心血管发育对三环PAH尤为敏感（上文已进行讨论），但很少有比较研究来评价此类规程对PAH的相对敏感性。进一步的研究表明由于仔鱼具有明显的敏感性，因此更适合使用仔鱼方案对石油碳氢化合物进行危害评价。例如，Schiff et al.（1992）发现美洲银汉鱼（*Menidia beryllina*）幼体在5种采出水检测试验方案中敏感性最高［紫色球海胆（*S. purpuratus*）受精试验 > 美洲银汉鱼（*M. beryllina*）幼体存活试验 > 拟糠虾（*A. bahia*）新生幼体存活试验 > 微生物毒性试验 > 沙蚕（*Neanthes arenaceodentata*）存活试验］。在评价化学分散剂和石油的相互作用时，Singer et al.（1998）发现，与红螺鲍（*H. rufescens*）胚胎和糠虾新生幼体（*H. costata*）相比，安芬拟银汉鱼（*A. affinis*）可能是对普拉德霍湾原油水溶性成分（WAF）最敏感的物种。当使用化学分散剂增强普拉德霍湾水溶性成分配置剂时，糠虾与其他2个物种相比更为敏感。应当注意的是，鱼类胚胎发育未包含在上述的比较之内，且鉴于近期PAH对心血管毒性的证据，这些终点应包括在分散剂对溢油影响的评价中（Carls et al.，1999；Couillard et al.，2005；McIntosh et al.，2010）。许多方案对非污染因素和自然存在的化合物都很敏感，因此可能会干扰毒性试验结果的解读。例如，糠虾、仔稚鱼以及某些情况下的海胆精子可能会受到高于或低于效应阈值离子浓度的影响（Pillard et al.，2000；Bay et al.，2003）。在环境水体可能受到采出水、农业废水或其他可能增加离子浓度水源（例如K^+、Ca^{2+}、Mg^{2+}、Br^-、SO_4^{2-}、HCO_3^-、$B_4O_7^{2-}$等）影响的情况下，这些离子含量和占比应被测定，并与已建立的效应模型相比较。此外，这些试验方案所选物种敏感性会随非离子氨浓度的升高而增强。因为所有此类规程都适用于TIE流程，有助于确认毒性原因，尤其是当非污染因素影响结果时。

如前文所述，由于此类规程对污染物的敏感度各不相同，美国EPA建议使用代表不同门和物种的复合方案进行测试，尤其当环境水体可能受到复杂化学混合物影响时。Schimmel et al. (1989) 使用了 5 种不同的大西洋沿岸生物的毒性测试方案 [环节藻（C. parvula），拟糠虾（A. bahia），红海胆（A. punctulata），美洲银汉鱼（M. beryllina），杂色鳉（C. variegatus）] 对 7 种不同污水及其受纳水体进行了毒性评价。对污水和受纳水的敏感性因各方案而异，没有一个试验方案是对每种污水及其受纳水体都最敏感。除在标准化规程中使用多种物种外，许多方案中还可以增加额外终点评价，以提供有关生态风险的额外信息。例如，可以用海胆精子和胚胎、鱼胚胎和仔鱼评价细胞遗传学终点（Anderson et al.，1994；Kocan et al.，1996）。

美国国家消除污染排放制度（NPDES）中的毒性试验、毒性鉴定评价和毒性消减评价（TIE/TRE）规程以及其他监测方法的结合，显著减少了来自点源污染的有毒输入（Norberg-King et al.，2005）。随着点源污染的减少，毒性试验在监测非点源污染造成的环境毒性项目中的应用是一个必然的趋势。下文将提供一些区域和州范围内的环境监测项目范例。这些项目包括了海水毒性试验，并讨论了加利福尼亚州雨水毒性的近期研究结果。

6.3.10.1　环境监控

海水毒性试验已被纳入切萨皮克湾和旧金山河口的区域环境监测以及加利福尼亚州的地表水监测项目。Hall & Alden（1997）于 1990—1994 年间在切萨皮克湾区域监测项目的综述中讨论了使用多种大西洋沿岸物种的毒性试验结果。其中包括了使用杂色鳉（C. variegatus）、草虾（Palaemonetes sp.）、近亲真宽水蚤（Eurytemora affinis）和拟糠虾（M. bahia）进行的 8 天生存及生长试验，以及使用侏儒蛤（M. lateralis）进行的 48 h 胚胎/幼体发育试验。毒性试验的结果整合为"毒性指数"（TOX – INDEX），以便筛选大量位点来对污染物的相对降解程度进行排序。水体毒性试验结果显示，在 5 年的测试期内，9 条河流和港口均存在一定程度的毒性。将水体毒性试验结果与沉积物毒性试验、水和沉积物化学分析相结合，证实了利用多种水和沉积物毒性试验以确定风险最高位点的实用性。该方法被用来推荐位点进行更全面的后续研究。除在切萨皮克湾的环境监测中使用之外，拟糠虾（A. bahia，或 M. bahia）还被用于进行实验室和野外暴露的环境毒性研究（Clark 1989；Schimmel et al.，1989；Kahn et al.，1993）。

在早期的一个使用软体动物的环境检测中，Woelke（1967）使用牡蛎（C. gigas）胚胎的原位暴露证明了在华盛顿州普吉特海湾制浆造纸厂附近的受纳水区具有毒性；而当废水停止排放时，环境水体的毒性快速消除。自 1993 年以来，在加利福尼亚州，海水毒性试验已被纳入"旧金山河口区域监测项目"，通过拟糠虾（A. bahia）生存和生长以及贻贝（M. galloprovincialis）胚胎发育情况来监测水体毒性。与其他监测项目一样，将该项测试的结果与水和沉积物化学分析以及沉积物毒性测试相结合，以提供河口水质变化的长期趋势。自检测开始以来，在河口中部几乎未观察到糠虾或双壳类动物胚胎水毒性案例。对糠虾的毒性试验表明，河口边缘的雨水输入毒性有所降低，这与进入河口流域的有机磷杀虫剂的使用减少一致（Anderson et al.，2007）。

为满足清洁水法案要求和提供全面的加利福尼亚地表水合理利用信息，州水资源管理委员会和地区水质管理委员会（统称为加利福尼亚州水委会）在 2001 年引入了地表水环境监测项目项目（SWAMP）（http：// www. waterboards. ca. gov/swamp/docs/cw102swampcmas. pdf）。该项目旨在协调全州范围内高质量、可持续、科学合理的方法和策略框架，以改进加利福尼亚的水质监测、评价和报告。该监测项目强调流域的特征，大多数毒性测试使用针对水体和沉积物物种的规程。海水毒性测试已纳入有淡水输入河口和海湾的 SWAMP 项目。该项目中最常用的海水毒性测试包括安芬拟银汉鱼（*A. affinis*）生存和生长以及地中海贻贝（*M. galloprovincialis*）胚胎/幼体发育测试。

6.3.10.2 雨水监测

近年来，加利福尼亚州的监管部门强调了非点源污染，将进入近岸水域生境（如海湾、河口）的雨水监测纳入海洋环境测试。已有很多使用太平洋沿岸物种的研究表明了雨水毒性。Schiff et al.（2003 年）采用紫色海胆（*S. purpuratusus*）受精试验监测进入加利福尼亚圣地亚哥湾雨水的毒性。将野外采集的样本送至实验室测试，并与现场测绘的雨水羽流相结合，以显示海湾的毒性程度。化学分析和 TIEs 结果表明毒性最有可能由锌引起。在一项使用海胆受精测试的类似研究中，Bay et al.（1999）绘制了巴罗纳河（Ballona Creek）进入南加州海洋的海洋表层水毒性程度。这些研究人员发现暴风雨过后的有毒羽流在圣莫尼卡湾延伸面积超过 4 km^2。化学分析和 TIEs 表明对海胆精子的毒性也可能与锌有关。这些研究者对大型底栖无脊椎动物群落进行了概要的研究，但未发现上述现象对近海环境的影响。

在加利福尼亚，有分别利用鱼胚胎和仔鱼来记录雨水毒性的两项研究。作为 NP-DES 雨水监测要求的一部分，Skinner et al.（1998 年）对美洲原银汉鱼（*M. beryllina*）和青鳉（*O. latipes*）的胚胎进行了试验，以调查圣地亚哥县沿海数条小溪的雨水影响。除记录卵死亡率和幼体孵化成功率之外，还评价了多项致畸终点。在 74% 的样品中观察到发育异常和/或幼体死亡。对鱼类胚胎的影响与样品中的金属有关，特别是铜、铅和锌。

Phillips et al.（2004 年）监测了进入加利福尼亚中部蒙特雷湾的 15 个观测站点样本，这是蒙特雷湾国家海洋保护区初期径流雨水采样项目的一部分。测试采用安芬拟银汉鱼（*A. affinis*）幼体和地中海贻贝（*M. galloprovincialis*）。进入系统的初期径流样本对安芬拟银汉鱼具有普遍毒性，通过 TIEs 和雨水化学分析，证明毒性是由高浓度的铜和锌引起。地中海贻贝（*M. galloprovincialis*）胚胎发育也受到大多数雨水样品的抑制，在经毒性鉴别评估的样品中发现金属（铜和锌）是导致毒性的原因。雨水毒性试验正在成为加利福尼亚州 NPDES 监测的必要组成部分。当结合化学分析和 TIEs 时，海水毒性测试可以提供关于雨水排放潜在危害的重要数据。这些信息可用于帮助识别目标污染物，是污染源控制前的关键一步。

6.3.10.3 过渡环境

沿海河口是世界上最重要的生态环境和受威胁最大的生境之一。过渡环境受到进入海洋系统的沿海流域淡水径流污染的问题对生态毒理学者们提出了特别的挑战。对河口水样品进行实验室测试的一个主要问题是确保样品盐度在被测试物种的盐度耐受范围内。试验具体的方法有多种，主要取决于研究目标。

例如，美国 EPA 制定的标准海水测试方案使用从冷冻海水中提取的高盐水或人工海盐来调整样本盐度（U. S. EPA，1995）。另一种方法是使用广盐性海洋物种来测试低盐度或不同盐度的水体。如银汉鱼（如 *Menidia* sp.、*A. affinis*），鲱鱼（*Clupea* sp.）和鳉鱼（*Fundulus* sp.）即为广盐性鱼类。某些银汉鱼可用于测试盐度为 0‰～34‰ 的水样。如拟糠虾（如 *A. bahia*）、双壳贻贝（如地中海贻贝 *M. galloprovincialis*）和一些牡蛎种（如 *C. virginica*）即为广盐性无脊椎测试物种。Stransky et al.（2014）使用实验室和野外暴露的糠虾和贻贝来监测城市雨水在海洋接收系统中引起的瞬时性毒性。Tait et al.（2014）在受雨水影响的岩质潮间带系统中进行了类似研究。在实验室中测试的样本变量少于野外暴露，而野外暴露更好地表征了接收系统中可变盐度和污染物浓度的相互作用。这些研究介绍了如何将标准的广盐性实验室试验生物应用于原位试验，以便更真实地评估偶发风暴事件对近海海洋系统的相对毒性。

可以想象，一些物种将有助于评估污染物和高盐度的相互作用。例如，从海水淡化厂排放高盐卤水到沿海水域时的环境。最近一项使用东太平洋物种地中海贻贝（*M. galloprovincialis*）和红螺鲍（*H. rufescens*）的研究表明，它们在胚胎-幼体阶段对盐水排放所造成的盐度升高特别敏感（Voorhees et al.，2013）。该研究还发现，2 种用于评价的河口物种（拟糠虾 *A. bahia* 和安芬拟银汉鱼 *A. affinis*）比严格意义上的海洋物种更能耐受高盐度。

评价河口生境毒性的另一种方法是使用耐高盐度的淡水物种。Werner et al.（2010）& Deanovic et al.（2013）的研究使用淡水端足类端足虫（*Hyalella azteca*）通过实验室暴露来评价被拟除虫菊酯杀虫剂高电导率实验室水和来自旧金山北部河口水污染的径流毒性。在受到来自农业径流的拟除虫菊酯和有机磷农药影响的加利福尼亚中部一个沿海河口，也有人使用端足虫（*H. azteca*）进行了类似研究（Anderson et al.，2014）。端足虫（*H. azteca*）的耐盐度高达 15‰，加上对杀虫剂的敏感性，特别适合用于监测非点源污染影响的过渡环境毒性。

6.3.10.4 展望

海水通过改变海水碳酸盐化学过程吸收大气中的二氧化碳，导致海洋酸化。该过程降低了海水的 pH 和碳酸根离子浓度。海水温度和碳酸盐预计将在未来几十年发生变化，海洋酸化被认为是海洋中最普遍的环境变化之一（Feely et al.，2009；Hofmann et al.，2014）。自然的海洋环流将深层低 pH 海水带到沿海表层，但导致 pH 降低可能是自然原因，也可能是人为原因（Boehm et al.，2015）。有证据表明沿海水域的酸化延伸到了近岸环境，并正向岩礁潮间带蔓延（Evans et al.，2013）。

海洋酸化有可能通过影响那些依靠吸收碳酸钙来发育和生长的物种而直接或间接地影响沿海水域。海洋酸化可能会影响珊瑚、双壳类、腹足类软体动物以及棘皮动物等海洋生物的骨骼或外壳发育（Hofmann et al.，2014）。对上述物种进行毒性试验适用于评价此类现象的影响。

近期对棘皮类和软体动物的研究已用于建立 pH 对胚胎及幼体发育影响的耐受阈值。目前已针对成年生物进行了大量研究（Miller et al.，2014；Sanford et al.，2014；Moulin et al.，2015），但也有人关注使用生物早期生命阶段进行毒性试验。紫色球海胆

(*S. purpuratus*) 幼体在当前及预测范围内的 pCO_2 条件下可正常发育（Kelly et al.，2013；Padilla-Gamino et al.，2013），但贻贝（*M. californianus*）幼体外壳的机械完整性在低 pH 时会受到显著影响（Gaylord et al.，2011）；而将奥林匹亚牡蛎（*Ostrea lurida*）置于 pH 为 7.8 的水中饲养时，其外壳生长速率较慢且体积较小（Hettinger et al.，2012）。一项对濒危的花斑鲍（*Haliotis kamtschatkana*）研究表明，当前大气中 CO_2 浓度增加 2 倍会显著增加其幼体死亡率，而幸存幼体中约 40% 外壳发育异常（Crim et al.，2011）。对海洋藻类物种的研究表明，巨藻（*M. pyrifera*）的孢子体阶段可能不受海洋碳化学变化的影响，更可能受温度和营养物质变化的影响（Hepburn et al.，2011；Fernandez et al.，2015）。

将软体动物胚胎-幼体发育初始阶段和变态过程结合的试验规程对于建立与生态学相关的效应极为有用（Crim et al.，2011）。例如，红螺鲍（*H. rufescens*）标准 48 h 胚胎-幼体发育测试可额外延长 10~12 天，以涵盖整个变态过程（Hunt 和 Anderson 1993；Conroy et al.，1996）。这已被证明是对污染物尤其是金属的敏感终点（Hunt et al.，2002）。因为该测试包含了浮游生物胚胎发育、壳形成、幼体变态和底栖生物补充等复杂的生理过程，所以该物种也可能对 pH 的变化较为敏感。由于低 pH 的影响也可能加剧人为污染物（如阳离子金属）的效应，未来的研究应利用对金属（和其他污染物）和低 pH 敏感的海洋物种来解决海洋酸化的交互效应。该效应可能会对珊瑚、软体动物和棘皮动物造成不同的影响。

科学家对调整毒性试验来解释淡水和海洋系统中的新兴化学物质（CEC）越来越感兴趣。近期用于生物分析工具的鱼类和无脊椎动物细胞系的研究进展鼓励了该方法在筛选化学物质对接收系统潜在影响的应用。例如，该方法在美国正用于筛选废水中内分泌干扰化学物质，以及研究其对淡水生态系统的潜在影响。此类方法正在扩大到旧金山湾的近岸海洋系统研究中（Jayasinghe et al.，2014）中，并被推广到加利福尼亚州的其他水域（Dodder et al.，2015）。由于这些方法被用于筛选 CECs，同时进行的研究则被推荐用于评价细胞筛选试验的结果与急性和慢性全生物毒性试验的相关性。例如，对海洋鱼类和无脊椎动物急性和慢性毒性试验的相关性研究已被推荐开展（Dodder et al.，2015）。

参考文献

ANDERSON B S, HUNT J W, 1988. Bioassay methods for evaluating the toxicity of heavy metals, biocides, and sewage effluent using microscopic stages of giant kelp *Macrocystis pyrifera* (Agardh): a preliminary report. Mar. Environ. Res.，26：113-134.

ANDERSON B S, HUNT J W, PIEKARSKI W J, et al.，1995. Influence of salinity on copper and azide toxicity to larval topsmelt *Atherinops affinis* (Ayres). Arch. Environ. Contam. Toxicol.，29，366-372.

ANDERSON B S, HUNT J W, TURPEN S L, et al.，1990. Copper toxicity to microscopic stages of *Macrocystis pyrifera*: interpopulation comparisons and temporal variability. Mar. Ecol. Prog. Ser.，68 (1-2)：147-156.

ANDERSON B S, MIDDAUGH D P, HUNT J W, et al.，1991. Copper toxicity to sperm, embryos, and larvae of topsmelt *Atherinops affinis*, with notes on induced spawning. Mar. Environ. Res., 31: 17 – 35.

ANDERSON B S, HUNT J W, PIEKARSKI W J, 1997. Recent advances in toxicity testing methods using kelp gametophytes // WELLS P G, LEE K, BLAISE C (Eds.). Microscale aquatic toxicology: advances, techniques and practice. Lewis Publishers, Boca Raton, FL: 255 – 268.

ANDERSON B S, HUNT J W, PHILLIPS B M, et al., 2007. Patterns and trends in sediment toxicity in the San Francisco estuary. Environ. Res., 105 (1): 145 – 155.

ANDERSON B S, PHILLIPS B M, HUNT J W, et al., 2014. Impacts of pesticides in a central California estuary. Environ. Monit. Assess., 186: 1801 – 1814.

ANDERSON S L, HOSE J E, KNEZOVICH J P, 1994. Genotoxic and developmental effects in sea urchins are sensitive indicators of effects of genotoxic chemicals. Environ. Toxicol. Chem., 13: 1033 – 1041.

APHA, 2012. Standard methods for the examination of water and wastewater. 22nd ed. Washington, DC.

ASTM, 2007. Standard guide for conducting acute toxicity tests on test materials with fishes, macroinvertebrates, and amphibians. E729 – 96 // Annual book of American society of testing and materials standards. vol. 11.06. West Conshohocken, PA, USA.

ASTM, 2008a. Standard guide for conducting acute toxicity tests on aqueous ambient samples and effluents with fishes, macroinvertebrates, and amphibians. E1192 – 97 // Annual book of American society of testing and materials standards. vol. 11.06. West Conshohocken, PA, USA.

ASTM, 2008b. Standard guide for conducting life-cycle toxicity tests with saltwater mysids. E1191 – 03a // Annual book of American society of testing and materials standards. vol. 11.06. West Conshohocken, PA, USA.

ASTM, 2012a. Standard guide for acute toxicity test with the rotifer Branchionus. E1440 – 91 // Annual book of American society of testing and materials standards. vol. 11.06. West Conshohocken, PA, USA.

ASTM, 2012b. Standard guide for conducting sexual reproduction tests with seaweeds. E1498 – 92 // Annual book of American society of testing and materials standards. vol. 11.06. West Conshohocken, PA, USA.

ASTM, 2012c. Standard guide for conducting static acute toxicity tests starting with embryos of four species of saltwater bivalve molluscs. E724 – 98 // Annual book of American society of testing and materials standards. vol. 11.06. West Conshohocken, PA, USA.

ASTM, 2012d. Standard guide for conducting static and flow-through acute toxicity tests with mysids from the west coast of the United States. 1463 – 92, vol. 11.06. West Conshohocken, PA, USA: 828 – 854

ASTM, 2012e. Standard guide for conducting static toxicity tests with microalgae. E1218 – 1304 // Annual book of American society of testing and materials standards. vol. 11.06. West Conshohocken, PA, USA.

ASTM, 2013. Standard guide for conducting early life-stage toxicity tests with fishes. E12411 – 1305 // Annual book of American society of testing and materials standards. vol. 11. 06. West Conshohocken, PA, USA.

BARRON M G, CARLS M G, HEINTZ R, et al., 2004. Evaluation of fish early lifestage toxicity models of chronic embryonic exposures to complex polycyclic aromatic hydrocarbon mixtures. J. Toxicol. Sci., 78: 60 – 67.

BAY S M, BURGESS R M, NACCI D, 1993. Status and applications of echinoid (Phylum Echinodermata) toxicity test methods // LANDIS W G, HUGHES J S, LEWIS M A (Eds.). Environmental toxicology and risk assessment. ASTM STP 1179. ASTM, Philadelphia, PA: 281 – 302.

BAY S, JONES B H, SCHIFF K, 1999. Study of the impact of stormwater discharge on Santa Monica Bay. Technical Publication USCSG – TR-02 – 99, Sea Grant Program, Wrigley Institute of Environmental Studies, University of Southern California, Los Angeles, CA: 16.

BAY S M, ANDERSON B S, CARR R S, 2003. Relative performance of porewater and solid-phase toxicity tests: characteristics, causes, and consequences // CARR R S, NIPPER M (Eds.). Porewater toxicity testing: biological, chemical and ecological considerations: methods, applications, and recommendations for future areas of research. SETAC Press, Pensacola, FL, USA: 11 – 36.

BOEHM A B, JACOBSON M Z, O'DONNELL M J, et al., 2015. Ocean acidification science needs for natural resource managers of the North American west coast. Oceanography, 28: 170 – 181.

BRANTON M, 1998. Microscale bioassays for corals // WELLS P G, LEE K, BLAISE C (Eds.). Microscale testing in aquatic toxicology: advances, techniques, and practice. CRC Press, Boca Raton, FL: 371 – 382.

BROOKS S J, BOLAM T, TOLHURST L, et al., 2008. Dissolved organic carbon reduces the toxicity of copper to germlings of the macroalgae, *Fucus vesiculosus*. Ecotoxicol. Environ. Saf., 70: 88 – 98.

CARLS M G, RICE S D, HOSE J E, 1999. Sensitivity of fish embryos to weathered crude oil: Part I. Low level exposure during incubation causes malformation, genetic damage, and mortality in larval pacific herring (*Clupea pallasi*). Environ. Toxicol. Chem., 18: 481 – 493.

CHERR G N, SHOFFNER-MCGEE J, SHENKER J M, 1990. Methods for assessing fertilization and embryonic-larval development in toxicity tests using the California (USA) mussel *Mytilus californianus*. Environ. Toxicol. Chem., 9: 1137 – 1146.

CLARK J R, 1989. Field studies in estuarine ecosystems: a review of approaches for assessing contaminant effects // COWGILL U M, WILLIAMS L R (Eds.). Aquatic toxicology and hazard assessment. vol. 12. STP 1027. American Society for Testing and Materials, Philadelphia, PA: 120 – 133.

CLARK J R, GOODMAN L R, BORTHWICK P W, et al., 1989. Toxicity of pyrethroids to marine invertebrates and fish: a literature review and test results with sediment-sorbed chemicals. Environ. Toxicol. Chem., 8: 393 – 401.

Code of Federal Regulations, 1990. Environmental effects testing guidelines (TSCA). Part 797-environmental effects testing guidelines, subpart B: aquatic guideline. 40 CFR: 298 – 392.

CONROY P T, HUNT J W, ANDERSON B S, 1996. Validation of a short-term toxicity test endpoint by comparison with longer-term effects on larval red abalone *Haliotis rufescens*. Environ. Toxicol. Chem., 15: 1245 – 1250.

COUILLARD C M, LEE K, LEGARE B, et al., 2005. Effect of dispersant on the composition of the water-accommodated fraction of crude oil and its toxicity to larval marine fish. Environ. Toxicol. Chem., 24: 1496 – 1504.

CRIM R N, SUNDAY J M, HARLEY C D G, 2011. Elevated seawater CO_2 concentrations impair larval development and reduce larval survival in endangered northern abalone (*Haliotis kamtschatkana*). J. Exp. Mar. Biol. Ecol., 400: 272 – 277.

CRIPE G M, 1994. Comparative acute toxicities of several pesticides and metals to *Mysidopsis bahia* and postlarval *Penaeus duorarum*. Environ. Toxicol. Chem., 13: 1867 – 1872.

DEANOVIC L, MARKEWICZ D, STILLWAY M, et al., 2013. Comparing the effectiveness of chronic water column tests with the crustaceans *Hyalella azteca* (Order: Amphipoda) and *Ceriodaphnia dubia* (Order: Cladocera) in detecting toxicity of current-use pesticides. Environ. Toxicol. Chem., 32: 707 – 712.

DINNEL P A, STOBER Q J, LINK J M, et al., 1983. Methodology and validation of a sperm cell toxicity test for testing toxic substances in marine waters. Final report FRI-UW-8306. Fish. Res. Inst., Schl. of Fish., Univ. of Washington, Seattle, WA: 208.

DINNEL P A, LINK J M, STOBER Q J, et al., 1989. Comparative sensitivity of sea urchin sperm bioassays to metals and pesticides. Arch. Environ. Contam. Toxicol., 18: 748 – 755.

DINNEL P A, FARREN H M, MARKO L, et al., 2005. *Clupea pallasi*, bioassay protocols: phase IV. Final technical report. Washington Department of Ecology, Olympia, WA: 26 (appendices).

DODDER N G, MEHINTO A C, MARUYA K A, 2015. Monitoring of constituents of emerging concern (CECs) in California's aquatic ecosystems: pilot study design and QA/QC guidance. Southern California coastal water research project technical report No. 854: 1 – 93.

Environment Canada, 2011. Biological test method: fertilization assay using echinoids (Sea Urchins and Sand Dollars). Method Development and Applications Unit Science and Technology Branch Environment Canada Ottawa. Ontario Report EPS 1/RM/27 2nd Ed. February 2011.

EVANS L H, BIDWELL J R, SPICKETT J, et al., 1996. Ecotoxicological studies in north west Australian marine organisms. Final technical report. Curtin University of Technology.

EVANS T G, CHAN F, MENGE B A, et al., 2013. Transcriptomic responses to ocean acidification in larval sea urchins from a naturally variable pH environment. Mol. Ecol., 22:

1609 – 1625.

FEELY R A, DONEY S C, COOLEY S R, 2009. Ocean acidification: present conditions and future changes in a high-CO_2 world. Oceanography, 22: 37 – 47.

FERNÁNDEZ P A, ROLEDA M Y, HURD C L, 2015. Effects of ocean acidification on the photosynthetic performance, carbonic anhydrase activity and growth of the giant kelp *Macrocystis pyrifera*. Photosynth. Res., 124: 293 – 304.

FINLAYSON B J, HARRINGTON J M, FUJIMURA R, et al., 1993. Identification of methyl parathion toxicity in Colusa Basin drain water. Environ. Toxicol. Chem., 12 (2): 291 – 303.

GAYLORD B, HILL T M, SANFORD E, et al., 2011. Functional impacts of ocean acidification in an ecologically critical foundation species. J. Exp. Biol., 214: 2586 – 2594.

GENTILE J H, SOSNOWSKI S L, 1978. Methods for the culture and short term bioassay of the calanoid copepod (*Acartia tonsa*) // EPA, U S (Eds.). Bioassay procedures for the ocean disposal permit program. EPA 600/9 – 78/010. United States Environmental Protection Agency, Gulf Breeze, FL.

GENTILE J H, JOHNS D M, CARDIN J A, et al., 1984. Marine ecotoxicological testing with crustaceans // PERSOONE G, JASPERS E, CLAUS C (Eds.). Ecotoxicological testing for the marine environment. State University of Ghent and Institute for Marine Scientific Research, Bredene, Belgium.

GIRLING J A, THOMAS K V, BROOKS S J, et al., 2015. A macroalgal germling bioassay to assess biocide concentrations in marine waters. Mar. Pollut. Bull., 91: 82 – 86.

GOODMAN L R, HANSEN D J, MIDDAUGH D P, et al., 1985. Method for early life-stage toxicity tests using three atherinid fishes and results with chlorpyrifos // CARDWELL R D, PURDY R, BAHNER R C (Eds.). Aquatic toxicology and hazard evaluation. ASTM STP 854. Americant Society for Testing and Materials, Philadelphia, PA, USA.

GRIES T H, 1998. Larval bioassay workshop summary workshop proceedings sponsored by the puget sound dredged materials management program. Washington Department of Ecology.

GROTHE D R, DICKSON K L, REED-JUDKINS D K (Eds.), 1996. Whole effluent toxicity testing: an evaluation of methods and prediction of receiving system impacts. SETAC Press, Pensacola, FL.

GULLY J R, BOTTOMLEY J P, BAIRD R B, 1999. Effects of sporophyll storage on giant kelp *Macrocystis pyrifera* (Agardh) bioassay. Environ. Toxicol. Chem., 18: 1474 – 1481.

HALL L W, ALDEN R W, 1997. A review of concurrent ambient water column and sediment toxicity testing in the Chesapeake Bay watershed: 1990 – 1994. Environ. Toxicol. Chem.

HARMON V L, LANGDON C J, 1996. A 7 – D toxicity test for marine pollutants using the Pacific mysid *Mysidopsis intii*. 2. Protocol evaluation. Environ. Toxicol. Chem., 15: 1824 – 1830.

HEPBURN C D, PRITCHARD D W, CORNWALL C E, et al., 2011. Diversity of carbon use strategies in a kelp forest community: implications for a high CO_2 ocean. Glob. Change Biol., 17: 2488 – 2497.

HETTINGER A, SANFORD E, HILL T M, et al., 2012. Persistent carry-over effects of planktonic exposure to ocean acidification in the Olympia oyster. Ecology, 93: 2758 – 2768.

HICKEN C E, LINBO T L, BALDWIN D H, et al., 2011. Sublethal exposure to crude oil during embryonic development alters cardiac morphology and reduces aerobic capacity in adult fish. Proc. Natl. Acad. Sci. USA, 108: 7086 – 7090.

HO K, CAUDLE D, 1997. Letter to the editor: ion toxicity and produced water. Environ. Toxicol. Chem., 16 (10): 1993 – 1995.

HO K T, MCKINNEY R A, KUHN A, et al., 1997. Identification of acute toxicants in New Bedford Harbor sediments. Environ. Toxicol. Chem., 16: 551 – 558.

HOFMANN G E, EVANS T G, KELLY M W, et al., 2014. Exploring local adaptation and the ocean acidification seascape studies: in the California current large marine ecosystem. Biogeosciences, 11: 1053 – 1064.

HOOK S E, LEE R F, 2004. Genotoxicant induced DNA damage and repair in early and late developmental stages of the grass shrimp *Paleomonetes pugio* embryo as measured by the comet assay. Aquat. Toxicol., 66: 1 – 14.

HOOTEN R L, CARR R S, 1998. Development and application of a marine sediment porewater toxicity test using *Ulva fasciata* zoopsores. Environ. Toxicol. Chem., 17: 932 – 940.

HUNT J W, ANDERSON B S, 1989. Sublethal effects of zinc and municipal effluents on larvae of the red abalone *Haliotis rufescens*. Mar. Biol., 101: 545 – 552.

HUNT J W, ANDERSON B S, 1993. From research to routine: a review of toxicity testing with marine molluscs // LANDIS W G, HUGHES J S, LEWIS M A (Eds.). Environmental toxicology and risk assessment, ASTM STP 1179. American Society for Testing and Materials, Philadelphia, PA, USA: 320 – 339.

HUNT J W, ANDERSON B S, TURPEN S L, et al., 1997. Precision and sensitivity of a seven-day growth and survival toxicity test using the west coast marine mysid crustacean *Holmesimysis costata*. Environ. Toxicol. Chem., 16: 824 – 834.

HUNT J W, ANDERSON B S, PHILLIPS B M, et al., 1999. Patterns of aquatic toxicity in an agriculturally dominated coastal watershed in California. Agric. Ecosystems Environ., 75: 75 – 91.

HUNT J W, ANDERSON B S, PHILLIPS B M, et al., 2002. Acute and chronic toxicity of nickel to marine organisms: implications for water quality criteria. Environ. Toxicol. Chem., 21: 2423 – 2430.

INCARDONA J P, COLLIER T K, SCHOLZ N L, 2004. Defects in cardiac function precede morphological abnormalities in fish embryos exposed to polycyclic aromatic hydrocarbons. Toxicol. Appl. Pharmacol., 196: 191 – 205.

INCARDONA J P, CARLS M G, TERAOKA H, et al., 2005. Aryl hydrocarbon receptor-independent toxicity of weathered crude oil during fish development. Environ. Health Perspect., 113: 1755 – 1762.

JAYASINGHE S, KROLL K, ADEYEMO O K, et al., 2014. Linkage of *In Vitro* assay results with *In Vivo* end points final report: phase 1. San Francisco Estuary Institute, Richmond. CA. Contribution #734.

JONES R J, MULLER J, HAYNES D, et al., 2003. Effects of herbicides diuron and atrazine on corals of the Great Barrier Reef, Australia. Mar. Ecol. Prog. Ser., 251: 153 – 167.

KAHN A A, BARBIERI J, KHAN S A, et al., 1993. Toxicity of ambient waters to the estuarine mysid *Mysidposis bahia*//LANDIS W G, HUGHES J S, LEWIS M A (Eds.). Environmental toxicology and risk assessment, ASTM STP 1179. American Society for Testing and Materials, Philadelphia, PA, USA: 405 – 412.

KARBE L, BORCHARDT T, DANNENBERG R, et al., 1984. Ten years of experience using marine and freshwater hydroid bioassays // PERSOONE G, JASPERS E, CLAUS C (Eds.). Ecotoxicological testing for the marine environment. State University of Ghent and Institute for Marine Scientific Research, Bredene, Belgium.

KELLY M W, GROSBERG R K, SANFORD E, 2013. Trade-offs, geography, and limits to thermal adaptation in a tide pool copepod. Am. Nat., 181: 846 – 854.

KOBAYASHI N, NOGAMI H, DOI K, 1972. Marine pollution bioassay by using sea urchin eggs in the inland Sea of Japan. Publ. Seto Mar. Biol. Lab., 19: 359 – 381.

KOCAN R M, 1996. Fish embryos as in situ monitors of aquatic pollution//OSTRANDER G K (Eds.). Techniques in aquatic toxicology. CRC Press, Boca Raton, FL, USA: 73 – 92.

KOCAN R M, HOSE J E, BROWN E D, et al., 1996. Pacific herring (*Clupea pallasi*) embryo sensitivity to Prudhoe Bay petroleum hydrocarbons: laboratory evaluation and in situ exposure at oiled and unoiled sites in Prince William Sound. Can. J. Fish. Aquat. Sci., 53: 2366 – 2375.

KUSK K O, PETERSEN S, 1997. Acute and chronic toxicity of tributyltin and linear alkylbenzene sulfonate to the marine copepod *Acartia tonsa*. Environ. Toxicol. Chem., 16.

LANGDON C J, HARMON V L, VANCE M M, et al., 1996. A 7 – D toxicity test for marine pollutants using the Pacific mysid *Mysidopsis intii*. 1. Culture and protocol development. Environ. Toxicol. Chem., 15: 1815 – 1823.

LASSUS P, LE BAUT C, LE DEAN L, et al., 1984. Marine ecotoxicological tests with zooplankton//PERSOONE G, JASPERS E, CLAUS C (Eds.). Ecotoxicological testing for the marine environment. vol. 2. State University of Ghent and Institute for Marine Scientific Research, Bredene, Belgium.

LEWIS M A, 1995. Algae and vascular plant tests//RAND G A (Eds.). Fundamentals of aquatic toxicology. 2nd ed. Taylor and Francis, Washington, DC: 135 – 169.

LUSSIER S M, KUHN A, CHAMMAS M J, et al., 1988. Techniques for the laboratory culture of *Mysidopsis* spp, Crustacea Mysidacea. Environ. Toxicol. Chem., 7: 969 – 978.

LUSSIER S M, KUHN A, COMELEO R, 1999. An evaluation of the seven-day toxicity test with *Americamysis bahia* (formerly *Mysidopsis bahia*). Environ. Toxicol. Chem., 18: 2888 – 2893.

MCINTOSH S, KING T, WU D, et al., 2010. Toxicity of dispersed weathered crude oil to early life stages of Atlantic herring (*Clupea harengus*). Environ. Toxicol. Chem., 29: 1160 - 1167.

MCKENNEY JR C L, 1998. Physiological dysfunction in estuarine mysids and larval decapods with chronic pesticide exposure//WELLS P G, LEE K, BLAISE C, et al., (Eds.). Microscale tesing in aquatic toxicology, advances, techniques, and practice. CRC Press, Boca Raton, FL: 465 - 478.

MCNULTY H R, ANDERSON B S, HUNT J W, et al., 1994. Age-specific toxicity of copper to larval topsmelt *Atherinops affinis*. Environ. Toxicol. Chem., 3: 487 - 492.

MIDDAUGH D P, ANDERSON B S, 1993. Utilization of topsmelt, Atherinops affinis, in environmental toxicology studies along the Pacific coast of the United States. Rev. Environ. Toxicol., 5: 1 - 49.

MILLER S H, ZARATE S, SMITH E H, et al., 2014. Effect of elevated pCO (2) on metabolic responses of porcelain crab (*Petrolisthes cinctipes*) larvae exposed to subsequent salinity stress. PLoS One 9: e109167.

MORRISON G, PETROCELLI E, 1990. Short-term methods for estimating the chronic toxicity of effluents and receiving waters to marine and estuarine organisms: supplement: test method for coot clam, *Mulinia lateralis*, embryo/larval test. Draft report. U. S. Environmental Protection Agency, Narragansett, RI.

MOULIN L, GROSJEAN P, LEBLUD J, et al., 2015. Long-term mesocosms study of the effects of ocean acidification on growth and physiology of the sea urchin *Echinometra mathaei*. Mar. Environ. Res., 103: 103 - 114.

NACCI D, COMELEO P, PETROCELLI E, et al., 1991. Performance evaluation of sperm cell toxicity test using the sea urchin, *Arbacia punctulata*//MAYES M A, BARRON M G (Eds.). Aquatic toxicology and risk assessment. vol. 14. American Society for Testing and Materials, Philadelphia, PA, USA: 324 - 336.

NACCI D, MORRISON G E, 1993. Standard operating procedures for conducting a sperm toxicity test using the Hawaiian sea urchin *Tripneustes gratilla*. Environmental Research Laboratory: Narragansett Contribution 1516. U. S. Environmental Protection Agency, Narragansett, RI.

NACCI D, COIRO L, KUHN A, et al., 1998. Non-destructive indicator of ethoxyresorufin-O-Deethylase activity in embryonic fish. Environ. Toxicol. Chem., 17: 2481 - 2486.

NEGRI A P, HEYWARD A J, 2000. Inhibition of fertilization and larval metamorphosis of the coral *Acropora millepora* (Ehrenberg, 1834) by petroleum products. Mar. Poll. Bull., 41: 420 - 427.

NIMMO D R, BAHNER L H, RIGBY R A, et al., 1977. *Mydisopsis bahia*, an estuarine species suitable for life cycle toxicity tests to determine the effects of a pollutant. Special technical publication. No. 634 // MAYER F L, HAMELINK J L (Eds.). Aquatic toxicology and hazard evaluation. Proceedings of the First Annual Sumposium American Society for Testing and Materials, Memphis, TN: 109 – 116.

NORBERG-KING T J, AUSLEY L W, BURTON D T, et al (Eds.), 2005. Toxicity reduction and toxicity identification evaluations for effluents, ambient waters, and other aqueous media. Society of Environmental Toxicology and Chemistry (SETAC), Pensacola, FL.

OBERDORSTER E, BROUWER M, HOEXUM-BROUWER T, et al., 2000. Long-term pyrene exposure of grass shrimp *Palaemonetes pugio*, affects molting and reproduction of exposed males and offspring of exposed females. Environ. Health Perspect., 108: 1 – 9.

PADILLA-GAMINO J L, KELLY M W, EVANS J M, et al., 2013. Temperature and CO_2 additively regulate physiology, morphology and genomic responses of larval sea urchins, *Strongylocentrotus purpuratus*. Proc. R. Soc. B, 280 (1759): 20130155.

PHILLIPS B M, NICELY P A, ANDERSON B S, et al., 2003a. Marine bioassay project eleventh report. State Water Resources Control Board, Sacramento, CA.

PHILLIPS B M, NICELY P A, HUNT J W, et al., 2003b. Toxicity of cadmium – copper-nickel-zinc mixtures to larval purple sea urchins (*Strongylocentrotus purpuratus*). Bull. Environ. Contam. Toxicol., 70: 592 – 599.

PHILLIPS B M, ANDERSON B S, HUNT J W, et al., 2004. Marine bioassay project twelfth report. State Water Resources Control Board, Sacramento, CA.

PHILLIPS B M, NICELY P A, HUNT J W, et al., 2005. Tolerance of five west coast marine toxicity test organisms to ammonia. Bull. Environ. Contam. Toxicol., 75: 23 – 27.

PILLARD D A, DUFRESNE D L, CAUDLE D D, et al., 2000. Predicting the toxicity of major ions in seawater to mysid shrimp (*Mysidopsis bahia*), sheepshead minnow (*Cyprinodon variegatus*), and inland silverside minnow (*Menidia beryllina*). Environ. Toxicol. Chem., 19: 183 – 191.

PROSPERI V A, BERTOLETTIA E, BURATINI S V, 1998. Toxicity tests with different age groups of *Mysidopsis juniae* // Conference proceeding. Society of environmental toxicology and chemistry (SETAC) 19th annual meeting. Charlotte, NC.

RAND G M, PETROCELLI S R, 1985. Fundamentals of aquatic toxicology: methods and applications. Taylor and Francis, Washington, DC.

RODGERS J H, DORN P B, DUKE T, et al., 1986. Mysidopsis sp.: life history and culture. Workshop report. Gulf Breeze, FL. American Petroleum Institute, Washington, DC.

ROSE W L, HOBBS J A, NISBET R M, et al., 2005. Validation of otolith growth rate analysis using cadmium-exposed larval topsmelt (*Atherinops affinis*). Environ. Toxicol. Chem., 24: 2612 – 2620.

SANFORD E, GAYLORD B, HETTINGER A, et al., 2014. Ocean acidification increases the vulnerability of native oysters to predation by invasive snails. Proc. R. Soc. Biol. Sci. Ser. B, 281: 20132681.

SCHIFF K C, GREENSTEIN D J, ANDERSON J W, et al., 1992. A comparative evaluation of produced water toxicity // RAY J P, ENGELHART F R (Eds.). Produced water. Plenum Press, New York, NY: 199 – 207.

SCHIFF K, BAY S M, DIEHL D W, 2003. Stormwater toxicity in Chollas creek and san Diego Bay, California. Environ. Monit. Assess., 81: 119 – 132.

SCHIMMEL S C, MORRISON G E, HEBER M A, 1989. Marine complex effluent toxicity test program: test sensitivity, repeatability and relevance to receiving water toxicity. Environ. Toxicol. Chem., 8: 739 – 746.

SCHIMMEL S C, THURSBY G B, 1996. Predicting receiving system impacts from effluent toxicity: a marine perspective//GROTHE D R, DICKSON K L, REED – JUDKINS D K (Eds.). Whole effluent toxicity testing: an evaluation of methods and prediction of receiving water impacts. SETAC pellston workshop on whole effluent toxicity. SETAC Press, Pensacola, FL, USA: 322 – 330.

SCHOLZ N L, INCARDONA J P, 2015. In response: scaling polycyclic aromatic hydrocarbon toxicity to fish early life stages: a governmental perspective. Environ. Toxicol. Chem., 34 (3): 459 – 461.

SHUGART L, BICKHAM J, JAKIM G, et al., 1992. DNA alterations//HUGGETT R J, KIMERLE R A, MEHRLE P M, et al (Eds.). Biomarkers: biochemical, physiological, and histological markers of anthropogenic stress. CRC Press, Boca Raton, FL: 125 – 153.

SINGER M M, GEORGE S, LEE I, et al., 1998. Effects of dispersant treatment on the acute aquatic toxicity of petroleum hydrocarbons. Arch. Environ. Contam. Toxicol., 34: 177 – 187.

SKINNER L, DEPEYSTER A, SCHIFF K, 1998. Developmental effects of urban stormwater in San Diego County, California in medaka (*Oryzias latipes*) and inland silverside (*Menidia beryllina*). Arch. Environ. Contam. Toxicol., 37: 227 – 235.

SNELL T W, JANSSEN C R, 1998. Microscale toxicity testing with rotifers//WELLS P G, LEE K, BLAISE C, et al (Eds.). Microscale testing in aquatic toxicology: advances, techniques, and practice. CRC Press, Boca Raton, FL: 409 – 422.

STEBBING A R D, BROWN B E, 1984. Marine ecotoxicological test with coelenterates //PERSOONE G, JASPERS E, CLAUS C (Eds.). Ecotoxicological testing for the marine environment. State University of Ghent and Institute for Marine Scientific Research, Bredene, Belgium.

STRANSKY C, ROSEN G, COLVIN M, et al., 2014. In situ storm water impact assessment in San Diego Bay, CA, USA//Proceedings, 35th annual meeting of the society of environmental toxicology and chemistry (SETAC). British Columbia, Vancouver.

TAIT K J, STRANSKY C, WELLS D, et al., 2014. Wet weather receiving water evaluation of a rocky intertidal area of special biological significance in La Jolla, CA, USA//Proceedings, 35th annual meeting of the society of environmental toxicology and chemistry (SETAC). British Columbia, Vancouver.

THURSBY G B, ANDERSON B S, WALSH G E, et al., 1993. A review of the current status of marine algal toxicity testing in the United States//LANDIS W G, HUGHES J S, LEWIS M A (Eds.). Environmental toxicology and risk assessment. ASTM STP 1179. American Society for Testing and Materials, Philadelphia, PA, USA: 362 – 377.

TSVETNENKO Y B, EVANS L H, GORRIE J, 1996. Toxicity of the produced formation water to three marine species. Final Technical Report Prepared for Ampolex Ltd, Curtin University of Technology.

TYLER-SCHROEDER D B, 1978a. Entire life-cycle toxicity test using grass shrimp (*Palaemonetes pugio* Holthuis) // EPA U S (Eds.). Bioassay procedures for the ocean disposal permit program. EPA 600/9 – 78/010. United States Environmental Protection Agency, Gulf Breeze, FL.

TYLER-SCHROEDER D B, 1978b. Static bioassay procedure using grass shrimp (*Palaemonetes* sp.) larvae // EPA U S (Eds.). Bioassay procedures for the ocean disposal permit program. EPA 600/9 – 78/010. United States Environmental Protection Agency, Gulf Breeze, FL.

TYLER-SCHROEDER D B, 1979. Use of the grass shrimp (*Palaemonetes pugio*) in a life-cycle toxicity test // MARKING L L, KIMERLE R A (Eds.). Aquatic toxicology. ASTM STP 667. American Society for Testing and Materials, Philadelphia, PA, USA: 153 – 170.

U. S. EPA, 1985a. Standard evaluation procedure: acute toxicity test for estuarine and marine organisms (Estuarine Fish 96 – hour Acute Toxicity). EPA-540/9 – 85 – 009.

U. S. EPA, 1985b. Standard evaluation procedure: acute toxicity test for estuarine and marine organisms (Mollusk 48 – hour Embryo Larvae Study). EPA-540/9 – 85 – 012.

U. S. EPA, 1985c. Standard evaluation procedure: acute toxicity test for estuarine and marine organisms (Mollusk 96 – hour Flow – Through Shell Deposition Study). EPA-540/9 – 85 – 011.

U. S. EPA, 1985d. Standard evaluation procedure: acute toxicity test for estuarine and marine organisms (Shrimp 96 – hour Acute Toxicity). EPA-540/9 – 85 – 010.

U. S. EPA, 1986a. Non-target plants: growth and reproduction of aquatic plants: phase 1 tiers 1 and 2. EPA-540/9 – 86 – 134. Hazard Evaluation Division.

U. S. EPA, 1986b. Standard evaluation procedure: fish early life-stage. EPA-540/9 – 86 – 138. Hazard Evaluation Division.

U. S. EPA, 1986c. Standard evaluation procedure: fish life-cycle toxicity tests. EPA-540/9 – 86 – 137. Hazard Evaluation Division.

U. S. EPA, 1991. Technical support document for water quality-based toxics control. EPA/505/2 – 90/001. Office of Water, Washington, DC.

U. S. EPA, 1995. Short-term methods for estimating the chronic toxicity of effluents and receiving waters to west coast marine and estuarine organisms. EPA/600/R-95/136. Office of Research and Development, Washington DC, USA.

U. S. EPA, 1996. Marine toxicity identification evaluation (TIE): phase I guidance document. EPA/600/R-95/054. Office of Research and Development, Washington, DC.

U. S. EPA, 2002a. Methods for measuring acute toxicity of effluents and receiving water to freshwater and marine organisms. EPA-821-R-02 – 012. Office of Research and Development, Washington, DC.

U. S. EPA, 2002b. Short-term methods for estimating the chronic toxicity of effluents and receiving waters to marine and estuarine organisms. EPA-821 – R-02 – 014. Office of Water, Washington, DC, USA.

U. S. EPA, 2012. Tropical collector urchin, *Tripneustes gratilla*, fertilization test method. EPA/600/R-12/022. Office of Research and Development, Washington, DC, USA.

VAZQUEZ L C, 2003. Effect of sperm cell density on measured toxicity from the sea urchin *Tripneustes gratilla* fertilization bioassay. Environ. Toxicol. Chem., 22: 2191 – 2194.

VOORHEES J P, PHILLIPS B M, ANDERSON B S, et al., 2013. Hyper-salinity toxicity thresholds for nine California ocean plan toxicity test protocols. Arch. Environ. Contam. Toxicol., 65: 665 – 670.

WARD G S, 1995. Saltwater tests // RAND G M (Eds.). Fundamentals of aquatic toxicology: effects, environmental fate, and risk assessment. 2nd ed. Taylor and Francis, Washington, DC.

WARD T J, RIDER E D, DROZDOWSKI D A, 1979. A chronic toxicity test with the marine copepod *Acartia tonsa* // MARKING L L, KIMERLE R A (Eds.). Aquatic toxicology. ASTM STP 667. American Society for Testing and Materials, Philadelphia, PA: 148 – 158.

WEIS P, WEIS J S, 1982. Toxicity of methylmercury, mercuric chloride, and lead in killifish (*Fundulus heteroclitus*) from Southampton, New York. Environ. Res., 28: 364 – 374.

WERNER I, DEANOVIC L A, MARKEWICZ D, et al., 2010. Monitoring acute and chronic water column toxicity in the northern Sacramento-San Joaquin Estuary, California, USA, using the euryhaline amphipod, *Hyalella azteca*: 2006 to 2007. Environ. Toxicol. Chem., 29: 2190 – 2199.

WOELKE C E, 1967. Measurement of water quality with the Pacific Oyster embryo bioassay. Special technical publication 416. American Society for Testing and Materials, Philadelphia, PA: 112 – 120.

WOODWORTH J G, KING C, MISKIEWICZ A G, et al., 1999. Assessment of the comparative toxicity of sewage effluent from 10 sewage treatment plants in the area of Sydney, Australia using an amphipod and two sea urchin bioassays. Mar. Pollut. Bull., 39: 174 – 178.

7　沉积物毒性试验

S. L. Simpson[①], 　O. Campana[②], 　K. T. Ho[③]

[①] 澳大利亚科学和工业研究组织。
[②] 美国纽约大学。
[③] 美国环保局。

7.1 引　言

沉积物是污染物进入城市径流、农业和工业等水体后的最终储存库。河口和沿海海洋环境中的沉积物是水生生态系统的重要组成部分，为底栖生物和远洋生物的早期生命阶段提供了重要栖息地，而这些生物支撑着广阔海洋的食物链。因此，我们需要了解沉积物中污染物的形态、归宿和效应。

沉积物毒性试验是在受控条件下将生物体暴露于沉积物，从而估计受污染沉积物对野外生物的毒性水平（ASTM，2008a；Greenstein et al.，2008；Kennedy et al.，2009；Simpson & Spadaro，2011；Rodrí-guez-Romero et al.，2013）。毒性水平只能通过化学分析推断。在化学分析中，可将测量到的污染物浓度与沉积物质量标准进行比较，以评价沉积物可能产生的有害效应。然而化学分析并不能评估未测量污染物、污染物混合物和非污染物（粒度或其他沉积物特征）的潜在效应，这些物质可能相互作用而产生影响。虽然化学分析可以提供污染物潜在生物利用度信息，但沉积物毒性试验则可反映污染物的生物利用率、污染物混合物的累积效应以及非污染应激源的相互作用，因为这些都直接影响到被测生物。由于群落终点难以确定，相比底栖生态评价（如群落分析、物种多样性、丰度和功能），沉积物毒性试验也往往能提供更多可量化证据来评价污染物影响（Johnston & Roberts，2009；Burton & John-ston，2010；Dafforn et al.，2012；Schleck-at et al.，2015）。

沉积物毒性试验是为了进行一系列的评价而进行的，且这些试验通常具有标准规范。常见应用包括：

（1）为在毒性数据库中具有匹配浓度数据的现有化学物和新型化学物制定基准（如指南、准则、标准），并在《欧盟化学品注册、评价、授权和限制规例》等框架下使用。

（2）调查污染物与环境变量之间的相互作用，如因营养过剩而引起的氨浓度升高，并预测气候条件（如温度）的变化。

（3）比较不同生物对污染物的敏感性和相对暴露途径。

（4）确定污染物浓度、暴露、生物利用度等和毒性之间的关系。

（5）为便于管理，确定受污染场地有毒沉积物的空间分布和等级，包括疏浚沉积物的处置。

（6）制定特异性沉积物中污染物的管理限制或可能的补救措施，并评价其有效性。

（7）实施生态风险评价，聚焦评价结果。

沉积物毒性试验在过去20年中取得了巨大的进展，现在的试验具有更大的环境相关性。这使得研究者更深入地了解物种对污染物的敏感性、生物行为和暴露途径，并提供了更多相关暴露条件的设计考虑。此外，还逐渐从以前占主导的急性致死性试验过渡到考虑多种可能的亚致死和慢性反应的试验方法（Scarlett et al.，2007a；Kennedy et al.，2009；Simp-son & Spadaro，2011；Fox et al.，2014；Simpson et al.，2016）。另外，还有更多方法可用于毒性评价，如使用生物标记物或基因组终点测定细胞反应。这些细胞反应可以单独使用，也可与传统的沉积物毒性试验评价相结合，以更好地了解生物的生存、发育和繁殖。由于毒性试验规程的步骤会对沉积物质量评价结果产生重大影响，因此要获得可靠的试验结果，就需要确保良好的试验设计、检测和报告（Harris et al.，2014）。有效应用沉积物毒性试验需要严谨的态度，了解各种方法的局限性（混淆因素和操作上定义的边界），以及认识到实验室中替代生物的暴露并非总能充分反映自然环境中物种的暴露条件。因此，利用沉积物毒性试验来区分或确定沉积物中个别化学物质的效应仍然具有挑战性。

本章描述了沉积物毒性试验的方法，主要评价污染沉积物对整个生物体的效应，包括生存、繁殖、发育和行为。针对多种生物体的试验，以及对生物标记物终点的评价，已经在第5章进行了讨论。此外，细胞系生物测定中使用分子和生化终点的方法（如以培养细菌和真核细胞为基础的检测和其他替代生物）也不在此进行阐述。这些亚生物水平的反应对于单个化学品的高通量检测越来越重要；但其并不能有助于尝试建立类似沉积物中生物的化学暴露条件和生物利用度。精心设计的生物体水平沉积物毒性试验提供了将各种终点锚定于生物体和种群水平反应的科学依据。本章的重点是介绍基于实验室的全沉积物毒性试验，围隔和现场毒性试验方法将在第8章中阐述。

7.1.1 健全生态毒理学的主要考虑因素

为了能够从沉积物毒性试验中获得一个健全的生态毒理学评价，应考虑一系列原则，从实验设计和测定的基本方面开始，以证明达到了所需的暴露条件，选择适当对照组、参考沉积物及终点，对结果进行无偏差分析。虽然不同的规范可能适用于不同的评价需求，但有些基本规范是实现实际评价所必需的。Harris et al.（2014）已经描述了一些生态毒理学评价所必需的规范，为使之与沉积物毒性评价相关的考虑因素更切合，本章对这些规范进行了修改。

7.1.1.1 充分规划和良好设计的保证

由于试验目的的不同，设计要求也不同。而且试验受到沉积物性质、相关污染物、预期暴露条件和数据分析方法等方面的影响［如确定沉积物中的存在、毒性机理或特定可测化学物质或终点的效应浓度（如 EC_{10}、EC_{50}）］。建议在初步规划过程中制定研究的数据质量目标，包括测试性能和验收标准（USEPA，2006a，b）（见第2.5节）。设计需要考虑如下内容：

（1）物种相关性。若选择一组试验，则需知道被试物种的数量和类型。

（2）需分析的终点，包括预期的灵敏度和允许终点完全表达所需测试持续时间。终点的选择会影响结果从个体水平到群体水平效应进行外推的有效性。

（3）评估特定沉积物是否引起毒性所需的重复次数，或者建立所需的每种浓度的暴露浓度和重复次数。各试验重复的次数可能会有所不同。例如，一个定性测试与一个已确定结果的测试，所需的重复次数就不一样。

（4）正在评价的暴露路径，试验期间预期暴露可能如何变化，以及维持特定暴露的必要步骤；分析与沉积物孔隙水或上覆水浓度相关的效应终点时，或分析暴露路径如何被 pH、酸挥发性硫化物（AVS）和有机碳（organic carbon，TOC）等因素改变时，需测量哪些暴露参数和测量频率。

（5）对食物添加的需求，及其对生物行为和污染物吸收的影响。

（6）确定试验是静态、半静态还是动态。因为不同设计都会影响总体结果。静态和半静态试验可能会高估上覆水中的暴露浓度，而清水流动的动态试验可能低估暴露。

（7）当创建和评价某一特定化学品的暴露试验时，采用哪些方法来评价添加的化合物是否已充分平衡，或生物有效态的浓度是否随时间而变化（如有机污染物的挥发、表层沉积物中 AVS 的氧化增加了金属的生物有效性）。

（8）通过监测手段协助识别混杂因素，这些因素可能与生物行为、超预期的压力源或无特征污染物有关。上述因素可能导致考虑毒性鉴别评价（TIE）操作。

7.1.1.2 反应终点基线的理解和定义

有必要定义未受到各类型沉积物暴露的生物的正常终点响应和可变性。由于更大的遗传变异性，野外采集的生物可能比实验室饲养的生物更具多变性。对于存在较大可变性的试验，可能需要更多的重复，以评价是否观察到毒性。基线值的考虑因素应包括以下几点：

（1）对照组沉积物特性对反应的影响方式（如粒度变化），并利用参考组沉积物对试验沉积物的理化特性分类。这些特性应与结果一起报告。

（2）生物大小、生命阶段、性别（如雌雄行为可能存在较大差异）和个体密度对试验终点的影响。适当地描述试验所使用生物的来源、生命阶段和历史背景很重要。

（3）了解摄食行为和生物营养要求。如提供过多或过少食物将改变生物行为、污染物暴露和反应（特别是生长）。

（4）理解密度和同类相食如何影响试验结果，以及未经试验因素（如寄生虫）如何影响生物健康。

（5）了解对非污染应激源（如粒度）的反应。

7.1.1.3 适当暴露途径和浓度（与所评价的环境相关）的使用，以及进行测量以定义暴露

暴露的环境相关性是一个关键考虑因素，既与试验生物在其自然环境中的预期行为有关，也与试验设计的暴露有关。多种暴露途径可能对沉积物中的生物产生影响。暴露注意事项应包括：

（1）暴露浓度与环境的相关性，例如，与环境赋存相比，暴露浓度不应过高，从而可被认为与环境有关。

（2）试验设计需要建立和维持一个理想的暴露，例如，沉积物是否会在评估地点自然地再悬浮？设计是否应该复制这一点？

（3）确定暴露量和达到评价所需的数据要求所必需的测量，包括在测试和评价目标污染物潜在生物可利用性部分期间出现的不同水－沉积物阶段的实测浓度。

(4) 模拟评价地点暴露条件所必需的半静态和流动试验的水更新率。

(5) 在试验过程中，主要暴露途径是否会因生物生命阶段、行为或污染物形式的变化而改变，以及这将会如何影响结果。

7.1.1.4 结果的统计分析及重复性

每次试验的重复次数，以及每个处理或重复的生物数量将影响试验结果，如是否具有足够的统计意义来判断是否发生了毒性（即假设是否得到统计上的支持或反对）。此外，试验的暴露次数将影响能否能确定如 EC_{10} 和 EC_{50} 效应浓度及其精度。这有利于进行定性试验和避免重复的大型试验。有必要将统计意义与生物或环境意义分开考虑。如果重复之间的方差或标准偏差较小，则结果可能具有统计学意义，但如果这些差异小于10%或15%（相对于对照组），则这些差异在生物学上可能并不显著。例如，如果对照组存活率为100%，标准偏差为0，试验组存活率为90%，标准偏差为5%，两者可能在统计上有所差异，但在生物学上不显著。统计设计的考虑因素包括：

(1) 试验旨在确定特定的暴露或浓度下是否发生显著的毒性效应。

(2) 为建立经验回归模型（如 EC_{10}、EC_{20}、EC_{50} 估算），需建立剂量-反应关系。

(3) 数据用于更复杂的暴露效应模型，考虑随时间变化的效应变化（如毒物代谢动力学/毒物效应动力学模型）。

(4) 如果效应是由生物的一个小变化所体现，如对比于对照组，生长的变化很微小，则需要更多的重复试验和加入更多的试验生物个体。需注意的是，应考虑与暴露和效应有关的重复性，在实验室（暴露量更可控）可能比在实地更容易实现。

(5) 范围毒性研究对效应阈值的推导是否有用，特别当需要进行多次重复试验来确定实际暴露特征，以及如果最初未达到合适的试验浓度范围导致无法重复试验时。

有关统计方面更加具体的描述，本书第 2 章已进行讨论。

7.1.1.5 混杂因素的考虑

许多因素会影响生物的反应和试验结果，虽然其中一些是设计所需的（如温度或盐度的控制），但不可控因素（如疾病、寄生虫）是不被希望出现的，如果可以的话，需要识别和解释。还有一些与试验样本相关的其他因素，这些因素可能会干扰任何浓度反应关系的解释（如暴露于多种未知的物质/应激源）。需要考虑的混杂因素包括：

(1) 特定于试验生物的因素，或来自其他生物的因素，如竞争、捕食、疾病和寄生虫（竞争和捕食相互作用被设计成群落试验，但应在基于实验室的单一生物全沉积物毒性试验中加以控制或消除）。

(2) 沉积物影响因素，如生物膜或真菌生长。

(3) 未知或变化的溶解或颗粒化学形态的暴露，这在计算和报告关于单一暴露途径的效应阈值时尤为重要［如 EC_{50} 以 mg/kg（指定湿重或干重）或 mg/L 为单位来表示］。

(4) 未知应激源（如氨、硫化物）或其他化学物质（数千种人为或自然的化学物质可能存在于沉积物中，通常只有主要污染物才被量化）。

(5) 对照组和参考组沉积物属性范围不匹配（如粒径极端值）。

(6) 试验条件的变化（如为生物提供的营养、光、溶解氧、温度、硬度、盐度等）。

7.1.1.6 结果的无偏差分析和汇报

其包括不同处理组以及处理方法和样本名称使用之间的生物样本随机分配等因素。这些因素有助于对终点和相关分析进行"盲"的无偏差评价。

有必要对试验一无预期或压力，以实现理想的剂量－反应关系或一个有特定结果的评价。在分析、总结和讨论结果时，其他需要考虑的因素包括：

(1) 系统讨论与可变性和可重复性有关的不确定性。

(2) 讨论已达到的暴露条件，及其在目标实地环境沉积物中可能具有的代表性。

(3) 实验室所观察到的剂量反应是否可能存在于实地。例如，野外环境中大于预期的水交换率或水流是否会导致不同的结果？

(4) 非单调剂量反应（如那些没有建立理想"S"形曲线的）是否可能是混杂因素（包括不良的实验设计或技术）导致的结果。

(5) 根据所有证据（暴露、条件、观察、测量、反应）讨论暴露浓度－反应是否合理。

(6) 考虑沉积物的效应是否来自于颗粒物[EC_{50}单位为 mg/kg（规定湿重或干重）]或溶解态（EC_{50}单位为 mg/L）暴露途径，以及所描述暴露是否会真实地发生于实地。当上覆水中的溶解污染物浓度高于受污染的实地环境时，这一点尤为重要。

应注意不要在试验或环境暴露之外进行过度外推。

7.1.2 试验生物选择的考虑因素

底栖生物几乎存在于所有的自然沉积物环境中，且行为各异，从而导致不同的污染物暴露途径和敏感性。由于这种多样性，没有一种底栖生物最适合于所有类型的污染沉积物的生态毒理学评价；理想情况下，应使用来自不同门和暴露途径的一系列生物。但是，由于试验生物和终点的选择，以及试验设计可能会对评价方案的结果产生重大影响，因此需要慎重选择试验生物。项目评价中选择的试验生物将作为沉积物生态系统中其他被保护生物的替代对象。使用替代物种进行环境风险评价的基础是其与目标生物具有相似的 DNA；因此，它们还应具有相似的污染物反应系统。但与生命阶段、行为和暴露途径敏感性相关的试验内容将影响替代生物在整个生态系统中的代表性。作为沉积物生物试验中生物选用的一般指南（ASTM，2008b），理想情况下，试验生物应符合以下条件：

(1) 正常行为状态下可与沉积物有直接接触，从而能通过水和沉积物暴露途径评价污染物的潜在影响，而非使用通常不与沉积物接触的生物或生物的生命阶段。

(2) 对已知的污染物种类，试验生物需有显著的敏感性，至少不会对已知的污染物不敏感。

(3) 能够耐受复杂的沉积物理化特性（如，粒度、盐度）。

(4) 与选定的暴露方法和终点兼容。

（5）具有标准化的应用和质保规程，以提供实验室间的比较方法。
（6）可通过培养或野外收集获得。
（7）易于在实验室维护。
（8）易于鉴定和识别（特别是从野外所采集的）。
（9）生命周期短或中等（数天至数周），从而无需进行长时间（数月）试验即可评价对生殖的影响。

这些标准很少有试验物种能够全部满足，但它们是进行沉积物生态毒理学评价设计时需要考虑的重要因素。前2个条件，即直接接触沉积物和影响终点对潜在目标污染物的敏感性，可能对评价结果的影响最大。由于使用的生物为替代物种，因此试验开展时，使用物种应为被评价地点的本土（现存的或历史上的）物种，或者与本土物种或其他在生态或经济上可能具高保护价值的生物有类似生态位（如行为和摄食）的物种。

7.1.2.1 污染物暴露途径和敏感性

底栖生物与溶解态和颗粒态污染物相互作用，并暴露于这些污染物中，其暴露方式包括从孔隙水和上覆水中吸收、直接的食物摄入、直接或无意中摄入沉积物（Rainbow，2007；Simpson & Batley，2007）。与沉积物颗粒相关的污染物浓度通常比孔隙水或上覆水中的多100～10000倍（Hassan et al.，1996；Simpson & Batley，2007）。由于具有更大的表面积，细颗粒上的污染物浓度往往大于粗颗粒上的污染物浓度（Chariton et al.，2010）。虽然溶解态污染物被认为是可能更具生物可利用性的污染物形式，且孔隙水是许多底栖生物接触污染物的重要途径，但通过沉积物颗粒摄入而暴露于污染物的可能性往往更大（Rainbow，2007；Casado-Martinez et al.，2010；Strom et al.，2011；Campana et al.，2012）。将孔隙水从沉积物中分离出来时，其污染物浓度通常会因挥发、沉淀或氧化而降低，因此在孤立的孔隙水中难以维持与实际环境相符的暴露条件。综上，使用沉积物栖息生物或生物群进行全沉积物毒性试验是获取所有暴露途径的最适合方法。

底栖生物常以沉积物为食，从沉积物表面或沉积物内部摄取生物体、有机和无机颗粒以获取营养。许多物种，尤其是多毛类蠕虫，会摄取地表下的沉积物，并将其转化为为粪粒输送到沉积物－水界面。这些不同的行为值得我们关注，因为这些行为既包括在进食过程中可能导致的不同暴露途径，也包括对沉积物化学性质的影响，以及在试验过程中可能发生的暴露。这些"生物扰动"活动改变了污染物在溶解态和颗粒态之间的分配，频繁地增加了污染物以溶解态（来自孔隙水）和颗粒态（细悬浮固体）从沉积物向水体中的释放（Aller et al.，2001；Ciutat & Boudou，2003；Belzunce et al.，2015）。生物扰动促进了沉积物与上覆水的相互作用，并使污染物以氧化形式存在于表面沉积物层和洞穴壁内（Peterson et al.，1996；Gerould & Gloss，1986；Simpson et al.，2012；Volkenborn et al.，2010）。毒性试验过程中溶解和颗粒通量的比例变化可能会改变污染物的暴露，影响试验结果，因此在试验设计期间需要加以考虑，并在可能的情况下进行监测。

底栖生物对沉积污染物的敏感性因物种和生物生命阶段的不同而差异显著。生物的穴居行为（如管栖生活或浮游生活、生物扰动活动、生物灌溉率）和摄食习惯（滤食或沉积摄食，以及它们对食物的选择性）会影响其暴露途径和敏感性。因此在条件允许的情况下，沉积物毒性评价应对具有不同行为和食性的生物进行一系列测试，以涵盖各种潜在的污染物暴露途径。有时评价方案可能需要使用本地物种；但使用本地物种来开发公认的标准测试规程通常非常耗时，且常常在许多试验中无法执行。此外，与替代物种相比，本地物种对潜在目标污染物质的敏感性可能更高或更低。

一般来说，没有一种物种对所有污染物都敏感，这也就解释了为何要建议对不同的门、食性和暴露方式进行试验。有些物种对许多污染物特别不敏感，如海洋蠕虫（*Nereis virens*）和卤虫（*Artemia* sp.）。这些生物可用于其他类型的测试，如海洋蠕虫（*N. virens*）由于其体积大，且能够生活在其他较敏感物种无法生存的受污染沉积物中，而常被用于生物累积试验（USEPA，1993）。卤虫（*Artemia* sp.）是一种易于培养的动物性饲料，而可被用于多种毒性和生物累计试验，因为其能在受污染的系统中生存，从而确保试验生物具有足够的营养供给。

7.1.2.2　潜在试验生物

大多数底栖生物物种已被应用于沉积物毒性试验中，包括细菌、藻类、甲壳类动物（如端足类、桡足类、糠虾类）、软体动物（双壳类/蛤类）、环节动物（多毛类动物）、腹足类动物（蜗牛）、线虫和棘皮动物（海胆、海参、海星）等（图7.1）。它们在底栖生物群落中扮演重要角色，其食物来源有细菌、藻类以及其他动植物碎屑，它们也是大型无脊椎动物、幼鱼和水鸟的重要食物来源。许多底栖生物直接摄取沉积物颗粒，因而直接暴露于沉积物结合污染物以及孔隙水和上覆水中的污染物。ASTM（2012a）讨论了选择本地物种作为试验生物的多种重要考虑事项。表7.1列出了经常被用于沉积物毒性试验的底栖生物种类。

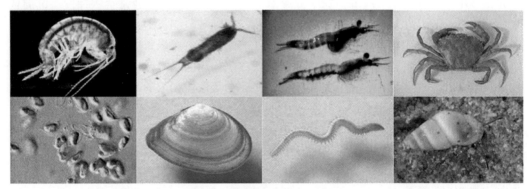

图7.1　全沉积物毒性试验物种

（上图，从左至右）端足类、桡足类、糠虾和蟹类（甲壳类动物）；（下图，从左至右）底栖藻类（植物）、双壳类/蛤类（软体动物）、多毛类蠕虫（环节动物）和蜗牛（腹足动物）。

细菌和藻类是所有沉积物的重要组成部分，可作为许多底栖生物的食物来源，并促进一些沉积物污染物向更高营养级转移。两者都会影响有机质降解、养分循环、有机化合物降解以及沉积物 – 水界面金属的再活化。细菌（与藻类）形成生物膜，此两门生物都可稳定表层沉积物和形成微生境。细菌或藻类无直接摄取颗粒的途径，因此毒性试验只评价孔隙水或上覆水中的污染物暴露。

对于河口和海洋沉积物，市售的 Microtox 试剂盒（Azur Environmental，1998）提供了细菌毒性测试的最佳示例。该方法使用海洋发光细菌费氏弧菌（*Vibrio fischeri*）作为试验生物，在暴露于沉积物 20 min 后，通过与对照组比较来评价其光输出的减少程度。该方法被认为是筛选和提供生态毒理学信息的实用工具，有助于对大量沉积物进行排序，从而快速确定潜在目标区域。该方法存在一个问题，即沉淀物浊度会干扰发光细菌的光输出，但这通常可以通过适当的控制变量来调控。大多数沉积物的藻类生物测定法的终点是评价对叶绿素产生（影响光合作用）和酶抑制的影响，而不是对藻类生长抑制作用的测量，因为藻类生长受沉积物中氨或营养物浓度的强烈影响（Adams & Stauber，2004）。热带河口常见的红树植物，由于试验困难，且对许多常见污染物的敏感性普遍较低，在毒性试验中的所占比例不足。

许多甲壳类动物已被成功地用于沉积物毒性评价，其中最典型的是端足类、小型哲水蚤和糠虾类。它们符合用于全沉积物毒性试验测试生物的多个选择标准，可采用一系列标准化或同行评审的方法（ASTM，2014；Greenstein et al.，2008；Perez-Landa & Simpson，2011）。由于小型甲壳类动物的生命周期通常很短（如许多桡足类生命周期为 20~30 天），而桡足类在达到完全形态之前要经历多个幼虫阶段和中期阶段（如无节幼体到桡足幼体到成体桡足繁殖），因此它们往往适用于进行短期亚慢性试验或生命周期全沉积物毒性试验（Chandler & Green，1996；Hack et al.，2008；Simpson & Spadaro，2011；Araujo et al.，2013）。试验终点通常包括生存、生殖和发育。蟹类虽然与沉积物密切相关，但尚未广泛应用于沉积物试验；然而，在仅考虑水污染物暴露的研究中，这些物种得到广泛应用（Rodrigues & Pardal，2014）。

由于生命周期长且复杂，使用软体动物、环节动物、腹足动物和棘皮动物进行的沉积物毒性试验不如使用甲壳动物常见。软体动物双壳类分布广泛，丰度高，易处理，其中许多物种在沉积物-水界面进食（Ringwood & Keppler，1998；Keppler & Ringwood，2002；King et al.，2010；Campana et al.，2013）。因此，与颗粒（食物）和水体中溶质相结合的污染物是重要的暴露途径（Griscom & Fisher，2004；King et al.，2010；Campana et al.，2013）。成年牡蛎未被用于全沉积物毒性试验（Ringwood，1992），但幼虫期和幼年期可被用于孔隙水试验，并可用于研究污染物从悬浮颗粒释放以及释放到水体的效应（Edge et al.，2012，2014）。多毛类（海底蠕虫）可能生活在人工建造的管道中，或在沉积物中自由生活，通过摄入的沉积物、孔隙水和上覆水，以及通过一些有机污染物的直接扩散与皮肤直接接触，通常暴露于与沉积物有关的污染物中（Bat & Raffaelli，1998；Morales-Caselles et al.，2008；Ramos-Gomez et al.，2011；Farrar & Bridges，2011）。腹足类动物、线虫和棘皮动物的行为或生命周期通常很复杂，并且难以评估比生存更敏感的终点，因此在沉积物毒性实验中所占比例较低（Ringwood，1992）。

表 7.1　河口和海洋全沉积物毒性试验

组织	试验物种	持续时间/终点	急性/慢性	文献
细菌	*Vibrio fischeri*（氨苯磺胺）	20 min 光反应	急性	Azur Environmental（1998），Environment Canada（2002）
原生动物	*Euplotes crassus*	浸提实验：8 h 细胞活力；24 h 反应	急性	Gomiero et al.（2013）
微藻	*Entomoneiscf punctulata*	24 h 酶（酯酶）抑制反应	急性	Adams & Stauber（2004），Adams（2016）
微藻	*Cylindrotheca closterium*（原名 *Nitzschia closterium*）	72 h 生长率	慢性	Moreno-Garrido et al.（2003a, b, 2007），Araujo et al.（2010）
桡足类	*Amphiascus tenuiremis* & *Microarthridion littorale*	14 d 存活和繁殖；21 d 全生命周期	慢性	Chandler & Green（1996），Kovatch et al.（1999），Kennedy et al.（2009）
桡足类	*Nitocra spinipes*	10 d 繁殖	慢性	Perez-Landa & Simpson（2011），Simpson & Spadaro（2011），Krull et al.（2014），Araujo et al.（2013），Spadaro & Simpson（2016a）
桡足类	*Robertsonia propinqua*	24 d 寿命周期试验	慢性	Hack et al.（2008）
桡足类	*Tisbe biminiensis*	7 d 繁殖	慢性	Araujo et al.（2013）
糠虾	*Americamysis bahia*	10 d 存活	急性	Kennedy et al.（2009）
端足类	*Ampelisca brevicornis*	28 d 存活，繁殖和生长	慢性	Costa et al.（1998, 2005）
端足类	*Corophium multisetosum*	10 d 存活；21 d 繁殖和生长	急性 慢性	Casado-Martinez et al.（2006），Castro et al.（2006）
端足类	*Corophium volutator*	28 d 存活和生长；28 d 和 76 d 生存，生长和繁殖	慢性	Scarlett et al.（2007a），Fox et al.（2014）
端足类	*Ampelisca brevicornis*, *Corophium volutator*, *Eohaustorius estuarius*, *Leptocheirus plumulosus*, *Rhepoxynius abronius*	10 d 存活	急性	Rodrıguez-Romero et al.（2013），ASTM（2014），Greenstein et al.（2008）
端足类	*Gammarus locusta*	28 d 存活，繁殖和生长	慢性	Costa et al.（1998, 2005）
端足类	*Hyalella azteca*（高至 15 ppt）	10 d 和 28 d 存活与成长；42 d 存活，生长和繁殖	慢性	ASTM（2010）
端足类	*Leptocheirus plumulosus*	28 d 繁殖和生长	慢性	ASTM（2008a），Kennedy et al.（2009）
端足类	*Melita plumulosa*	幼年期 10 d 存活	急性	Spadaro et al.（2008），Strom et al.（2011）

续表7.1

组织	试验物种	持续时间/终点	急性/慢性	文献
端足类	*Melita plumulosa*	10 d 繁殖	慢性	Mann et al.（2009），Simpson & Spadaro（2011），Spadaro & Simpson（2016b）
双壳类	*Mercenaria mercenaria*	幼体 7 d 生长	亚致死	Ringwood & Keppler（1998），Keppler & Ringwood（2002）
双壳类	*Tellina deltoidalis*	10 d 存活	急性	King et al.（2010）
双壳类	*Tellina deltoidalis*	30 d 存活和生长	慢性	Campana et al.（2013），Spadaro & Simpson（2016c）
多毛类蠕虫	*Arenicola marina*	10 d 和 21 d 存活	急性	Bat & Raffaelli（1998），Morales-Caselles et al.（2008）
多毛类蠕虫	*Neanthes arenaceodentata*	20 d 存活，28 d 生长	慢性	Bridges & Farrar（1997），Farrar & Bridges（2011）
多毛类蠕虫	*Nereis virens*	7 d 回避行为，身体机能受损	急性	Van Geest et al.（2014a, b）
贻贝	*Mytilus galloprovincialis*	胚胎 2 d 在沉积物-水界面中发育	亚致死	Anderson et al.（1996），Greenstein et al.（2008）
蜗牛	*Hydrobia ulvae*	暴露后 48 h 喂食，24 h 回避	亚致死	Krell et al.（2011），Araujo et al.（2012）

7.1.3 潜在试验终点

为了最大限度地利用沉积物毒性试验结果来帮助受污染沉积物控制方案的管理决策，必须考虑所评价效应的生物学和生态学相关性，以选择最合适的试验终点。沉积物毒性试验最常用于提供污染物在生物水平效应的信息。这些效应包括生存、生长、发育、繁殖和行为。通过将这些试验终点纳入模型，许多试验终点可用于预测化学品可能在种群水平构成的风险（Kuhn et al.，2000，2002）。越来越多的亚生物体水平终点被开发用来预测生物体水平的效应（如生物标记物反应或细胞毒性、遗传毒性、免疫毒性和内分泌效应的测定）。当然，我们有必要理解毒性更高层次的机制，以确定从亚生物体到生物体再到种群水平的效应；其中一种方法是有害结局路径（AOP）模型（Ankley et al.，2010）。AOP 是毒物和种群变化之间的一种形式化联系。AOP 力求阐明并尝试通过关系将关键事件联系起来，直到达到不良结果为止。关键事件可以发生在组织的关键水平，包括分子、细胞、器官、生物体和种群水平。AOP 中所有关键事件的记录和说明可能非常复杂，而且很难完成。在本章中，我们主要讨论生物水平的终点。第5章讨论的是生物标志物的使用，第6章讨论的是细胞毒性终点。

测试终点通常被称为急性或慢性，尽管有时也常使用亚慢性、亚致死和致死等术语。一项试验是急性还是慢性取决于所考虑的物种和终点，该定义并不严格。慢性毒性试验可定义为物种在至少一个完整生命周期内暴露于受污染沉积物,或物种在一个或多个关键和敏感生命阶段暴露于受污染沉积物。而该定义的后半部分更好地描述了亚慢性毒性试验,在该试验中,暴露时间不包括整个生命周期,但至少发生在生物体生命周期的一个关键阶段(如生殖繁育:Mann et al.,2009;Perez-Landa & Simpson,2011)。许多生物体有相对较长的生命周期(如超过4个月),在实际应用中,暴露期超过该生物体生命周期的10%就通常被认为足以描述一种慢性试验。因此除细菌(低于24 h)和藻类(48~72 h为生长速率终点)外,10天的暴露时间对大多数生物而言通常被认为是慢性的。28~60天的持续暴露时间通常用于评价污染物对端足类、双壳类和蠕虫物种生存、生长和生殖的慢性效应(ASTM,2008a)。对于生命周期较短(如20~60 d)的端足类和桡足类生物,可以使用持续时间为10天的亚慢性试验来评价对其生殖和发育的效应(Mann et al.,2009;Perez-Landa & Simpson,2011)。急性毒性通常是指相对于生物体整个生命周期较短的暴露时间所产生的有害效应;急性致死试验是使用最广泛的试验终点。急性沉积物毒性试验通常在暴露4~10天后评价存活,不过对行为终点(如回避行为和暴露后摄食)的评价期可能短至48 h。

沉积物毒性评价中最常用的试验终点是暴露期为10天的幼年或青年成体的急性存活(USEPA,1994)。然而现在人们普遍认识到,这些试验可能忽视了长期暴露或对其他敏感生命阶段潜在效应的观测(Simpson & Spadaro,2011)。成体生物对污染物的敏感度远低于其处于胚胎期或早期生命阶段(Williams et al.,1986;Hutchinson et al.,1998)。亚致死试验是为了评价毒物对生物体的慢性或亚慢性效应,包括生物量(如个体平均体重)、生长(如平均体重的变化,有时为体长或表面积的变化)、发育(幼体-幼体-成年期)和生殖(如生育力、繁殖力或所产生子代数量)的变化,通常比急性生存试验更敏感。亚致死试验终点为预测对群落水平的长期效应提供了更多信息,所以往往更适合于风险评价(Kuhn et al.,2000,2002)。评价亚生物体水平效应(如生理或生化反应的生物标志物)和行为反应(如回避行为和摄食)的试验终点为识别潜在的个体健康损害提供了有用的信息,但出于风险评价的目的,这些终点提供的实际信息通常被认为更难与潜在的种群水平效应联系起来。许多研究者比较了致死和亚致死试验终点的敏感性(Scarlett et al.,2007a;Greenstein et al.,2008;Kennedy et al.,2009;Simpson & Spadaro,2011)。

7.1.3.1 行为试验终点

生物对污染物和非污染物压力的反应可能会表现出行为变化,这些行为变化可作为沉积物毒性试验的终点(Scarlett et al.,2007b;Hellou,2011)。行为终点通常被认为是"早期预警"反应,因为快速反应可以提供即时保护,以规避压力源,并可能代表早期反应(如一旦检测到污染物时,停止滤水或进食);而一些较长期的行为反应可能会有不良的后果,如底栖生物(端足类或双壳类)无法钻洞、运动障碍(如蛇尾类的僵直)或交配和繁殖失败而导致对污染物的摄食的增加。矛盾的是,生物如不能规避污染物则增加了不良效应产生的可能性(Ward et al.,2013a)。污染物暴露后的摄食减少是沉积物毒性试验中使用最广泛的行为反应终点(Moreira et al.,2005,2006;Krell et al.,2011;Rosen & Miller,2011),其次是回避行为反应(Ward et al.,2013a,b)。

与回避行为相比，与摄食有关的行为反应可能更容易与潜在的机体水平效应联系起来，如生物体生长、发育和繁殖。

7.1.3.2 生物标志物、基因表达、遗传毒性和基于细胞的生物试验终点

第5章已详细地讨论了亚生物体水平试验终点［评价对生理和生化反应（如生物标记物或基因表达变化）的效应］。虽然理论上发生在分子水平的变化可能比在更高生物体水平的反应更敏感、更具体，但目前尚未有证据证明，与全沉积物试验中确定的亚致死终点相比，这些分子水平的反应能否提供更高的敏感性或降低可变性（Martín-Díaz et al.，2004；Simpson & Spadaro，2011；Edge et al.，2014）。最常被评价的生物标志物是有关于生物转化和抗氧化酶活性以及细胞氧化损伤的生化指标，如溶酶体的不稳定性（Monserrat et al.，2007；Edge et al.，2012；Martins et al.，2012；Taylor & Maher，2016）。对整个生物体进行更大范围的生物标记物检测可提供更多证据说明生物体暴露于污染物，但这些并非是直接的效应证据，尽管后者正在不断增加（Boldina-Cosqueric et al.，2010；Taylor & Maher，2010；Hook et al.，2014a；Regoli & Giuliani，2014）。

基于分子的生物标志物正越来越多地被提议作为试验终点，包括在评价基因表达变化（RNA提取）时被称为生态毒理基因组学和转录组学的生物标记物（目的是识别开启或关闭的基因类别），或用来识别蛋白质或代谢物产生率变化的蛋白质组学和代谢组学（Biales et al.，2013；Hook et al.，2014a）。这些终点具有提供特定化学物质生物可利用部分的分子指纹的潜力；但这些终点需要通过AOPs（Ankley et al.，2010；Lee et al.，2015）锚定到生物体水平（如繁殖结果）和/或种群水平（如种群增长）的响应。开发详细和有用的AOP还不是很常见，本章后面也将有所讨论。

基于生物标记物和分子反应的试验终点的可变性通常很高，且可能对许多非污染因素更敏感，如盐度和光照条件的变化。然而这些技术的进一步发展可能识别出对单种污染源和非污染压力的更具体响应。

污染物与生物体细胞内遗传物质的相互作用可能导致一系列遗传毒性/致突反应，从而可能导致发育异常（致畸性）和癌症形成（致癌性）效应。基于细胞的生物检测可用于化学物质内分泌干扰潜力的高通量筛选（例如，CALUX；USEPA（2014）Method 4435-59）。但是对于沉积物毒性评价，如用于沉积物化学提取的评价（Li et al.，2013；Gao et al.，2015），由于污染物的生物利用度没有很好地反映在所观察到的响应中，因此许多评价终点的相关性大大降低。

7.1.3.3 时间依赖性终点

生物体对毒物的作用效应不仅取决于剂量，还取决于暴露时间。当生物体内一系列生物扰动产生不良反应时，毒性就发生了。何时能观察到结果取决于毒物的类型、环境条件、种类和终点的选择。

诸如存活、回避或静止等终点是质反应的，这意味着我们看不到一个渐进的效应，而只能观察反应的存在与否（如被观察到反应个体的比例），这取决于暴露的时间长短。生死显然是不可逆的，但如果消除压力，行动可能会恢复。恢复时间是一个非常有

价值的反应，特别是可以在生态风险评价中用来测定许多终点，以调查生物体/种群的"可塑性"。可塑性，即在压力消失后恢复到原始状态（在干扰事件之前，如对测定终点不再产生有害效应的毒物浓度）的能力。

分级终点是指随时间推移以不同程度量化的效应，如生长、生殖、摄食速率、生化或遗传生物标记。选择终点时需着重考虑的是测量压力响应的时间尺度。时间尺度从几分钟到几天不等。没有一种时间尺度适用于所有终点，因为某些终点需要比其他终点更长的暴露时间才能观察到效果。敏感性是每个依赖于时间成分终点的特征。若选择的时间尺度太短，则响应将无法被检测到，因为无法从自然变化（如酶的激活或抑制、DNA改变）中识别信号（Regoli et al., 2004）。同样，若等所有效应都显示出来后再进行测定终点，那么潜伏反应可能被遗漏或变得不明显。例如在评价生殖终点时，毒性试验中并不会对子代的反应进行常规观察，在许多典型脉冲暴露（泄漏事件、偶发径流事件或周期性农药处理）的情况下，也未考虑暴露时间（Zhao & Newman, 2007）。

7.1.4 干扰因素

沉积物中生物的响应会受多种非生物和生物因素的影响，因此在试验设计和结果解释时，考虑这些因素很重要。干扰因素可能会产生协同和拮抗影响，从而影响对某些沉积物污染物效应的归因。当污染程度较低时，可能会产生大于预期的毒性（相对于控制组反应的负面效应）；或者当污染物达到可产生预期毒性的浓度时，却未观察到明显的毒性（Bridges et al., 1997; Spadaro et al., 2008）。

在毒性试验的研究和解释中，对试验物种的潜在干扰因素和耐受限度的了解是一个重要因素。非生物因素包括沉积物特性（如粒径）、试验条件（如温度、盐度、光照）及其可变性（如沉积物异质性）的影响，以及自然产生的压力因素（如氨和硫化物）的影响。生物因素包括疾病、捕食、同类相食和营养需求等，这些因素可能受到食物添加、有机物、细菌、藻类和/或沉积物中的小型生物（如小型底栖动物）数量和质量的影响。

通过适当的测试设计，包括提供足够的、优选的猎物和避免试验生物密度过大，可以避免实验系统中的捕食或同类相食等干扰因素。每个试验容器中生物的数量应进行优化，以提供最大的统计功效，同时确保生物不会受到捕食和非化学污染负面反应的胁迫。可通过适当控制检测到养殖或野外采集动物的疾病，通常存活率为80%～100%，生长或繁殖率为70%～100%。试验生物，特别是人工培养的，还应进行季度对照试验，以确保其健康。应保留每季度培养反应的参照表，以确保生物体对化学刺激的反应一致。

物理化学条件，包括沉积物粒度分布，应在所选试验生物已知的耐受限度内（Moore et al., 1997）。一些生物对沉积物粒度分布的耐受范围较窄，其阈值往往存在于粉质或砂质沉积物范围的两端。这些生物的耐受性可能受生物体穴居行为的影响，也可能因食物的供应而改变（在基本营养物质非常低的沙质沉积物中，生物的耐受性可能会增加）。含有致密粘土的沉积物可能会对生物体造成压力，因为生物体很难挖洞。控制组和对照组的沉积物应涵盖试验沉积物的粒径和盐度范围。

溶解态氨和硫化物自然存在于大多数沉积物中，但人为输入（如营养物质和有机碳）会大大增加其浓度。许多营管栖的底栖无脊椎动物通过它们的洞穴进行上覆水循环，以提供氧气，并降低它们暴露在包括氨和硫化氢在内的孔隙水污染物中的风险（Knezovich et al., 1996；Wang & Chapman, 1999）。但在利用物种不栖息于孔隙水中的生命阶段（如双壳类和海胆幼体试验）进行孔隙水试验, 孔隙水中氨和硫化物浓度可能导致有毒暴露（ASTM, 2012a）。对氨和硫化物的耐受性取决于生物体的敏感性和行为，应在试验可接受标准中规定耐受范围。试验过程中，上覆水 pH、温度、曝气方式的差异会影响氨和硫化氢的毒性，随着 pH 的增加和温度的升高，NH_3 中毒性较强的非离子态的相对比例增加，而 H_2S 中毒性较弱的非离子态的相对比例降低（Miller et al., 1990；Wang & Chapman, 1999）。

在试验设计时，应测定孔隙水和上覆水中的氨和硫化物浓度，并考虑实验室毒性试验在上覆水中的暴露情况。为达到所需条件，可以适当修改试验设计，如可以提高上覆水的更新速度以降低氨浓度（Word et al., 2005）。然而在一项对 30 个沉积物进行全沉积物 TIE 分析的综述中，未有全沉积物试验显示硫化物毒性，仅有 2 个显示出氨毒性（Ho & Burgess, 2013）。早期文献强调氨是沉积物中一种重要的毒物（Ankley et al., 1990），这可能是孔隙水试验的人为结果。为减少如氨之类干扰因素的影响而对沉积物进行操作时，应考虑到这些操作也可能会改变其他水溶性污染物浓度或生物利用度。由于硫化物的挥发性，及其与氧气结合后可产生毒性较小的硫酸盐，正常处理沉积物（在曝气的沉积水试验系统中的混合、分配和 24 h 平衡）可能会将硫化物中等浓度水平降低到毒性阈值以下。pH 和盐度的变化也可能改变其他污染物（如金属）的生物利用度和毒性（Ho et al., 1999a；Riba et al., 2004）。如果氨是一种主要毒物，可以使用 TIE 规程识别并将其除去（见第 7.3 节）。

食物（营养）的可用性可能经常通过改变摄食行为而影响试验结果，受试生物有可能选择摄食人为添加的纯净食物，而不是从试验沉积物中自然搜寻和获取食物。但对于有机物含量很少的沙质沉积物，可能有必要补充食物以使生物体在长期暴露的情况下能够在正常范围内生长发育。用无污染的食物喂养受试生物，可能会因选择性摄食而减少受试生物暴露于沉积物中污染物的风险，需要对此进行进一步评估（McGee et al., 2004；Spadaro et al., 2008）。McGee et al.（2004）观察到急性和慢性毒性的差异可能与试验中所使用的食物有关，添加食物可能会改善慢性暴露中的毒性效应。相反，如果已知污染物，且食物的加标方法确定可靠，则可将受污染或加标的食品添加到系统中。然而，确保提供特征良好的加标食品增加了毒性试验步骤的复杂性。如果能达到试验沉积物类型范围的可接受终点反应，一般建议不提供额外的食物；当然应提供足够的营养，以免损害受试生物体的健康。非典型小型底栖动物的存在会影响受试生物的食物可利用性，如当提供额外的食物时，线虫数量会迅速增加，并可能导致对受试物种的竞争胁迫。此外，过量的食物也可能导致氨浓度升高、藻类或细菌生长。

光致毒性是控制沉积物中某些污染物（如某些多环芳烃）毒性的重要因素之一（Ankley et al., 1997；Fathallah et al., 2012）。由于光致毒性可能会在野外自然发生，因此实验室试验期间使用的光线条件、水深和浊度需尽可能接近野外条件。即使阳光不能穿透混浊的水体，但沉积物中的生物幼虫可能会依然受到影响（Pelletier et al., 1997）。

7.2 如何进行沉积物毒性试验

以下章节的目的是依次说明在进行沉积物毒性试验时所采取的每个步骤和所记录的信息。本节假设已经根据上述考虑因素选择了试验物种和终点。具体涉及引言中讨论的许多关于生态毒理学的诸多考虑因素，并强调了设计时的考虑因素。这些设计考虑因素可以应用于符合环境实际的暴露条件下，对生物对污染沉积物的反应进行统计学上的可靠评价。

7.2.1 沉积物收集和准备

7.2.1.1 收集与储存

为沉积物毒性评价而收集和试验的样品数量可通过一个更广泛的沉积物质量评价规程确定，该规程将详细说明实现数据质量目标所需的诸多质量保证（见第 7.2.5 节）。此外，该规程还将作为确定用于评价试验准确性和控制沉积物数量和类型的参考。在试验之前，样品采集和所有形式处理（运输、储存和操作）所用的方法将改变样品物理化学性质，可能改变污染物的总浓度和/或改变污染物生物利用度，以及非污染压力（如氨），从而影响毒理学分析结果。因此需要提供不同层次的定量信息，以说明试验沉积物与从现场收集的沉积物的相似性。现场记录的信息应包括现场的位置、深度、水质参数和其他相关观测结果（ASTM，2008c；Batley & Simpson，2016）。

根据评价的目的，沉积物上部几厘米通常是关注的范围（如 0~2 cm 或 0~10 cm 的表层沉积物）。该沉积物层通常代表了可供许多底栖无脊椎动物生活的污染区，同时也是可以很容易地进行重悬浮并使底栖动物和水柱生物可利用的深度。诸多文献已详细地描述了可用的收集沉积物设备和技术（Mudroch & Azcue，1995；USEPA，2001）。抓取采样装置通常用于许多采样项目。如果对历史污染感兴趣，可以收集沉积物芯，将其均质或切片，以获得特定深度的沉积物用于毒性试验。所需的沉积量取决于试验的特定性（种类和规程），但一般 3 kg 沉积物湿重就足以进行大多数具平行组的毒性试验。如果要进行阈值、化学、孔隙水试验或 TIEs，就需要收集更多的沉积物。如果要对同一沉积物进行沉积物化学分析，则应考虑不同的处理和储存需要。大量文献已详细描述了向沉积物中加入污染物以达到所需的暴露浓度的流程（Northcott & Jones，2000；Simpson et al.，2004；Hutchins et al.，2008）。

用于毒性试验的沉积物应进行冷藏（非冷冻）（USEPA，1993；ASTM，2008c）；然而，对于合适的储存时间看法各异。采集样本后就立即开展试验不太现实，合理的看法是：若已知沉积物毒物是挥发性或不稳定的，试验一般应在 2 周内进行（ASTM，2008c）；如果已知毒物相对稳定（如 PCB、PAH），似乎适度的甚至长期储存也可能不会改变毒性（DeFoe & Ankley，1998）。尽快对沉积物进行试验是合理的，但似乎即使长期储存也不会导致沉积物毒性失效。为了尽量减少沉积物扰动而导致影响沉积物毒性的因素，建议不要对沉积物进行细筛，除非有必要去除会干扰试验终点的掠食性或当地的物种。

已有文献详细描述了从沉积物中收集孔隙水的方法，并讨论了预防措施以及人工制备的影响（Carr & Nipper，2003；Chapman et al.，2002）。流程、测定的报告，以及结果诠释对于孔隙水试验至关重要。因为一旦孔隙水从沉积物中被分离出来，相关污染物的溶解浓度可能在试验前和测试期间发生较大变化（Chapman et al.，2002；Simpson & Batley，2003）（有关孔隙水试验的更多讨论见第 7.3 节）。

7.2.1.2 控制和参考沉积物

很少有生物体的反应会不受沉积物性质（如粒径）的影响。控制沉积物试验提供了有助于实现试验生物接近最优反应的试验条件。控制沉积物可包括野外收集的自然沉积物（从生物体收集点收集）或通过混合沉积物基质（如沙子、淤泥和粘土）以达到所需条件的配制沉积物。当沉积物特性（如粒度、有机碳）影响试验反应时，可通过使用控制沉积物来匹配这些条件（如粉质和砂质粒度控制）。这些控制沉积物通常被称为"参考沉积物"，一般是从污染评价现场附近收集，但不在污染源影响范围内或已知不含目标污染物的污染区域收集。采用参考沉积物进行的平行试验提供了评价区域内变化的非污染因素（包括目标化学物质以外的自然压力和其他人为毒物）如何影响试验终点的信息。根据被试验沉积物的性质范围，可将有显著意义的生物反应与对照组或参考沉积物进行统计比较。由于这些沉积物起到"阴性对照"的作用，它们应该是相对无污染物和无毒的。控制流程图有助于监控试验规程内部和规程之间试验终点的性能（ASTM，2013）。

7.2.1.3 特性鉴定

由于污染物的生物可利用性影响其在沉积物中的毒性，因此必须考虑沉积物的收集、处理和储存如何改变沉积物的化学和物理特性（Bull & Williams，2002；Simpson & Batley，2003）。在毒性试验之前，许多评价规程将分析沉积物中的目标污染物，以及 pH、氧化还原电位、含水量、粒度、有机碳、AVS、颗粒铁和锰，以及孔隙水成分（如氨、硫化物、金属）（Vandegehuchte et al.，2013）。对于大多数沉积物毒性试验规程，在开始之前，沉积物将被彻底均质化，并在平行处理组间进行分配，从而改变其生物利用度，甚至可能改变某些污染物的总浓度。对于需要对沉积物进行筛分以去除大颗粒和碎片或分离原生生物或以其他方式严重干扰的试验规程，建议在提供试验结果的同时，提供一份说明，说明沉积物扰动如何潜在地改变了试验沉积物中污染物浓度或生物利用度。

将沉积物和上覆水置于试验容器后，可能需要几天至几周时间使沉积物重新平衡和重建氧化还原分层。建议在试验前对沉积物的处理如何影响污染物浓度（如挥发性物质的损失）、生物利用度（如孔隙水和 AVS）或其他可能影响其毒性的因素进行一些评估。

7.2.2 试验生物

生物可以来源于人工培养（实验室或水产养殖）或从野外采集。培养方法因生物体而异，培养规程的变化可能会影响试验中所用生物的敏感性。应在试验前的数天至数周内确定性别（雄性/雌性）、生命阶段或状态（如幼体/成体、妊娠期、大小），以及喂食模式和环境暴露条件（如盐度、温度、光照）。同其他可能影响生物生存条件的因素一样，这些因素应在沉积物平行组和实验组间保持一致。在试验准备阶段死亡、目测不健康或表现不典型的生物应被清除和丢弃。不健康生物的迹象包括：双壳类动物的开口、颜色异常、身体"过于松弛"或对触碰无反应，或肠道内无食物。

使用实验室培养的生物比野外采集的生物更易实现试验生物之间较低的变异性，因为要了解野外生物的生活史和背景较为困难。为了最大限度地减少生物从野外采集点或培养设施到实验室的运输过程中以及在预试验阶段所受到的压力，在运输途中和在实验室存放时应为生物提供一个栖息的基质（如 2~4 cm 厚具上覆水的野外沉积物）。生物的保持密度应类似于试验中使用的密度，并应在试验开始之前提供一个适应期（如 2~10 天），以便其可以调整适应实验室试验条件（ASTM，2012b）。当试验条件与野外现场有显著差异时（如盐度），则生物对试验条件的适应操作应缓慢进行，如以 5‰ 盐度和 3 ℃/24 h 阶梯调整（ASTM，2012b）。同样，若 pH、盐度和温度等培养条件与试验条件不一致时，应通过改变生物的培养条件使其逐渐适应试验条件。对于特定生命阶段（如生殖终点）效应评价试验，试验开始前的几天至几周，需将雄性和雌性用托盘隔离，并准备好受精雌性或幼体。

7.2.3 试验暴露条件和设置

试验设计将影响预期暴露条件的实现和期望终点反应的检测。暴露室的尺寸和配置是首要考虑因素，包括沉积物和上覆水的体积，以及实现预期暴露所需的水交换率。为了在大多数表层沉积物中形成原位含氧沉积层，以及深层沉积物的某些平衡，建议至少在添加试验生物之前的 24 h 将沉积物置于毒性试验室（USEPA，2000；ASTM，2014）。

每天换水 1 次或频率更少的静态或半静态设计通常比流通设计更容易配置，其优缺点通常取决于试验生物种类和持续时间。由于沉积物中污染物的流通性，静态试验常常会由于上覆水缺乏交换或替换而导致过高估计生物暴露于上覆水溶解污染物的情况。渗透设计通常可以更好地模拟现场可能存在的上覆水条件，但往往需要大量的洁净上覆水来维持。如果评价生物的生命阶段非常短，且经常出现在水体，则这种设计可能不实用。如果野外的上覆水是沉积物污染来源，那么渗流试验中洁净上覆水可能会低估试验中的毒性。应规定静态更新和渗流规程的水更新频率。

在最终确定试验设计之前，应确定孔隙水中的氨浓度特征。为减少氨等修正因子的影响而对沉积物进行的任何操作，还应考虑它们如何改变暴露污染物及其他可溶性污染物的浓度或生物利用度（在"干扰因素"一节中讨论）。虽然理想的试验生物能够容忍环境条件的大范围变化，以便试验能够适用于不同物理化学性质的沉积物，但大多数生物对某些变量的容忍限度较窄。对大多数生物来说，温度（如 21°±1°）、光照（如 12 h 暗/12 h 光照周期和规定的光照强度）、pH（如 8.0±0.2）、盐度（如 30±2‰）和溶解氧浓度（如 80%～110% 溶解氧饱和度）必须保持在特定范围内。试验的"坚固性"是一个术语，常用来描述一种试验方法偏离指定的试验或环境条件的不敏感性。应规定并报告这些条件的预期范围和限值，以评估试验性能和终点结果的可接受性。

7.2.4 暴露条件监测

相关污染物的浓度和生物可利用度可能会因试验操作、试验设置所需时间以及试验过程中生物相互作用导致的物理化学性质变化而变化。为了将任何生物所受影响与特定污染物的暴露联系起来，可在试验开始和结束时采集沉积物样本进行分析，从沉积物表层（如 0～0.5 cm 深度）和更深的沉积物中提取样本，以便对暴露情况进行更详细的分析。分析可能包括污染物浓度，以及提供有关污染物生物利用度信息的测定（如 pH、氧化还原电位、粒度、TOC、AVS、孔隙水污染物浓度、氨和硫化物）（Simpson et al.，2016；Vandege-huchte et al.，2013）。

暴露期间需要监测标准水质参数（温度、pH、盐度、溶解氧），以确保它们保持在可接受的试验范围内。特别是静态和静态更新试验中，对监测上覆水中溶解氨和目标污染物很有用。如果试验室足够大，可容纳带有上覆水和孔隙水的沉积物，那么被动取样技术就可以用于监测其中污染物的浓度（USEPA，2012；Perron et al.，2013；Lydy et al.，2014；Peijnenburg et al.，2014；Amato et al.，2014）。

除物理化学测定外，还应观察记录生物行为，包括回避沉积物、不打洞、停止捕食或滤食，或任何异于对照/参考组的行为。

7.2.5 试验可接受性、数据分析和诠释

从取样收集阶段到最后的统计分析阶段，都应考虑沉积物毒性试验反应所产生实验和数据的质量。这意味着需要将第 7.1.1 节中概述的所有通用数据质量目标和原则考虑在内。虽然试验数据分析的目的是量化相对于控制/参考组沉积物的效应变化，但在风险评价规程中使用的数据还将考虑试验条件如何接近于可能发生在野外的生物暴露。

7.2.5.1 试验可接受标准

需要制定和报告试验可接受性标准，以确认达到所需暴露的条件和暴露的准确反应。确定试验可接受性的标准主要针对于试验生物物种、试验条件、规程和终点，以及观察和测定。本章前面讨论了包括沉积物和水的特性以及其他干扰因素在内的测定标准。

阳性和阴性对照品的使用有助于确定试验生物的反应是否与试验条件和实验室内/外的表现相一致。对照和参考沉积物起阴性对照作用。阴性对照通常用于每一批试验，并常用于确定观测效果的假设检验。阳性对照的目的是识别受试生物是否行为或反应异常。参考毒物试验是最常见的阳性对照形式，它将试验生物暴露于一种终点反应已知

（在特定范围内）的毒物中。定期使用它们来确定培养或现场采集生物的健康状况或反应性。阳性对照试验的 EC_{50}/LC_{50} 值应在该试验生物和试验条件的质量控制图平均值的 2 个标准差（SD）范围内。参考毒物旨在为毒性试验方法的重现性（精确度）和培养或收集的试验生物随时间变化提供一个通用的测定方法。如果结果超出标准范围，可能需要进行调查。建议使用控制图来跟踪阴性对照和阳性对照随时间变化的表现（图 7.2）。

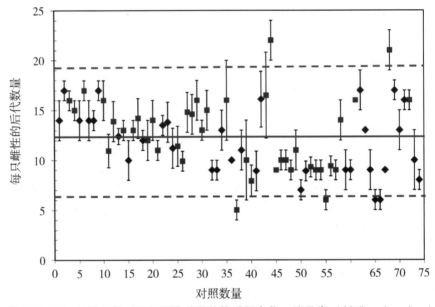

图 7.2 控制图示例，追踪阴性对照和阳性对照随的时间变化：端足类（Melita plumulosa）的总生殖量

图中线段表示平均值（实线）和平均值的两个标准差（虚线）。方形表明泥质沉积物更多（50% <63 mm），菱形表明沙质沉积物更多。

（本图例已修改，包含其他数据，引自 Spadaro, D. A., Simpson, S. L., 2016b. Appendix E. Protocol for 10 – day whole-sediment sub-lethal (reproduction) and acute toxicity tests using the epibenthic amphipod Melita plumulosa. In: Simpson, S. L., Batley, G. E. (Eds.), Sediment Quality Assessment: A Practical Handbook. CSIRO Publishing, Canberra, Australia, 265 – 275.）

参考毒物试验通常仅在水中短期进行，如 96 h ~ 7 d 的静态试验，而且通常暴露于单一化学品的单一或梯度稀释浓度中，不过也可使用加标沉积物暴露。理想情况下这些试验应在开始全沉积物试验的几天内，使用同一批野外采集的试验生物进行。对于培养生物，可常规进行参考毒物试验，而不是对试验中使用的每一批试验生物进行试验。如果沉淀物中添加了参考毒物，则应仔细考虑添加周期和平衡期（Simpson et al., 2004）。

7.2.5.2 毒性测定

如果符合试验可接受标准，确定毒性的统计分析通常较简单。这些分析将确定试验沉积物中平行组生物的效应是否与对照或参考沉积物中的平行组生物发生的效应存在显著差异（ASTM，2013）。试验设计需规定必要的重复程度以提供统计能力，用来正确检测毒性效应并评价反应的大小。应计算每种处理的基本统计终点，如存活率（平均 SD）或损伤百分比（例如生长、生殖、行为），并与对照和参考沉积物结果进行比较。

最初数据的正态性和方差齐性应该通过处理与对照（或参考）组数据的两两比较来检验［如通过 t 检验（student's t-test）或方差分析（ANOVA），然后进行 Dunnett 或 Tukey 检验］。如果不满足正态性和方差齐性要求，可以对数据进行转换和重新检验，如果仍然不合格，可以采用非参数检验（如 Wilcoxon Rank Sum 检验）进行统计比较。一般采用假设检验来分析试验生物对对照和处理沉积物反应的统计差异，显著水平通常为 $\alpha = 0.05$（即导致 5% "假阳性"的概率）。值得注意的是，在生物试验设计过程中，通过增加重复组往往可以将"假阳性"（第 I 类错误——检测到一个不存在的效应）和"假阴性"（第 II 类错误——未检测到存在的效应）的可能性降至最低。然而统计重复的需求要与实验室可行性相匹配。更详细的毒性试验数据分析的统计方法描述可见 ASTM（2008a，2013）、OECD（2006），以及本书第 2 章。

除统计分析之外，确定试验沉积物是否有毒的标准通常规定了效应的量级，例如，存活率或繁殖率比对照或参考沉积物低 20%。这个量级一般会考虑过去试验（实验室间和实验室内）的表现与反应可变性的关系，这些反应是被试验沉积物类型的典型特征（Simpson & Spadaro，2011）。例如，反应的变异性可能允许检测到显著差异，但沙质沉积物比粉质沉积物更大。如果将沉积物分类为轻微或高毒性，则可能需要考虑变异性差异。除此之外，毒性的附加量级还可以任意设定（如 20%～50% 的差异为中毒性，大于 50% 的差异为高毒性）。当对多重生物测定的一系列毒性试验结果进行排序时，毒性量级可能变得很重要。

7.2.5.3 效应阈值的计算（LC_{50}、EC/IC_{10}）

如果研究的目的是提供污染物的浓度梯度，则可以使用效应数据计算污染物的效应阈值，如 EC/IC_{10} 和 EC/IC_{50} 值。效应阈值和相关的不确定量应根据化学物质的测得浓度来计算，如 $EC_{50} = 50$（40～60）mg 铜/kg 干重（括号中的值是置信区间）。化学品的标称浓度只能在无法测量的情况下使用，并且必须说明标称浓度变化导致的不确定度。除报告效应阈值外，还应描述化学物质在溶解相和颗粒相之间的分配（如分配系数 K_d），以提供导致效应的有关暴露途径信息。

当效应数据是用于建立沉积物污染物的物种敏感性分布（SSDs）（Simpson et al.，2011；Vangheluwe et al.，2013）或建立特异性地点的管理限制时（Simpson et al.，2013），关于污染物种类和暴露途径的信息变得更加重要。对于沉积物污染物，SSD 的理想输出量通常为百分比危害或物种保护浓度，单位为 mg/kg 干重。SSD 中每个物种的无效应阈值将特定于沉积物特性，因为它们会影响污染物形态以及溶解态或颗粒态污染物暴露的相对贡献。这对于推导可应用于野外沉积物的指导值仍然是一个重大挑战，这些沉积物的性质与用于推导效应阈值的沉积物的性质（如不同的粒径、AVS、TOC）明显不同。

7.3　全沉积物毒性鉴定评价

现今使用 TIE 规程处理天然水中溶解毒物的技术已经很成熟（例如，USEPA，1996）。沉积物（全沉积物或孔隙水）（USEPA，2007）的 TIE 规程在持续扩展，并正在考虑用于监管。很大程度上是因为对影响水生生态系统（特别是底栖生物健康）的毒物类别的识别正在成为沉积物质量评价规程中一个日益重要的部分（NFESC，2003；USEPA，2007；Ho & Burgess，2009；Ho et al.，2009，2013；Burgess et al.，2011；Araujo et al.，2013；Camargo et al.，2015）。TIEs 涉及对沉积物或沉积物成分（如孔隙水）的操作，以去除或改变孔隙水或全沉积物中单个毒物（如氨）或一类污染物（如疏水性有机物、金属）的生物利用度，从而利于识别诱导可观察到毒性效应的单个毒物或毒物类别（Ankley & Schubauer-Berigan，1995；Burgess et al.，2003、2004、2007、2011；Ho et al.，2002、2004；USEPA，2007）。在 USEPA（2007）的框架中，TIE 方法分为三个阶段：表征、鉴定和确认。

选择使用孔隙水还是全沉积物进行 TIEs，与选择孔隙水或全沉积物进行毒性试验的原因有许多相同之处。TIEs 孔隙水试验的优点是能够使用许多用于废水 TIEs 开发的既定方法。其他优点包括：假设孔隙水被认为是许多毒物暴露的主要途径，能够使用与沉积物基质不相容的试验生物；可很好地理解样品操作对水化学的影响。

然而，许多因素不利于孔隙水 TIE 规程，包括许多样品操作工序会影响污染物形态和孔隙水毒性试验生物利用度。对孔隙水隔离以进行试验或 TIEs 会导致许多化学物质的平衡发生变化，如金属可能被氧化、孔隙水中有机污染物和有机碳可能沉淀、因孔隙水从系统中移除导致孔隙水和沉积物颗粒之间的污染物平衡被破坏并无法重建。此外，某些生物（穴居的端足类或蠕虫）的暴露途径是孔隙水和上覆水的混合暴露，现在却暴露在 100% 孔隙水中，从而增加了它们暴露于氨、金属和硫化物等水溶性污染物的机会。这可能解释了为何早期孔隙水 TIE 研究比近期全沉积物 TIE 研究更频繁地表明氨是一种毒物。一般来说，如果可以成功实施，则全沉积 TIE 是首选方法。而不能成功实施的原因包括：只在孔隙水中发现毒性，而不是全沉积物相；或相对于全沉积物，水相毒物鉴别的方法更先进。

为了将孔隙水和全沉积物毒性试验方法间的规程差异（包括样品制备）与生态现实主义结合起来，我们希望采用适用于全沉积物的 TIE 规程。全沉积物 TIEs 有望更准确，并为生物提供更真实的暴露途径。迄今为止，3 种主要的 TIE 分离方法已被应用：

（1）在沉积物中添加金属螯合树脂被发现是一种实用的全沉积 TIE 方法，此方法可降低金属的浓度和毒性，而对沉积物中存在的氨和非极性毒物的毒性影响很小（Burgess et al.，2000）。暴露后能够从测试系统中分离出树脂和累积的金属，从而可以启动 TIE 规程的识别阶段。

（2）在全沉积物毒性试验中，通过添加海藻（石莼，*Ulva lactuca*）或沸石来去除氨毒性（Besser et al.，1998；Ho et al.，1999b；Burgess et al.，2004）。

（3）添加椰子炭粉可有效去除有机污染物（如多环芳烃、多氯联苯及农药）的毒性（Ho et al.，2004）。

污染物暴露途径对选定污染物敏感性信息的纳入可以进一步改进未来的 TIE 方法。此外，基因组生物标志物的发展为 TIEs 对特定毒物和途径（而非仅是毒物类别）的识别提供了潜力（Biales et al.，2013；Hook et al.，2014b）。

7.4 展　　望

化学品生态风险评价的关键挑战是确定生态系统中发生有害效应的概率和程度，其最终目标是保护种群、群落和生态系统的长期续存。目前，风险评价方案是基于代替物实验：在恒定且通常有利的实验室条件下进行标准测试，仅提供相对少量物种的生物水平终点数据。就化学品风险评价而言，每次只对一种化学品进行试验；但联合国环境规划署的数据表明，目前有超过 10 万种不同的化学物质在被使用（UNEP，2010）。对于受许多化学品影响的现场沉积物质量的评价，在预测混合物的影响、生物利用度修正因子和非污染压力如何改变化学品污染带来的风险等方面仍然存在相当大的挑战。目前我们已知的海洋物种总数为 21.2 万种；但据估计，地球上有 140 万～160 万种海洋物种（Bouchet，2006）。人类愈发地希望从更偏远、特征不明确的环境中获取资源，如在深海（Collins et al.，2013）或极地地区进行采矿作业。这些活动将导致全新的物种暴露在化学品中，因而可能需要开发新的沉积物生态毒理学方法来评价这些环境中的风险。例如，需要开发适合水深超过 2000 m（如生活在无光和高压下的生物）或夏季较短近冰冻水域的方法。毋庸置疑，在无数可能的环境条件下试验所有物种和化学组合是不切实际的。

7.4.1　基于生物利用度的沉积物质量指南的推导

自从将全沉积物毒性试验引入生态风险评价框架以来，不同类型沉积物的污染物生物利用度差异带来了挑战（Chapman et al.，1998；Simpson & Batley，2007；Maruya et al.，2012）。公认的污染物生物利用度考虑因素是非离子疏水性有机污染物浓度（hydrophobic organics contaminant concentrations，HOCs）与沉积物有机碳浓度的标准化（Di Toro & McGrath，2000；USEPA，2003，2012），以及利用酸挥发性硫化物（acid-volatile sulfide，AVS）同步提取金属（simultaneously extracted metals，SEM）理论预测有效硫化物（available sulfide，AVS）摩尔浓度超过 SEM 时的无毒性。然而，这两种考虑都有局限性；由于黑碳的存在（如热解碳：烟灰、煤焦油、煤，不完全燃烧的残留物：木炭）（USEPA，2003，2012），HOCs 的生物利用度预测变得较为复杂，AVS-SEM 理论（USEPA，2003，2012）通常预测的是无毒性，而不是有毒性，因为它没有解释 AVS 计算中未考虑的粘土/淤泥、有机碳、铁和锰氧氢氧化物相部分对金属生物利用度的额外影响（Strom et al., 2011；Simpson et al., 2011；Campana et al., 2012, 2013；Besser et al., 2013）。对生物利用度修正因子的进一步了解有助于生物利用度沉积物质量指南的制定。预测无效应浓度（predicted no effects concentrations, PNEC）沉积物所使用的一系列标准化慢性生态毒性试验可从物种敏感性分布（SSDs）中获得，另外生物利用度模型提供了针对所有沉积物类型和条件定制PNEC沉积物的方法。

建立沉积物污染物的PNEC沉积物值可能首先要考虑对单个物种的慢性影响（如Fox et al., 2014）。特别对于金属，现在认为有必要将生物利用度纳入PNEC沉积物的推导中（Simpson et al., 2011；Campana et al., 2013；Schlekat et al., 2015）。最好的例子是镍和淡水沉积物（Schlekat et al., 2015）。此研究产生了9种底栖生物的慢性生态毒性数据，通过检查其中许多物种的生物利用度关系，可以开发出适用于各种沉积物类型的底栖生物物种敏感性分布（SSD），包括合理的最坏情况下的镍PNEC沉积物。对于海洋生态毒理学，目前可能尚无必要的标准化慢性检验数量；只要遵守合理的生态毒理学原则，表7.1中所列的许多方法均可作为标准化方法，以提供适当水平的环境和统计质量的效应数据。以具有不同性质的海洋沉积物中的铜为例，单一沉积物类型（Simpson et al., 2011）的SSDs结合模型（Simpson & King, 2005；Simpson, 2005）或详细的生物有效性修饰因子研究（Strom et al., 2011；Campana et al., 2012），也可能适用于基于生物利用度的PNECs推导，以用于风险评价项目（Simpson et al., 2013）。使用加标沉积物的全沉积物试验是建立必要生物利用度关系的一个基本部分（Campana et al., 2012；Besser et al., 2013），确保通过加标和平衡产生的污染物暴露提供所需的沉积物和孔隙水之间的理想分配，以便建立各种底栖无脊椎动物物种的浓度-毒性反应关系（Simpson et al., 2004；Hutchins et al., 2008；Brumbaugh et al., 2013）。

7.4.2 评价生态毒理学效应的沉积物暴露新方法

遵循合理的生态毒理学原理（第7.1.1节），开发的全沉积物生态毒理学标准方法已经可用于各种环境中的多种生物。但评价仍存在一个主要难题：对于大多数评价环境，我们没有大部分其他物种的效应数据，如大于99%的现存物种。在获取PNEC沉积物值的同时，如何确保我们正在保护上述的其他物种？

基于实验室的污染物效应阈值比基于野外的底栖生态评价更易推导，主要因为我们可以控制实验室沉积物毒性试验规程所提供的暴露条件。目前的难题在于基于实验室的沉积物毒性试验无法提供有关构成群落的许多底栖生物信息，也不允许可能影响污染物暴露和动物健康的生物相互作用。相反，野外实验（如在有自然生物聚集的围隔或野外定殖实验中，Olsgard, 1999；Chariton et al., 2011；Hill et al., 2013）可提供许多底栖生物敏感性信息，并允许群落相互作用；然而，由于许多污染物在野外暴露环境中的暴露难以被控制和测量，由此产生的效应关系难以解释。

Ho et al.（2013）表明，解释群落终点的一些困难可以通过使用一种新颖的混合暴露方法来解决：将完整的沉积物芯收集至实验室，然后通过向沉积物芯表面添加有毒物质来暴露现有的底栖生物群落。这种方法改进自Chandler et al.（1997）所描述方法：

利用沉积物芯将完整的小型和大型底栖动物群落带进实验室，并将它们暴露在2 cm厚的带有加标污染物的沉积物中，"群落"必须垂直迁移到含氧层才能生存；暴露于污染层两周后，加入一层2 cm厚的干净"无DNA"沉积物，那些在污染层中存活下来的微生物必须再次垂直迁移到氧化的表层；1周后对幸存的群落进行评价。利用无DNA沉淀物创建只有活的生物体才可以迁移到的层，可将基因组终点（因该终点无法区分活的和死的DNA）被包括在内（Chariton et al.，2014）。该方法已成功应用于检验三氯生（Ho et al.，2013；Chariton et al.，2014）、联苯菊酯和纳米铜（研究文献待发表）添加沉积物对河口小型和大型底栖生物群落的效应。

7.4.3 动态毒性（模拟走出黑盒的方法）

影响我们当前风险评价方法的另一点是无法完全模拟现实环境，从而使得对生态系统环境污染的实际后果产生了高度的不确定性。这种不确定性的根源在于生态因素（如竞争、捕食、资源限制）、环境条件（如生境质量、物理压力）和化学应激本身（如化学物质的混合物）在时间和空间上的高度变异性。因此，当外推超出试验条件时，许多化学品生态效应的可能性和程度往往仍然不确定。

在所有必要环境中对各种化学物质进行评价并不切实际，我们需要做的是改进现有方法的应用范围，对沉积物中化学物质的长期影响进行稳健评价，且有必要构建可精确推断其他环境和条件效应（和风险）的模型。只有通过更深入地了解影响暴露途径的因素和化学品对底栖生物造成毒性的机制，提升预测毒性的能力，才能做到这一点。

目前对水或沉积物毒性测试方法的弱点是我们对单一暴露期终点的评估。我们应当把生物看作动态系统，摆脱传统的以动物为基础的"黑盒"模式，即给生物体注入化学物质，然后观察最终结果。传统的试验侧重于收集有关少数最终不良结局（如生存、生长或生殖）的信息，这些信息虽然有价值，但主要为描述性的，没有对调节这些结局的机制进行任何深入了解。毒性途径是动态的。在低浓度的化学物质暴露下，毒性可能是可逆的，生物可能通过适应性反应恢复；在较高浓度暴露下，这些变化可能是不可逆的，最终导致生物功能受损或死亡。无论是哪种途径发生，描述生活史特征（如生存、生长、繁殖）的质量数据的生成和分析，对于准确地描述与种群动态相关的个体过程以及化学应激对这些过程的效应都至关重要。

基于对毒性基础途径相关的强大科学知识，科学的进步导致了预测毒性的新愿景。后文将向读者介绍不同的新兴技术和概念，这些技术和概念将在近期为更现代化的生态毒理学创造了惊人的机会。

7.4.3.1 机理效应模型

生态和机理效应模型可以有助于实现上述愿景。机理效应模型包括毒物代谢动力学-毒物效应动力学（toxicokinetic-toxicodynamic，TK-TD）、动态能量预算（dynamic energy budget，DEB）或基于个体的种群模型（individualbased models，IBMs）等。下文将进一步介绍它们是一种怎样有价值的工具，如何明确表示随时间推移的生物和化学过程，以及将对个体的效应转化为对诸如种群和群落生态系统的效应。

以下概述了一些机理效应模型。本书的第3章详细介绍了生态毒理学建模方法。

TK-TD模型模拟了一种毒物（在水或沉积物中）的外部浓度随时间变化的推移对暴露生物生活史特征的影响。TK-TD两步建模应用了毒物代谢动力学和毒物效应动力学的概念。毒物代谢动力学包括毒物在生物体内的吸收（摄取，生物富集）、分布、生物转化和消除的过程（生物对化学物质的反应）。毒物效应动力学逐步定量地将生物体内毒物浓度与个体生物水平上的时间效应联系起来（化学物质对生物体的影响）。

与其为每个物种设计一个新模型，不如构建一个通用模型来捕捉动物共有的过程。强有力的证据表明调节与种群动态相关的个体特征的代谢过程在物种间是保守的。例如，尽管物种之间存在巨大的形态、发育和行为差异，但几乎所有的物种都遵循典型的Von Bertalanffy生长模式（Von Bertalanffy，1957）。DEB理论（Kooijman et al.，1989）建立在这一前提之上，并试图通过一组基于基本原理（如质能守恒）的共享过程来解释关键生理生命特征的多样性（即随着时间推移的生长和繁殖量）。关注共性而非差异，使得通用模型（如基于DEB的模型）可以在物种间保留模型结构；因为物种间差异是通过控制能源获取和分配的模型参数的变化来表征。能量分配模式的变化对毒物的反应揭示了"生理作用模式"。了解毒物的生理作用模式细节的能力突出了基于DEB模型的另一个重要潜在用途：外推到种群水平的效应。例如，累积繁殖的降低可能是由于在胚胎阶段胚胎的死亡率或母体喂养率减少（即可用较少的能源配予繁殖），但这两种不同的生理行为模式可能会对种群结构和动态产生截然不同的影响。

IBMs（Martin et al.，2012，2013）基于这样一个前提：种群水平的过程产生于个体的行为。这些模型将个体生物表示为彼此不同的独特实体，并在其生命周期中发生变化。种群动态是通过个体之间或其非生物环境的相互作用而产生的。

机理效应模型能够将生态因素和环境条件与化学效应相结合，有助于缩小对个体的实验室试验与真实环境中的生态系统之间的差距。

7.4.4　生态毒理学新技术与新视野

毒性试验的新愿景是发展更好预测性、更高处理量和更低成本的方法，同时减少生物的使用和化学试验所需时间。目前的生态毒理学的自动化还处于最低水平。

7.4.4.1　实验室芯片技术

微流控芯片，也被称为芯片实验室或微全分析系统（mTAS），在一块微芯片上集成了大量生物和化学操作。相应设备作为自动化成像系统与基于显微镜的读数相结合，为分析小型化、自动化和并行化提供了新的工具，同时提高准确性和分辨率。对于毒性试验，这些集成的微系统可以减少实验成本和分析时间。它们采用无毒、透明和廉价的聚合物（如聚二甲基硅氧烷）制成，可以通过模塑获得微结构，以模拟微生物维持的环境，以便研究人员用来研究生理功能（Zheng et al.，2014）。在一个微流体装置中使用尺寸为几十微升的通道或腔室阵列、阀门、混合器和其他构件，可以产生多个有毒浓度梯度，同时对不同化合物进行高通量筛选，并可保持良好控制的微环境参数。

已有研究人员创建的微流控平台通过3D环境扫描电子显微镜以获得斑马鱼幼虫

（*Dano rerio*）形态学特征的高清成像，无需任何染色程序（Akagi et al.，2014），或采用微型摄像机和全自动化分析方法检测海洋端足类动物（*Allorchestes compressa*）游泳活动的亚致死变化（Cartlidge et al.，2015）。Zheng et al.（2013）证明利用集成微流控装置，可以在芯片上成功培养和暴露海洋微藻，提供多种生物反应（如细胞分裂率、自发荧光、酯酶活性）的在线测量。

7.4.4.2 有害结局途径

如前所述，AOP 是一个概念性框架，用于组织和描述有关生物组织各层次的毒性机制和结局的现有知识。AOP 代表了从分子启动事件（MIE；一种毒物与其分子靶点的直接相互作用）到触发生物扰动并发生不良后果的过程（AO）（Ankley et al.，2010）。在生态毒理学中，AO 可在个体水平上描述对生存、生长或繁殖的影响，或在更高水平上（如种群水平）描述生物的衰退。由 MIE 引发 AO 所必需的一系列事件是在细胞、组织、器官或个体水平上一系列可测量/可观察到的必要的生物学变化（定义为关键事件，key events，KEs），而关键事件间的联系（key event relationships，KERs）描述了 MIEs、KEs 和 AOs 之间的联系（定性和定量）（Groh et al.，2015）。

AOP 开发的基本原则是将 AOP 视为通用性的而不是特定于化学的。KEs 和 KERs 独立描述，并通常以一个非分支序列连接以定义单个 AOP。AOPs 也可以连接在一个更广泛的网络背景下，考虑潜在的相互作用，多重扰动的累积影响，以及作为生命阶段、性别、分类群等的函数响应的相似性和差异性（Villeneuve et al.，2014a，b）。通过 KERs 的鉴定和定量定义来验证 AOP，将允许基于试验开始时的 KERs 测定来预测 AO，而无需进行整个生物体毒性试验或野外研究来直接观察 AO，后者试验可能非常昂贵甚至无法进行（Groh et al.，2015）。

AOP 路径上所有关键事件的记录和说明可能非常复杂且难以完成。OECD（2013）发布了一份关于制定和评价不良结局途径的指导文件，其目的是为 AOP 的制定和描述提供最佳实践。人们期望作为 AOP 开发基础的测量和科学支持会随时间的推移而发展，从而增加 AOP 知识库。

向基于 AOP 范式过渡的化学安全性评价也侧重于现有体内数据与体外和生物信息学方法的集成。体外方法广泛用于药物在靶组织中的吸附、分布、代谢和排泄（adsorption，distribution，metabolism，and excretion；ADME）的时间过程研究。然而体外试验的主要缺点是其预测体内毒性的能力有限，因为细胞培养过程中产生的微环境与体内环境截然不同。微流控技术可能最终使先进的体外系统的制造成为可能，使体外制剂（组织或细胞）的培养能够长期保存其体内原始来源的所有特性，提供反映体内毒性动力学的方法，建立体外终点与体内有害效应之间的明确关系。例如，为绕过动物实验，考虑到不同器官和组织之间复杂的相互作用，基于多组织的微流控装置在单个芯片上模拟了生物体内描述毒代动力学模型的一些生理相关过程（ADME）（Baker，2011；van Midwoud et al.，2011）。将这些过程与环境生物利用度因素相结合，即整合这些分析与整个生物沉积物毒性评价将是未来的挑战。

7.4.5 结论

这些新技术和方法在生态毒理学方面,特别是在沉积物毒性试验上的应用,呈现出明显的局限性和挑战。机理模型应用的复杂性、模型输入所需的时间以及密集型沉积物毒性试验所需的资源,为其在生态毒理学中的应用带来了较大障碍。然而机理建模方法已经应用于不同的研究,如利用沉积物毒性试验分析毒物对软体动物(Ducrot et al., 2007)、昆虫(Beaudouin et al., 2012)、多毛类动物(Jager & Selck, 2011)和海洋微藻(Miller et al., 2010)的效应。关于 AOP,一个重要的限制是其自身定义为非化学特定实体,即迄今为止不需要外部或内部暴露的知识。这阻止了可提供关键信息的生物利用度数据或毒代动力学过程的使用,例如,触发 MIE 的毒物的外部或内部浓度是多少?暴露频率和持续时间如何影响 MIE 的激活?显然,芯片实验室技术应用于沉积物毒性试验仍是未来的发展方向。模拟因沉积物特性或生物本身所改变的生物利用度数据的问题是微流控技术短期内可能无法解决的挑战。然而它们以更高分辨率和准确度对微环境进行时空控制的创新能力可用于从体外系统推断毒代动力学数据,以预测体内结果,并结合成熟的常规试验来评价外部暴露,减少了对动物试验的需要。

参考文献

ADAMS M S, STAUBER J L, 2004. Development of a whole-sediment toxicity test using a benthic marine microalga. Environ. Toxicol. Chem., 23: 1957 – 1968.

ADAMS M S, 2016. Appendix D. Protocol for whole-sediment bioassay using the marine microalga *Entomoneis cf punctulata*//SIMPSON S L, BATLEY G E (Eds.). Sediment quality assessment: a practical handbook. CSIRO Publishing, Canberra, Australia: 55 – 264.

AKAGI J, ZHU F, HALL C J, et al., 2014. Integrated chip-based physiometer for automated fish embryo toxicity biotests in pharmaceutical screening and ecotoxicology. Cytometry, 85A: 537 – 547.

ALLER J Y, WOODIN S A, ALLER R C, 2001. Organism-sediment Interactions. University of South Carolina Press, Columbia, SC, USA.

AMATO H D, SIMPSON S L, JAROLIMEK C, et al., 2014. Diffusive gradients in thin films technique provide robust prediction of metal bioavailability and toxicity in estuarine sediments. Environ. Sci. Technol., 48: 4485 – 4494.

ANDERSON B S, HUNT J W, HESTER M, et al., 1996. Assessment of sediment toxicity at the sediment-water interface//OSTRANDER G K (Eds.). Techniques in aquatic toxicology. CRC Press, Boca, Raton, FL, USA: 609 – 624.

ANKLEY G T, SCHUBAUER-BERIGAN M K, 1995. Background and overview of current sediment toxicity identification evaluation procedures. J. Aquat. Ecosyst. Healt., 4: 133 – 149.

ANKLEY G T, KATKO A, ARTHUR J W, 1990. Identification of ammonia as an important sediment-associated toxicant in the lower Fox River and Green Bay, Wisconsin. Environ. Toxicol. Chem., 9: 312-322.

ANKLEY G T, ERICKSON R J, PHIPPS G J, et al., 1995. Effects of light intensity on the phototoxicity of flouranthene to a benthic invertebrate. Environ. Sci. Technol., 29: 2828-2833.

ANKLEY G T, BENNETT R S, ERICKSON R J, et al., 2010. Adverse outcome pathways: a conceptual framework to support ecotoxicology research and risk assessment. Environ. Toxicol. Chem., 29: 730-741.

ARAUJO C V M, TORNERO V, LUBIAN L M, et al., 2010. Ring test for whole-sediment toxicity assay with-a-benthic marine diatom. Sci. Tot. Environ., 408: 822-828.

ARAUJO C V M, BLASCO J, MORENO-GARRIDO I, 2012. Measuring the avoidance behaviour shown by the snail *Hydrobia ulvae* exposed to sediment with a known contamination gradient. Ecotoxicology, 21: 750-758.

ARAUJO G S, MOREIRA L B, MORAIS R D, et al., 2013. Ecotoxicological assessment of sediments from an urban marine protected area (Xixova-Japui State Park, SP, Brazil). Mar. Pollut. Bull., 75: 62-68.

ASTM (American Society for Testing and Materials), 2008a. Standard guide for conducting 10-day static sediment toxicity tests with marine and estuarine amphipods (E1367-03 (2008)). Annual Book of ASTM Standards, Vol 11.06. West Conshohocken, PA, USA. http://www.astm.org/Standards/E1367.htm.

ASTM, 2008b. Standard guide for designing biological tests with sediments (E1525-02 (2008)). Annual Book of ASTM Standards, Vol 11.06, West Conshohocken, PA, USA. http://www.astm.org/Standards/E1525.htm.

ASTM, 2008c. Standard guide for collection, storage, characterization, and manipulation of sediments for toxicological testing and for selection of samplers used to collect benthic invertebrates (E1391-03 (2008)). Annual Book of ASTM Standards, Vol 11.06, West Conshohocken, PA, USA. http://www.astm.org/Standards/E1391.htm.

ASTM, 2010. Standard test method for measuring the toxicity of sediment-associated contaminants with freshwater invertebrates (E1706-05 (2010)). Annual Book of ASTM Standards, Vol 11.06, West Conshohocken, PA, USA. http://www.astm.org/Standards/E1706.htm.

ASTM, 2012a. Standard guide for conducting static acute toxicity tests starting with embryos of four species of saltwater bivalve molluscs (E724-98 (2012)). Annual Book of ASTM Standards, Vol 11.06. West Conshohocken, PA, USA. http://www.astm.org/Standards/E724.htm.

ASTM, 2012b. Standard Guide for Selection of Resident Species as Test Organisms for Aquatic and Sediment Toxicity Tests (E1850-04 (2012)). Annual Book of ASTM Standards, Vol 11.06, West Conshohocken, PA, USA. http://www.astm.org/Standards/E1850.htm.

ASTM, 2013. Standard Practice for Statistical Analysis of Toxicity Tests Conducted Under ASTM Guidelines (E1847-06 (2013)). Annual Book of ASTM Standards, Vol 11.06,

West Conshohocken, PA, USA. http://www.astm.org/Standards/E1847.htm.

ASTM, 2014. Standard Test Method for Measuring the Toxicity of Sediment-associated Contaminants with Estuarine and Marine Invertebrates (E1367 – 03 (2014)). Annual Book of ASTM Standards, Vol 11.06. West Conshohocken, PA, USA. http://www.astm.org/Standards/E1367.htm.

Azur Environmental, 1998. Microtox acute toxicity solid phase test. Microtox® Manual, Carlsbad, CA, USA.

BAKER M, 2011. Tissue model: a living system on a chip. Nature, 471: 661 – 665.

BAT L, RAFFAELLI D, 1998. Sediment toxicity testing: a bioassay approch using the amphipod *Corophium volutator* and the polychaete *Arenicola marina*. J. Exp. Mar. Biol. Ecol., 226: 217 – 239.

BATLEY G E, SIMPSON S L, 2016. Sediment sampling, sample preparation and general analysis // SIMPSON S L, BATLEY G E (Eds.). Sediment quality assessment: a practical handbook. CSIRO Publishing, Canberra, Australia: 15 – 46.

BEAUDOUIN R, DIAS V, BONZOM J M, 2012. Individual-based model of *Chironomus riparius* population dynamics over several generation to explore adaptation following exposure to uranium – spiked sediments. Ecotoxicology, 21: 1225 – 1239.

BELZUNCE-SEGARRA M J, SIMPSON S L, AMATO E D, et al., 2015. Interpreting the mismatch between bio-accumulation occurring in identical sediments deployed in field and laboratory environments. Environ. Pollut., 204: 48 – 57.

BESSER J M, INGERSOLL C G, LEONARD E N, et al., 1998. Effect of zeolite on toxicity of ammonia in fresh-water sediments: implications for toxicity identification evaluation procedures. Environ. Toxicol. Chem., 17: 2310 – 2317.

BESSER J M, BRUMBAUGH W G, INGERSOLL C G, et al., 2013. Chronic toxicity of nickel-spiked freshwater sediments: variation in toxicity among eight invertebrate taxa and eight sediments. Environ. Toxicol. Chem., 32: 2495 – 2506.

BIALES A D, KOSTICH M, BURGESS R M, et al., 2013. Linkage of genomic biomarkers to whole organism end points in a toxicity identification evaluation (TIE). Environ. Sci. Technol., 47: 1306 – 1312.

BOLDINA-COSQUERIC I, AMIARD J – C, AMIARD – TRIQUET C, et al., 2010. Biochemical, physiological and behavioural markers in the endobenthic bivalve *Scrobicularia plana* as tools for the assessment of estuarine sediment quality. Ecotoxicol. Environ. Saf., 73: 1733 – 1741.

BOUCHET P, 2006. The magnitude of marine biodiversity // DUARTE C (Eds.). The exploration of marine biodiversity: scientific and technological challenges. Fundación, BBVA, Bilbao, Spain: 31 – 62.

BRIDGES T S, FARRAR J D, 1997. The influence of worm age, duration of exposure and endpoint selection on bioassay sensitivity for *Neanthes arenaceodentata* (Annelida: Polychaeta). Environ. Toxicol. Chem., 16: 1650 – 1658.

BRIDGES T S, FARRAR J D, DUKE B M, 1997. The influence of food ration on sediment toxicity in *Neanthes arenaceodentata* (*Annelida*: Polychaeta). Environ. Toxicol. Chem., 16: 1659 – 1665.

BRUMBAUGH W G, BESSER J M, INGERSOLL C G, et al., 2013. Preparation and characterization of nickel spiked fresh-water sediments for toxicity tests: toward more environmentally realistic nickel partitioning. Environ. Toxicol. Chem., 32: 2482 – 2494.

BULL D C, WILLIAMS E K, 2002. Chemical changes in an estuarine sediment during laboratory manipulation. Bull. Environ. Contam. Toxicol., 68: 852 – 861.

BURGESS R M, CANTWELL M G, PELLETIER M C, et al., 2000. Development of a toxicity identification evaluation procedure for characterizing metal toxicity in marine sediments. Environ. Toxicol. Chem., 19: 982 – 991.

BURGESS R M, PELLETIER M C, HO K T, et al., 2003. Removal of ammonia toxicity in marine sediment TIEs: a comparison of *Ulva lactuca*, zeolite and aeration methods. Mar. Pollut. Bull., 46: 607 – 618.

BURGESS R M, PERRON M M, CANTWELL M G, et al., 2004. Use of zeolite for removing ammonia and ammonia-caused toxicity in marine toxicity identification evaluations. Arch. Environ. Contam. Toxicol., 47: 440 – 447.

BURGESS R M, PERRON M M, CANTWELL M G, et al., 2007. Marine sediment toxicity identification evaluation methods for the anionic metals arsenic and chromium. Environ. Toxicol. Chem., 26: 61 – 67.

BURGESS R, HO K, BIALES A, et al., 2011. Recent developments in whole sediment toxicity identification evaluations: innovations in manipulations and endpoints // BRACK W (Eds.). Effect-directed analysis of complex environmental contamination. Springer, Berlin, Germany: 19 – 40.

BURTON G A, JOHNSTON E L, 2010. Assessing contaminated sediments in the context of multiple stressors. Environ. Toxicol. Chem., 29: 2625 – 2643.

CAMARGO J B D A, CRUZ A C F, CAMPOS B G, et al., 2015. Use, development and improvements in the protocol of whole-sediment toxicity identification evaluation using benthic copepods. Mar. Pollut. Bull., 91: 511 – 517.

CAMPANA O, SPADARO D A, BLASCO J, et al., 2012. Sublethal effects of copper to benthic invertebrates explained by changes in sediment properties and dietary exposure. Environ. Sci. Technol., 46: 6835 – 6842.

CAMPANA O, BLASCO J, SIMPSON S L, 2013. Demonstrating the appropriateness of developing sediment quality guidelines based on sediment geochemical properties. Environ. Sci. Technol., 47: 7483 – 7489.

CARR R S, NIPPER M J, 2003. Porewater toxicity testing. Society of Environmental Toxicity and Chemistry (SETAC), Pensacola, FL, USA.

CARTLIDGE R, NUGEGODA D, WLODKOWIC D, 2015. Gammarus Chip: innovative lab – on-a-chip technology for ecotoxicological testing using the marine amphipod *Allorchestes* compressa. Proc. SPIE 9518, Bio-MEMS and Medical Microdevices II 951812. http://dx. doi. org/10.1016/B978 - 0 - 12 - 803371 - 5.00007 - 2.

CASADO-MARTINEZ M C, BEIRAS R, BELZUNCE M J, et al., 2006. Interlaboratory assessment of marine bioassays to evaluate the environmental quality of coastal sediments in Spain. IV. Whole sediment toxicity test using crustacean amphipods. Ciencias Mar., 32: 149 - 157.

CASADO-MARTINEZ M C, SMITH B D, LUOMA S N, et al., 2010. Metal toxicity in a sediment-dwelling polychaete: threshold body concentrations or overwhelming accumulation rates? Environ. Pollut., 158: 3071 - 3076.

CASTRO H, RAMALHEIRA F, QUINTINO V, et al., 2006. Amphipod acute and chronic sediment toxicity assessment in estuarine environmental monitoring: an example from Ria de Aveiro, NW Portugal. Mar. Pollut. Bull., 53: 91 - 99.

CHANDLER G T, GREEN A S, 1996. A 14 - day harpacticoid copepod reproduction bioassay for laboratory and field contaminated muddy sediments // OSTRANDER G K (Eds.), Techniques in Aquatic Toxicology. CRC, Boca Raton, FL, USA: 23 - 39.

CHANDLER G T, COULL B C, SCHIZAS N V, et al., 1997. A culture-based assessment of the effects of chlorpyrifos on multiple meiobenthic copepods using microcosms of intact sediments. Environ. Toxicol. Chem., 16: 2339 - 2346.

CHAPMAN P M, WANG F, JANSSEN C, et al., 1998. Ecotoxicology of metals in aquatic sediments: binding and release, bioavailability, risk assessment, and remediation. Can. J. Fish Aquat. Sci., 55 (10): 2221 - 2243.

CHAPMAN P M, WANG F Y, GERMANO J D, et al., 2002. Pore water testing and analysis: the good, the bad, and the ugly. Mar. Pollut. Bull., 44 (5): 359 - 366.

CHARITON A A, ROACH A C, SIMPSON S L, et al., 2010. The influence of the choice of physical and chemistry variables on interpreting the spatial patterns of sediment contaminants and their relationships with benthic communities. Mar. Freshwater Res., 61: 1109 - 1122.

CHARITON A A, MAHER W A, ROACH A C, 2011. Recolonisation of translocated metal-contaminated sediments by estuarine macrobenthic assemblages. Ecotoxicology, 20: 706 - 718.

CHARITON A A, HO K T, PROESTOU D, et al., 2014. A molecular-based approach for examining responses of microcosm - contained eukaryotes to contaminant-spiked estuarine sediments. Environ. Toxicol. Chem., 33: 359 - 369.

CIUTAT A, BOUDOU A, 2003. Bioturbation effects on cadmium and zinc transfers from a contaminated sediment and on metal bioavailability to benthic bivalves. Environ. Toxicol. Chem., 22: 1574 - 1581.

COLLINS P C, CROOT P, CARLSSON J, et al., 2013. A primer for the Environmental Impact Assessment of mining at seafloor massive sulfide deposits. Mar. Policy, 42: 198 - 209.

COSTA F O, CORREIA A D, COSTA M H, 1998. Acute marine sediment toxicity: a potential new test with the amphipod *Gammarus locusta*. Ecotoxicol. Environ. Saf., 40: 81 - 87.

COSTA F O, NEUPARTH T, CORREIA A D, et al., 2005. Multi-level assessment of chronic toxicity of estuarine sediments with the amphipod *Gammarus locusta*：Ⅱ. Organism and population-level endpoints. Mar. Environ. Res. ，60：93 – 110.

DAFFORN K A，SIMPSON S L，KELAHER B P，et al. ，2012. The challenge of choosing environmental indicators of anthropogenic impacts in estuaries. Environ. Pollut. ，163：207 – 217.

DEFOE D L，ANKLEY G T，1998. Influence of storage time on the toxicity of freshwater sediments to benthic macroinvertebrates. Environ. Pollut. ，99：123 – 131.

DI TORO D M，MCGRATH J A，2000. Technical basis for narcotic chemicals and polycyclic aromatic hydrocarbon criteria. Ⅱ. Mixtures and sediments. Environ. Toxicol. Chem. ，19：1971 – 1982.

DUCROT V，PERY A R R，MONS R，et al. ，2007. Dynamic energy budget as a basis to model population-level effects of zinc-spiked sediments in the gastropod *Valvata piscinalis*. Environ. Toxicol. Chem. ，26：1774 – 1783.

EC（European Commission）Report，2012. Addressing the new challenges for risk assessment. SCENIHR（Scientific Committee on Emerging and Newly Identified Health Risks），SCHER（Scientific Committee on Health and Environmental Risks），SCCS（Scientific Committee on Consumer Safety）. http：// ec. europa. eu/health/scientific_committees/emerging/docs/scenihr_o_037. pdf.

EDGE K，JOHNSTON E，ROACH A，et al. ，2012. Indicators of environmental stress：cellular biomarkers and reproductive responses in the Sydney rock oyster（*Saccostrea glomerata*）. Ecotoxicology，21：1 – 11.

EDGE K，DAFFORN K A，ROACH A C，et al. ，2014. A biomarker of contaminant exposure is effective in large scale study of ten estuaries. Chemosphere，100：16 – 26.

Environment Canada，2002. Biological test method：reference method for determining the toxicity of sediment using luminescent bacteria in a solid-phase test. Report EPS 1/RM/42. Ottawa，ON，Canada.

FARRAR J D，BRIDGES T S，2011. 28 – Day chronic sublethal test method for evaluating whole sediments using an early life stage of the marine polychaete *Neanthes arenaceodentata*. US Army Corps of Engineers，Vicksburg，MS，USA. Report ERDC TN-DOER-R14.

FATHALLAH S，MEDHIOUB M N，KRAIEM M M，2012. Photo-induced toxicity of four polycyclic aromatic hydrocarbons（PAHs）to embryos and larvae of the carpet shell clam *Ruditapes decussatus*. Bull. Environ. Contam. Toxicol. ，88：1001 – 1008.

FOX M，OHLAUSON C，SHARPE A D，et al. ，2014. The use of a *Corophium volutator* chronic sediment study to support the risk assessment of medetomidine for marine environments. Environ. Toxicol. Chem. ，33：937 – 942.

GAO J J，SHI H H，DAI Z J，et al. ，2015. Variations of sediment toxicity in a tidal estuary：a case study of the South Passage，Changjiang（Yangtze）Estuary. Chemosphere，128：7 – 13.

GEROULD S，GLOSS S P，1986. Mayfly-mediated sorption of toxicants into sediments.

Environ. Toxicol. Chem. , 5: 667 – 673.

GOMIERO A, DAGNINO A, NASCI C, et al. , 2013. The use of protozoa in ecotoxicology: application of multiple endpoint tests of the ciliate *E. crassus* for the evaluation of sediment quality in coastal marine ecosystems. Sci. Tot. Environ. , 442: 534 – 544.

GREENSTEIN D, BAY S, ANDERSON B, et al. , 2008. Comparison of methods for evaluating acute and chronic toxicity in marine sediments. Environ. Toxicol. Chem. , 27: 933 – 944.

GRISCOM S B, FISHER N S, 2004. Bioavailability of sediment-bound metals to marine bivalve molluscs: an overview. Estuaries, 27: 826 – 838.

GROH K J, CARVALHO R N, CHIPMAN J K, et al. , 2015. Development and application of the adverse outcome pathway framework for understanding and predicting chronic toxicity: challenges and research needs in ecotoxicology. Chemosphere, 120: 764 – 777.

HACK L A, TREMBLAY L A, WRATTEN S D, et al. , 2008. Toxicity of estuarine sediments using a full life-cycle bioassay with the marine copepod *Robertsonia propinqua*. Ecotox. Environ. Saf. , 70: 469 – 474.

HARRIS C A, SCOTT A P, JOHNSON A C, et al. , 2014. Principles of sound ecotoxicology. Environ. Sci. Technol. , 48: 3100 – 3111.

HASSAN S M, GARRISON A W, ALLEN H E, et al. , 1996. Estimation of partition coefficients for five trace metals in sandy sediments and application to sediment quality criteria. Environ. Toxicol. Chem. , 15: 2198 – 2208.

HELLOU J, 2011. Behavioural ecotoxicology, an "early warning" signal to assess environmental quality. Environ. Sci. Pollut. Res. , 18: 1 – 11.

HILL N A, SIMPSON S L, JOHNSTON E L, 2013. Beyond the bed: effects of metal contamination on recruitment to bedded sediments and overlying substrata. Environ. Pollut. , 173: 182 – 191.

HO K T, BURGESS R M, 2009. Marine sediment toxicity identification evaluations (TIEs): history, principles, methods, and future research // KASSIN T A, BARCELO D (Eds.). Contaminated Sediments. Springer, Berlin, Germany: 75 – 95.

HO K T, BURGESS R M, 2013. What's causing toxicity in sediments? Results of 20 years of toxicity identification and evaluations. Environ. Toxicol. Chem. , 32: 2424 – 2432.

HO K T, KUHN A, PELLETIER M C, et al. , 1999a. pH dependent toxicity of five metals to three marine organisms. Environ. Toxicol. Chem. , 14: 235 – 240.

HO K T, KUHN A, PELLETIER M C, et al. , 1999b. Use of *Ulva lactuca* to distinguish pH-dependent toxicants in marine waters and sediments. Environ. Toxicol. Chem. , 18: 207 – 212.

HO K T, BURGESS R M, PELLETIER M C, et al. , 2002. An overview of toxicant identification in sediments and dredged materials. Mar. Pollut. Bull. , 44: 286 – 293.

HO K T, BURGESS R M, PELLETIER M C, et al. , 2004. Use of powdered coconut charcoal as a toxicity identification and evaluation manipulation for organic toxicants in marine sediments. Environ. Toxicol. Chem. , 23: 2124 – 2131.

HO K T, GIELAZYN M L, PELLETIER M C, et al., 2009. Do toxicity identification and evaluation laboratory-based methods reflect causes of field impairment? Environ. Sci. Technol., 43: 6857-6863.

HO K T, CHARITON A A, PORTIS L M, et al., 2013. Use of a novel sediment exposure to determine the effects of triclosan on estuarine benthic communities. Environ. Toxicol. Chem., 32: 384-392.

HOOK S E, GALLAGHER E P, BATLEY G E, 2014a. The role of biomarkers in the assessment of aquatic ecosystem health. Integr. Environ. Assess. Manag., 10: 327-341.

HOOK S E, OSBORN H L, GOLDING L A, et al., 2014b. Dissolved and particulate copper exposure induce differing gene expression profiles and mechanisms of toxicity in a deposit feeding amphipod. Environ. Sci. Technol., 48: 3504-3512.

HUTCHINS C M, TEASDALE P R, LEE S Y, et al., 2008. Cu and Zn concentration gradients created by dilution of pH neutral metal-spiked marine sediment: a comparison of sediment geochemistry with direct methods of metal addition. Environ. Sci. Technol., 42: 2912-2918.

HUTCHINSON T H, SOLBE J, KLOEPPER-SAMS P J, 1998. Analysis of the Ecetox Aquatic Toxicity (EAT) database Ⅲ: Comparative toxicity of chemical substances to different life stages of aquatic organisms. Chemosphere, 36 (1): 129-142.

JAGER T, SELCK H, 2011. Interpreting toxicity data in a DEB framework: a case study for nonylphenol in the marine polychaete *Capitella teleta*. J. Sea Res., 66: 456-462.

JOHNSTON E L, ROBERTS D A, 2009. Contaminants reduce the richness and evenness of marine communities: a review and meta-analysis. Environ. Pollut., 157: 1745-1752.

KENNEDY A J, STEEVENS J A, LOTUFO G R, et al., 2009. A comparison of acute and chronic toxicity methods for marine sediments. Mar. Environ. Res., 68: 118-127.

KEPPLER C J, RINGWOOD A H, 2002. Effects of metal exposures on juvenile clams, *Mercenaria mercenaria*. Bull. Environ. Contam. Toxicol., 68: 43-48.

KING C K, DOWSE M C, SIMPSON S L, 2010. Toxicity of metals to the bivalve *Tellina deltoidalis* and relationships between metal bioaccumulation and metal partitioning between seawater and marine sediments. Arch. Environ. Contam. Toxicol., 58: 657-665.

KNEZOVICH J P, STEICHEN D J, JELINSKI J A, et al., 1996. Sulfide tolerance of four marine species used to evaluate sediment and pore-water toxicity. Bull. Environ. Contam. Toxicol., 57: 450-457.

KOOIJMAN S A L M, VAN DER HOEVEN N, VAN DER WERF D C, 1989. Population consequences of a physiological model for individuals. Funct. Ecol., 3: 325-336.

KOVATCH C E, CHANDLER G T, COULL B C, 1999. Utility of a full life-cycle copepod bioassay approach for assessment of sediment-associated contaminant mixtures. Mar. Pollut. Bull., 38: 692-701.

KRELL B, MOREIRA-SANTOS M, RIBEIRO R, 2011. An estuarine mud snail in situ toxicity assay based on postexposure feeding. Environ. Toxicol. Chem., 30: 1935-1942.

KRULL M, ABESSA D M S, HATJE V, et al. , 2014. Integrated assessment of metal contamination in sediments from two tropical estuaries. Ecotoxicol. Environ. Saf. , 106: 195 – 203.

KUHN A, MUNNS W R J, POUCHER S, et al. , 2000. Prediction of population-level response from mysid toxicity test data using population modeling techniques. Environ. Toxicol. Chem. , 19: 2364 – 2371.

KUHN A, MUNNS W R J, SERBST J, et al. , 2002. Evaluating the ecological significance of laboratory response data to predict population-level effects for the estuarine amphipod *Ampelisca abdita*. Environ. Toxicol. Chem. , 21: 865 – 874.

LEE J W, WON E-J, RAISUDDIN S, et al. , 2015. Significance of adverse outcome pathways in biomarker-based environmental risk assessment in aquatic organisms. J. Environ. Sci. , 35: 115 – 127.

LI J Y, TANG J Y M, JIN L, et al. , 2013. Understanding bioavailability and toxicity of sediment-associated contaminants by combining passive sampling with in vitro bioassays in an urban river catchment. Environ. Toxicol. Chem. , 32: 2888 – 2896.

LYDY M J, LANDRUM P F, OEN A, et al. , 2014. Passive sampling methods for contaminated sediments: state of the science for organic contaminants. Integr. Environ. Assess. Manag. , 10: 167 – 178.

MANN R M, HYNE R V, SPADARO D A, et al. , 2009. Development and application of a rapid amphipod reproduction test for sediment quality assessment. Environ. Toxicol. Chem. , 28: 1244 – 1254.

MARTIN B T, ZIMMER E I, GRIMM V, et al. , 2012. Dynamic Energy Budget theory meets individual-based modelling: a generic and accessible implementation. Meth. Ecol. Evol. , 3: 445 – 449.

MARTIN B T, JAGER T, NISBET R M, et al. , 2013. Extrapolating ecotoxicological effects from individuals to populations: a generic approach based on dynamic energy budget theory and individual-based modeling. Ecotoxicology, 22: 574 – 583.

MARTINS M, COSTA P M, RAIMUNDO J, et al. , 2012. Impact of remobilized contaminants in *Mytilus edulis* during dredging operations in a harbour area: bioaccumulation and biomarker responses. Ecotoxicol. Environ. Saf. , 85: 96 – 103.

MARTÍN-DÍAZ M L, BLASCO J, SALES D, et al. , 2004. Biomarkers as tools to assess sediment quality: laboratory and field surveys. Trends Anal. Chem. , 23: 807 – 818.

MARUYA K A, LANDRUM P F, BURGESS R M, et al. , 2012. Incorporating contaminant bioavailability into sediment quality assessment frameworks. Integr. Environ. Assess. Manag. , 8: 659 – 673.

MCGEE B L, FISHER D J, WRIGHT D A, et al. , 2004. A field test and comparison of acute and chronic sediment toxicity with the marine amphipod *Leptocheirus plumulosus* in Chesapeake Bay, USA. Environ. Toxicol. Chem. , 23: 1751 – 1761.

MILLER D C, POUCHER S, CARDIN J A, et al. , 1990. The acute and chronic toxicity of ammonia to marine fish and a mysid. Arch. Environ. Contam. Toxicol. , 19: 40 – 48.

MILLER R J, LENIHAN H S, MULLER E B, et al. , 2010. Impacts of metal oxide nanoparticles on marine phytoplankton. Environ. Sci. Technol. , 44: 7329 – 7334.

MONSERRAT J M, MARTINEZ P E, GERACITANO L A, et al. , 2007. Pollution biomarkers in estuarine animals: critical review and new perspectives. Comp. Biochem. Physiol. C Toxicol. Pharmacol. , 146: 221 – 234.

MOORE D W, BRIDGES T S, GRAY B R, et al. , 1997. Risk of ammonia toxicity during sediment bioassays with the estuarine amphipod *Leptocheirus plumulosus*. Environ. Toxicol. Chem. , 16: 1020 – 1027.

MORALES – CASELLES C, RAMOS J, RIBA I, et al. , 2008. Using the polychaete *Arenicola marina* to determine toxicity and bioaccumulation of PAHS bound to sediments. Environ. Monit. Assess. , 142: 219 – 226.

MOREIRA S M, MOREIRA-SANTOS M, GUILHERMINO L, et al. , 2005. Short-term sublethal in situ toxicity assay with *Hediste diversicolor* (Polychaeta) for estuarine sediments based on postexposure feeding. Environ. Toxicol. Chem. , 24: 2010 – 2018.

MOREIRA S M, LIMA I, RIBEIRO R, et al. , 2006. Effects of estuarine sediment contamination on feeding and on key physiological functions of the Polychaete *Hediste diversicolor*: laboratory and in situ assays. Aquat. Toxicol. , 78: 186 – 201.

MORENO-GARRIDO I, HAMPEL M, LUBIAN L M, et al. , 2003a. Marine benthic microalgae *Cylindrotheca closterium* (Ehremberg) Lewin and Reimann (*Bacillariophyceae*) as a tool for measuring toxicity of linear alkylbenzene sulfonate in sediments. Bull. Environ. Contam. Toxicol. , 70: 242 – 247.

MORENO-GARRIDO I, HAMPEL M, LUBIAN L M, et al. , 2003b. Sediment toxicity tests using benthic marine microalgae *Cylindrotheca closterium* (Ehremberg) Lewin and Reimann (*Bacillariophyceae*). Ecotoxicol. Environ. Saf. , 54: 290 – 295.

MORENO-GARRIDO I, LUBIAN L M, JIMENEZ B, et al. , 2007. Estuarine sediment toxicity tests on diatoms: sensitivity comparison for three species. Estuar. Coastal Shelf Sci. , 71: 278 – 286.

MUDROCH A, AZCUE J M, 1995. Manual of aquatic sediment sampling. CRC Press, Boca Raton, FL, USA.

NEWMAN C M, 2010. Fundamentals of ecotoxicology. 3rd ed. CRC Press, Boca Raton, FL, USA.

NFESC (Naval Facility Engineering Services Center), 2003. Using sediment toxicity identification evaluations to improve the development of remedial goals for aquatic habitats. Special Publication SP-2132 – ENV, Port Hueneme, CA, USA.

NORTHCOTT G L, JONES K C, 2000. Spiking hydrophobic organic compounds into soil and sediment: a review and critique of adopted procedures. Environ. Toxicol. Chem. , 19: 2418 – 2430.

OECD (Organisation for Economic Cooperation and Development), 2006. Current approaches in the statistical analysis of ecotoxicity data: a guidance to application // OECD series on testing and assessment No. 54. ENV/JM/MONO (2006) 18. Environment Directorate, Paris, France.

OECD (Organisation for Economic Cooperation and Development), 2013. Guidance document on developing and assessing adverse outcome pathways // OECD series on testing and assessment, No 184. ENV/JM/MONO (2013) 6. Environment Directorate, Paris, France.

OLSGARD F, 1999. Effects of copper contamination on recolonisation of subtidal marine soft sediments-an experimental field study. Mar. Pollut. Bull., 38: 448 – 462.

PEIJNENBURG W J, TEASDALE P R, REIBLE D, et al., 2014. Passive sampling methods for contaminated sediments: state of the science for metals. Integr. Environ. Assess. Manag., 10: 179 – 196.

PELLETIER M C, BURGUESS R M, HO K T, et al., 1997. Phototoxicity of individual polycyclic aromatic hydrocarbons and petroleum to marine invertebrate larvae and juveniles. Environ. Toxicol. Chem., 16: 2190 – 2199.

PEREZ-LANDA V, SIMPSON S L, 2011. A short life-cycle test with the epibenthic copepod *Nitocra spinipes* for sediment toxicity assessment. Environ. Toxicol. Chem., 30: 1430 – 1439.

PERRON M M, BURGESS R M, SUUBERG E M, et al., 2013. Performance of passive samplers for monitoring estuarine water column concentrations: 1. Contaminants of concern. Environ. Toxicol. Chem., 32: 2182 – 2189.

PETERSON G S, ANKLEY G T, LEONARD E N, 1996. Effect of bioturbation on metal-sulfide oxidation in surficial fresh-water sediments. Environ. Toxicol. Chem., 15: 2147 – 2155.

RAINBOW P S, 2007. Trace metal bioaccumulation: models, metabolic availability and toxicity. Environ. Int., 33: 576 – 582.

RAMOS – GÓMEZ J, MARTINS M, RAIMUNDO J, et al., 2011. Validation of *Arenicola marina* in field toxicity biomass using benthic cages biomarkers as tools for assessing sediment quality. Mar. Pollut. Bull., 62: 1538 – 1549.

REGOLI F, GIULIANI M E, 2014. Oxidative pathways of chemical toxicity and oxidative stress biomarkers in marine organisms. Mar. Environ. Res., 93: 106 – 117.

REGOLI F, FRENZILLI G, BOCCHETTI R, et al., 2004. Time-course variation of oxyradical metabolism, DNA integrity and lysosomal stability in mussels, *Mytilus galloprovincialis*, during a field translocation experiment. Aquat. Toxicol., 68: 167 – 178.

RIBA I, DELVALLS T A, FORJA J M, et al., 2004. The influence of pH and salinity values in the toxicity of heavy metals in sediments to the estuarine clam *Ruditapes phillipinarum*. Environ. Toxicol. Chem., 23: 1100 – 1107.

RINGWOOD A H, KEPPLER C J, 1998. Seed clam growth: an alternative sediment bioassay developed during EMAP in the Carolinian Province. Environ. Monit. Assess., 51: 247 – 257.

RINGWOOD A, 1992. Comparative sensitivity of gametes and early developmental stages of a sea urchin species (*Echinometra mathaei*) and a bivalve species (*Isognomon californcum*) during metal exposures. Arch. Environ. Contam. Toxicol., 22: 288 – 295.

RODRIGUES E T, PARDAL M Â, 2014. The crab *Carcinus maenas* as a suitable experimental model in ecotoxicology. Environ. Int., 70: 158 –182.

RODRIGUEZ-ROMERO A, KHOSROVYAN A, DEL VALLS T A, et al., 2013. Several benthic species can be used interchangeably in integrated sediment quality assessment. Ecotoxicol. Environ. Saf., 92: 281 –288.

ROSEN G, MILLER K, 2011. A postexposure feeding assay using the marine polychaete *Neanthes arenaceodentata* suitable for laboratory and in situ exposures. Environ. Toxicol. Chem., 30: 730 –737.

SCARLETT A, ROWLAND S J, CANTY M, et al., 2007a. Method for assessing the chronic toxicity of marine and estuarine sediment-associated contaminants using the amphipod Corophium volutator. Mar. Environ. Res., 63: 457 –470.

SCARLETT A, CANTY M N, SMITH E L, et al., 2007b. Can amphipod behaviour help to predict chronic toxicity of sediments? Hum. Ecol. Risk Assess., 13: 506 –518.

SCHLEKAT C E, GARMAN E R, VANGHELUWE M L U, et al., 2015. Development of a bioavailability-based risk assessment approach for nickel in freshwater sediments. Integr. Environ. Assess. Monit. http://dx.doi.org/10.1016/B978 –0 –12 –803371 –5.00007 –2.

SIMPSON S L, BATLEY G E, 2003. Disturbances to metal partitioning during toxicity testing of iron (II) -rich estuarine porewaters and whole-sediments. Environ. Toxicol. Chem., 22: 424 –432.

SIMPSON S L, BATLEY G E, 2007. Predicting metal toxicity in sediments: a critique of current approaches. Integr. Environ. Assess. Manag., 3: 18 –31.

SIMPSON S L, KING C K, 2005. Exposure-pathway models explain causality in whole-sediment toxicity tests. Environ. Sci. Technol., 39: 837 –843.

SIMPSON S L, KUMAR A, 2016. Sediment ecotoxicology//SIMPSON S L, BATLEY G E (Eds.). Sediment quality assessment: a practical handbook. CSIRO Publishing, Canberra, Australia: 77 –122.

SIMPSON S L, SPADARO D A, 2011. Performance and sensitivity of rapid sublethal sediment toxicity tests with the amphipod *Melita plumulosa* and copepod *Nitocra spinipes*. Environ. Toxicol. Chem., 30: 2326 –2334.

SIMPSON S L, ANGEL B M, JOLLEY D F, 2004. Metal equilibration in laboratory-contaminated (spiked) sediments used for the development of whole-sediment toxicity tests. Chemosphere, 54: 597 –609.

SIMPSON S L, BATLEY G E, HAMILTON I, et al., 2011. Guidelines for copper in sediments with varying properties. Chemosphere, 85: 1487 –1495.

SIMPSON S L, WARD D, STROM D, et al., 2012. Oxidation of acid-volatile sulfide in surface sediments increases the release and toxicity of copper to the benthic amphipod *Melita plumulosa*. Chemosphere, 88: 953 –961.

SIMPSON S L, SPADARO D A, O'BRIEN D, 2013. Incorporating bioavailability into management limits for copper and zinc in sediments contaminated by antifouling paint and aquaculture. Chemosphere, 93: 2499 – 2506.

SIMPSON S L, BATLEY G E, MAHER W A, 2016. Chemistry of sediment contaminants // SIMPSON S L, BATLEY G E (Eds.). Sediment quality assessment: a practical handbook. CSIRO Publishing, Canberra, Australia: 47 – 75.

SIMPSON S L, 2005. An exposure-effect model for calculating copper effect concentrations in sediments with varying copper binding properties: a synthesis. Environ. Sci. Technol., 39: 7089 – 7096.

SPADARO D A, SIMPSON S L, 2016a. Appendix F. Protocol for whole-sediment sub-lethal (reproduction) toxicity tests using the copepod *Nitocra spinipes* (harpacticoid) // SIMPSON S L, BATLEY G E (Eds.). Sediment quality assessment: a practical handbook. CSIRO Publishing, Canberra, Australia: 276 – 284.

SPADARO D A, SIMPSON S L, 2016b. Appendix E. Protocol for 10 – day whole-sediment sub-lethal (reproduction) and acute toxicity tests using the epibenthic amphipod *Melita plumulosa* // SIMPSON S L, BATLEY G E (Eds.). Sediment quality assessment: a practical handbook. CSIRO Publishing, Canberra, Australia: 265 – 275.

SPADARO D A, SIMPSON S L, 2016c. Appendix G. Protocols for 10 – day whole-sediment lethality toxicity tests and 30 – day bioaccumulation tests using the deposit-feeding benthic bivalve *Tellina deltoidalis* // SIMPSON S L, BATLEY G E (Eds.). Sediment quality assessment: a practical handbook. CSIRO Publishing, Canberra, Australia: 285 – 293.

SPADARO D A, MICEVSKA T, SIMPSON S L, 2008. Effect of nutrition on toxicity of contaminants to the epibenthic amphipod, *Melita plumulosa*. Arch. Environ. Contam. Toxicol., 55: 593 – 602.

STROM D, SIMPSON S L, BATLEY G E, et al., 2011. The influence of sediment particle size and organic carbon on toxicity of copper to benthic invertebrates in oxic/sub – oxic surface sediments. Environ. Toxicol. Chem., 30: 1599 – 1610.

SWARTZ R C, FERRARO S P, LAMBERSON J O, et al., 1997. Photoactivation and toxicity of mixtures of polycyclic aromatic hydrocarbon compounds in marine sediment. Environ. Toxicol. Chem., 16: 2151 – 2157.

TAYLOR A M, MAHER W A, 2010. Establishing metal exposure-dose-response relationships in marine organisms: illustrated with a case study of cadmium toxicity in *Tellina deltoidalis* // PUOPOLO K, MARTORINO L (Eds.). New oceanography research developments: marine chemistry, ocean floor analyses and marine phytoplankton. Nova Science, New York, USA: 1 – 57.

TAYLOR A M, MAHER W A, 2016. Biomarkers // SIMPSON S L, BATLEY G E (Eds.). Sediment quality assessment: a practical handbook. CSIRO Publishing, Canberra, Australia: 157 – 193.

UNEP (United Nations Environment Programme), 2010. Harmful substances and hazardous waste: factsheet. www.unep.org/pdf/brochures/HarmfulSubstances.pdf. [2015 – 10].

USEPA (US Environmental Protection Agency), 1993. Guidance manual: bedded sediment bioaccumulation test. EPA-600-R-93 – 183. Office of Research and Development, Washington, DC, USA.

USEPA, 1994. Methods for assessing the toxicity of sediment-associated contaminants with estuarine and marine amphipods. 600-R-94 – 025. Office of Research and Development, Washington, DC, USA.

USEPA, 1996. Marine toxicity identification evaluation (TIE) procedures manual. Phase I guidance document. EPA-600-R-96 – 054. Office of Research and Development, Washington, DC, USA.

USEPA, 2000. Methods for assessing the toxicity and bio-accumulation of sediment-associated contaminants with freshwater invertebrates. EPA-600-R-99-064. 2nd ed. Office of Research and Development, Washington, DC, USA.

USEPA, 2001. Methods for collection, storage and manipulation of sediments for chemical and toxicological analyses. Technical manual EPA-823-B-01-002. Office of Water, Washington, DC, USA.

USEPA, 2003. Procedures for the derivation of equilibrium partitioning sediment benchmarks (ESBs) for the protection of benthic organisms: PAH mixtures. EPA-600-R-02 – 013. Office of Research and Development, Washington, DC, USA.

USEPA, 2005. Procedures for the derivation of equilibrium partitioning sediment benchmarks (ESBs) for the protection of benthic organisms: metal mixtures (Cadmium, Copper, Lead, Nickel, Silver, and Zinc). EPA-600-R-02 – 011. Office of Research and Development, Washington, DC, USA.

USEPA, 2006a. Guidance on systematic planning using the data quality objectives process. EPA-240-B-06-001. Office of Environmental Information, Washington, DC, USA.

USEPA, 2006b. Data quality assessment: statistical methods for practitioners. EPA-240-B-06-003. Office of Environmental Information, Washington, DC, USA.

USEPA, 2007. Sediment toxicity identification evaluation (TIE). Phases Ⅰ, Ⅱ, and Ⅲ guidance document. EPA-600-R-07 – 080. Office of Research and Development, Washington, DC, USA.

USEPA, 2012. Equilibrium partitioning sediment benchmarks (ESBs) for the protection of benthic organisms: procedures for the determination of the freely dissolved interstitial water concentrations of nonionic organics. EPA-600-R-02 – 012. Office of Research and Development, Washington, DC, USA.

USEPA, 2014. Screening for dioxin-like chemical activity in soils and sediments using the CALUX® bioassay and TEQ determinations. Method 4435 – 59. Test methods for evaluating solid waste. SW-846 On-line. US Environmental Protection Agency, Office of Solid Waste, Economic, Methods, and Risk Analysis Division, Washington, DC, USA. http://www.epa.gov/waste/hazard/testmethods/sw846/online/index.htm.

VAN GEEST J L, BURRIDGE L E, KIDD K A, 2014a. The toxicity of the anti-sea lice pesticide AlphaMax® to the polychaete worm *Nereis virens*. Aquaculture, 430: 98 – 106.

VAN GEEST J L, BURRIDGE L E, KIDD K A, 2014b. Toxicity of two pyrethroid-based anti-sea lice pesticides, AlphaMax® and Excis®, to a marine amphipod in aqueous and sediment exposures. Aquaculture, 434: 233 – 240.

VAN MIDWOUD P M, VERPOORTE E, GROOTHIUS G M M, 2011. Microfluidic device for in vitro studies on liver drug metabolism and toxicity. Integr. Biol., 3: 509 – 521.

VANDEGEHUCHTE M B, NGUYEN L T H, DE LAENDER F, et al., 2013a. Whole sediment toxicity tests for metal risk assessments: on the importance of equilibration and test design to increase ecological relevance. Environ. Toxicol. Chem., 32: 1048 – 1059.

VANGHELUWE M L U, VERDONCK F A M, BESSER J M, et al., 2013b. Improving sediment-quality guidelines for nickel: development and application of predictive bioavailability models to assess chronic toxicity of nickel in freshwater sediments. Environ. Toxicol. Chem., 32: 2507 – 2519.

VILLENEUVE D L, CRUMP D, GARCÍA-REYERO N, et al., 2014a. Adverse outcome pathways (AOP) development I: strategies and principles. Toxicol. Sci., 142: 312 – 320.

VILLENEUVE D L, CRUMP D, GARCÍA-REYERO N, et al., 2014b. Adverse outcome pathways (AOP) development II: best practices. Toxicol. Sci., 142: 321 – 330.

VOLKENBORN N, POLERECKY L, WETHEY D S, et al., 2010. Oscillatory porewater bioadvection in marine sediments induced by hydraulic activities of *Arenicola marina*. Limnol. Oceanogr., 55: 1231 – 1247.

VON BERTALANFFY L, 1957. Quantitative laws in metabolism and growth. Q. Rev. Biol., 32: 217 – 231.

WANG F Y, CHAPMAN P M, 1999. Biological implications of sulfide in sediment: a review focusing on sediment toxicity. Environ. Toxicol. Chem., 18: 2526 – 2532.

WARD D J, SIMPSON S L, JOLLEY D F, 2013a. Slow avoidance response to contaminated sediments elicits sub-lethal toxicity to benthic invertebrates. Environ. Sci. Technol., 47: 5947 – 5953.

WARD D J, SIMPSON S L, JOLLEY D F, 2013b. Avoidance of contaminated sediments by an amphipod (*Melita plumulosa*), a harpacticoid copepod (*Nitocra spinipes*) and a snail (*Phallomedusa solida*). Environ. Toxicol. Chem., 32: 644 – 652.

WILLIAMS K, GREEN D W J, PASEOE D, et al., 1986. The acute toxicity of cadmium to different larval stages of *Chironomus riparius* (Diptera: Chironomidae) and its ecological significance for pollution regulation. Oecologia, 70: 362 – 366.

WORD J Q, GARDINER W W, MOORE D W, 2005. Influence of confounding factors on SQGs and their application to estuarine and marine sediment evaluations // WENNING R J, BATLEY G E, INGERSOLL C G, et al (Eds.). Use of sediment quality guidelines and related tools for the assessment of contaminated sediments. society of environmental toxicology and chemistry. Pensacola, FL, USA: 633 – 686.

ZHAO Y, NEWMAN M C, 2007. The theory underlying dose-response models influences predictions for intermittent exposures. Environ. Toxicol. Chem., 26: 543 – 547.

ZHENG G, WANG Y, WANG Z, et al., 2013. An integrated microfluidic device in marine microalgae culture for toxicity screening application. Mar. Pollut. Bull., 72: 231 – 243.

ZHENG G, LI Y, LIU X, et al., 2014. Marine phytoplankton motility sensor integrated into a microfluidic chip for high-throughput pollutant toxicity assessment. Mar. Pollut. Bull., 84: 147 – 154.

8 海洋环境围隔与实地毒性试验

A. C. Alexander[①], **E. Luiker**[②], **M. Finley**[③], **J. M. Culp**[④]

---

① 加拿大，新布伦瑞克大学。
② 加拿大，新布伦瑞克大学。
③ 美国，威斯康辛州政府。
④ 加拿大，新布伦瑞克大学。

8.1 引　　言

8.1.1 什么是围隔实验？

围隔是指生物、栖息地和化学条件可控的室外封闭设施（1～10000 L 以上）（图 8.1），用于在实际野外条件下模拟复杂的暴露动态（Culp & Baird，2006）。围隔结合了实地和实验室试验技术，在允许控制一些目标参数（如包括栖息环境和物种）的同时，实现接近自然环境条件（如昼夜温度循环），以创建较实验室试验更真实的暴露情境。按照时间跨度，围隔实验可分为短期（<1 个月）、中期（1 个月至 1 年）和长期（>1 年），通常根据研究对象的生命周期、种群或群落来选择时间轴。

目前，海洋围隔已得到了广泛应用（参见 Grice & Reeve，1982；Clark & Noles，1994；Oviatt，1994；Peterson et al.，2009；Stewart et al.，2013）。此系统被认为有助于加深了解生物多样性与生态系统功能（Emmerson et al.，2001）、人类活动和生境影响（Renick 人，2015）之间的关系，最近也用于对气候变化的复杂反应（Lejeusne et al.，2010；Stewart et al.；2013）。海洋围隔在实验过程中面临着独特的挑战，其设备结构必须坚固，能够抵抗海水腐蚀；在进行潮下实验时，还应足够结实，以承受波浪作用。此外，海洋围隔系统规模可以很大（每个实验池容量大于 1000 L），这个规模会带来自身的挑战，并常常会阻碍可移动围隔设施的构建。

与实地试验相比，围隔具更强的可控性和可复制性，同时比实验室生物测试更贴近现实情境（表 8.1）。围隔虽然可能会很昂贵且需要复杂的后勤保障，但能够帮助降低试验不确定性。基于围隔的实验往往可产生高质量、可重复的数据，且数据比传统的实地研究更易收集和解释。

表 8.1　海洋环境中围隔研究的优劣势、机遇与威胁

优势	劣势	机遇	威胁
·可长期对 2 个或 2 个以上的营养级进行研究 ·可通过自然（如上升流）或非自然（如引入一种污染物）手段来控制环境 ·控制和复制试验	·价格昂贵，需后勤保障 ·与实地试验相比，失去真实性 ·封闭的混杂效应（如：壁效应，混合效应）	·可在历史和现今数据集的基础上进行推论 ·将来自不同学科的科学家聚集在一起进行一项研究的能力	·知识缺口，即对目标系统的真正了解有限；此缺点普遍存在于所有试验案例中

图 8.1 现代海洋围隔示例

(A、B) 为基于陆地的海洋围隔设施，(C、D) 为将试验水体与周围水体相分开的封闭式柱。(A)(B) 分别位于荷兰瓦赫宁根和美国康涅狄格大学，用于评估潮汐系统中复杂污水和对比试验海岸群落物种入侵的管理方案。(C)(D) 的原位技术是利用大型的聚氨酯漂浮袋，位于 2013 年部署在瑞典一处湾峡，用来测试海洋酸化率。

尽管有反对声音认为与整个生态系统研究相比，围隔的有效性受到了空间尺度和持续时间的约束（Carpenter，1996，1999；Carpenter et al.，1998），但一些最重要的全生态系统实验只有在进行了大量的围隔研究之后，才有可能实施（Schindler，1998）。此外，与整个生态系统研究不同，围隔可以在不破坏自然系统的情况下提供模拟过程的方法（Guckert，1993；McIntire，1993；Lawton，1996；Boyle & Fairchild，1997；Drenner & Mazumder，1999；Clements et al.，2002）。

围隔在模拟一个过程或因果关系时最为实用（Lamberti & Steinman，1993）。de Lafontaine & Leggett（1987）认为，理想的围隔围护系统应满足4个标准：①良好的时间与空间重复性；②自然环境的再现性；③构建具代表性的群落；④考虑到在自然密度下捕食者和被捕食者之间的关系。在最好的情况下，围隔所解决的是基本问题或混杂问题，用以澄清野外实地数据的解读。此外，该类研究对于隔离极端事件、物种入侵和需要与周围景观部分隔离的栖息地破碎化的影响至关重要（Stewart et al.，2013）。Odum（1984）提出的"揭露整体的基本属性"认为，围隔是切实可行的实验工具，可以同时监测种群和生态系统。

8.1.2 实地生物监测与围隔实验的结合

由于工业和城市活动排除的污水往往是漫射的、复杂的混合物，可能具有刺激作用（如营养掩蔽）或抑制效应（如污染物毒性），实地调研难以轻易地将因果相联系。在这方面，原位实验（如围场、湖沼围栏）具有实用性，因为可以在生物测定中分离出混合因素。例如，可以将自然应激源（如捕食）的效应与污染物的效应分开。因此，多种相互作用应激源的混杂因素可以被分离，从而揭示生物反应的原因。

与实地生物监测或封闭实验相比，围隔能够控制更多的变量，从而分离出影响生物反应的相互作用条件和污染物。围隔相对于实验室毒性试验而言，能够容纳更高程度的复杂性，且可产生关于污染物对水生生物群落亚致死和慢性效应的重要信息。围隔研究与实地生物监测的结合有益于生态风险评价（Norton et al.，2014），且该种结合已被用于进行证据权重风险评价（Culp et al.，2000a）。下一节的具体内容是：①阐述围隔研究在分析复杂应激源和生物学效应方面的实用性；②描述不同海洋围隔设计的基本方法学；③讨论围隔实验中较大生物体的需求；④概述海洋围隔研究的具体问题。

8.2 围隔——保护海洋生态系统的管理工具

环境管理者需借助工具对来自人为应激源影响的环境风险进行评价和管理。围隔作为一个独特的实验系统，允许研究人员在涉及不同食物链水平的现实环境条件来控制受体、风险和暴露途径（Perceval et al.，2009）。这些围隔特性可提高可用信息的深度，以支撑生态风险评价和制定环境标准和指南。

"风险评价是评价因暴露于一种或多种应激源而可能发生或正在发生的有害生态影响的一个可能性过程"（US EPA，1992）。围隔实验可以支持环境风险评价，因为它们允许在半自然条件下控制危害-受体相互作用，这对确定风险至关重要。在加拿大环境部开发的人工河流围隔中（Culp & Baird，2006），有害暴露的数量和浓度可以通过稀释和流速来控制，有害暴露时间控制尺度可以从脉冲到延长压力期。该围隔中受体为河流底栖生物群落，暴露途径/生境为天然的河流底层，应激源包括市政施工污染、纸浆厂生产污染和矿山尾矿废水、农药、营养盐和沉积物。

围隔研究同样可用于建立多种应激源情境下的因果关系。Culp et al.（2003）通过设置4个处理组调查了2种来源的废水对加拿大亚伯达的一条河流的因果关系：含1%城市污水排放物（MSE）组、含3%纸浆厂排放物（PME；比例基于河流中典型污染物浓度）组（作为对照组），以及含2种排放物的混合物组。基于这样的实验设计，可以确定MSE和PME都是营养盐主要来源，其中MSE是氮的主要来源，PME是磷的主要来源。这2种污染源结合后对初级生产者产生了协同效应，在下游创造了富营养化条件。这项研究为流域管理者提供了相应信息，以便观察监测整个河流系统及其排放源，并进行相应的管理。Clark & Clements（2006）也使用了类似的方法研究了金属矿对阿肯色河的影响。

为了在多种应激/累积效应情境下建立一个强有力的因果关系，需要结合各种来源信息的证据权重方法（Lowell et al.，2000；Culp & Baird，2006）。根据实地结果可以测定污染物或其他指标和生态系统成分（然而暴露水平和生态交互作用未知），通过基于实验室的生物分析研究（无菌环境特定浓度下的单物种研究），围隔研究可以弥补实地结果的不足。围隔可以根据实地数据验证受控条件下的应激效应，并提供陡峭的暴露-响应曲线（Lowell et al.，2000）。

美国国家海洋和大气管理局（National Oceanic and Atmospheric Administration，NOAA）在海洋环境中使用了类似的方法，开发了一种模块化的海水围隔试验系统，用于预测来自污染物尤其是石油烃和化学污染物对盐沼动植物的影响（Scott et al.，2013）。如上文所述的淡水围隔案例，这些围隔系统可控制危害参数（集中或结合其他应激源），并利用盐沼生境及其自然生物区系进行暴露。该模块化围隔系统所得结果可被用于泄漏前应急规划和生态风险评价预测（Scott et al.，2013）。

对于新的或其他未评估的物质，围隔研究可有效地确定实际的NOEC（无观察效应浓度）或无观察效应的物质释放量。经过验证，该方法在淡水和海洋生态系统中都已被证明有效。围隔可在真实的环境背景中（包括生态系统某组成部分）测试物质。因为对非目标/未测试类群的效应可能会被忽略，所以单物种生物测定试验无法复制生态系统中释放物质的效应。这点在杀虫剂混合物的研究中得到了证明，不同的底栖无脊椎动物类群对暴露处理表现出不同的敏感性（Alexander & Culp，2013）。Foekema et al.（2015）也使用了类似方法，通过室外围隔测定了海洋底栖和浮游生物群落中溶解铜的NOEC，并与基于实验室单物种试验的物种敏感度分布结果进行比较。海洋潮间带围隔研究中，采用2种底栖生物［沙蚕（*Hediste diversicolor*）；双壳软体动物（*Scrobicularia plana*）］对银纳米颗粒的环境归宿和毒性进行测定（Buffett et al.，2014）。此方法的特别之处在于通过在围隔内进行抽放水来模拟潮汐作用，潮汐周期为6 h。终点包括银纳米粒子的环境归宿、银生物富集和生物标记物响应（Buffett et al., 2014）。

此外，在围隔研究中目标物质可单独或与其他物质/应激源结合后试验，从而确定潜在的累加、协同或拮抗效应。Alexander et al.（2013）对单种和多种农药与营养剂混合效应的研究证明了这一点。结果表明，当暴露于低浓度农药时，营养剂以一种刺激作用"掩盖"了农药对底栖生物群落的效应；暴露于高浓度农药时，营养剂具有拮抗作用，对底栖生物群落的整体毒性增加（Alexander et al., 2013）。

8.3 水生围隔设计通则

8.3.1 陆基围隔

研究人员已使用各种路基围隔系统来评估应激源的归宿和效应，并量化多物种环境下的因果关系（图 8.1A 和 B）。围隔方法的选择取决于研究问题（如包括肉食性鱼类），但由于常规的围隔往往需要大量的时间和成本投入，因而研究人员通常更偏向于使用小型围隔。幸运的是，较小的系统可在节省了大量的成本的情况下生成与大型系统类似的因果信息。例如，Howick et al.（1994）发现尽管玻璃纤维槽比土池便宜 80%，但产生的农药对无脊椎动物效应相同。路基围隔在评价农药生态风险方面有着悠久的历史，其规模在 100～1000 m^2 之间，并已被纳入美国环保署农药注册的高级测试规程（Graney et al., 1994）。部分陆地围隔为人造池塘，深度通常不高于 3 m，边坡接近于 2∶1（Christman et al., 1994; Howick et al., 1994; Johnson et al., 1994）。试验水一般来自于地下井水或附近地表水。

装配式水箱属于要求最为严苛的路基围隔，其中平行组沉积物和水源须相同，生物群落也需精心受控。Rand et al.（2000）就建立此类系统提供了详细的示例和方法步骤。水箱的大小介于 2000～20000 L 之间，足以代表简单的食物网（通常不包括鱼类），并具有周围环境的温度、光照和风条件（Graney et al., 1994）。水箱围隔主要优点在于成本低，还可防止目标化合物污染自然环境。然而小尺寸围隔往往限制了较大捕食者的纳入，并可能导致不良的壁效应。但即使是试验过的某些最大围隔（1300 m^3）也被发现在长期实验（如大于 50 天）中形成相当大的壁效应（Grice et al., 1980）。另外还需格外注意的是采样技术的选择，因为破坏性取样（如沉积物取样）会影响此类小型系统的实验结果（Graney et al., 1995）。

MERL 是一个典型的路基海洋围隔（海洋生态系统研究实验室，美国罗得岛州纳拉甘西特市）。此系统建立于 1976 年，位于罗德岛大学，起初由美国环保署资助建设（Grice & Reeve, 1982; Odum, 1984）（图 8.2）。与实验室或野外围隔研究相比，MERL 温度、混合、交换率和生物材料均受到管控，且这些控制处于中等水平。该试验系统包含 14 个小型钢化玻璃纤维水箱（直径 2 m，深 6 m，沉积物表面 0.76 m^2，总体积 13 m³），以及由野外采集的浮游生物和底栖生物群落，它们的规模比例与附近的纳拉甘塞特湾相匹配。每个 MERL 水箱中包含 1 个适合长期试验的多营养级底栖生物群落。大量试验表明，MERL 围隔在进行的试验中充分复制了足够的野外变量，并且一旦储存起来，仍然与野外群落相似（Oviatt et al.，1980；Pilson & Nixon，1980）。该系统已进行过一些长期常规实验（大于 1 年）。如最近的实验研究了沿海酸化的季节性模式，以及营养富集如何改变这些模式（Nixon et al.，2015）。

8.3.2　围隔研究

围隔方法还包括湖沼、沿海和远洋围隔。湖泊围隔利用固定在底部沉积物上的分隔侧壁隔离出水环境的重复部分，并用过滤（Forrest & Arnott，2006）或未经过滤（Thompson et al.，1994）的远洋海水填充。湖泊围隔与其他围隔方法有着类似的优势（如可重复的设计、标准化的理化环境、多物种的交互）。湖泊围隔必须考虑到由受限的垂直和水平混合而导致的壁效应。壁效应可影响营养和化学动力学、物理化学性质和物间相互作用（如捕食），从而导致相对于周围水体的条件随时间发生变化（Graney et al.，1995）。滨岸围隔与湖泊围隔的不同之处在于，它们具有一个天然的海岸线和三堵嵌在沉积物中的塑料墙（Graney，1995）。尽管它们有着与湖泊围隔相仿的优势，但其局限性包括顶层捕食者密度的高重复变化，以及水深波动可能会损害围隔结构的完整性。

目前，已建立的滨海围隔较少（见位于挪威 Solbergstrand 的海洋科学研究站和瑞典卡尔斯克罗纳的波罗的海海洋实验室）。在挪威的混泥土滨海围隔（23 m³）采用防波堤模拟奥斯陆峡湾潮间带（Gray，1987；Bokn et al.，2001）。这些采用硬底围隔的实验开始前需提前准备 3 年，此期间让底栖生物群落自由生长（Bakke，1990）。位于 Solbergstrand 的滨海围隔研究首先提出，应注意确保封闭系统的影响被记录在案，因为某些物种的活动会受到封闭（特别是软体动物）的强烈影响（Gray，1987）。此后，研究者对此进行了严谨的试验，描述了这些围隔试验中的 50 种肉眼可见藻类和动物与奥斯陆峡湾典型中等被遮蔽海岸对应物种有何相似之处（Bakke，1990；Bokn et al.，2001）。

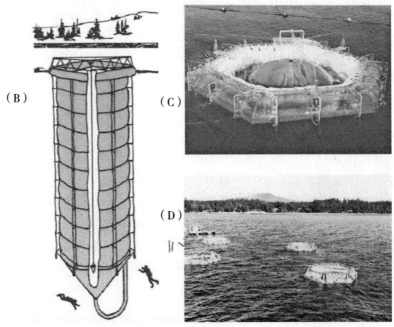

图 8.2 路基（A）和远洋围隔（B、C、D）实验的典型示例

路基实验（A），如纳拉甘西特湾的海洋生态系统研究实验室（Marine Ecosystems Research Laboratory，MERL），是由罗德岛大学和美国环保署运营共同管理。MERL 包含 14 个小型钢化玻璃纤维容器（直径 2 m，深 6 m，沉积物表面 0.76 m^2，总体积 13 m^3），按与附近纳拉甘西特湾相匹配的比例，播种野外收集的浮游生物和底栖生物群落。远洋围隔实验（B）（C）（D）位于加拿大不列颠哥伦比亚省萨尼奇湾的生态系统种群受控实验室（Controlled Ecosystems Populations Experiment，CEPEX）。最初的 CEPEX 由 3 个韧性锥形袋组成，总体积 68 m^3，顶部直径 2.5 m，底部直径 0.2 m，深 16 m（见 Parsons，1987）。(C) 为带压克力圆顶的早期 CEPEX 设计图，圆顶的作用是收集海面气体。(B)(D) 为之后演变的更大型版本，总体积 1300 m^3，顶部直径 10 m，底部直径 0.2 m，深 24 m。图片来源：Parsons, T. R., 1978. Controlled Ecosystems Populations Experiment, Mar. Pollut. Bull. 9, 203–205; and Odum, E. P., 1984. The mesocosm. Bioscience 34, 558–562. （已授权）。

一个经典的远洋围隔研究是 CEPEX（受控生态系统种群实验），但也可参照苏格兰的艾维海湾实验（Gamble et al.，1977）。CEPEX 是一个用于调查加拿大不列颠哥伦比亚省萨尼奇湾一带浮游生物动态的系泊远洋围隔（图 8.2）。初始实验系统由 3 个体积 68 m^3、顶部直径 2.5 m、底部直径 0.2 m、深 16 m 的锥形袋组成（Parsons，1978）。随后发展成为更大型的版本，体积 1300 m^3，顶部直径 10 m，底部直径 0.2 m，深 24 m。前后两种版本均捕获静水体，未进行潮汐交换稀释或补充（Grice et al.，1980；Menzel & Case，1977）。CEPEX 建立的最初目的是研究天然未受污染的浮游植物动态。在试验过程中发现 CEPEX 可为铜暴露对种群选择性摄食和捕食及其恢复情况的影响提供开创性见解（Grice et al.，1980；Menzel & Case，1977）。近年来，挪威（Riebesell et al.，2008）和韩国（Kim et al.，2008）的远洋围隔已被用于检测海洋酸化，而且欧盟委员会对其装置和部署提供了大量指导（Riebesell et al.，2009）。现今关于海洋酸化的大部分远洋研究工作是基于早期对珊瑚礁的围隔研究（见 1996 特刊，*Ecological Engineering* v6：1-225）。对这些研究的广泛回顾表明，只要有适当的资源，围隔研究有可能提高模型精度，是研究全球珊瑚礁生态恢复的独特工具（Luckett et al.，1996；Kleypas et al.，2006）。

8.3.3 人工河流方法

通过人工河流方法已广泛探寻了克服混合和水体运动技术问题的方法，所以此法可以为海洋方面的相关研究提供有用见解。河流围隔应用于淡水生态和生态毒理学研究已有 50 多年历史（McIntire，1933）。它们的尺寸范围从大型、具直流式人工渠道（＞100 m^3）到较小（＜1 m^3）、由部分再循环或直流式的装配式储罐组成的系统（Swift et al.，1993；Culp & Baird，2006）。大型河流围隔（长度 ＞50 m，＞100 m^3）是非常稀有的资源（Swift et al.，1993）。高昂的建造和维护成本极大地限制了复制流的数量和实验设计的多样性。此外，鉴于其成本和后勤的复杂性，往往只有政府机构或大型企业才拥有运作此类系统的能力。虽然大型实验河流有助于对目标种群进行从压抑扰动中恢复的长期、多代检查，但这些实验很少超过 12 个月（Swift et al.，1993）。在意识到需要更大、更符合实际生态的复制系统时，Mohr et al.（2005）开发出了一种新围隔设施，该设施提供了高度灵活的设计，可分割为 8 条复制流（每条最长 106 m），或连接为一条长度 850 m 的单一流。

相较于大型河流围隔，小型河流围隔则被生态学家和生态毒理学家广泛采用（Lamberti & Steinman，1993；Culp et al.，2000bc；Dubé et al.，2002；Schulz et al.，2002；Crane et al.，2007）。小型河流围隔规模从流动水槽（Bothwell，1993），到可用于交通运输的河滨围隔（Culp & Baird，2006），再到带有自然定殖基质的温室系统（Clark & Clements，2006）。水源包括地下水和地表水，水的流速可通过搅拌系统、水泵或实验单元的坡度控制。通常情况下，水温不被刻意控制。底栖生物群落的建立各不相同，包括自然形成、移播参考河段底栖生物群落（Alexander et al.，2008，2013）以及直接移植参考河段基底（Clark & Clements，2006）。由于规模较小，研究人员通常进行短期实验（＜30 天）。短期研究降低了壁效应发生的概率，并减少了由于缺乏如来自上游无脊椎动物栖息地迁徙的生物过程而导致的与参考河流结构之间不良偏差。总而言之，小规模河流围隔实验与野外研究和实验室研究相结合时，已被证明可提供丰富的信息（Culp et al.，2000a；Clements et al.，2002；Clements，2004）。

8.4 拓展：为大型鱼类开发系统

　　大型鱼类对实验设计提出独特的挑战。它们的存在可以决定实验的规模、复杂性和成本。然而，使用大型物种是利大于弊的，如可增加生态真实性，提供更多用于生化试验的组织样本数和进行重复检测的可操作性。在过去的20～30年间，研究中鱼类的使用频率大大增加。鱼类生命力强、适应性强且多样性丰富。此外，它们的形态和生理系统（Pohlenz & Gatlin，2014；Bolon & Stoskopf，1995；Sunyer，2013）、基本生化路径和响应通常类似于哺乳动物（Bolon & Stoskopf，1995）。鱼类对疼痛刺激也表现出与哺乳动物相似的行为、生理和激素应激反应（Sneddon，2009；Braithwaite & Huntingford，2004；Baker et al.，2013）。某些鱼类和其他脊椎动物之间的相似性使它们成为许多研究学科的合适模型。随着我们对鱼类生物学、行为和饲养知识的拓展，适当饲养和使用鱼类作为实验动物的指导方针也在不断更新，如加拿大动物管理委员会（Canadian Council on Animal Care，CCAC）、美国兽医协会（American Veterinary Medical Association，AVMA）和国际实验动物评估和认可委员会（Association for Assessment and Accreditation of Laboratory Animals International，AAALAC）。这些适当的鱼类福利可确保为它们提供最佳的健康和福祉（食物、栖息地，以及表达正常行为的能力），并确保它们不会遭受过度的痛苦或折磨。

　　与许多现有的围隔设计相比（如某物种数量 > 10），大型鱼类需要更大容量和合适的水源供应。水作为鱼类的生命维持系统，可提供氧气并清除有毒代谢废物。某些物种（如冷水鲑鱼种）适应了水流湍急、温度范围狭窄、富氧的水体，更容易受到水质细微变化的影响。而其他物种（如暖水鲤科物种）则可以忍受更宽泛的环境条件，如水流、溶氧量和温度。保持适当的水化学和充足的氧气供应，可在一个给定的水族箱中最大限度地增加鱼的数量，从而减少实验占用面积。在保持温度、溶解氧水平和水化学参数以及最大限度减少有毒代谢废物的堆积方面，具有高循环率的大型水箱比小型水箱更有弹性。

　　传统的鱼类养殖系统属于静态系统，通常使用较小的水箱（1～300 L），水输入较少，有时会附带小型过滤装置。此类系统承载能力较低，极易受到水质变化的影响（毒素堆积，低溶氧量）。更复杂的系统已被开发出来，通常用于小型鱼类。对于大型鱼类来说，建议采用直流式或再循环系统。这样的系统除了更有弹性，还为鱼类提供了足够的空间来表现出更正常的游泳行为。直流式系统需要持续的高质量水供应。虽然再循环系统所需水量相对较少，但往往需要更多的专业知识、时间和维护，且水质可能更多变。当生物过滤器的负载无法承受高种群密度或饲喂率时，可能会出现一个常见的水质问题，即有毒氨的积累（Yanong，2003）。其他水质问题，如低溶氧量、高二氧化碳含量、pH下降和水温升高也会随着时间推移而出现。

水箱的设计也可根据研究物种的特性而定。较活跃的鱼类（如鲑鱼）需要更高的水流量和持续游泳的空间。圆形水箱具有高水流量和更高的承载能力，因鱼可持续游动，从而具有自洁的额外益处。但这种圆形设计不适用于不太活跃的鱼类（如狗鱼科）或易受高水流量过度冲击的小型鱼类。瑞典池塘为圆形边缘的方形水池，被认为是最适合那些需要更大水体表面积和不垂直聚集于水体鱼类（如大西洋鲑鱼 *Salmo salar*；Piper et al., 1982）的池塘。圆角边缘使水池具有与圆形水池相同的自清洁机制。矩形槽或跑道不能反映正常的环境条件。它们的水流模式很差，无流动庇护点，也无法达到实现自清洁的高流速（Piper et al., 1982）。因此，矩形水道的水质较差，水道流出端的水质明显恶化，经常导致鱼群聚集于流入端附近。在这种情况下，鱼类的疾病发生率和传播增加（图8.3）。矩形设计最适合低密度的小型鱼类。土质池塘可用于不需要高水质和/或在实验设计中强化环境控制的生存能力较强鱼类（如鲫鱼）。土质池塘的优势在于能够产生天然饲料（如浮游动物和水生无脊椎动物），并模拟更多的自然环境。然而，此类池塘往往会有过多植被，增加了清洁、分配饲料以及收集鱼类等方面的困难。此外，土池也往往需要进行曝气，而清洁减少则会增加病原体暴露的风险。

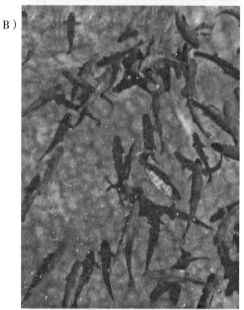

图8.3 水道中鱼类状态恶化

（A）虹鳟（*Oncorhynchus mykiss*）因疾病出现黑化，图为健康状态的虹鳟；（B）圣克洛伊岛基地捕获的溪红点鲑（*Salvelinus fontinalis*）因骨骼畸形导致神经损伤，表现出异常的旋转游动行为；（C）美国威斯康星州奥西奥拉州鱼类孵化场。

> 围隔：是否越大越好？
>
> 　　围隔的物理尺寸从人工建造的系统（人造水箱、人造水池、水渠）到天然生境中隔离出来的分区（湖泊围隔、沿海围隔）（Boyle & Fairchild, 1997）。基于以往相关文献的回顾，发现研究平均持续时间为 49 天，空间为 1.7 m³（图 8.4）（Petersen et al., 1999, 2009）。此类试验比生态过程研究时间更短、更小，但两者都是对重复和受控宏观生态实验时间和成本的让步。对污染物的潜在生态效应和归宿的回顾表明，扰动研究与野外实地数据结合能够得出数月和数十米尺度上合理可信的种群和群落数据（Sanders, 1985）。然而，确定生态真实性的程度取决于研究问题和可用于进行实验的资金。较大型系统由于成本限制，为了大规模的、单一干预因素的围隔，往往会倾向于牺牲可复制性（Riebesell et al., 2008）。在整个围隔领域的文献中，普遍采用组合方法，即不同尺度的证据权重为同一种科学叙述提供重叠的、独特的视角（围隔与野外研究相结合：Schindler, 1998；野外研究：Culp & Baird, 2006；实验室试验：Sanders, 1985；建模方法：Stewart et al., 2013）。

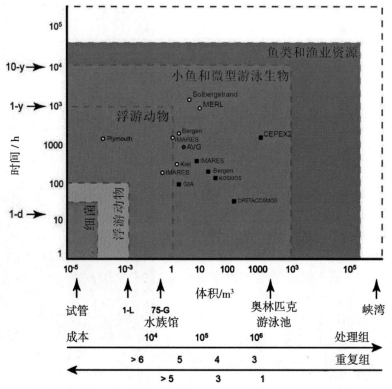

图 8.4　海滨和远洋围隔生物群落营养等级与围隔规模和试验持续时间之间的关系

8.5　海洋围隔设计的思考

海洋围隔最初于1970年代流行。到了1980年代，围隔方法需要对规模效应进行深入研究，才能将研究结果应用到实际系统中（Petersen et al., 2009）。众所周知，小型围隔和短期实验会限制试验中营养级的数量（图8.4）。围隔的尺寸还会产生改变自然暴露的人为干扰（如壁效应、动物运动、混合减少和水交换率改变）。所有这些都与薄弱的实验设计有关——这个问题并非只在围隔研究中才存在（Kuiper, 1982; Bloesch et al., 1988）。

开放式系统的应用近年来在淡水研究中受到青睐。在淡水研究中，水生无脊椎动物种群被允许定居或离开池塘围隔生境。但在海洋环境中，开放的系统允许内外界水体交换，让营养物质和新资源流入更接近自然系统。封闭系统也有着不同的优势，如更适合于有毒物质测试，因为此系统能够防止化学物质释放或暴露后的生物体回到自然环境；另外还可以模拟一个广泛的非点源污染，为管理和建模提供一个状况更糟糕的场景。

封闭水体会改变混合状态。湍流会影响物种间的相互作用、生长以及微粒的运动。根据Sanford（1997）观点，湍流混合与光、温度、盐度和营养盐同样重要，应被直接测量和记录。目前，在描述最好的混合系统中，可找到的是多尺度实验生态系统研究中心的底栖-远洋围隔（Petersen et al., 2009），它被广泛地应用工程学原理来开发合适的混合机制（Tatterson, 1991）。相反，早期的柔性围隔（如CEPEX）通常不混合（图8.2b）。经teele et al.（1977）估计，CEPEX的垂直湍流扩散速度为$0.1\ cm^2/s$，相较周围的表层水慢10倍。近年来的研究开始通过使用鼓泡和水平栅格增加了垂直混合，从而在自然系统范围内制造湍流（Svensen et al., 2001; Nerheim et al., 2002）。

8.6　展　　望

随着海洋和水生环境新问题的出现，未来围隔研究面临新的挑战和机遇。我们预计与气候变化有关的海洋围隔和淡水围隔实验将会增加，未来可能会倾向于通过河流围隔、建模和野外调查多种方法研究压力源对河流生态系统的累积影响。累积效应的调查也与海洋围隔研究有关，在海洋围隔研究中，自然（如季节性）和人为（如城市化、资源开发）梯度的重要性影响着海洋群落。

另一个受到人们持续关注并被不断扩大的领域是海洋温度上升和海水酸化。这将是一个从多个方面影响海洋群落的重大问题。其他一些未来的研究问题可能包括气候变化对北极冰川融化或海平面上升的影响。例如，融冰期和持续时间的变化如何影响水生群落的结构或功能？河口地区是否会暴露于不同的盐度水平，以及会给河口群落造成怎样的影响？

利用围隔研究物种入侵是另一个有趣的问题。为什么物种或疾病会入侵新的生境？为什么有些地区比其他地区更易受到入侵？入侵者是否改变了生态系统各组成部分之间的相互作用（如捕食者-被捕食循环、资源利用、碳通量）？本地物种是否与低纬度入侵者共存？食物网是否会被破坏？保护工作是否可以保护到气候变化敏感地区？

无论研究的主题是什么，围隔研究往往依赖于创新。最具影响力的围隔研究将是跨学科的努力，采用创新的方法来研究问题，而非简单地在室外进行实验室实验。它们还将结合自然变化的相关来源（如温度、混合、移植）来理清物种和环境之间重要的相互作用。与任何实验一样，了解研究的局限性和适当规模至关重要。

8.7 结论：围隔试验在海洋生态毒理学中的应用

围隔是跨学科研究的有效工具，并在历史上产生关于海景功能和扰动做出反应的新概念。这类研究往往会产生高质量的数据，对于梳理复杂因素以揭示生态系统的潜在模式特别有用。围隔研究通过揭示生物体与生态系统之间关系的复杂性和简单性，使得研究人员有时也意识到我们理解的局限性（Benton et al., 2007）。在监管背景下，如加拿大的环境影响监测计划认为，围隔是一种公认的鱼类研究替代监测方法。尽管大型鱼类对实验设计提出了挑战，但使用大型生物体的受益更好。越来越多的围隔用于探索非常复杂的全球问题，包括气候变化影响（Davis et al., 1998；Petchey et al., 1999）、生物多样性丧失（Balvanera et al., 2006；Cardinale et al., 2006）、环境污染、渔业过度捕捞（Micheli, 1999），以及极端生境研究（Stark et al., 2014）。鉴于气候变化、海洋酸化、沿海发展和城市化对全球海洋构成的威胁，未来将需要跨学科研究，而围隔研究可以成为任何研究项目的有力工具。

参考文献

ADAMS S M, 2003. Establishing causality between environmental stressors and effects on aquatic ecosystems. Hum. Ecol. Risk Assess., 9: 17-35.

ALEXANDER A C, CULP J M, 2013. Predicting the effects of insecticide mixtures on non-target aquatic communities. Chapter 3 // TRDAN S (Eds.). Insecticidese development of safer and more effective technologies. InTech, Rijeka, Croatia: 83-101. http://www.intechopen.com/books/insecticides-development-of-safer-and-moreeffective-technologies/predicting-the-effects-of-insecticidemixtures-on-non-target-aquatic–communities.

ALEXANDER A C, HEARD K S, CULP J M, 2008. Emergent body size of mayfly survivors. Freshwater Biol., 53: 171-180.

ALEXANDER A C, LUIS A T, CULP J M, et al., 2013. Can nutrients mask community responses to insecticide mixtures? Ecotoxicology, 22: 1085 – 1100.

BAKER T R, BAKER B B, JOHNSON S M, et al., 2013. Comparative analgesic efficacy of morphine sulfate and butorphanol tartrate in koi (Cyprinus carpio) undergoing unilateral gonadectomy. J. Am. Vet. Med. Assoc., 243 (6): 882 – 890.

BAKKE T, 1990. Benthic mesocosms: II. basic research in hard-bottom benthic mesocosms // LALLI C M (Eds.). Coastal and estuarine studies. Springer, New York, NY, USA: 122 – 135.

BALVANERA P, PFISTERER A B, BUCHMANN N, et al., 2006. Quantifying the evidence for biodiversity effects on ecosystem functioning and services. Ecol. Lett., 9: 1146 – 1156.

BENTON T G, SOLAN M, TRAVIS J M J, et al., 2007. Microcosm experiments can inform global ecological problems. Trends Ecol. Evol., 22: 516 – 521.

BLOESCH J, BOSSARD P, BUHRER H, et al., 1988. Can results from limnocorral experiments be transferred to in situ conditions? Hydrobiologia, 159: 297 – 308.

BOKN T L, HOELL E E, KERSTING K, et al., 2001. Methods applied in the large littoral mesocosms study of nutrient enrichment in rocky shore ecosystems-EULIT. Cont. Shelf Res., 21: 1925 – 1936.

BOLON B, STOSKOPF M K, 1995. Fish // ROLLIN B E (Eds.). The experimental animal in biomedical research: care, husbandry, and well-being-an overview by species, vol. II. CRC Press: 15 – 30.

BOTHWELL M L, 1993. Artificial streams in the study of algal/nutrient dynamics // LAMBERTI G A, STEINMAN A D (Eds.). Research in artificial streams: applications, uses, and abuses: 327 – 333. J. N. Am. Benthol. Soc., 12: 313 – 384.

BOYLE T P, FAIRCHILD J F, 1997. The role of mesocosm studies in ecological risk analysis. Ecol. Appl., 7: 1099 – 1102.

BRAITHWAITE V A, HUNTINGFORD F A, 2004. Fish and welfare: do fish have the capacity for pain perception and suffering? Anim. Welfare, 13 suppl.: 87 – 92.

BUFFET P-E, ZALOUK-VERGNOUX A, CHÂTEL A, et al., 2014. A marine mesocosm study on the environmental fate of silver nanoparticles and toxicity effects on two endobenthic species: the ragworm *Hediste diversicolor* and the bivalve mollusk *Scrobicularia plana*. Sci. Total Environ., 470 – 471 (1): 1151 – 1159.

CARDINALE B, SRIVASTAVA D, DUFFY J, et al., 2006. Effects of biodiversity on the functioning of trophic groups and ecosystems. Nature, 443: 989 – 992.

CARPENTER S R, COLE J J, ESSINGTON T E, et al., 1998. Evaluating alternative explanations in ecosystem experiments. Ecosystems, 1: 335 – 344.

CARPENTER S R, 1996. Microcosm experiments have limited relevance for community and ecosystem ecology. Ecology, 77: 677 – 680.

CARPENTER S R, 1999. Microcosm experiments have limited relevance for community and ecosystem ecology: reply. Ecology, 80: 1085 – 1088.

CHRISTMAN V D, VOSHELL J R, JENKINS D G, et al., 1994. Ecological development and biometry of untreated pond mesocosms // GRANEY R L, KENNEDY J H, RODGERS J H (Eds.). Aquatic mesocosm studies in ecological risk assessment. CRC Press, Boca Raton, FL: 105-129.

CLARK J L, CLEMENTS W H, 2006. The use of in situ and stream microcosm experiments to assess population-and community-level responses to metals. Environ. Toxicol. Chem., 25 (09): 2306-2312.

CLARK J R, NOLES J L, 1994. Contaminant effects in marine/estuarine systems: field studies and scaled simulations // GRANEY R L, KENNEDY J H, RODGERS J H (Eds.). Aquatic mesocosm studies in ecological risk assessment. CRC Press, Boca Raton, FL, USA: 47-60.

CLEMENTS W H, CARLISLE D M, COURTNEY L A, et al., 2002. Integrating observational and experimental approaches to demonstrate causation in stream biomonitoring studies. Environ. Toxicol. Chem., 21 (6): 1138-1146.

CLEMENTS W H, 2004. Small-scale experiments support causal relationships between metal contamination and macroinvertebrate community responses. Ecol. Appl., 14: 954-967.

CRANE M, BURTON G A, CULP J M, et al., 2007. Review of aquatic in situ approaches for stressor and effect diagnosis. Integr. Environ. Assess. Manage., 3: 234-245.

CULP J M, BAIRD D J, 2006. Establishing cause-effect relationships in multi-stressor environments // HAUER F R, LAMBERTI G A (Eds.). Methods in Stream Ecology. 2nd ed. Academic Press, Burlington, MA, USA: 835-854.

CULP J M, LOWELL R B, CASH K J, 2000a. Integrating in situ community experiments with field studies to generate weight-of-evidence risk assessment for large rivers. Environ. Toxicol. Chem., 19: 1167-1173.

CULP J M, PODEMSKI C L, CASH K J, 2000b. Interactive effects of nutrients and contaminants from pulp mill effluents on riverine benthos. J. Aquat. Ecosyst. Stress Recovery, 8 (1): 67-75.

CULP J M, PODEMSKI C L, CASH K J, et al., 2000c. A research strategy for using stream microcosms in ecotoxicology: integrating experiments at different levels of biological organization with field data. J. Aquat. Ecosyst. Stress Recovery, 7: 167-176.

CULP J M, CASH K J, GLOZIER N E, et al., 2003. Effects of pulp mill effluent on benthic assemblages in mesocosms along the Saint John River, Canada. Environ. Toxicol. Chem., 22: 2916-2925.

DAVIS A J, JENKINSON L S, LAWTON J H, et al., 1998. Making mistakes when predicting shifts in species range in response to global warming. Nature, 391: 783-786.

DE LAFONTAINE Y, LEGGETT W C, 1987. Evaluation of in situ enclosures for larval fish studies. Can. J. Fish Aquat. Sci., 44: 54-65.

DRENNER R W, MAZUMDER A, 1999. Microcosm experiments have limited relevance for community and ecosystem ecology: comment. Ecology, 80: 1081-1085.

DUBÉ M G, CULP J M, CASH K J, et al., 2002. Artificial streams for environmental effects monitoring (EEM): development and application in Canada over the past decade. Water Qual. Res. J. Can., 37: 155 – 180.

EFSA PPR Panel (European Food Safety Authority Panel on Plant Protection Products and their Residues), 2013. Guidance on tiered risk assessment for plant protection products for aquatic organisms in edge-of-field surface waters. EFSA J., 11 (7): 3290. http://dx.doi.org/10.1016/B978 – 0 – 12 – 803371 – 5.00008 – 4, 268. http://www.efsa.europa.eu/sites/default/files/scientific_output/files/main_documents/3290.pdf.

EMMERSON M C, SOLAN M, EMES C, et al., 2001. Consistent patterns and the idiosyncratic effects of biodiversity in marine ecosystems. Nature, 411: 73 – 77.

Environment Canada, 2010. Pulp and paper environmental effects monitoring technical guidance document: Chapter 8 alternative monitoring methods. https://www.ec.gc.ca/esee-eem/3E389BD4 – E48E-4301 – A740 – 171C7A887EE9/PP_full_versionENGLISH [1] – FINAL-2.0.pdf.

Environment Canada, 2012. Metal mining technical guidance for environmental effects monitoring: Chapter 9 alternative monitoring Methods. https://ec.gc.ca/Publications/default.asp?lang = En&xml = D175537B – 24E3 – 46E8 – 9BB4 – C3B0D0DA806D.

FOEKEMA E M, KAAG N H, KRAMER K J M, et al., 2015. Mesocosm validation of the marine No Effect Concentration of dissolved copper derived from a species sensitivity distribution. Sci. Total Environ., 521 – 522: 173 – 182.

FORREST J, ARNOTT S E, 2006. Immigration and zooplankton community responses to nutrient enrichment: a mesocosm experiment. Oecologia, 150: 119 – 131.

GAMBLE J C, DAVIES J M, STEELE J H, 1977. Loch Ewe bag experiment, 1974. Bull. Mar. Sci., 27: 146 – 175.

GRANEY R L, KENNEDY J H, RODGERS J H, 1994. Introduction // GRANEY R L, KENNEDY J H, RODGERS J H (Eds.). Aquatic mesocosm studies in ecological risk assessment. CRC Press, Boca Raton, FL, USA: 1 – 4.

GRANEY R L, GIESY J P, CLARK J R, 1995. Field Studies // RAND G M (Eds.). Fundamentals of aquatic toxicology: effects, environmental fate, and risk assessment. Taylor & Francis, Washington, DC, USA: 257 – 305.

GRAY J S, 1987. Oil pollution studies of the Solbergstrand mesocosms. Philos. Trans. R. Soc. B, 316: 641 – 654.

GRICE G D, REEVE M R, 1982. Marine mesocosms. Springer, New York, NY, USA.

GRICE G D, HARRIS R P, REEVED M R, 1980. Large-scale enclosed water-column ecosystems an overview of foodweb I, the final CEPEX experiment. J. Mar. Biol. Assoc. UK, 60: 401 – 414.

GUCKERT J B, 1993. Artificial streams in ecotoxicology // LAMBERTI G A, STEINMAN A D (Eds.). Research in artificial streams: applications, uses, and abuses: 350 – 356. J. N. Am. Benthol. Soc., 12: 313 – 384.

HOWICK G L, DENOYELLES F, GIDDINGS J M, et al., 1994. Earthen ponds vs. fiberglass tanks as venues for assessing the impact of pesticides on aquatic environments: a parallel study with Sulprofos//GRANEY R L, KENNEDY J H, RODGERS J H (Eds.). Aquatic mesocosm studies in ecological risk assessment. CRC Press, Boca Raton, FL, USA: 321-336.

JOHNSON P C, KENNEDY J H, MORRIS R G, et al., 1994. Fate and effects of Cyfluthrin (Pyrethroid insecticide) in pond mesocosms and concrete microcosms//GRANEY R L, KENNEDY J H, RODGERS J H (Eds.). Aquatic mesocosm studies in ecological risk assessment. CRC Press, Boca Raton, FL, USA: 337-371.

KIM J M, SHIN K, LEE K, et al., 2008. In situ ecosystem-based carbon dioxide perturbation experiments: design and performance evaluation of a mesocosm facility. Limnol. Oceanogr.: Methods, 6 (6): 208-217.

KLEYPAS J A, FEELY R A, FABRY V J, et al., 2006. Impacts of ocean acidification on coral reefs and other marine calcifiers: a guide for future research. Report of workshop held 18-20 April 2005, St. Petersburg, FL: 88. http://www.ucar.edu/communications/Final_acidification.pdf.

KUIPER J, 1982. Ecotoxicological experiments with marine plankton communities in plastic bags//GRICE G D, REEVE M R (Eds.). Marine mesocosms. Springer, New York, NY, USA: 181-193.

LAMBERTI G A, STEINMAN A D, 1993. Conclusions//LAMBERTI G A, STEINMAN A D (Eds.). Research in artificial streams: applications, uses, and abuses: 370. J. N. Am. Benthol. Soc., 12: 313-384.

LAWTON J H, 1996. The ecotron facility at silwood park: the value of "big bottle" experiments. Ecology, 77: 665-669.

LEJEUSNE C, CHEVALDONNÉ P, PERGENT-MARTINI C, et al., 2010. Climate change effects on a miniature ocean: the highly diverse, highly impacted Mediterranean Sea. Trends Ecol. Evol., 25: 250-260.

LOWELL R B, CULP J M, DUBÉ M G, 2000. A weight-of-evidence approach for northern river risk assessment: integrating the effects of multiple stressors. Environ. Toxicol. Chem., 4 (2): 1182-1190.

LUCKETT C, ADEY W H, MORRISSEY J, et al., 1996. Coral reef mesocosms and microcosms-successes, problems, and the future of laboratory models. Ecol. Eng., 6: 57-72.

MCINTIRE C D, 1993. Historical and other perspectives of laboratory stream research//LAMBERTI G A, STEINMAN A D (Eds.). Research in artificial streams: applications, uses, and abuses: 318-323. J. N. Am. Benthol. Soc., 12: 313-384.

MENZEL D W, CASE J, 1977. Concept and design: controlled ecosystem pollution experiment. Bull. Mar. Sci., 27: 1-7.

MICHELI F, 1999. Eutrophication, fisheries, and consumerresource dynamics in marine pelagic ecosystems. Science, 285: 1396 – 1398.

MOHR S, FIEIBICKE M, OTTENSTROER T, et al., 2005. Enhanced experimental flexibility and control in ecotoxicological mesocosm experiments: a new outdoor and indoor pond and stream system. Environ. Sci. Pollut. Res. Int., 12: 5 – 7.

NERHEIM S, STIANSEN J E, SVENDSEN H, 2002. Gridgenerated turbulence in a mesocosm experiment. Hydrobiologia, 484: 61 – 73.

NIXON S W, OCZKOWSKI A J, PILSON M, et al., 2015. On the response of pH to inorganic nutrient enrichment in well-mixed coastal marine waters. Estuaries Coasts, 38: 232 – 241.

NORTON S B, CORMIER S M, SUTER G W, 2014. Ecological causal assessment. CRC Press, Boca Raton, FL, USA.

ODUM E P, 1984. The mesocosm. Bioscience, 34: 558 – 562.

OVIATT C A, WALKER H, PILSON M E Q, 1980. An exploratory analysis of microcosm and ecosystem behavior using multivariate techniques. Mar. Ecol. Progr. Ser., 2: 179 – 191.

OVIATT C A, 1994. Biological considerations in marine enclosure experiments: challenges and revelations. Oceanography, 7: 45 – 51.

PARSONS T R, 1978. Controlled aquatic ecosystem experiments in ocean ecology research. Mar. Pollut. Bull., 9: 203 – 205.

PARSONS T R, 1982. The future of controlled ecosystem enclosure experiments // GRICE G D, REEVE M R (Eds.). Marine mesocosms biological and chemical research in experimental ecosystems. Springer, New York, NY, USA: 411 – 418.

PERCEVAL O, CAQUET T, LAGADIC L, et al., October 2009. Mesocosms: their value as tools for managing the quality of aquatic environments // Recap prepared from the meeting of the ecotoxicology symposium in le croisic, France.

PETCHEY O L, MCPHEARSON P T, CASEY T M, et al., 1999. Environmental warming alters food-web structure and ecosystem function. Nature, 402: 69 – 72.

PETERSEN J E, CORNWELL J C, KEMP W M, 1999. Implicit scaling in the design of experimental aquatic ecosystems. Oikos, 85: 3 – 18.

PETERSEN J E, KENNEDY V S, DENNISON W C, et al., 2009. Enclosed experimental ecosystems and scale: tools for understanding and managing coastal ecosystems. Springer, New York, NY, USA.

PILSON M E Q, NIXON S W, 1980. Annual nutrient cycles in a marine microcosm // GIESY J P (Eds.). Microcosms in ecological research, DOE symposium series, 52. Springfield, VA, USA: 753 – 778. CONF – 781101.

PIPER R G, MCELWAIN I B, ORME L E, et al., 1982. Fish hatchery management. United States Fish and Wildlife Service, Washington, DC.

POHLENZ C, GATLIN III D M, 2014. Interrelationships between fish nutrition and health. Aquaculture, 431: 111 – 117.

RAND G M, CLARK J R, HOLMES C M, 2000. Use of outdoor freshwater pond microcosms: Ⅱ. Responses of biota to pyridaben. Environ. Toxicol. Chem., 19 (2): 396 – 404.

RENICK V C, ANDERSON T W, MORGAN S G, et al., 2015. Interactive effects of pesticide exposure and habitat structure on behavior and predation of a marine larval fish. Ecotoxicology, 24: 391 – 400.

RIEBESELL U, LEE K, NEJSTGAARD J C, 2009. Pelagic mesocosms. Chapter 6 // RIEBESELL U, FABRY V J, HANSSON L (Eds.). Guide to best practices for ocean acidification research and data reporting. Report for the European Commission EUR 24328 EN.

RIEBESELL U, BELLERBY R G J, GROSSART H-P, et al., 2008. Mesocosm CO_2 perturbation studies: from organism to community level. Biogeosciences, 5: 1157 – 1164.

SANDERS F S, 1985. Use of large enclosures for perturbation experiments in lentic ecosystems: a review. Environ. Monit. Assess., 5: 55 – 99.

SANFORD L P, 1997. Turbulent mixing in experimental ecosystem studies. Mar. Ecol. Progr. Ser., 161: 265 – 293.

SCHINDLER D W, 1998. Whole-ecosystem experiments: replication versus realism: the need for ecosystem-scale experiments. Ecosystems, 1: 323 – 334.

SCHULZ R, THIERE G, DABROWSKI J M, 2002. A combined microcosm and field approach to evaluate the aquatic toxicity of azinphosmethyl to stream communities. Environ. Toxicol. Chem., 21 (10): 2172 – 2178.

SCOTT G I, FULTON M H, DELORENZO M E, et al., 2013. The environmental sensitivity index and oil and hazardous materials impact assessments: linking prespill contingency planning and ecological risk assessment. J. Coastal Res., 69: 100 – 113.

SNEDDON L U, 2009. Pain perception in fish: indicators and endpoints. ILAR J., 50: 338 – 342.

STARK J, JOHNSTONE G, RIDDLE M, 2014. A sediment mesocosm experiment to determine if the remediation of a shoreline waste disposal site in Antarctica caused further environmental impacts. Mar. Pollut. Bull., 89 (1 – 2): 284 – 295.

STEELE J H, FARMER D M, HENDERSEN E W, 1977. Circulation and temperature structure in large marine enclosures. J. Fish Res. Board Can., 34: 1095 – 1104.

STEWART R I, DOSSENA M, BOHAN D A, et al., 2013. Mesocosm experiments as a tool for ecological climate-change research. Chapter 2. Adv. Ecol. Res., 48: 71 – 181.

SUNYER J O, 2013. Fishing for mammalian paradigms in the teleost immune system. Nat. Immunol., 14: 320 – 326.

SVENSEN C, EGGE J K, STIANSEN J E, 2001. Can silicate and turbulence regulate the vertical flux of biogenic matter? A mesocosm study. Mar. Ecol. Progr. Ser., 217: 67 – 80.

SWIFT M, TROELSTRUP JR N H, DETEBECK N E, et al., 1993. Large artificial streams in toxicological and ecological research. J. N. Am. Benthol. Soc., 12: 359 – 366.

TATTERSON G B, 1991. Fluid mixing and gas dispersion in agitated tanks. McGraw – Hill, New York, NY, USA.

THOMPSON D G, HOLMES S B, PITT D G, et al., 1994. Applying concentration-response theory to aquatic enclosure studies//GRANEY R L, KENNEDY J H, RODGERS J H (Eds.). Aquatic mesocosm studies in ecological risk assessment. CRC Press, Boca Raton, FL, USA: 129-156.

US EPA, 1992. Framework for ecological risk assessment. US Environmental Protection Agency, Washington, DC, USA. Risk Assessment Forum Report N. EPA/630/P-04/068B. http://www.epa.gov/raf/publications/pdfs/FRMWRK_ERA.PDF.

YANONG R P E, 2003. Fish health management considerations in recirculating aquaculture systems: Part 1: Introduction and general principles. Circular FA-120. UF IFAS Cooperative Extension Service. http://edis.ifas.ufl.edu/fa099.

9 生态风险和证据权重评价

P. M. Chapman[①]

[①] 加拿大 Chapema 环境战略有限公司。

9.1 引　　言

前几章对海洋生态毒理学的各个方面进行了论述，从污染物与实验设计到建模与监测，包括专业内容和应用（生物标志物、水和沉积物测试、微生态和中生态）。本章将提供基于应用海洋生态毒理学的知情决策框架。

知情决策并不包含完全的知识和绝对的确定性。相反，它包含最好的可用知识，加上最好的专业判断，并合理地减少不确定性。不确定性是必需被承认和接受的，尤其是考虑到全球气候变化这一现实（Landis et al., 2012；参见本书第10章）。如前文所述，它可以被减少和限制，但无法被完全消除。

本章将介绍两种互补的决策框架：生态风险评价（ecological risk assessment，ERA）和证据权重（weight of evidence，WoE）评价；并探讨海洋生态毒理学在这两种框架内的作用（毒性试验和生物富集评价）；最后对探讨这些作用的可能性和概率。

9.2　生态风险评价

9.2.1　生态风险评价导论

ERA包含一个收集数据并评价其对知情决策充分性的框架。它的基础是评价和分配不同活动或压力源对环境不利影响的概率。风险指概率（probability）；危险则指可能性（possibility）。换而言之，如果水环境中一定量铜具有对鱼类产生有害效应的风险，就是有一定概率会发生；如果水环境中一定量的铜有可能对鱼类产生有害效应，就是有一定的可能性，且这个可能性是否存在还有待确定。ERA框架通过识别、计算和报告来估计应激源有害效应的不确定性。

ERA及其所包含的框架不能提供100%的确定性；实际也不存在绝对的准确性。一个正确执行的ERA会衍生出两种产物：与风险有关的信息和与知情决策相关的不确定性，以及应处理/解决的关键不确定性，以提高此类决策的准确性。

ERA侧重于应激源，包括生物可利用的污染物（换句话说，它们会影响生物系统）。极端情况下，暴露于应激源可能导致3种有害结局而影响生态系统结构和功能：①种群水平的变化（包括种群瓶颈和应激源诱导的选择）；②遗传多样性的变化；③进化轨迹的变化。

根据个人观点的不同，上述变化可以是消极的、中性的或积极的。例如，虹鳟是一种入侵物种，在 19 世纪 50 年代中期之前，只在北美落基山脉以东被发现。如今，人类活动使其除在南极洲外其他大洲均有分布。它作为一种游钓和商业性鱼类受到珍视和保护，但它的引入显然改变了原有的水生生态系统。从人类需求的角度来看，虹鳟的引入所导致的变化是积极的。然而，这种引入对一些水生生态系统可能产生积极和消极的影响，这取决于当地物种及其在食物链中的作用。

导致全球范围内环境变化的五大环境应激源按重要性排序如下：①全球气候变化；②生境变化；③外来物种引入／入侵；④富营养化，包括有害藻华；⑤化学污染。

实施 ERA 的原因是为了预测不同应激源和活动产生的生态后果，评价行动的潜在影响，以及确定其现在和未来的意义：

（1）追溯性风险评价，用来评价已发生事情的重要性。这是 ERA 最常见用途，但不一定是最有用的。更好的方式应是在环境破坏发生之前进行预测和预防，而不是在事后尝试补救，因为补救本身也可能有风险（Moriarty, 2015；参见第 9.2.3 节）。

（2）主动性风险预测，用来评价未来可能发生事情的重要性。这是在 ERA 中不常见，却越来越重要的手段。

（3）比较不同行动或发展带来的风险，以评价不同活动的风险并确定其优先级，包括降低风险等管理措施（参见第 9.2.3 节）。人类的一切行为对环境均具有负面影响，人类的存在本身就改变了生态系统（Chapman, 2013a）。例如，虽然废水处理将减少废水中污染物，但方法是将废水中的污染物浓缩为半液态残留物，即污泥。而废水处理和污泥处置过程会耗费能源，产生温室气体，以及占用栖息地资源（Chapman, 2013b）。

（4）对不同应激源或行为所造成的风险进行排序，类似于前述观点，但仅排序而不进行比较，并包括前文列出的五大应激源。

（5）根据必要性和适当性来制定保护性（有时为特定性）基准。可以制定并使用数字基准来区分需要管理行动（如补救）的领域和不构成重大环境风险的领域。以实现采取有针对性的管理行动，从而减少这些行动对环境的破坏（Chapman & Smith, 2012）。例如，政府通常会为一般且保守的保护（如水生和陆地环境）制定环境质量基准，这些政府基准可以特定性使用或修改。

判定风险存在需满足两个条件：首先，应激源（即化学的、生物的或物理实体的）具有引起一个或多个有害效应的内在能力；其次，应激源与受体［即潜在受应激源影响的生态功能资源（个体、种群、群落、生境）］在足够的持续时间和程度上共存或接触，从而引发已确定的有害效应。

ERA 有 4 个科学组成部分，它们均具有生态毒理学成分，在第 9.2.2 节（图 9.1）中会进行讨论：

（1）问题阐述。定义问题、评价和测量终点［即分别为需要保护的环境价值（如商业性渔业），其他可外推表示评价终点的可测量的方面（如关键鱼种的生存、生长和繁殖）］；总结可用信息；制定概念模型（第 9.2.2 节）。

（2）暴露评价。确定暴露浓度或水平、生物利用度（对于化学品）、敏感物种／种群。

（3）效应评价。确定直接或间接的可能对个体（如与更高层次生物相关的生物标记物)、种群和群落产生的效应。

(4) 风险特征。根据暴露与效应的比较来估计风险。

ERA 的非科学组成部分 (图 9.1) 是指通过与风险管理者进行初步和最终讨论,以及从管理者那获得的信息,这包括所有利益相关者与科学家沟通后的初步意见,以及与决策者沟通后的最终意见。风险管理是一个非科学过程。科学证据仅是其中的一个考虑因素,但并不总是最重要的考虑因素,还应考虑其他 3 个因素:经济、社会和政治。

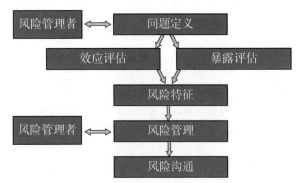

图 9.1 生态风险评价的科学 (红色字体) 和非科学部分 (白色字体)

9.2.2 海洋生态毒理学在生态风险评价中的作用

在 ERA 的 4 个组成部分中,海洋生态毒理学具有 3 个重要作用:问题阐述、效应评价和风险特征。海洋生态毒理学在暴露评价中不起作用。暴露评价只与应激源有关,而与应激源的潜在效应无关。

问题阐述涉及以下组成部分,其中海洋生态毒理学在了解潜在效应方面具有作用,无论是普遍性的还是特定性的:①应激源的识别和特征 (物理的、化学的、生物的);②识别受体 (潜在的处于风险的生物群);③潜在生态效应的识别;④评价和试验终点的确定;⑤开发概念模型,以示意图形式或卡通形式说明应激源与受体之间的关系 (一个简化概念模型范例如图 9.2 所示);⑥制定环境风险假设和保护目标 (人类价值,例如生态系统服务功能),以确定将在下一 ERA 环节 (暴露和效应评价) 中的评价/测试内容。

效应评价描述了应激源和受体之间的关系,并用于确定应激源与生物反应和 (或) 生态构成之间的关系。毒性试验和物种敏感度分布的计算 (图 9.3,详述见下一段) 是 ERA 构成的关键因素。相对于根据生物群中化学污染物浓度确定基准,联合毒性试验和生物富集评价的重要性在进一步提高 (Meador et al., 2014)。确定基准不仅对了解生态系统功能重要本地生物中污染物剂量具有重要意义,同样也对了解毒性试验生物中污染物剂量至关重要。因为决定毒性的是生物体内的剂量 (即浓度和形态),而不是环境中的浓度。

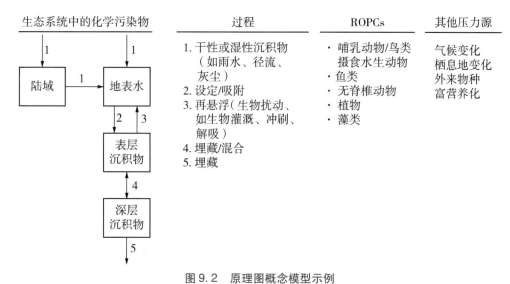

图 9.2 原理图概念模型示例

ROPCs（receptors of potential concern），关注的潜在受体。

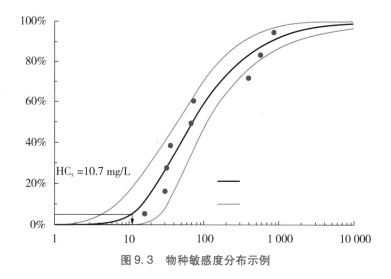

图 9.3 物种敏感度分布示例

将被试验生物百分比与毒性试验结果进行作图，即可得到物种敏感度分布。具体而言，将每个被测物种不受到影响或受到可忽略的影响时［10%～20%效应水平，如抑制浓度（IC）为 10～20］的应激源浓度（在图 9.3 中的示例，锶；McPherson et al.，2014）与试验生物受影响比率进行比较，可得出测试物种从最高耐受到最低耐受的曲线。绘制一条最佳拟合线，当它与 5% 的物种测试线相交时，意味着 95% 的该物种群体不受此浓度的影响。这一交点被称为 HC_5，它是建立在 Posthuma et al.（2002）的基础上，即"在水生生物群落中存在足够的冗余，以允许一些损失"（CCME，2007，第二部分，第 3.1—3.5 节）。正如上述 CCME 中所指出的，考虑到可能损失高达 5% 的物种群体，因此"确定低或无效应水平数据不太重要了"，重点应放在无或可忽略的效应数据上（如10%～20%效应水平）。

风险特征是 ERA 的最后一层，整合了 ERA 的其他部分。从独立证据（lines of evidence，LoE）得出的结论是相互关联的，类似于使用证据权重（WoE）确定的（第9.3节）多重证据间的一致性，而且不同的观点有助于确定可能的机制。风险特征可定性或定量地估计效应的大小和概率。

9.2.3 海洋生态毒理学在风险评价管理中的作用

如第9.2.2节所述，海洋生态毒理学提供了应激源潜在有害效应的信息。这些信息为非科学家人员的风险管理决策相关提供了相关背景知识依据（图9.1）。

一旦做出风险管理的决策，就需要评价其有效性。具体来说，需要确定是否做出了正确决策，以减少或最小化应激源的潜在有害效应。此外，由于人类的每个行动都有其潜在的负面效应，因此需要对风险管理本身进行评价，以确定是否正在发生出乎意料的有害效应，这些效应比正在处理的效应更值得关注，即风险管理行为本身是否弊大于利（Moriarty，2015）。

因此，在初步评价后，海洋生态毒理学具有持续作用。确切的作用将取决于地点和情况，但可能需要一些最初为确定风险管理效能而进行的相同测试和评价。由于可能出现意想不到的有害效应，因此需要根据新的信息和最佳的专业判断进行不同的测试和评价。

9.3 证据权重

9.3.1 证据权重简介

证据权重（WoE）指综合多学科环境研究产生的数据，涉及多个独立证据，通常包括化学和生物测量。这是一种基于多重独立证据确定可能的生态影响的方法。

证据权重评价提供了3种类型的信息：①应激源对环境有害效应的相对确定性；②可能的因果关系；③关键的不确定性，若得到解决，将改进管理决策。加拿大环境部（2012）将证据权重定义为"从不同的科学证据中收集信息以得出关于危害的概率和大小结论的所有程序"。ECHA（2014）指出："在风险评价中，衡量和预测效应的最可靠方法包括精心设计的观察性研究和以一种证据权重方式进行实验的组合，旨在证明关键机制和联系。

证据权重评价为环境评价数据（例如 Chapman & Anderson 2005；McDonald et al.，2007；加拿大环境部和安大略省环境部 2008；Cormier et al.，2010；Suter et al.，2010；

ECHA 2010；Suter & Cormier 2011；Chapman & Smith 2012；Cormier & Suter 2013；ECHA 2014；Hope & Clarkson 2014）和其他数据（如 Borgert et al.，2011）的整合提供了一种公认的方法。Linkov et al.（2015）特别提到了在证据权重方法论定量、透明和客观方面的规范所做出的巨大努力；他们指出"整合独立证据是环境评价的重要组成部分，环境评价应标准化，以建立类似工作间的一致性和可比性"。

在证据权重分析中，决策的基础是将一个或多个统计分析、评分系统和定性判断汇入一个逻辑系统。定性证据权重方法通常包括对结果的分类（如低、中、高），然后根据最佳专业判断对结果进行集成，以获得一个全面的结论（如沉积物质量三元法：Chapman，1990；Chapman et al.，2002；Chapman & McDonald，2005）。半定量方法以相同的方式开始，但随后是将结果的分类转换为数值分数，并将数值加权因子应用于此分数，以获得总体证据权重"分数"（如框架的提出，Menzie et al.，1996；Hope & Clarkson 2014）。此外，基于数字评级和权重的系统应用，也可采用混合方法；但总体证据权重分数是定性的，而非绝对的（如 McDonald et al.，2007）。美国环保署因果分析/诊断决策信息系统使用证据权重和环境流行病学原则一起评价因果关系（CADDIS；http://www.epa.gov/caddis/）。

大多数证据权重评价都有 3 个共同的考虑因素：①终点的相对权重；②对各终点响应程度（定性或定量地）评级；③相关终点的并发反应（Menzie et al.，1996）。使证据权重评价在技术上可行的关键就是透明度，以确保结果的可重复性（Linkov et al.，2009）。为实现这种透明性和可重复性，由独立证据提供的证据质量通常是加权先验的（即在结果已知之前），以防止在应用加权前出现与已知反应相关的潜在偏差，以及在某些情况下应用后验（即事后）调整，以考虑相关证据之间反应的一致性（或一致性缺乏）。如在 Haake et al.（2010）& Wiseman et al.（2010）的案例研究中，即使面对不确定的因果关系和重大的不确定性，结构化、透明的证据权重评价也能提供足够的确定性来为决策提供依据。

学者建议对不同的证据进行加权，以进行证据权重评价（Menzie et al.，1996；McDonald et al.，2007；Suter & Cormier 2011；Hope & Clarkson 2014）。对解决具体环境问题权重的考虑通常分为 4 类：①证据与受保护生态价值之间的关联强度；②典型性（空间、时间、特异性）；③研究设计和执行（针对特异性调查）；④数据质量。为保持清晰度和透明性，首选方法是根据这些事先考虑因素对权重进行标准化，这样才能以系统、无偏差的方式应用于证据。实际上，环境评价数据的性质很复杂，可能会出现先验无法预料的反应组合；必要时允许后验判断（即考虑实际证据反应的性质和方向）来微调证据权重结果（Menzie et al.，1996；McDonald et al.，2007；SABCS，2010）。加权应包括先验的考虑，对变化或反应方向的考虑，以及基于结果的性质、杂性和确定性的后验考虑。Suter & Cormier（2011）以及 Hope & Clarkson（2014）认为对证据权重的权重因子和分数的定量是通过最佳专业判断得出的，因此应该定性地解释权重因子和分数。

9.3.2 海洋生态毒理学在证据权重中的作用

证据权重评价包括对暴露和效应的单独证据。海洋生态毒理学提供了有关效应的信息，以补充和帮助了解应激源暴露下生物群落的状态信息。尽管生态毒理学研究从定义上看并非完全贴合现实（贴近现实研究排序：室内毒性研究 < 微宇宙试验 < 围隔实验 < 原位暴露），但它们提供了有关应激源可能做出反应的必要信息。对暴露群落评价虽然现实，但由于某的随机性（自然可变性）而可能缺乏说服力，除非应激源的效应是极端的。

在证据权重中使用海洋生态毒理学的最佳方法是根据一个特异性问题建立概念模型。该模型旨在说明与效应（即变化，可正面或负面）相关的潜在关注应激源（stressors of potential concern，SOPCs）、暴露途径和潜在关注受体（receptors of potential concern，ROPCs）之间的相互作用。图 9.4 中提供了一个将矿业废水排放至海湾的示例。在这种情况下，潜在关注应激源是指废水带来的潜在毒物和营养物质。潜在关注受体是指水体中的附石藻类（旧称为固着植物）、浮游植物和浮游动物、沉积物中的底栖无脊椎动物，以及深海和远洋鱼类等。受纳环境的毒性试验包括水和沉积物毒性测试。生物富集评价更有可能涉及足够大的潜在关注受体，以便相对容易地收集足够的生物量用于化学分析；但根据潜在关注物质及其通过食物网的途径，可能涉及诸如无脊椎动物和浮游生物等较小的潜在关注受体。

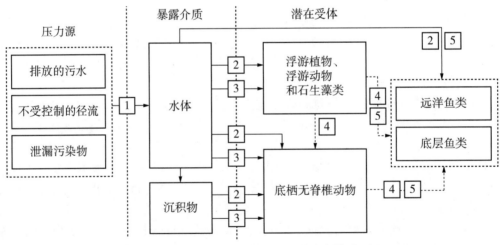

图 9.4　矿业废水排入海湾的特异性概念模型示例

1—污染物（具潜在毒性）和营养物（潜在富集）排放；2—潜在的直接毒性；3—潜在营养物富集，或营养物平衡改变；4—基于食物供应变化的潜在间接效应；5—组织化学的潜在改变。

表 9.1 中的示例说明了在证据权重评价中如何结合毒性测试、化学分析和生物群落评价，来判断是否存在低、中或高水平的潜在影响（即化学品对生物群体产生有害效应）。该例子较简单，只涉及 3 个证据；当然其他证据也可能存在（Chapman & Hollert，2006）。例如，少数具有生物放大效应的有机化学物质（如甲基汞、2,3,7,8-TCDD、

滴滴涕、多氯联苯）可被视为单独证据。从生物富集评价中获得的临界体内残留（即对受影响生物造成有害效应的浓度）可以基准的形式提供证据：低于该基准时几乎不可能诱导效应；高于时则很有可能诱导效应；还可用来确定与化学应激源相关的因果关系。效应生物标记物（在个体生物体中测量的从分子到细胞、代谢和生理反应到行为变化的特定生物反应）不仅与暴露有关，还会提供额外的证据。化学污染以外的应激源（如生境变化）也可以提供额外的证据。

表9.1 沉积物中化学应激源、毒性和群落的证据权重分类示例（依据关注程度的分类）

证据	高	中	低
体相化学（与SQG相比）	预期有害效应：一项或多项SQG-high超标	有害效应可能会或不会发生：一项或多项SQG-low超标	无预期有害效应：所有污染物浓度均低于SQG-low
毒性终点（相对于参考值）	主要：在一项或多项毒理学终点中，统计上显著降低幅度>50%	轻微：在一项或多项毒理学终点中，统计上显著降低幅度>20%	忽略不计：所有毒理学终点指标降低幅度≤20%
总体毒性（基于一项以上的试验）	重要：多个试验/终点显示出主要毒性效应	潜在：多个试验/终点显示出较小的毒性效应和/或一个试验/终点显示出主要效应	忽略不计：在不超过一个终点观察到较小毒性效应
生物群落的变化（单变量和多变量的评价）	与参考"不同"或"非常不同"	与参考"可能不同"	与参考"相同"
整体证据权重评价	显著的有害效应：化学物浓度升高 一个或多个毒理学终点指标降低幅度超过50% 生物群落的结构和功能与参考不同	潜在的有害效应：化学物浓度升高 两个或多个毒理学终点降幅超过20% 生物群落的结构和功能可能与参考有区别	无显著的有害效应：不超过一个毒理学终点的轻微减少 生物群落的结构和功能与参考无区别

合理独立证据的关键决定因素取决于SOPC、ROPC和需回答的问题（即待检验的假设）。具体问题具体分析，无通用的公式。然而对于受化学污染的沉积物而言，组成沉积物质量"三合一"（化学、毒性和生物群落）的证据可能有用且必要（Chapman 1990；Chapman et al.，2002；Chapman & McDonald，2005）。表9.2提供了这3个沉积物质量"三合一"证据的决策矩阵。

表 9.2 化学应激源、毒性和生物群落证据权重分类的决策矩阵

	化学	毒性	群落变化	评价
1	低	低	低	无需采取进一步措施
2	中－高	低	低	无需采取进一步措施；化学污染物无生物利用度
3	低	低	中－高	生物群落改变原因的确定
4	低	中－高	低	毒性与化学物质浓度升高或生物群落变化无关——可能基于实验室工序或未测得的压力源；需调查
5	中－高	中－高	低	毒性可能与化学物质有关，但无证据表明与生物群落变化有关——可能基于实验室工序或未测得的压力源；需调查
6	中－高	低	中－高	确定生物群体变化原因——缺少证据表明此变化与化学物质有关，需调查[①]
7	低	中－高	中－高	确定毒性和生物群体变化的原因[①]
8	中－高	中－高	中－高	需采取管理措施[②]

[①]生物群落的变化可能由化学应激源以外的因素导致，可能是自然因素（如竞争/捕食、生境差异）、人为因素（如生境差异、物种入侵）或两者兼有（如有害藻华、入侵物种）。

[②]可能性的确定。理想情况下，化学物质浓度变化因与所观察到的生物效应（即因果效应）紧密相关，以确保管理措施可解决问题。可能需要进一步的调查来确定因果关系，如毒性鉴定评价和/或污染物残留分析。

单独的终点可包含在每个独立证据中（如金属、PAHs、PCBs 等化学品；生存、生长、繁殖毒性；生物群落的丰度、多样性、优势）。

9.4 生态风险评价中的海洋生态毒理学和证据权重

海洋生态毒理学在生态风险评价和证据权重中有两个主要作用：毒性和生物富集试验（统称为生物测定测试）。毒性测试需要侧重于确定急性（死亡率）和慢性反应（如生长、繁殖力、行为）的无或可忽略效应水平。尽管无效应（即 10% 效应）水平是首选，但 CCME（2007）还是把 20% 效应水平作为"负面效应的阈值水平"。换句话说，超过 20% 效应水平则可能发生负面效应。CCME（2003）& CCME（2007）都不认为 10% 效应水平值得信赖。实际上，加拿大关于钴的国家水质基准是基于 25% 效应水平。USEPA（2002）整体废水毒性测试指南也依据 IC_{25}（25% 效应水平）来评价废水毒性。Bruce & Versteeg（1992）估算 20% 效应浓度作为一个与环境相关的浓度时，与自然变异相比，可最大限度地减少对种群的有害效应。Barnthouse et al.（1987）& Norberg-King（1993）将此最小有害效应阈值设置为 25%。许多学者认为 10% 和 20% 的效应阈值是无或最小效应浓度的替代品（如 Dyer et al., 1997；Grist et al., 2003；Versteeg &

Rawlings 2003；Barnthouse et al.，2007；Oris et al.，2012）。USEPA 也接受以 20% 效应水平来制定国家水质标准（如 USEPA 氨水水质标准，http：// water. epa. gov/scitech/swguidance/standards/criteria/aqlife/ammonia/upload/AQUATIC-LIFE-AMBIENT-WATER-QUALITY – CRITERIA-FOR-AMMONIA-FRESHWATER-2013. pdf）。

除受威胁和濒危物种（如北美西海岸的鲑鱼）以外，对种群造成 10%～25% 的慢性效应不会对该种群产生不利影响。所以了解一个物种的种群动态和生活史也很重要，而不是仅仅依靠毒性试验来预测效应（Stark et al.，2004；Stark 2005；Hanson & Stark 2011，2012）。

以北美西海岸受威胁的鲑鱼为例，低水平的毒性会影响幼鱼的生长，从而降低其种群生存能力。鱼入海前在河流中的生长速度对其生存能力和洄游能力有很大的影响；较大的鱼类在返回其出生地产卵的过程中具有较强的生存能力（Spromberg & Meador，2005）。一个鲑鱼群与其他鲑鱼群争夺栖息地的间歇效应会减少整个种群的规模（Spromberg & Scholz，2011），本质上是"远距离作用"（Spromberg et al.，1998）。因此，正如在生态风险评价和证据权重中一样，海洋生态毒理学只是确定和规避有害环境效应和影响的依据之一。

9.5 展　　望

9.5.1　五大环境应激源

如前所述（第 9.2.1 节），造成环境变化的主要环境压力有 5 种，化学污染只是其中一种，而且相对而言可以说是最不重要的。在生态风险评价和证据权重中，海洋生态毒理学需要适应和发展，以提供与所有 5 种应激源相关的有用成分信息，不仅仅是单独的信息，而是要整合的信息。物种之间或更高层次生物组织之间的共性可能被发现（Sulmon et al.，2015）。

关于气候变化（第 10 章将详细讨论），需要从毒性和生物累积两方面评价温度和海洋酸化对海洋动物的直接影响。温度和海洋酸化也是改变化学毒性的因素（第 10 章），有着间接影响。

生境变化会改变水流状态和泥沙运动/负荷的时空格局（第 10 章），从而影响海洋动物，同样需要评价生境变化的直接和间接影响。

我们也需要评价富营养化的直接和间接影响，包括有害藻华。关于物种引入和入侵，需要在生态风险评价和证据权重范围内进行生态毒理学研究，以确定这些物种是否以及如何影响当地生态系统，并且要选取代表性的引进/入侵物种（即具有增强生态系统服务功能潜力的物种）进行适当的毒性测试和生物富集评价。

未来海洋生态毒理学的作用可能需要考虑获得性耐受性和相关能量需求的影响。需要控制种群间耐受性变异（Sun et al., 2015）。

9.5.2 良好的生态系统状况

欧盟根据其海洋战略框架指令（MSFD, 2008），整合了环境保护和可持续利用的概念，并要求成员国获得良好环境状况（good environmental status, GES）。在 MSFD（2008）附件 I 中，有 11 个定性描述定义了 GES，其中 7 个要求进行包括海洋生态毒理学在内的生态风险评价和/或证据权重评价，包括：①维护生物多样性；②入侵物种/引进物种不会对生态系统造成不利影响；③商业用鱼类和贝类保持健康；④海洋食物网得以维持；⑤海床的完整性不会对底栖生物或其他生态系统产生不利影响；⑥污染物不是有害污染物；⑦鱼和其他海产品可以安全食用。其他 4 个描述是：①尽量减小人类引起的富营养化，包括有害藻华；②水文的永久改变不会对海洋生态系统产生不利影响；③海洋垃圾不对海岸和海洋环境造成危害；④包括水下噪音在内的能量不会对海洋环境造成不利影响。

上述定性描述没有明确说明生态系统不能改变的方面，如物种不能消失或被取代，只是强调仍然要保持生态系统多样性、健康和可持续发展，以及维持人类获取食用海产品的现状。重点是生态系统功能和生态系统服务，而不是生态系统结构。因此，要真正确定 GES 是否已实现或可维持，就需要改变我们监测和评价海洋（和其他）生态系统的方式。

GES 的确定需要可靠的工具；Lyons et al.（2010）& Robinson et al.（2012）强调了结合化学污染物浓度监测和生物效应（如海洋生态毒理学）测量的重要性。现在的化学污染物不仅指传统意义上的金属和有机化学等无机物质，还包括越来越多的新兴化学品（如微塑料、药品和个人护理产品、阻燃剂），而这些新兴化学品作用方式和潜在的生物效应仍有待充分确定。因此，化学和生物学的结合尤其重要。此外，化学污染物和其他应激源（如气候变化和海洋酸化、生境变化、入侵/引进物种、有害藻华、富营养化）的共同作用，使得它们对生态系统功能的影响乃至最终对生态系统服务的定位变得更为复杂，加大了做出对应决策的难度。

对 GES 的评价不应基于它是否存在于个体物种的存在或缺失评测（即非结构），而应基于确定什么对维持生态系统功能至关重要。这可能包括入侵/引进物种，只要它们"处于不会对生态系统造成不利影响的水平"（MSFD, 2008）。

对不可接受的生态系统功能变化和可能改变生态系统功能的污染物或其他应激源影响，都需要进行早期预警。早期预警既需要生物标记物，也需要生物指示物（如整体生物毒性试验），但也未必需要按照目前制定和使用的基准。

关键生态系统功能的生物标记物是必需的。因为它将为某些化学和/或其他应激源（最终可能会破坏特定生态系统功能，并对生态系统服务产生不利影响）提供早期预警。目前尚不清楚有哪些主要针对暴露而非效应的生物标记物真正适用于上述目的。组学技术可能有助于开发必要的基于效应的生物标记物。

目前海洋毒性试验一般是在实验室使用自然生物进行，通常测量生存、生长和繁殖的终点。同样地，试验的焦点需要从保护结构转向保护功能。这可能意味着毒性测试针对特定的官能团，也可能是在对应激源影响更不敏感的生物体上进行。例如，保护食物链中一个虽较敏感但对生态系统功能却非必需的物种可能意义不大。一些公认的毒性测试可能不再适用。

需要为无机、有机和新兴污染物制定基于生物中污染物浓度（即关键的体内残留）的基准（Meador et al., 2014）。此类基准不仅是为了了解对生态系统功能很重要的生物中污染物剂量的重要性，而且也是为了了解毒性试验生物体的重要性。

所有信息都应被整合到不依赖于简化指标的多应激源、特定生态系统的 ERA/WoE 评价中（Green & Chapman, 2011）。最终目标必须是通过确定和预测污染物和/或其他应激源对特定生态系统功能的作用阈值/临界点，从而为实现和维持 GES 的决策提供可靠的信息。

参考文献

BARNTHOUSE L W, SUTER G W, ROSEN A E, et al., 1987. Estimating responses of fish populations to toxic contaminants. Environ. Toxicol. Chem., 6: 811 – 824.

BARNTHOUSE L W, MUNNS JR W R, SORENSEN M T, 2007. Population-level ecological risk assessment. CRC Press, Boca Raton, FL, USA.

BORGERT C J, MIHAICH E M, ORTEGO L S, et al., 2011. Hypothesis-driven weight of evidence framework for evaluating data within the US EPA's Endocrine Disruptor Screening Program. Regul. Toxicol. Pharmacol., 61: 185 – 191.

BRUCE R D, VERSTEEG D J, 1992. A statistical procedure for modeling continuous toxicity data. Environ. Toxicol. Chem., 11: 1485 – 1494.

CCME (Canadian Council of Ministers of the Environment), 2003. Canadian water quality guidelines for protection of aquatic life: guidance for site-specific application of water quality guidelines in Canada and procedures for deriving numerical water quality objectives. Winnipeg, MB, Canada.

CCME, 2007. A protocol for the derivation of water quality guidelines for the protection of aquatic life 2007 // Canadian environmental quality guidelines. Winnipeg, MB, Canada. http://ceqg-rcqe.ccme.ca/.

CHAPMAN P M, ANDERSON J, 2005. A decision-making framework for sediment contamination. Integr. Environ. Assess. Manag., 1: 163 – 173.

CHAPMAN P M, HOLLERT H, 2006. Should the sediment quality triad become a tetrad, a pentad, or possibly even a hexad? J. Soils Sed., 6: 4 – 8.

CHAPMAN P M, MCDONALD B G, 2005. Using the sediment quality triad in ecological risk assessment // BLAISE C, FÉRARD J-F (Eds.). Small-scale freshwater toxicity investigations: hazard assessment schemes. vol. 2. Netherlands: Kluwer Academic Press: 305 – 330.

CHAPMAN P M, SMITH M, 2012. Assessing, managing and monitoring contaminated aquatic sediments. Mar. Pollut. Bull., 64 (10): 2000 – 2004.

CHAPMAN P M, MCDONALD B G, LAWRENCE G S, 2002. Weight of evidence frameworks for sediment quality and other assessments. Hum. Ecol. Risk Assess., 8:

1489 – 1515.

CHAPMAN P M, 1990. The sediment quality triad approach to determining pollution-induced degradation. Sci. Total Environ., 97 – 98: 815 – 825.

CHAPMAN P M, 2013a. Polluting to pollute or "polluting" to protect? Mar. Pollut. Bull., 70: 1 – 2.

CHAPMAN P M, 2013b. Treatment by any other name would be more environmentally friendly. Mar. Pollut. Bull., 77: 1 – 2.

CORMIER S M, SUTER II G W, 2013. A method for assessing causation of field exposure-response relationships. Environ. Toxicol. Chem., 32: 272 – 276.

CORMIER S M, SUTER II G W, NORTON S B, 2010. Causal characteristics for eco-epidemiology. Hum. Ecol. Risk Assess., 16: 53 – 73.

DYER S, LAUTH J, MORRALL S, et al., 1997. Development of a chronic toxicity structure-activity relationship for alkyl sulfates. Environ. Toxicol. Water Qual., 12: 295 – 303.

ECHA (European Chemicals Agency), 2010. Practical guide 2: how to report weight of evidence. Helsinki, Finland. http://echa.europa.eu/documents/10162/13655/pg_report_weight_of_evidence_en.pdf.

ECHA, 2014. Principles for environmental risk assessment of the sediment compartment// Proceedings of the Topical Scientific Workshop, Helsinki, Finland, 1 – 8 March, 2013. http://echa.europa.eu/documents/10162/13639/environmental_risk_assessment_final_en.pdf.

Environment Canada and Ontario Ministry of the Environment, 2008. Canada-ontario decision-making framework for assessment of great lakes contaminated sediment. Ottawa, ON, Canada. http://publications.gc.ca/site/archivee-archived.html?url=http://publications.gc.ca/collections/collection_2010/ec/En164 – 14 – 2007-eng.pdf.

Environment Canada, 2012. Federal contaminated sites action plan (FCSAP): ecological risk assessment guidance. Prepared by Azimuth Consulting Group, Vancouver, BC, Canada. http://www.ec.gc.ca/Publications/default.asp?lang=En&xml=D86920CE-2DFE-40CB – 985F – 60A3EA6069A9.

GREEN R, CHAPMAN P M, 2011. The problem with indices. Mar. Pollut. Bull., 62: 1377 – 1380.

GRIST E P, WELLS N C, WHITEHOUSE P, et al., 2003. Estimating the effects of 17α-ethinylestardiol on populations of the fathead minnow *Pimephales promelas*: are conventional endpoints adequate? Environ. Sci. Technol., 37: 1609 – 1616.

HAAKE D M, WILTON T, KRIER K, et al., 2010. Causal assessment of biological impairment in the Little Floyd River, Iowa, USA. Hum. Ecol. Risk Assess., 16: 116 – 148.

HANSON N, STARK J D, 2011. Utility of population models to reduce uncertainty and increase values relevance in ecological risk assessments of pesticides: an example based on acute mortality data for daphnids. Integr. Environ. Assess. Manag., 8: 262 – 270.

HANSON N, STARK J D, 2012. Comparison of population level and individual level endpoints to evaluate ecological risk of chemicals. Environ. Sci. Technol., 46: 5590 – 5598.

HOPE B K, CLARKSON J R, 2014. A strategy for using weight-of-evidence methods in ecological risk assessments. Hum. Ecol. Risk Assess., 20: 290 – 315.

LANDIS W G, DURDA J L, BROOKS M L, et al., 2012. Ecological risk assessment in the context of global climate change. Environ. Toxicol. Chem., 32: 1 – 14.

LINKOV I, LONEY D, CORMIER S, et al., 2009. Weight-of-evidence evaluation in environmental assessment: review of qualitative and quantitative approaches. Sci. Total Environ., 407: 5199 – 5205.

LINKOV I, MASSEY O, KEISLER J, et al., 2015. From "weight of evidence" to quantitative data integration using multicriteria decision analysis and Bayesian methods. Altex, 32: 3 – 8.

LYONS B P, THAIN J E, STENTIFORD G D, et al., 2010. Using biological effects tools to define good environmental status under the European Union marine strategy framework directive. Mar. Pollut. Bull., 60: 1647 – 1651.

MCDONALD B G, DE BRUYN A M H, WERNICK B G, et al., 2007. Design and application of a transparent and scalable weight-of-evidence framework: an example from Wabamun Lake, Alberta, Canada. Integr. Environ. Assess. Manag., 3: 476 – 483.

MCPHERSON C, LAWRENCE G, ELPHICK J, et al., 2014. Development of a strontium chronic effects benchmark for aquatic life in freshwater. Environ. Toxicol. Chem., 33: 2472 – 2478.

MEADOR J P, WARNE ST J M, CHAPMAN P M, et al., 2014. Tissue-based environmental quality standards. Environ. Sci. Pollut. Res., 21: 28 – 32.

MENZIE C, HENNING M H, CURA J, et al., 1996. Special report of the Massachusetts weight-of-evidence work group: a weight-of evidence approach for evaluating ecological risks. Hum. Ecol. Risk Assess., 2: 277 – 304.

MORIARTY P, 2015. Reliance on technical solutions to environmental problems: caution is needed. Environ. Sci. Technol., 49: 5255 – 5256.

MSFD (Marine Strategy Framework Directive), 2008. Directive 2008/56/EC of the European parliament and the council of 17 June, 2008: establishing a framework for community action in the field of marine environmental policy. http://ec.europa.eu/environment/marine/eu-coast-and-marine-policy/marine-strategy-framework-directive/index_en.htm.

NORBERG – KING T J, 1993. A linear interpolation method for sublethal toxicity: the inhibition concentration (ICp) approach. version 2.0 // National effluent toxicity assessment center technical report 39: 3 – 93. United States Environmental Protection Agency, Environmental Research Laboratory, Duluth, MN, USA.

ORIS J T, BELANGER S E, BAILER A J, 2012. Baseline characteristics and statistical implications for the OECD 210 fish early life stage toxicity test. Environ. Toxicol. Chem., 31: 370 – 376.

POSTHUMA L, SUTER II G W, TRAAS T (Eds.), 2002. Species sensitivity distributions in ecotoxicology. CRC Press, Boca Raton, FL, USA.

ROBINSON C D, GUBBINS M J, LYONS B P, et al., 2012. Assessing good environmental status for descriptor 8: an integrated assessment of contaminants and their biological effects across multiple matrices in the Firth of Forth, Scotland. ICES CM 2012/G, 16. https://www.ices.dk/sites/pub/CM20Doccuments/CM-2012/G/G1612.pdf

SABCS (Science Advisory Board for Contaminated Sites in British Columbia), 2010. Guidance for a weight of evidence approach in conducted detailed ecological risk assessments (DERA) in British Columbia. Ministry of Environment, Victoria, BC, Canada.

SPROMBERG J A, MEADOR J P, 2005. Relating results of chronic toxicity responses to population-level effects: modeling effects on wild Chinook salmon populations. Integr. Environ. Assess. Manag., 1: 9-21.

SPROMBERG J A, SCHOLZ N L, 2011. Estimating the future decline of wild Coho salmon populations resulting from early spawner die-offs in urbanizing watersheds of the Pacific Northwest, USA. Integr. Environ. Assess. Manag., 7: 648-656.

SPROMBERG J A, JOHNS B M, LANDIS W G, 1998. Metapopulation dynamics: indirect effects and multiple discrete outcomes in ecological risk assessment. Environ. Toxicol. Chem., 17: 1640-1649.

STARK J D, BANKS J E, VARGAS R, 2004. How risky is risk assessment: the role that life history strategies play in susceptibility of species to stress. PNAS, 101: 732-736.

STARK J D, 2005. How closely do acute lethal concentration estimates predict effects of toxicants on populations? Integr. Environ. Assess. Manag., 1: 109-113.

SULMON C, VAN BAAREN J, CABELLO-HURTADO F, et al., 2015. Abiotic stressors and stress responses: what commonalities appear between species across biological organization levels? Environ. Pollut., 202: 66-77.

SUN P Y, FOLEY H B, BAO V W W, et al., 2015. Variation in tolerance to common marine pollutants among different populations in two species of the marine copepod *Tigriopus*. Environ. Sci. Pollut. Res., 22: 16143-16152.

SUTER II G W, CORMIER S M, 2011. Why and how to combine evidence in environmental assessments: weighing evidence and building cases. Sci. Total Environ., 409: 1406-1417.

SUTER II G W, NORTON S B, CORMIER S M, 2010. The science and philosophy of a method for assessing environmental causes. Hum. Ecol. Risk Assess., 16: 19-34.

USEPA (US Environmental Protection Agency), 1992. Framework for ecological risk assessment. EPA/630/R-92/Washington, DC, USA.

USEPA, 2002. Short-term methods for estimating the chronic toxicity of effluents and receiving water to freshwater organisms. EPA-821-r-02-13. Office of Water, Washington, DC, USA.

USEPA, 2007. Framework for metals risk assessment. EPA/20/R-07/001. Washington, DC, USA.

VERSTEEG D, RAWLINGS J, 2003. Bioconcentration and toxicity of docecylbenzene sulfonate (C12LAS) to aquatic organisms exposed in experimental streams. Arch. Environ. Contam. Toxicol., 44: 237-246.

WISEMAN C D, LEMOINE M, CORMIER S, 2010. Assessment of probable causes of reduced aquatic life in the Touchet River, Washington, USA. Hum. Ecol. Risk Assess., 16: 87-115.

10　全球变化

J. L. Stauber，　A. Chariton，　S. Apte[①]

---

① 澳大利亚联邦科学与工业研究组织（CSIRO），水土研究所。

10.1 引　　言

人类活动对沿海和海洋生态系统结构和完整性的影响日益加剧。土地资源利用变化、沿海城市化和工业化的加剧、人口的增长、水资源可利用度和质量的改变，以及气候变化，对海洋栖息地、生态过程和群落以及沿海城市的宜居性产生了重大影响。

全球变化可能由人为压力或自然破坏性事件引起。虽然自然事件不可预测且无法受控，但是可以对人为引起的全球变化进行管理和预测，尽管其中仍然存在较大的不确定性（Duarte，2015）。本章的重点为人为引起的全球变化，Duarte（2015）将其定义为：人类活动对生物圈主要功能调节过程的影响所导致的全球范围变化。

这里只考虑全球范围的变化，这些变化可能是正面的，也可能是负面的。

海洋拥有地球上最大的生物群落，也是人类压力源的主要受体（Duarte，2015）。Halpern et al.（2008）绘制了一幅显示 17 种人为因素驱动全球海洋生态系统变化的累积影响示意图。他们预计累积影响最大的海洋生态系统是大陆架、岩礁和珊瑚礁。全球气候变化是一个重要因素，尤其是对近海生态系统而言。

了解和预测化学物质对海洋和河口生物种群和群落的影响是全球生态毒理学研究的重点。生态毒理学家面临的一个特殊挑战是预测全球变化和污染物对生物个体的联合效应将如何在种群和群落水平表现出来。污染物是环境中浓度高于自然背景值的物质，它们可能是物理类（如盐度）、化学类（如金属）或生物类（如微生物致病体）。

全球范围内，我们在对多种环境压力源以及直接和间接压力源如何与污染物相互作用并影响海洋和河口生态系统方面取得了有限的成就。生物体对化学胁迫存在一定的敏感性，而多重胁迫对生态系统健康的影响在未来可能变得越来越重要。

本章将讨论全球变化的驱动因素，包括流域土地利用变化、沿海开发（包括港口活动和工业化）以及气候变化。我们考虑了多种压力源以及它们可能如何相互作用对现在和未来 50～100 年人类时代海洋和河口生态系统的潜在影响。

10.2　流域土地利用变化

10.2.1　人口增长与景观变迁

2012 年，世界人口达 70 亿（UNFPA，2014）。即使人口增长率总体下降，但在 1990 年到 2010 年间世界人口增长了 30%，预计到 2083 年将达到 100 亿（UNFPA，2014）。为了满足这种增长及其对资源日益增长的需求，地球表面 1/3～1/2 发生了变化，现在25%的陆地作为农业用地。虽然海岸占地球陆地面积的比例不足5%，但它们是人口增长的中心，39%的世界人口生活在沿海100 km以内的地区。如在澳大利亚，由于那里大部分的陆地景观无法支撑高人口密度，所以这个数字更高：85%的澳大利亚人居住在距海岸50 km以内（ABS，2003）。人口向沿海地区迁移在发展中国家也很常见。全球39个人口超过500万的大都市中，60%都位于距海岸100 km以内，包括12个人口超过1000万的城市（全球共有16个）（Nicholls et al.，2007）。

人口增长，特别是沿海地区的人口增长，加上沿海土地利用的日益增加和变化，导致了关键海洋环境的显著损失。例如，目前全球海草场面积低于19世纪80年代前估计的29%。在20世纪的最后20年中，全世界20%的珊瑚礁已消失，同样比例的珊瑚礁正在退化。在同时期，估计有35%的红树林也已消失（Valiela et al.，2001）。尽管这些环境的丧失和退化可以归结于许多原因，但最明显的是生境破坏，环境污染物也直接或间接地促成了这些趋势。据估计，21世纪初，270万 km^2 的海洋生境受到非点源有机和无机污染物的影响，此外，还有160万 km^2 受到营养盐输入的影响（Halpern et al.，2008）。

基于农业、水产养殖、采矿、工业发展、疏浚工程、大规模制造和城市化等人类活动释放的化学污染物正在威胁世界范围内的沿海生态系统。沿海和海洋环境的污染主要是人类活动造成的。联合国环境规划署（United Nations Environment Programme，UNEP）估计市场上约有5万种工业、农业和家用化学品（UNEP，2013）。这些污染物可通过面源输入（如雨水径流和大气沉降）、污水处理厂和经处理工业废水的直接排放而进入水环境。污染物可能包括自然产生的成分（如金属/类金属）、营养物质或人造化学品（如工业化学品、杀虫剂和药品）。

污染和非污染压力源的共存，产生了复杂的压力源和效应，并随着我们在沿海地区活动的增加而加剧。因此，我们研究的重点是了解海洋生态系统的风险，预测影响并制定缓解措施。

水环境中的污染物可能来自点源或面源（图10.1）。点源来自特定的固定地点，如与生产工厂、污水处理厂和雨水管道相关的排污出口。虽然污染物来源的类型各不相同，但一般来说，与特定活动相关的污染物通常有特定的类别，如采矿作业的金属、污水处理厂的营养物质、药品和个人护理产品（PCPs）。因此在某些情况下，可以使用沉积物的化学特征来确定污染物的可能来源（Costanzo et al.，2001；Hajj-Mohamad et al.，2014）。但因为河口地区的污染输入大量并多样化，且污染物的扩散可能强烈地被水动力条件改变，所以这种鉴定方法在河口地区难以实施。污染物的面源通常与流域径流有关，这些污染物的负荷往往为偶发性，其成分反映了周围景观的组成，如农药来自农业，石油碳氢化合物来自道路。

其他的面源包括大气污染物（如通过烟囱、汽车尾气排放和挥发性污染物）的沉降，以及与过去生产活动相关的残留污染物（如工业旧址）。残留污染物可以来自点源和面源，在许多情况下包含持久性有机污染物（persistent organic pollutants，POPs），其生产已被逐步淘汰或大幅减少。例如，在澳大利亚悉尼港的一个前化学工业基地周围的沉积物中，仍能检出高浓度的二噁英和呋喃（如polychlorinated dibenzodioxins，TCDD），该基地曾被用来生产多种化学物质，包括木材防腐剂、除草剂、杀虫剂和塑料。这些污染物随后被栖息于沉积物的生物所吸收,并通过食物链传递给鱼类、甲壳类和海洋爬行动物/哺乳动物。鉴于二噁英对健康的严重影响以及在当地几种鱼类中高浓度二噁英的检出,自该地化学生产活动停止的30年以来,港口的大部分区域仍禁止捕鱼(Manning & Ferrell, 2007)。基于有机氯、多氯联苯和其他残留有机污染物的持久性,它们仍可能在改变河口和海洋系统的生态中发挥着重要作用(Chariton et al., 2010)。尽管研究的重点是调查正在进入环境的污染物,但考虑这些残留污染物可能对环境产生的作用也很重要。

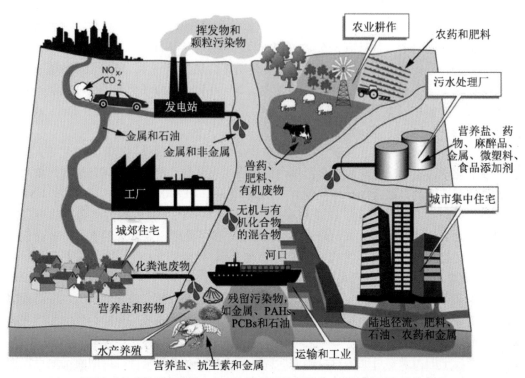

图10.1 土地利用与进入河口和海洋环境污染物来源和类型之间的关系

虽然各种的污染物已被证明对海洋环境有广泛的影响,但本章关注的重点是营养盐、杀虫剂和各种新兴污染物。

10.2.2 营养盐和富营养化

河口营养盐输入的自然来源为上升流、有机物质的细菌分解和地质风化。氮和磷对于维持河口和海洋环境的初级生产力和自然功能至关重要,此外也提供了生命的基本组成部分(如氨基酸和核酸)。虽然以 N_2 形式存在的氮是地球大气的最高组成部分,但是元素氮必须被转化(或固定)为还原态才可被生物利用。地球上只有 0.002% 的氮存在于生物物质中(如生物体和碎屑物质),而在自然存在的无机形式中,如硝酸盐、亚硝酸盐和铵的比例要少几个数量级(Howarth, 2008; Vitousek et al., 1997a)。但正是这些无机态氮对水生植物的生长至关重要,从而限制了世界上大多数温带地区河口和海洋环境的初级生产力。

尽管全球的氮总量是固定的，但人类活动（尤其是合成氮肥料的生产、鼓励固氮的农业实践以及通过化石燃料燃烧产生的活性氮）导致氮不成比例地向更活泼的氮形态转移（Galloway et al.，2004）。据估计，21世纪初全球活性氮的形成量增加了33%~55%，人类活动过程中产生的活性氮已超过自然界形成的量（Howarth & Marino，2006）。预计到2030年，活性氮的年产量将在1990年8000万吨的基础上增长1.7倍（Vitousek et al.，1997b）。

磷在水生系统中以溶解态和颗粒态形式存在。颗粒态磷可通过吸附于沉积物颗粒和有机物质（如腐烂物质）或与蛋白质结合而进入海洋环境。溶解态磷的种类众多，包括无机正磷酸酯、低分子量磷酸酯、焦磷酸酯和长链多磷酸酯。磷酸盐化合物在水中可以被酶解或化学水解成正磷酸盐，然后被植物、藻类和细菌同化（Correll，1998）。通过同化和沉积的共同作用，沉积物可以成为磷的储存库；这些磷在有氧条件下相对稳定，但在缺氧条件下则重新进入水体（Correll，1998）。虽然人们普遍认为磷限制淡水系统的初级生产力，而氮限制海洋系统的初级生产力，但一些研究对这种过于简单化的观点提出了质疑。例如，澳大利亚西部Peel-Harvey河口初级生产力限制和营养盐有效性之间的关系被证明具有季节性，夏季表现为氮限制，冬季则为磷限制（McComb & Davis，1993）。

沿海地区向城市化、近郊化（分散的城市和农村特色的混合景观）和农业集约化的转变导致流入河口的营养盐负荷显著增加。据估计，每年有超过6000万吨的氮流入海洋，是19世纪中期估计量的2倍（Boyer & Howarth，2008；Galloway et al.，2004）。氮负荷的增加在农业最为发达的北半球温带地区极为明显。例如，在韩国和美国东北部，每年流入海洋的氮负荷量分别由基线100 N/km^{2}增加到1700 N/km^2 & 1000 kg N/km^2（Howarth & Marino，2006）。

磷的增加趋势与氮相似，1950—1995年间预计有6亿吨磷被使用于陆地环境（Brown et al.，1998）。虽然大部分磷作为肥料被人工矿化，但在许多工业化程度较低的地区，磷主要以粪肥的形式获得（Carpenter et al.，1998）。目前存在的主要问题是磷在许多系统中的输入大大超过了农业的需求，导致大量的磷积累在土壤中，最终流入水域（Carpenter et al.，1998）。

海洋水域可以根据其限制生长的营养盐输入进行分类（图10.2），富营养化这一术语通常用于定义生态系统对营养盐供过于求的反应。与大多数以直接急性或慢性毒性为生态毒理学终点的污染物相比，主要关注的问题应为富营养化的间接影响，而不是营养盐的实际毒性。

氮和磷都限制植物的生长，而这些营养盐的过量输入主要通过增加海洋浮游植物和附生藻类的生物量来导致初级生产力的显著增加（Smith，1998）。这些变化所产生的影响远不止是水质透明度和其他审美价值的下降。初级生产者的增殖阻挡了阳光，导致包括珊瑚礁和海草床等关键生态系统的底栖生物生境质量大规模下降（Bellwood et al.，2004）。这些环境完整性的降低可能导致生物多样性的进一步下降。更重要的是，本地植物生物量的过度增长以及额外有机物质（即有机碳）导致微生物活性的增加，会耗尽上覆水域的溶解氧（DO），导致低氧（DO < 2 mg/L）或缺氧条件（无溶解氧），从

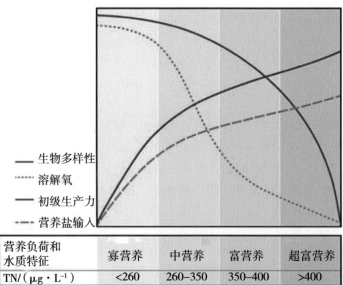

营养负荷和 水质特征	寡营养	中营养	富营养	超富营养
TN/($\mu g \cdot L^{-1}$)	<260	260–350	350–400	>400
TP/($\mu g \cdot L^{-1}$)	<10	10–30	30–40	>40
叶绿素a/($\mu g \cdot L^{-1}$)	<1	1–350	3–5	>5
透明度/m	>6	3–6	1.5–3	<1.5

图 10.2 基于营养负荷的海水分类和关键变化

而可能致使鱼类和其他生物大量死亡。

富营养化还会导致生态系统最小组成部分（真核微生物和浮游植物）的组成发生变化，从而影响到大型藻类和鱼类（Brodie et al., 2005; Chariton et al., 2015a; Smith, 1998）。以河口微型底栖无脊椎动物为例，富营养化通常导致体型较小但数量丰富的机会主义类群成为优势群落，如 capetillid 和 spionid polychates。总的来说，上述变化反映了生物多样性、均匀性和生物量下降（Warwick, 1986）。在浮游植物群落中，机会主义类群占据优势可能会对环境产生严重的影响，为 60~80 种有毒物种爆发成为优势种创造适宜条件（Smayda, 1997）。最重大的有毒浮游植物爆发事件之一发生于波罗的海，富营养化导致了 75000 km² 的 *Chrysochromulina polylepis* 爆发，造成大量鱼类、植物和大型无脊椎动物的死亡（Rosenberg et al., 1988）。近年来，越来越多的证据表明富营养化正在导致以珊瑚为食的棘冠海星（Crown of Thorns starfish，COT）的爆发，此物种对澳大利亚大堡礁的衰退起到了重要作用（Brodie et al., 2005; 专栏 10.1）。然而 COT 的爆发与营养盐之间的关系较为复杂。研究表明，营养盐能通过促进更大的浮游植物物种增殖间接推动 COT 种群增殖，因为这些浮游植物优先被 COT 幼虫摄食（Brodie et al., 2005）。

富营养化的生态影响既广泛又严重；然而与污染输入停止后浓度还能长时间保持稳定的持久性污染物（如有机氯和多环芳烃）相比，越来越多的研究表明通过减排计划可有效地抵消富营养化，如减少营养盐使用和输入，并促进营养盐向海洋转移扩散（Boesch et al., 2001）。例如，美国佛罗里达州的坦帕湾在 20 年间总氮量减少了 50%，使水体透明度增加了 50%、海草场恢复了 27 km² （Greening & Janicki，2006）。更详细的关于营养盐化学和生态毒理学信息，可参见以下文章：Smith，1998；Howarth & Marino，2006；Kennish & de Jonge，2011。

大堡礁案例研究

澳大利亚昆士兰的大堡礁（GBR）是地球上最大的生物群落，全长 2300 km，覆盖面积 344400 km²。大堡礁作为世界自然遗产，有 600 多种软/硬珊瑚、100 多种水母、3000 多种软体动物、500 多种蠕虫、1625 种鱼类、133 种鲨鱼和鳐鱼以及 30 多种鲸鱼和海豚。珊瑚礁群落具有重要的经济和生态意义，并提供了重要的生态系统服务，包括渔业、海岸保护、旅游业和新型活性化合物药物等（Moberg，1999）。人为活动导致的气候变化以及其他如沉积物、营养盐和污染物压力因素正在对这些珊瑚礁群落的长期恢复力产生重大影响。这些人为压力源与其他大规模干扰（特别是热带风暴和以珊瑚为食的刺冠海星种群爆发）相互作用（图 10.3）。Déath et al.（2012）估计大堡礁的珊瑚覆盖率在 1985—2012 年间从 28% 下降到 14%。热带气旋、COTs 对珊瑚的捕食以及珊瑚白化分别占损失的 48%、42% 和 10%。大堡礁正日益受到多重压力源的影响：

（1）沿海流域的泥沙径流增加了水体浊度，并减少了光穿透。沉积作用通过沉积物中有机物触发微生物的过程杀死珊瑚；微生物的呼吸作用导致缺氧和 pH 降低，从而引发珊瑚种群退化（Weber et al.，2012）。

（2）如敌草隆的污染物是一种常用于农业（如甘蔗种植）和船用防污涂料的除草剂。该污染物是一种有效的光合作用抑制剂，对热带系统中的珊瑚虫黄藻、壳状珊瑚藻、海草和浮游植物有潜在的影响（Haynes et al.，2000；Harrington et al.，2005）。大堡礁潮下沉积物中敌草隆的浓度高达 10 mg/kg ww，而 Harrington et al.（2005）的研究表明微量敌草隆的存在显著增强了珊瑚藻的沉积胁迫。

（3）来自流域营养盐的输入导致潜在富营养化。

（4）航运活动造成的石油泄漏。

（5）水温升高导致珊瑚白化。

（6）海洋酸化。

在过去的 20 年里，可能由于气候变化导致珊瑚钙化减少，如滨珊瑚这样的大堡礁造礁珊瑚的生长速度已经下降了 21%，表现为骨骼密度和线性扩展率的降低（Hoegh-Guldberg et al.，2007）。珊瑚骨骼密度的降低也可能导致珊瑚的侵蚀率增加，增加它们应对捕食和风暴破坏的脆弱性，从而影响生境的结构、多样性和生态系统服务（如海岸保护）。珊瑚在不利的条件下维持骨骼生长和密度的能量消耗也可能影响其繁殖。

除对珊瑚造成影响外，珊瑚藻作为珊瑚的主要定居基质，对 pH 很敏感，它们需要镁和钙来形成外骨骼。如果珊瑚藻丰度下降，则为与它们争夺空间和光线的大型藻类提供理想的栖息地，从而可能影响珊瑚的补充。大型藻类形成稳定的群落，会削弱珊瑚的重建能力，从而影响生态恢复力，将生态系统推向另一种状态（Mumby et al.，2007）。

Hoegh-Guldberg et al.（2007）在综述中总结预测了来自这些多重压力源变化的场景，旨在为未来的适应性管理提供一个框架。对于珊瑚礁的管理，首先应该关注当地的压力源，如水质，这可能会减少 COTs 爆发。此外，还需要对珊瑚礁附近及珊瑚礁上的渔业活动进行管理。管理干预措施的目标应该是协助珊瑚群落适应包括气候变化在内的当前和未来多重压力源。减少 CO_2 的排放，使大气中的 CO_2 含量保持在 500 ppm 以下，对于珊瑚礁以及现在和将来依赖它们的人类至关重要。

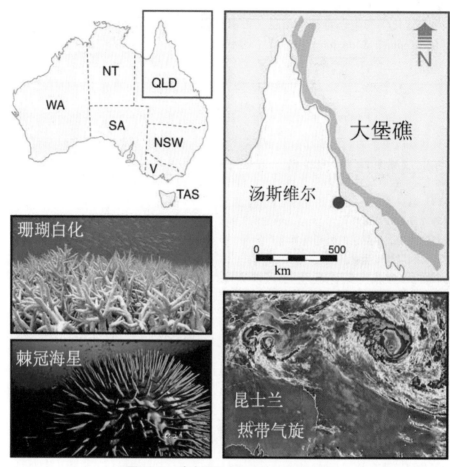

图 10.3 澳大利亚大堡礁的特定压力源

预测营养物质过剩对海洋环境生态的长期影响较为困难。显然水产养殖将在海洋和沿海系统的富营养化方面发挥越来越大的作用；预计 2090 年分配给水产养殖的面积将比 1990 年增加 1800 倍（Duarte et al.，2009），导致农业来源氮和磷负荷的预测变得愈加不确定。我们可以合理地假设随着人口增长和沿海开垦面积的增加，农业径流将继续按比例增加；不过经济和环境的成本将日益受到关注。例如，谷物作物施氮量在 1960—1995 年间增加了 7 倍，但作物产量仅增加了 1 倍，氮肥利用率从 70 kg 谷物/kg 氮降至 25 kg 谷物/kg 氮（Keating et al.，2010）。在欧洲，包括减排措施和更广泛的社会资源损失在内的氮损失环境成本为每年 700 亿～3200 亿欧元，超过了农业施氮的经济效益（Brink et al.，2011）。虽然氮明显被过度使用，但由于它不断地被固定，所以不会被耗尽。然而对于磷来说，情况并非如此：与石油相似，磷肥是从经过数百万年形成的矿床中提取而来。虽然未来对磷肥的需求会不断增加，但其供应量将日益受限。总的来说，这表明水产养殖在沿海和海洋环境富营养化中的作用呈日益重要的趋势。

10.2.3 农药

农药是一种可用于杀死、驱除或减少害虫种类的化学混合物。自20世纪40年代中期大规模使用以来，农药的成分和属性发生了几次重大变化。第一代合成农药是包括艾氏剂、狄氏剂、DDTs和硫丹的有机氯农药。最初这些化学物质被证明是高效的杀虫剂，并被广泛用于减少农业害虫、斑疹伤寒和疟疾等疾病的传播媒介。在其使用高峰时期（1950—1980年），每年使用的DDT超过40000 t（Geisz et al.，2008）。然而到了20世纪60年代，人们越来越清楚地认识到由于有机氯的持久性、过度使用和通过食物网实现生物放大的特点，有机氯农药对环境产生了显著而深远的影响。早在1968年，匈牙利就禁止了DDT的使用，到20世纪70年代初，瑞典、挪威和美国已经禁止使用大多数的有机氯农药。尽管全球禁止使用，但每年仍有高达4000 t DDT被生产，主要用于病媒控制，而印度和朝鲜（可能）仍用于农业用途（van den Berg et al.，2012）。

第二代农药是有机磷农药，包括二嗪磷、毒死蜱、马拉硫磷等化学物质。与有机氯相比，有机磷通常只在环境中残留几天或几周；但它们的毒性远远超过有机氯。有机磷的毒性是通过抑制乙酰胆碱酯酶作用引起的，乙酰胆碱酯酶是一种支持神经活动的酶。这一机制使得有机磷暴露对包括人类、海洋鱼类和无脊椎动物在内得多种生物具有潜在毒性（Bouchard et al.，2011；Janaki Devi et al.，2012；Canty et al.，2007）。鉴于只有1%喷洒农药是有效的，其余99%则用于非目标环境，如土壤、水体和大气，因此高毒性且非特异性有机磷的潜在生态影响非常显著。然而，由于它们在环境中的持久性较低，将海洋系统的长期变化与有机磷的使用联系起来颇具挑战性。

最近开发的杀虫剂包括拟除虫菊酯，它主要成分是菊的天然衍生物——除虫菊酯。更新的除草剂包括一些广泛使用的化学物质，如草甘膦，俗称农达。草甘膦含有一个膦酰基团，可以抑制酪氨酸、色氨酸和苯丙氨酸的合成。另一类对海洋环境影响日益严重的除草剂是敌草隆、阿特拉津和三嗪类。这些除草剂通过破坏光合作用来发挥作用。重要的是目前正在开发和使用一些非农药方法以最大限度提高植物对害虫的抗性，其中最引人关注的是通过RNA干扰（RNA interference，RNAi）对农作物进行基因修饰（Castel & Martienssen，2013）。

10.2.3.1 海洋中的农药

环境中有机氯农药的残留问题依然存在。21世纪初，世界海洋环境中仍能检出浓度较低的DDT及其代谢物。由于有机氯具有疏水性，DDT及其代谢物在沉积物和生物中的浓度更高，因此在许多情况下其浓度仍然值得关注（Galanopoulou et al.，2005）。DDT的半衰期为10～14年，近年来海洋生物体内的DDT总浓度（身体负荷）总体上有所下降（Sericano et al.，2014）。瑞典的一项长期调查证明了这一趋势，研究人员发现在1969—2012年间，海洋生物（海鳗卵和鲱鱼）中的DDT总浓度下降了96%～99%（Nyberg et al.，2015）。研究人员在多氯联苯（PCBs）、六氯环己烷（HCHs）和六氯苯（HCBs）中也发现了类似趋势。

尽管农药的使用量在不断增加，但使用的农药类型却在全球范围内发生了变化，除草剂（47.5%）现在占农药使用的最大比例，其次是杀虫剂（29.5%）、杀菌剂（17.5%）和其他杀虫剂（5.5%）（Pimentel，2009）。尽管除草剂的降解速度可能相对较快，但它们在沿海地区的使用（如农业及城市用途）广泛且多样化，并随着沿海地区经济的发展而不断增加。即使在高度保护的环境中，如澳大利亚的卡卡杜世界遗产地，敌草隆也可以在河口沉积物中被检出，因为其不断地被用于控制河滩地的杂草入侵（Chariton 未发表）。

由于敌草隆和阿特拉津等除草剂的设计目的是用于干扰光合作用，其流入沿海环境可能会对藻类和海草等非目标的海洋光合作用物种产生广泛影响。这些物种的消失可能产生级联效应：例如海草床既是重要的栖息地，也是儒艮、海牛和绿海龟等标志性物种的食物来源（Reich & Worthy，2006；Bjorndal，1980；Bayliss & Freeland，1989）。同样地，珊瑚礁附近沿海水域的除草剂浓度经常被证明足以引起亚致死效应，如抑制海洋微藻的生长和光合作用（Magnusson et al.，2008）。鉴于微藻是底栖生物和远洋食物网构成的基础，农药对微藻和其他初级生产者造成大规模不利影响的可能性值得关注。越来越多的证据表明除草剂通过影响共生藻（一种甲藻，通过互惠关系为石珊瑚提供营养）导致珊瑚礁的消失（Jones，2005；Veron，1995）。有趣的是，不管除草剂的作用模式如何，其毒性反应并不局限于光合作用系统，它们对一系列河口无脊椎动物和其他生物也有毒性作用（Macneale et al.，2010）。

10.2.4 新兴污染物

海岸景观和沿海区域社会经济结构的显著变化导致了越来越复杂的污染物混合物进入海洋系统（图10.1）。药物、PCPs、微塑料、纳米材料和麻醉剂等统称为"受关注的新兴污染物"。对于许多生物群体来说，这并不是新问题，如自20世纪70年代中期以来，关于抗生素在水生环境中的问题就已在文献中被提出（Hignite & Azarnoff，1977）。然而我们之前对这些污染物的分布和生态毒理学的理解受到了我们鉴定和量化环境相关浓度能力的限制（通常要求检测限在 ng/L 级到 mg/L 级范围内）。

许多新兴污染物的作用方式和生态毒理学效应与农药和金属有关的污染物存在很大差异。例如，药物具有特定的生理作用方式（如抗炎和脂质调节），用于靶向特定的生化途径或生物系统。相反，农药通常被设计成对特定类群内的大多数物种具有致命性。但也有例外，如广谱抗生素，其目的是杀死或抑制广泛的细菌类群。

农药的种类和使用频率通常由作物、害虫和季节决定，与之相反，海洋环境可能持续暴露在各种药物的低浓度混合物中。其中最普遍的包括非甾体抗炎药、抗生素、降血脂药、性激素和抗癫痫药（Santos et al.，2010）。

许多受人关注的新兴污染物缺乏与药物相似的特定作用方式。例如，三氯生（triclosan）是一种抗菌剂，是牙膏和液体肥皂等一系列PCPs的添加剂。由于其目标类群广泛（细菌），且具持久性和生物蓄积能力，因此它可能与传统污染物（如多氯联苯和金属）有更多的共同之处。此外，三氯生的毒性并不局限于其目标类群，高浓度的三氯生对许多小型和大型动物也具有毒性（Chariton et al.，2014）。尽管三氯生在许多产品中仍有应用，但欧盟正在逐渐淘汰三氯生，美国面临类似趋势的压力日益增加。

一些新型污染物的结构特性（尤其是塑料）给海洋环境造成了重大的生态毒理学问题。全球塑料生产始于20世纪50年代，2010年的产量为2.65亿t，自20世纪70年代以来增长了5倍（Cózar et al.，2014）。塑料的生态效应很大程度上取决于其尺寸和材料类型。例如，较大的漂浮塑料碎片可能会对海鸟、海龟和鱼类造成结构性损伤（如窒息和堵塞消化道），而较小的塑料碎片（如用于面部磨砂的塑料颗粒）可能会引起无脊椎动物的类似反应。一项关于海洋塑料碎片分布的Meta分析表明，塑料碎片积聚与洋流和陆地使用之间存在着很强的相关性（Cózar et al.，2014）。例如，在占据亚洲东海岸大部分地区的北太平洋，塑料垃圾总量占全球海洋塑料垃圾（估值为7000～35000 t）的33%～35%。北太平洋"东部垃圾区"的海洋塑料碎片的堆积最为明显，该区域的面积虽然难以量化，但据估计在70万～150万平方千米，相当于太平洋面积的0.41%～8.1%（Moore，2003）。

许多塑料还含有潜在的有毒化学物质，如邻苯二甲酸酯、阻燃剂和双酚A。因此无论是通过塑料分解还是海洋生物摄入的化学物质富集都引起了人们的关注（Tanaka et al.，2013）。即使是惰性塑料也能吸附疏水性污染物，为污染物的富集提供了额外的途径。对东太平洋垃圾区塑料碎片的随机调查发现超过50%的塑料碎片样品含有农药、PCBs和PAHs，尤其是在具高度耐降解性的聚乙烯塑料上（Rios et al.，2010）。微塑料也存在类似问题（Teuten et al.，2007）。

虽然进入海洋系统的污染物混合物的多样性和复杂性带来了一些巨大的挑战，但通过适当的监管可以有效地减少其对环境的影响。这不仅在有机氯农药中得到了证明，在很多的污染物中也得到证明。例如，自2004年欧盟禁止使用五溴联苯醚（penta-PBDE）以来，格陵兰环斑海豹脂肪中溴化二苯醚的浓度总体上有所下降（Law，2014）。2008年国际海事组织（International Maritime Organization，IMO）禁止在大型海船上使用三丁基锡（tributyltin）防污涂料，在后期也观察到了类似的趋势（Law，2014）。然而对于生态毒理学家来说，新兴污染物所带来的挑战不仅是要理解特定污染物的生态影响，还要了解多种污染物之间的相互作用，包括遗留污染物，它们的浓度和混合物随时间的推移发生较大变化。本章后面将讨论环境基因组学在应对这些挑战方面的潜在作用。

10.3 港口和工业相关变化

海洋对于全球粮食安全、人类健康和气候调节至关重要。全球超过30亿人的生计依赖于海洋和沿海生物多样性的服务，随着人口的增长，对海洋食物和资源的需求必然会增加（EASAC，2015）。

20世纪以来，人类对海岸的利用急剧增加。世界上沿海许多三角洲、岛屿和河口的人口增长导致了沿海自然景观（包括沿海森林、湿地和珊瑚礁）广泛转向农业、水产养殖、工业和住宅用途（Valiela，2006）。城市化的中心往往发生在具有重要生态意义的沿海生境附近。例如，世界上58%的珊瑚礁位于人口超过10万的主要城市中心50

km 以内，64%的红树林和62%的主要河口位于这些中心附近（Agardy et al.，2005）。

了解未来50年我们对海洋环境及其资源利用的变化将为规划提供了参考依据，因此对工业部门和政府至关重要。本节将探讨未来可能影响海洋环境的全球变化。了解未来的发展轨迹也将有助于海洋生态毒理学的发展，以便了解需要付出更大努力的方向。

10.3.1 对沿海水域的要求

沿海地区面临多重压力会导致生态系统退化，对海岸带群落产生影响。19世纪中期和20世纪，人类在河流流域、城市和沿海开发以及渔业过程中不受管制的活动导致海岸、海湾和河口的严重退化。沿海生态系统面临的最大威胁是与发展相关的生境和服务丧失。许多沿海地区已退化或改变，人类正面临日益严重的海岸侵蚀、洪水、水质下降和日益增加的健康风险。如筑坝、渠化和沿海水道改道的工程结构改变了循环模式，改变了淡水、沉积物和营养盐的输送。

提高预测沿海生态系统变化的能力是一个重大的挑战。对生态系统反应的预测仍然模糊，很大程度上仅为定性，因此应用十分有限（Duarte，2015）。原来认为生态系统对压力的反应是平滑的、线性的，但大量证据表明多个相互作用的元素组成的复杂自然系统往往对压力表现出非线性反应，当压力超过一个极限，即临界值或临界点时，对压力的最初平稳、渐进的反应被状态的突然变化所取代（Andersen et al.，2009；Duarte et al.，2009，2012a，b）。减压后系统的恢复轨迹通常会遵循不同的路径，从而达到不同的平衡状态。

10.3.2 沿海工业

全球范围内成熟工业的地理足迹正在发生变化，传统重工业和化学工业正在向发展中国家转移。随着经济合作与发展组织（OECD）国家越来越依赖从中国和印度等国的进口，材料生产已经转移（World Ocean Review，2010）。据预测，该种变化将持续下去，因为全球工业产品需求到2050年将增加1倍以上。目前，中国是氨、水泥、钢铁和甲醇的最大生产国。随着中国经济向服务业转移，中国的生产将遵循OECD的发展趋势，有望于2050年趋于稳定（World Ocean Review，2010）。相比之下，印度、非洲和中东的工业活动预计到2050年将显著增加。随着3D打印、先进机器人技术、轻质材料、纳米技术和海洋生物技术等技术的应用，现有的海洋行业将在未来几十年发生重大变化。预测突破性技术的影响非常困难，并且可能会与基于当前发展轨迹的简单外推预测发生巨大偏差。

炼油厂主要位于沿海地区。全球最大的10家炼油厂大多位于亚太地区，其中印度拥有全球最大的炼油厂，其次是委内瑞拉和韩国。由于全球石油行业的经济整合和对大型炼油厂的日益依赖，预计1/5的炼油厂将在未来5年内停止运营。

石化工业通过蒸汽裂解或催化裂解将石脑油和天然气组分（如丁烷、乙烷、丙烷）等原料转化为乙烯、丙烯、苯、二甲苯等石化原料。这些化学品经过进一步加工，产生如油漆、轮胎、洗涤剂、农用化学品和塑料制品等最终产品。大约80%的石化产品的制造成本与能源以及作为原料的石油和天然气有关。目前石化工业主要集中在北美、西欧和亚洲。然而由于拥有丰富的天然气储量，中东在未来几年很可能成为一个重要的天然气生产国。由于生产设备陈旧和生产成本较高，欧洲石化产品的产量正在下降。相比之下，巴西在生物衍生化学品和燃料方面处于世界领先地位。

10.3.3 油气勘探

自从19世纪中期开始工业石油开采以来，全球已开采了1470亿吨石油，超过一半的量是在1990年以后开采（World Ocean Review，2010）。全球老油田产量在未来20年将下降，对油气资源的依赖将加大（图10.4、图10.5）。

目前最高产的近海地区是北海、墨西哥湾、巴西、西非大西洋海域、阿拉伯湾和东南亚的海域。据估计剩余的可采常规石油近50%在近海，25%在深海（IEA，2012）。目前海上石油开采占全球石油产量的37%。随着浅海（深度小于400 m）油气储量的持续枯竭，人们越来越关注深海（500～1500 m）和超深海（大于1500 m）海域油气储量的勘探开发。

人们对北极的兴趣与日俱增，据估计世界上约30%未探明的天然气和13%未探明的石油可能存在北极圈以北的海域（USGS，2008）。虽然大多数钻探将在水深不到500 m的海上进行，但北极的条件极其恶劣，导致开采难度较大。随着北极海冰因气候变化而融化，开发北极地区的石油和天然气储备将变得越来越可行。而南极洲的油气储量则尚未明确。

因海上原油的意外泄漏而产生的浮油经常会漂向海岸，杀死海鸟和海洋哺乳动物。然而油轮事故只占全球海洋石油污染的10%左右。大多数石油通过不太明显的途径进入海洋，这使得很难准确估计全球石油进入海洋环境的状况。约有5%的石油污染来源于自然，35%来自油轮运输和其他航运业务，包括非法排放和油罐清洁（World Ocean Review，2010）。

目前海上天然气产量为65万亿立方米，占全球天然气产量的1/3。当前全球28%的天然气产量来自海洋。北海是目前最重要的天然气产区，但在不久将会被其他地区所取代，如中东、印度、孟加拉国、印度尼西亚和马来西亚。

包括煤层气、页岩气和致密气在内的非常规天然气是一种相对较新的天然气来源，它极大地降低了国际市场的天然气价格。非常规天然气储量丰富，遍布各大洲。据估计目前页岩气和煤层气资源约占全球天然气资源的51%。廉价页岩气的供应改变了美国石化行业的面貌。乙烷裂解炉正在建设中，美国生产的乙烯产品将出口。

液化天然气（liquefied natural gas，LNG）在天然气行业中扮演着至关重要的角色，因为用大型油轮运输冷却液化天然气比管道运输便宜。液化天然气已占当今全球天然气贸易的1/4。天然气在未来更可能通过船舶运输，而不是通过管道运输。

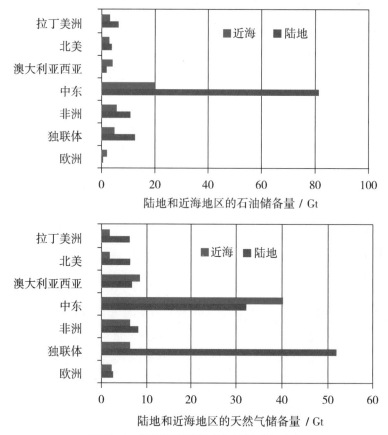

图 10.4　按区域划分的油气储量地理分布

来源：World Ocean Review, 2010. Living with the Oceans, vol. 1. Maribus, Hamburg, Germany。

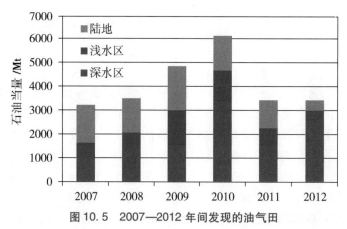

图 10.5　2007—2012 年间发现的油气田

来源：World Ocean Review, 2014. Marine Resources-Opportunities and Risks, vol. 3. Maribus, Hamburg, Germany。

10.3.4 航运与港口

航运占全球商业贸易的90%,并以每年约10%的速度增长(IMO,2012)。海运可以分为两个次级市场:液体货物(如石油、石油产品和液化天然气)和干货(物)。干货由散装货物组成,其中最重要的5种是铁矿石、煤炭、谷物、磷酸盐和铝土矿。全球最重要的一种货物是原油,仅原油就约占海运货物总量的25%。因此世界上主要航线是从石油生产中心如阿拉伯湾绕过好望角或通过苏伊士运河,由非洲向北和向西延伸到欧洲和北美。其他的航线连接阿拉伯海湾至东亚和加勒比海至美国墨西哥湾沿岸。

就数量而言,铁矿石和煤炭是重要的干散货。铁矿石通过超大型船舶进行长途运输,主要从巴西运往西欧和日本,以及从澳大利亚运往日本。最重要的煤炭运输路线是从主要出口国澳大利亚和南非到西欧和日本,从哥伦比亚和美国东海岸到西欧,以及从印度尼西亚和美国西海岸到日本。航道横穿一些最具生态敏感性的海域,常见的海上搁浅和事故给海洋环境带来了额外的压力。许多环境生态影响与航运有关,其中包括石油泄漏以及在压舱水或船体上"搭便车"的入侵物种转移。此外,船只撞击海洋哺乳动物的频率同样值得关注。防污涂料(主要是铜基杀菌剂)也会降低船舶密度较高的港口或码头水质。此外,海洋商用船只动力系统排放在最终到达海洋的全球空气污染物和温室气体排放总量中占大部分(Blasco et al., 2014)。

即使对经济增长只做了适度的假设,港口货运量预计到2030年也将增长57%。船舶的平均规模已大幅增加,港口当局必须通过扩大港口基础设施和改善港口通道(如通过加深航道)来应对船舶规模的增加。预计到2060年,港口将位于海上的人工岛上,布局可以得到优化。这些港口将由浮动的供给装置/河道端口支撑,可以根据不断变化的需求进行移动。

20世纪60年代,集装箱运输被首次引入美国,被认为是20世纪运输的最关键的技术革命之一。标准化集装箱的使用节省了成本,货物只需包装一次就可以通过各种运输方式进行长途运输——卡车、铁路或轮船。自1985年以来,全球集装箱运输量以每年约10%的速度增长,在2008年达到13亿t,并将继续保持增长,到2030年,集装箱运量将增长2倍(World Ocean Review,2010)。在未来,基于相同规格和标准化格式的集装箱将继续使用,并可能内置智能,以传递目的地、内容和旅程细节;下一代超大型船舶将装载18000个集装箱(McKinnon et al., 2015)。

10.3.5 疏浚

疏浚工程包括挖掘或清除海床上的沉积物和/或岩石,是港口作业和沿海及海洋基础设施发展的常规部分。疏浚作业主要有两种类型(McCook et al., 2015):资本疏浚是为了开发新的项目,如码头、港口或拓宽现有通道;维护疏浚则是将先前疏浚的区域保持在所需的深度。定期(如每年)进行维护疏浚工作,持续时间较短(数天至数周),一般会清除含有较高比例细颗粒的沉积物。通过疏浚清除的沉积物可用于填海,或弃置于陆地。然而大部分在港口、河口和海上疏浚的材料被倾倒在海上,只有少量的疏浚材料得到利用。大多数发达国家对可能有毒的疏浚沉积物的处置有严格控制，这些沉积物需要在和/或安置在安全的填埋场之前进行处理。

疏浚作业几乎总会使沉积物再悬浮并增加浊度，再悬浮的程度和影响取决于沉积物的物理和化学特性、场地条件、设备类型和疏浚方法。释放到水体中的沉积物可直接影响海洋生物，如珊瑚、海绵和贝类，并可导致鳍鱼的膜刺激和鳃磨损（OSPAR，2004，2008b）。悬浮泥沙水平的升高可以吸收、散射或降低光照，对许多光合底栖生物生物包括硬珊瑚和软珊瑚、海草、红树林、大型藻类（包括海藻）和藻类等造成极为重要的影响。沉积物也可以从悬浮物中沉淀下来，可能导致海底生物窒息。此外沉积物中营养盐的释放可能导致富营养化和氧消耗的增加（OSPAR，2004，2008a，b）。

10.3.6 填海

填海造地是指从海洋中创造新土地的过程。最简单的填海方法是用大量的重岩石和/或水泥填满该地区，再用粘土和土壤填充，直到达到所需的高度。排干淹没湿地海水的方法常被用于开垦农业用途用地。

第一次大规模填海工程发生在 20 世纪 70 年代，当时荷兰的鹿特丹港正在扩建（OSPAR，2008a）。这是当今填海造地时代的开始，并在全球范围内迅速蔓延。1975 年，新加坡政府开始在新加坡东端兴建新机场。樟宜机场由从海底回收的 4000 多万立方米海床沙建成。填海造地的著名例子还包括香港、新加坡、荷兰（OSPAR，2008a，b；Hilton & Manning，1995）和中国大陆的大部分海岸线（An et al.，2007）。人工岛屿是填海造地的一个例子，荷兰的弗莱福圩田（970 km^2）是世界上最大的填海人工岛，由艾瑟尔湖围填而成。日本大阪的关西国际机场和香港国际机场也是填海造地的例子。

填海造地使海洋栖息地永久消失。据估计，填海造地导致中国近 51% 的滨海湿地消失（An et al.，2007）。填海造地还可能影响海岸和陆地起源的栖息地类型，如沙丘或淡水系统。沉降会成为一个问题，一方面是因为填土上的土壤压实，另一方面是因为湿地被堤坝包围，并排水以形成圩田（即围海造地或河堤保护的低洼土地）（Hoeksema，2007）。

10.3.7 海滩修复

海滩修复是利用来自内陆或近海的沙或泥等材料修复海滩的过程。这可以用于建造遭受海滩贫瘠或海岸漂移侵蚀的海滩（Nordstrom，2000；Hamm & Stive，2002）。尽管这不是一个长期的解决方案，但它比其他类型的海岸防护便宜。长期以来，海滩修复一直被视为海岸保护的必要条件，同时也是一种拓展生活和娱乐的形式。这些措施改善了数百万人的生活质量。如澳大利亚的海岸线和沙滩是重要的娱乐和旅游资源。澳大利亚黄金海岸的可伦宾海滩在修复之前被严重侵蚀。同样的情况也适用于西班牙的地中海和大西洋沿岸以及许多其他沿海地区。美国、荷兰和比利时的东部及西部海岸每年也得到了修复。

10.3.8 酸性硫酸盐土

长期以来，许多沿海地区的酸性硫酸盐土壤对土地所有者和环境来说是一个问题。富含硫化铁的土壤在被水覆盖时是良性的，但是当它们变干时，氧气与硫化物结合产生硫酸，使土壤酸化。雨后，酸性物质进入水体，造成水体酸化和氧气减少。这种酸可以溶解金属（如铝），如果被排放到河流和河口，金属与酸的结合可以杀死植物和动物。酸性硫酸盐土壤的污染效应表现为观察到河口水域中鱼类因缺氧、溶解铝中毒和鱼病（如红斑溃烂）而死亡（Sammut et al., 1995）。从地理分布上来看，大多数酸性硫酸盐土分布在沿海地区，由近代或半近代沉积物形成。它们通常局限于相对靠海地区，形成了海相和河口沉积。许多沿海地区均发现了酸性土壤，在澳大利亚沿海地区尤为普遍。管理硫酸酸性土的成本包括了更换受损的基础设施所产生的成本，可能是巨大的。仅在昆士兰，每年的费用约为 1.89 亿美元，这还不包括对渔业和农业的直接损失（Ozcoasts, 2010）。

沿海地区的排水影响地下水位，并可能引发酸性硫酸盐土的问题。酸性硫酸盐土的扰动通常与运河、住宅和码头开发期间的疏浚、挖掘和排水活动有关。

干旱还会导致酸性硫酸盐土的暴露和酸化。预防性管理措施包括保持较高水位以防止土壤干燥，或通过填塞排水管或在排水管中蓄水。另一种技术是用潮水冲刷酸性废水，因为碱性海水可以中和酸。

10.3.9 渔业和水产养殖

预计未来几年内世界各地对鱼和其他渔业产品的需求将增加，野生鱼类资源将面临巨大压力。根据 2009 年收集的全球数据，全球约三分之一的鱼类资源被过度开发、耗尽或正在从枯竭中恢复，超过一半的鱼类资源被认为已得到充分开发（FAO, 2012）。因此，野生捕捞渔业在未来不太可能产生更高的产量。人们将主要通过水产养殖来满足日益增长的需求（图 10.6）。

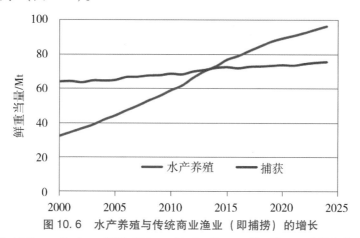

图 10.6 水产养殖与传统商业渔业（即捕捞）的增长

改编自：World Ocean Review, 2013. Living with the Oceans. The Future of Fish-the Fisheries of the Future, vol. 2. Maribus, Hamburg, Germany。

全球40%以上的鱼类和贝类消费由水产养殖业提供。在过去的20年里，没有任何其他食品生产部门增长得如此之快（图10.7）。当今全世界每年养殖的鱼、贻贝、蟹和其他水生生物约6000万吨。亚洲（尤其是中国）是最重要的水产养殖地区，目前供应了全球89%的产量。在一些工业化和发展中国家（特别是东南亚、东亚以及拉丁美洲部分地区）的沿海水产养殖在不断增长，其他地区还有巨大的发展和扩大空间（FAO，2012）。

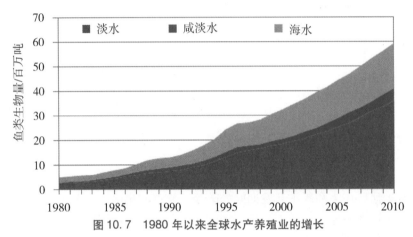

图10.7　1980年以来全球水产养殖业的增长

改编自：World Ocean Review, 2013. Living with the Oceans. The Future of Fish e the Fisheries of the Future, vol. 2. Maribus, Hamburg, Germany。

水产养殖发展的制约因素有很多，包括适宜水资源的日益短缺、日益拥挤的沿海地区有限的新业务开展机会、环境对人口较多沿海地区营养盐和污染有限的承载能力以及更严格的环境法规。未来水产养殖生产能力的大部分扩展可能发生在海洋，其中一些将越来越多地向近海转移，以摆脱沿海水域的限制。

水产养殖的可持续集约化和开发是全球海产品安全面临的重大挑战。水产养殖饲养的许多鱼类是肉食性鱼类，依赖于其他鱼类的供应作为食物。尽管数量因种类而异，但养殖1 kg的鱼平均需要约5 kg的鱼粉和鱼油。然而水产养殖的一个优势是养殖鱼类和海鲜所需的饲料要比饲养家畜（如肉牛和猪）少得多。生产1 kg牛肉所需的饲料是生产1 kg鱼所需的15倍（World Ocean Review, 2010）。然而水产养殖仍然需要大量的野生鱼类以加工成鱼粉和鱼油作为饲料。因此，水产养殖仍然可能是造成过度捕捞的一个原因。

许多环境问题与废物的处理以及工业上广泛使用的杀菌剂和兽药有关。来自集约化养殖场产生的食物、粪便和代谢废物会导致水体富营养化。许多养鱼场比猪或牛的集约化养殖更环保。后者从用于施肥的泥浆和肥料中释放出大量的氮和磷，水产养殖的氮和磷排放量则低得多（World Ocean Review, 2013）。

是为获得最大产量而集约化养殖的鱼类比野生鱼类更易患病，因此抗生素和其他药物被广泛使用，特别是在东南亚。已有迹象表明这些措施不再有效。近年水产养殖中使用的抗生素导致了多耐药病原体的传播，大多数现有抗生素对这些病菌无效。加速养殖鱼类到野生鱼类的疾病传播。沿海水域的水产养殖已导致重大疾病爆发，影响了当地物种的生态（State of Environment，2011）。

为了满足未来的需求，将需要进行可持续的水产养殖集约化。陆地系统中可持续的集约化被定义为一种"产量增加而不对环境产生不利影响，也不耕种更多土地"的生产形式（Garnett & Godfray，2012）。这是对世界人口日益增长的粮食需求带来的挑战的一种回应，因为在这个世界上，土地、水、能源和其他投入的供应都是有限的，而且无法得到可持续的利用。

10.3.10 采矿和矿山废物处理

10.3.10.1 海底采矿

海底采矿尚处于初级阶段，然而在相对较浅的水域已经有成功的采矿案例。20世纪60年代，海洋钻石公司在纳米比亚海岸不到20 m深的水域发现了超过100万克拉的钻石。如今作为全球领先的钻石公司戴比尔斯（De Beers），其大部分钻石产量都来自南部非洲大陆架水深小于300 m的潜水地带。

近年来，人们对深海采矿（尤其是金属开采）的兴趣与日俱增，对锰结核和海底块状硫化物（seafloor massive sulfides，SMS）的商业兴趣尤为强烈。海底块状硫化物是在热液喷口与周围较冷的海水相互作用下，从热液流体中沉淀出来的富含硫的金属矿床。硫化物矿床含有如银、金、铜、锰、钴和锌等贵重金属。块状硫化物和硫化物泥浆形成于板块边界附近的火山活动区域，深度为500～4000 m（World Ocean Review，2010）。

沿着海底山脉边缘的1000～3000 m发现了钴结壳。锰结核通常位于4000 m以下，主要由锰、铁以及具有经济价值的元素（包括钴、铜和镍，约占总重量的3%）组成（Margolis & Burns，1976）。此外还发现了其他重要元素的踪迹，如铂或碲。锰结核覆盖了深海的大面积区域，质量密度可达75 kg/km^3，大小从"土豆"到"足球"不等。结核密度最大的区域是墨西哥西海岸、秘鲁盆地和印度洋。由于从海底表面收集结核相当容易，实际的采矿过程不存在任何重大技术问题。

可以使用液压泵或铲斗系统将矿石运到地表进行开采。深海采矿作业设备包括：采矿支撑平台或船舶；发射和回收系统；带采矿头、离心泵和垂直运输系统的履带；还有电气、控制、仪表和可视化系统。采矿业一直在开发专门的挖泥船、泵、履带爬行器、钻机、平台、切割机和取芯器，其中许多设计都是用于深海恶劣条件下工作的机器人。正在开发中的潜艇车可以在5000 m深度以下作业。

《联合国海洋法公约》中的《国际深海采矿条例》已在1994年开始生效（Glasby，2000）。公约设立了国际海底管理局（International Seabed Authority，ISA），负责管理各国专属经济区（沿海国家周围370 km区域）以外的深海采矿企业。迄今为止，国际海底管理局已与13个承包商签订了17份为期15年的深海多金属结核和多金属硫化物勘探合同。其中11份合同是在太平洋的克拉里昂-克利伯顿大断裂带勘探多金属结核，2份合同是在西南印度洋脊和大西洋中脊勘探多金属硫化物。

鹦鹉螺矿业公司（Nautilus Minerals Inc.）在巴布亚新几内亚海域获得了首个多金

属 SMS 矿床的采矿租约，命名为索瓦拉 1 号（Solwara 1）。索瓦拉 1 号项目位于俾斯麦海（太平洋西南部），是世界上第一个旨在开采铜、金和银的深海采矿项目（Batker & Schmidt, 2015）。索瓦拉 1 号工地面积为 0.11 km^2，水深为 16000 m，预计最大扰动面积为 0.14 km^2。它位于巴布亚新几内亚海岸 30 km 处，毗邻一个被称为"珊瑚三角"的地区。该区域约占地球海底面积的 2%，却包含世界上 76% 的珊瑚和 37% 的珊瑚鱼。

深海采矿的潜在影响可能与沉积物的扰动和悬浮有关。当开采的尾矿（通常是细颗粒）被沉积回海洋时，可能会产生浑浊的羽流。这些羽状物会影响浮游动物和光线的穿透，进而影响该地区的食物网（Ahnert & Borowski, 2000；Nath & Sharma, 2000）。部分海底的移除会对底栖生物的栖息地造成严重的干扰（Ahnert & Borowski, 2000）。在某些情况下，废物将占泵入海面物质总量的 90%。因此海底作业将在海底沉积大量废物。深海群落的特征和分布图尚不明确，目前还不知道如何（甚至是否会）恢复挖掘区，需要开展研究以了解受深海采矿影响后的生态修复和重建。

10.3.10.2 甲烷水合物

甲烷水合物是由甲烷和水组成的白色冰状固体。它们是一种尚未被开发的潜在能源。甲烷分子被包裹在由水分子组成的微型结构中。甲烷气体主要是由生活在深层沉积物的微生物产生，它们将有机物缓慢地转化为甲烷。甲烷水合物只有在压力大于 3500 Kpa 和在海洋底部以及深海海床的低温下（0～4 ℃之间）才稳定。在水深 350 m 以下时，压力足以稳定水合物。因此，甲烷水合物主要出现在水深为 350～5000 m 的大陆边缘。由于深海沉积物中有机质不足，在广阔的海洋盆地底部几乎未发现任何水合物。

10.3.10.3 矿山废弃物处理

全球大约有 2500 个矿山在运营（Vogt, 2012）。几乎所有的尾矿都是在陆地上处理，通常是在尾矿库（也称为尾矿坝）。然而在某些地方，地面处理可能不是技术上最可行的选择。例如在印度尼西亚和巴布亚新几内亚，山区地形、高降雨和地震的问题结合在一起使得有效尾矿库的开发非常困难；在挪威等国家，可用土地的缺乏也是一个问题（Vogt, 2012）。

另一种废物的管理策略是海洋处置，适用于相对靠近沿海地区或通过管道进入的矿场。早期的作业基本上是无计划的，尾矿和其他废料直接排入大海。随着时间的推移，尾矿排放口和尾矿沉积池的设计修改将它们深度逐渐增加（Ellis & Ellis, 1994；Ellis et al., 1995）。目前世界上至少有 15 个采矿和矿物加工作业在海洋环境中使用海底尾矿处置方式（Vogt, 2012）。

海洋尾矿处置一般是指海洋浅层环境中（地表排放）的尾矿处置；海底尾矿处置或安置是指将沉积的尾矿沉降至 100～1000 m 深处。深海尾矿处置（deep-sea tailings placement，DSTP）是一种较新的做法，即将尾矿埋设在深度大于 1000 m 处。图 10.8 展示了 DSTP 的运行组成部分和潜在环境影响的概念图。为了避免尾矿再悬浮，尤其是影响到具有生物生产的透光带时，尾矿从深埋的管道中排出。

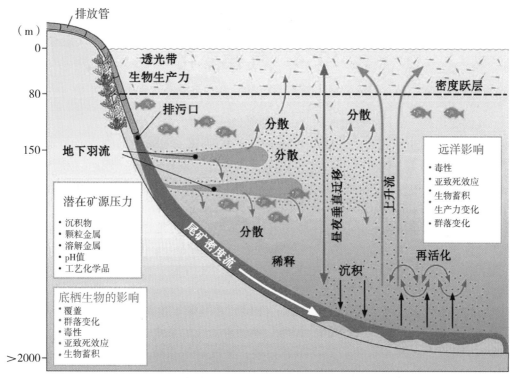

图 10.8　深海尾矿处置（DSTP）对环境的影响

深度未按比例显示。生物放大作用是一种罕见的现象，仅限于几种有机化学物质，其定义为 3 个或 3 个以上营养等级生物通过摄食导致体内污染物浓度的增加。CSIRO。

DSTP 的基本前提是尾矿可以在透光带以下的深度以稳定的羽流形式排放到海底。尾矿固体最终以足迹的形式沉积在海底，当地的海底地形决定了沉积物的形状。DSTP 要求尾矿管道排放深度大于表层混合层、透光层和上升流层的最大深度，将尾矿置于这些区域之下可最大限度地使其稳定沉积在海底。尾矿液和固体都是海洋环境的压力源，可能会影响远洋和底栖生物。图 10.8 描述了 DSTP 可能会导致透光层表层水受污染的过程。海床上的强洋流能使沉积的尾矿重新流动，这些尾砂可能因上升流被带到表层，供透光层的生物群利用。类似地，一些生物（如浮游动物和小型鱼类）每天在水体上下移动，可能作为有毒污染物的潜在携带者，将污染物带给表层的捕食者。适合的地点仅限于一些海洋岛屿和群岛，那里的海岸附近有很深的水。在珊瑚礁边缘之外的海底峡谷和自然切割的水道被认为是特别合适的地点。

10.4 气候变化

气候变化是一个日益紧迫的问题，可能对地球上的生命产生深远的影响。气候变化指的是气候在几十年来的长期趋势；不同于气候变异，气候变异指的是每年的变化（Mapstone，2011）。在 2014 年全球风险年度报告中，世界经济论坛（World Economic Forum，WEF，2014）将气候变化列为全球 31 大风险中的第 5 大风险，全球风险被定义为"在长达 10 年的时间范围内对若干国家和行业造成重大负面影响的事件"。与气候变化相关的潜在环境风险包括导致物种（包括害虫）的生物地理分布发生重大变化及生物多样性重大损失的气温升高、降雨模式改变、海平面上升、海洋酸化、洪水和干旱等极端天气事件数量的增加，以及龙卷风、火灾和山体滑坡等自然灾害的数量和级别的增加。

政府间气候变化专门委员会（IPCC，2014）在其第 5 次评估报告（Intergovernmental Panel on Climate Change，IPCC，2014）中指出，气候系统的变暖是明确的，且许多观测到的变化是前所未有的。自 1880 年以来，全球陆地和海洋表面平均温度上升了近 1 ℃。在北半球，1983—2012 年是 1400 年以来最热的 30 年。自 1910 年以来，澳大利亚的平均气温上升了 0.9 ℃；与 1951—1980 年相比，在过去 15 年中，非常温暖的月份出现的频率增加了 5 倍（State of the Climate，2014）。

人们普遍认识到气候变化的速度和轨迹及其影响在区域和地方尺度上存在很大差异。除了影响污染物的输入、运输和归宿外，气候变化还可能影响生态系统的结构和功能，以及对化学污染物和其他压力源的敏感性。

在接下来的章节中，我们将首先描述与海洋生态系统相关的单个气候变化压力源，然后探讨这些压力源和化学污染物之间可能的相互作用。评价这些多重压力源的影响，需要改变生态风险评价方法和适应性管理。

10.4.1 气候变化影响的环境变量

10.4.1.1 温度

气候变化的预测模型表明，到 2100 年，全球平均地面气温将比 1986—2005 年的参考值上升 1～5 ℃，但这在很大程度上取决于地理位置（IPCC，2014）。在同一时期，100 m 水深的海洋温度预计会提高 0.6～2.0 ℃，其中热带和北半球亚热带地区的升温幅度最大。在更深的区域，南大洋变暖最为明显（IPCC，2014）。

大多数海洋生物都是恒温动物，因此温度成为控制它们生理过程的重要变量。许多动物已经适应了日常和季节性的温度波动。然而对于压力源的组合，可能会超出生态系统的弹性范围。虽然对温度胁迫的遗传适应可以让种群在相对较强的选择压力下生存，但缺点是这通常也会导致种群遗传多样性降低，因此适应一组环境胁迫会产生适应成本，并会增加对其他胁迫的易感性（Moe et al.，2013）。

热带物种对温度变化的耐受性通常低于温带物种，在温度波动期间，热带物种的共生关系可能不如温带物种稳定（Przeslawski et al.，2008）。珊瑚及其甲藻共生体对海洋温度上升的敏感性已是很好的证据（Hoegh-Guldberg，1999）。据预测，全球气温上升 1 ℃，澳大利亚大堡礁 65% 的珊瑚将出现白化（Hennessy，2011）。当温度长时间（3～4 周）超过夏季最高温度几度时，虫黄藻就会被排出，导致珊瑚白化和潜在的死亡。虽然珊瑚在轻度热应激后可能会恢复，但它们通常会表现出生长缓慢、钙化和繁殖力下降，并可能遭受更严重的疾病（Hoegh-Guldberg et al.，2007）。

水温还会影响许多海洋无脊椎动物的繁殖时间。海洋温度的上升可能会导致温带地区动物产卵时间模式的变化（Przeslawski et al.，2008）。许多无脊椎动物每年会根据月球运动和温度变化而大量产卵，如棘皮动物、软体动物、珊瑚和多毛类。热变化可能会对这些产卵事件有直接或间接的影响（Przeslawski et al.，2008）。例如，如果产卵与浮游植物作为食物来源的可用性不一致，幼虫的存活和生存可能会受到抑制。除了影响繁殖时间，温度升高还可能减少或增加热带无脊椎动物（如海绵）的繁殖能力（Ettinger-Epstein et al.，2007）。

10.4.1.2　全球水循环和盐度

气候变化对全球水循环、海洋和河口系统均有影响。例如，气候变化对沿海地区的盐度变化影响很大。在全球范围内，气候变暖将导致水体蒸发增加，大气中的水蒸气和降水量增加；但预计这种现象的区域差异较大。最近在一些集水区发现极端降水频率和降雨量呈上升趋势，这意味着沿海地区发生洪水和风暴潮的风险更高。

在一些地区，如澳大利亚东南部，沿海降雨量减少、云量减少、蒸发量增加及水位下降导致了净水分不足。如墨累河和河口遭遇了长达十年的干旱。气候变化直接导致副热带山脊气压系统南移造成的长期干旱，再加上上游水资源的过度分配，导致 2005—2010 年间的水量非常少（Stauber et al.，2008）。因此，墨累河和河口及其邻近湿地受到低水位以及酸性硫酸盐土壤和沉积物中含硫物质的双重影响。

海洋表面盐度变化的观测结果也间接为全球海洋水循环的变化提供了证据。很可能自 20 世纪 50 年代以来，以蒸发为主的高盐度海区的海水变得更咸，以降水为主的低盐度海区的海水变得更淡（IPCC，2014）。

棘皮动物（尤其是海胆）对盐度降低的耐受性非常有限，已有许多淡水径流导致的成年个体死亡的报道（Lawrence，1996）。由于整个季节的补充可能会消失，因此对淡水不耐受物种的幼虫会变得非常脆弱。盐度降低事件的时间对于预测种群反应至关重要。

10.4.1.3　低氧

低氧即溶解氧低，是气候变化的另一个压力源，可能会越来越多地影响敏感的物种。据报道，波罗的海、墨西哥湾和切萨皮克湾等大片沿海地区都出现了低氧现象，预计这将因气候变化而恶化（Hooper et al.，2013）。水温升高降低了氧气的可用性，再加上增加的降水量会将富含营养的暖水带到敏感地区，导致富营养化和有机物负荷的增加。低氧会降低生物对 PAHs 和二噁英等污染物的解毒能力，干扰内分泌系统和生殖（Wu，2002）。暴露于这些化学物质也可能会阻碍物种对气候变化下低氧增加的反应能力。

在无脊椎动物中，甲壳类动物被认为是最容易受到低氧和有机物负荷影响的生物（Gray et al.，2002）。例如，在过去的 40 年里，波罗的海片脚类动物 *Monoporeia affinis* 的数量大幅度下降，与 20 世纪 70 年代相比，如今的种群丰度不足 10%（Eriksson-Wiklund & Sundelin，2004）。波罗的海的 *Monoporeia affinis* 卵子生成过程对低氧和温度极为敏感，主要发生在 8—11 月，这期间的氧气含量很低，而海面温度最高。因此，这一物种可能会被困在一个氧气枯竭的深水和过于温暖的浅水之间日益缩小的栖息地。取而代之的是更耐受的物种，如多毛类正在入侵，由于它们是沉积物中良好的生物扰动器，这可能会导致更多的污染物再悬浮和顺势效应，从而对其他敏感物种产生影响（Hedman，2006）。

10.4.1.4　海平面上升和风暴潮

1901—2010 年，海平面温度升高和南北极冰盖融化导致全球海平面平均上升 0.19 m（0.17～0.21 m）。19 世纪中叶以来，海平面上升的速度超过了此前 2000 年的平均速度（IPCC，2014）。自 1979 年以来，在每个季节和每一个连续十年中，北极地区海冰的面积都以每十年 3.5%～4.1% 的速度减少。南极洲的地区差异很大，某些地区的海冰面积增加，而其他地区则减少（IPCC，2014）。

相对于 1986—2005 年，全球平均海平面预估到 2100 年将上升 0.26～0.82 m（IPCC，2014），相对海平面上升的局部差异很大（由潮汐、风和大气压力模式、海洋环流变化、大陆垂直运动造成）（IPCC，2014）。因此预计海平面上升的影响是局部的。在最坏的情况下，将要受影响的大多数人生活在中国（7200 万人）、孟加拉国（1300 万人，损失了 16% 的全国水稻产量）和埃及（600 万人，损失了 12%～15% 的农业用地）（Nicholls & Leatherman，1995）。基里巴斯和马绍尔群岛等地势低洼的太平洋珊瑚环礁正在受到严重威胁。比海平面上升造成的土地直接损失更重要的是相关间接损害，包括侵蚀模式和沿海基础设施的破坏、水田盐碱化、沿海城市污水系统功能欠佳、沿海生态系统的损失以及生物资源的损失。

如果海平面上升速度很快，或如防波堤等人为屏障限制了红树林群落的扩张，则无脊椎动物群落的组成可能会由于潮间带栖息地的丧失而发生重大变化，这将取决于潮差和当地地貌。海平面的变化也会改变幼虫的传播模式（Przeslawski et al.，2008）。

由于气候变化，预计风暴的频率和严重程度都将会增加，可对海洋生境造成重大的物理干扰。Fabricius et al.（2008）发现近海珊瑚礁比近海珊瑚礁更容易受到风暴的破坏，特别是存在海岸径流污染时（Fabricius et al.，2008）。波浪的作用可以将固着的无脊椎动物从基质中分离出来，如果新生幼体供应不足或缺乏足够完整的沉降基质，那么它们的生存风险就特别大。

10.4.1.5　海洋酸化

自工业时代开始，海洋对 CO_2 的吸收导致了海洋酸化。以氢离子浓度来衡量，海洋表层水的 pH 下降了 0.1，相应的酸度增加了 26%（IPCC，2014）。根据情景模拟，预计到21世纪末，全球海洋pH将下降0.06～0.32；到2050年，碳酸盐饱和度将不足以维持珊瑚礁生长所需（IPCC，2014）。

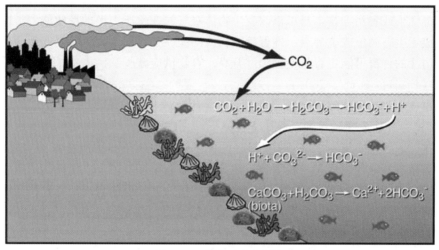

图10.9　海洋酸化的过程

来源：Hoegh-Guldberg, O., Mumby, P. J., Hooten, A. J., Steneck, R. S., Greenfiled, P., Gomez, E., Harvell, C. D., Sale, P. F., Edwards, A. J., Caldeira, K., Knowlton, N., Eakin, C. M., Iglesias-Prieto, R., Muthiga, N., Bradbury, R. H., Dubi, A., Hatziolos, M. E., 2007. Coral reefs under rapid climate change and ocean acidification. Science 318, 1737–1742。

全球大约有25%的人为CO_2排放进入海洋，其与水发生反应产生碳酸，碳酸分解成碳酸氢盐和H^+（图10.9）。然后质子与更多的碳酸盐离子结合，产生更多的碳酸氢盐，进一步降低了碳酸盐浓度，使那些具有碳酸钙外壳的海洋生物（如珊瑚、牡蛎、海胆和有孔虫类）无法获取碳酸盐（图10.9）。碳酸还与贝壳中的碳酸钙发生反应，导致贝壳溶解。壳层形成的削弱会降低对捕食者、物理损伤和干燥的防护能力，危及珊瑚礁无脊椎动物浮游和底栖生命阶段的生存（Przeslawski et al., 2008）。表层水中CO_2的增加会导致无脊椎动物发生酸中毒，降低代谢率，随后影响无脊椎动物的摄食、生长和繁殖。但大多数研究都是基于室内试验，且实验中的CO_2含量的设定变化较大，因此我们对于逐渐增加的海洋酸化对无脊椎动物的效应尚不清楚（Przeslawski et al., 2008）。

pH的变化也会影响生物地球化学过程，如金属形态的改变，可能会产生实质性的生物效应。海洋酸化也可能通过改变离子通道直接或间接地影响藻类的离子和养分同化，或通过养分有效性的变化间接影响藻类的离子和养分同化（Li et al., 2013）。研究者认为海洋酸化与氮限制的联合效应可以影响海洋硅藻和潜在的海洋食物网。

10.4.2　多重压力源：气候变化和污染物的相互作用

尽管全球气候变化越来越为科学界、监管界以及知情公众所接受，但迄今为止的讨论并没有将污染物作为环境中额外压力源的作用包括在内。

Noyes et al.（2009）首次发表了气候变化与化学污染物之间潜在相互作用的论述。

气候变化所改变的环境变量可以影响污染物的环境归宿和行为、化学吸附、分布和代谢的毒物动力学，以及污染物与目标分子和受体之间的毒理－动力学相互作用。

多重压力源可以通过两种方式相互作用：气候变化压力源可以增加或减少污染物对生物的毒性，或者污染物本身可以改变生物体对气候变化压力源的反应能力。这可能最终导致接近生态阈值或临界点，即群落结构或功能在小扰动下发生突变。生活在生理耐受范围边缘的种群可能更容易受到温度升高、食物供应减少和污染物暴露等多重压力源的影响（Heugens et al., 2001）。

多重压力源对生物的效应估计较为复杂，因为这些效应可能是直接的（如减少繁殖）、间接的（如改变捕食者－被捕食者关系）或诱导的（与物理或生态变化有关，而不是直接归因于化学压力源）。越来越多的证据表明，多重压力源会影响生物的生存、生长、繁殖、新陈代谢、行为和增殖，尤其是在生命早期阶段。水温可能影响生物繁殖的时间，如产卵时间和浮游时间。生物会对这些压力源的反应表现出物种特异性，并且为了预测这些效应，每种压力源的暴露规模和持续时间都十分重要（Przeslawski et al., 2008）。暴露的时机也很重要，若压力源（脉冲事件）暴露与生物敏感的生命阶段（如产卵或成熟）相吻合，则可能会产生更大的效应。

气候变化的压力源、污染物和生态系统之间的相互作用如图10.10所示。气候变化的影响包括温度升高、降雨模式改变、海洋酸化、缺氧和极端事件的增加。气候效应作为共同的压力源会直接影响污染物的迁移、转化、生物富集等过程，并最终影响污染物的生物利用度和毒性效应。气候变化也会对生物产生直接的影响。例如，当环境条件超出生物的耐热性和耐盐性范围，会导致其生物地理分布发生变化和受到更具耐受性的外来物种入侵。污染物也可以直接或间接作用于生物，使它们对气候变化的影响更加敏感。但迄今为止，我们尚未对这些影响进行充分的研究。

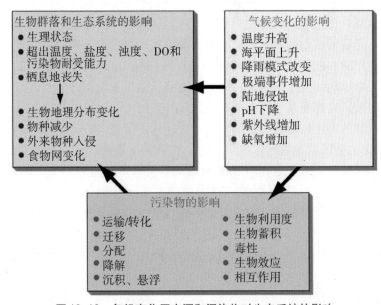

图10.10　气候变化压力源和污染物对生态系统的影响

修改自：Schiedek, D., Sundelin, B., Readman, J.W., Macdonald, R.W., 2007. Interactions between climate change and contaminants. Mar. Pollut. Bull. 54, 1845－1856。

收集气候变化影响的环境变量与相关污染物之间所有潜在相互作用的经验数据是不切实际的,因此有必要制定预测方法,将机械性数据纳入风险评价过程(Hooper et al., 2013)。另一个挑战是污染物很少单独出现,因而会存在多种污染物间协同或拮抗作用的未知影响。为简单起见,一般认为毒性作用是加成的,而气候共同压力源可能会对不同污染物产生不同的影响。

10.4.2.1 对污染物归宿、迁移和生物富集的影响

气候变化以及相关的土地利用变化可能会影响未来化学物质的使用模式。病虫害和病媒的增加将导致杀虫剂、杀菌剂和药物使用频率和时间的增加,可能导致排放到沿海海洋环境的量增加。气候变化将通过改变大气、水、土壤和生物之间的物理、化学和生物驱动因素(包括反应速率),从而影响污染物的环境归宿和行为。例如,温度升高和随后的冰雪融化可以使得持久性有机污染物重新进入水和大气中,并通过远距离运输将污染物输送至距其源头数千公里以外的高纬度地区(Noyes et al., 2009)。对于沉积物中的 Hg 等遗留化学物质,温度升高可能会加快 Hg 甲基化和挥发,导致 Hg 再活化和排放增加(Bogdal & Scheringer, 2011)。

污染物的脉冲释放和扩散源在飓风和洪水等极端事件中尤为重要,这些事件的频率预计会随着气候变化而增加。热带海洋和河口生态系统可能比温带地区面临更大的由地表径流和季节性强降雨事件带来的污染物输入风险,而热带系统也能更快地从干扰中恢复。

与气候有关的食物网的变化也会改变污染物的生物富集和迁移。储存在高脂肪(如鲸鱼和北极熊)中的生物富集污染物可能会在压力时期被释放出来。例如当北极的冰融化时,北极熊的食物来源——海豹就会变得稀缺,北极熊不得不利用储存的脂肪作为能量储备来抵抗饥饿。但反过来这会导致 POPs 从能量储存组织中释放,POPs 和饥饿的复合效应已被证明会影响其甲状腺功能,进而影响行为和认知功能,降低狩猎能力(Stirling et al., 1999)。因此,食物网中的微小变化会对食物链的上层产生潜在影响。

与温度依赖性的化学归宿和生物能模型联合应用的生物富集模型是研究气候变化对污染物生物富集影响的主要工具(描述污染物在生物系统中的吸收和分布)(Gouin et al., 2013)。至今大多数据都集中在单一地区的化学物质分布上,很少有关于食物网变化等间接影响的数据。Gouin et al.(2013)研究表明气候变化对化学物质长期归宿和迁移的影响估计相对较小,对于数量有限的有机化学物质而言,预测的变化与基线值相差不到 2 倍(Gouin et al., 2013)。气候变化对水生食物链中生物富集的直接影响在很大程度上取决于化学物质分配特性和生物转化速率常数。如 POPs 等中性有机化学物质,通常被认为在生物积累方面有问题,但也在基线值的 2 倍以内(Gouin et al., 2013)。然而与化学物质物理化性质有关的不确定性可能导致暴露的变异性比归宿和生物富集模型所预测的更大。据预测,污染物使用和排放的变化对污染物的暴露影响大于气候效应对污染物分配和生物富集的影响。因此,需要建模和监测数据以便更好地预测气候变化对污染物暴露的影响,尤其是在南半球。

10.4.2.2 对污染物毒性的影响

1. 温度

温度是气候变化参数,海洋表面温度的升高,不仅改变了污染物的环境分布,也改

变了它们的毒性。通常更高的温度会导致更高的污染物毒性，但此现象一般针对非常特定的物种和污染物。水温升高会增加污染物的吸收和生物富集。例如，鳃通气速率和代谢的增加可导致污染物的高组织浓度。而生物对污染物的净化和解毒能力也会随着温度的升高而增强。例如，加拿大底鳉（*Fundulus heterclitus*）在 25 ℃ 时，清除毒杀酚同系物的速度是 15 ℃ 的 2 倍（Maruya et al.，2005）。然而，这通常是以机体的能量消耗作为代价。较高的温度也会增加海洋生物体内金属的生物富集，如甲壳类、棘皮动物和软体动物（Marques et al.，2010）。较高温度还可以促进生物将污染物转化为毒性更强的化合物，如将 PCBs 转化成毒性更强的羟基化 PCB 代谢物。遗憾的是，许多关于温度对污染物毒性影响的研究都使用了非常高的污染物浓度和大的温度变化，这些在自然系统中不太可能发生。相反，生物可能会逐渐适应因全球变暖导致的温度变化，因此真正的问题不在于温度升高本身，而是在于温度升高如何与其他压力源相互作用（尤其是污染物）。

Negri（数据未发表）发现 Cu 和温度都抑制了幼体鹿角珊瑚（*Acropora millepora*）的变态率。在 Cu 浓度为 0.5～75 mg/L 的条件下，温度从 28 ℃ 上升至 34 ℃ 时，变态率下降；温度从 28 ℃ 上升至 32 ℃ 时，EC_{50}（即抑制 50% 个体变态的有效浓度）由 35 μg Cu/L 下降至 10 μg Cu/L。在 34 ℃，即使没有 Cu，幼体的变态行为也被完全抑制。

污染物暴露也会改变生物的热耐受性，通常用临界高温（critical thermal maximum，CTM）来衡量。水生动物 CTM 可能因有毒污染物的暴露而发生变化。鱼、爬行动物和两栖动物等变温动物尤其脆弱，因为它们无法调节自己的体温。热带物种（如珊瑚）已经生活在接近 CTM 的环境，可能比温带物种更脆弱（Heugens et al.，2001）。生物的生理条件也可以受温度影响，如热应激蛋白的诱导可能会影响生物对毒物的敏感性。

Adams et al.（未发表）研究了温度和 Cu 对固定化热带桡足动物 *Acartia sinjiensis* 的联合急性毒性效应（图 10.11）。结果表明当温度从 30 ℃ 上升至 34 ℃ 时，Cu 的毒性呈温度依赖性增加，48 h 的 EC_{50} 值从 44 μg Cu/L 下降至 10 μg Cu/L。它们同时还研究了 Cu 对桡足类热耐受性的影响，在 0～38 μg Cu/L 范围内，随着 Cu 浓度的增加，热耐受性有不同程度的降低（图 10.11）。

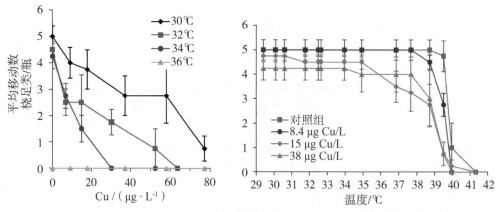

图 10.11　Cu 和温度 48 h 内对海洋桡足动物 *Acartia sinjiensis* 生存的联合效应

2. 盐度

河口地区生物暴露于污染物和盐度波动。盐度和污染物之间的相互作用较为复杂，因为盐度既可以影响污染物的化学形态，也可以影响生物的生理过程，从而影响毒性。对于金属而言，由于金属络合作用的增加，盐度的增加通常会降低其生物利用度和毒性，但这取决于金属的吸收途径和作用方式。例如，Wildgust & Jones（1998）发现河口糠虾（*Neomysis integer*）在 20‰ 的最适盐度条件下对 Cd 的耐受性最强，在较高或较低盐度，糠虾的死亡率较高。Cd 毒性的差异不能完全用 Cd 形态的差异来解释，渗透胁迫的增加导致了 Cd 耐受性降低。

相比之下，许多有机化合物（如有机磷酸盐）的毒性通常随盐度的增加而增强，部分原因是由于溶解度降低、持久性增强和生物富集增加。在 3～4 倍于盐水虾（*Artemia* sp.）等渗盐度条件时，暴露于有机磷农药乐果，其死亡率增加（Song & Brown., 1998）。污染物也可以改变生物的渗透调节。例如，当阿特拉津暴露浓度为 5 mg/L 时，通常耐受多种盐度的鱼苗在高盐度和低盐度环境下表现出更高的死亡率（Fortin et al., 2008）。

洪水事件导致的水体盐度降低，加上杀虫剂和营养物质等污染物的输入，已被证明会对珊瑚产生有害影响，使它们更容易受到真菌感染、藻类和藤壶的寄生，导致死亡率上升（Smith et al., 1996）。

紫外线

在海洋系统中，增加紫外线辐射可使 PAHs 的毒性增强多达 100 倍。实验室研究表明，在 UVA（320～400 nm）辐射下，采用先前认为的无效应暴露浓度，蒽、芘和荧蒽对海洋无脊椎动物幼体和胚胎的毒性显著大于 PAHs 单独暴露（Pelletier et al., 1997）。毒性机制是通过形成与各种大分子相互作用的活性氧，在急性暴露下引起细胞损伤和致死（Hooper et al., 2013）。

10.4.3 对环境风险评价的影响

气候因素驱动的复杂性、不确定性和可变性对影响的预测环境管理项的实施带来了重大挑战。污染物影响与气候变化的相互作用能在不同的时空尺度上发生。在纳入包括气候变化在内的多种压力源影响的风险评价中，都需要考虑使用和释放的化学物质种类和数量的变化、污染物在环境中的运输、归宿和富集以及污染物对生物的效应（Stauber et al., 2012）。

目前的生态风险评价（ecological risk assessment，ERA）框架是为了研究小地理范围内特定压力源（通常是化学压力源）作用于特定受体的风险，而忽略了其他非污染压力源（物理和生物压力源）。传统 ERA 评估相对于参考地点或条件是否有生态系统服务的变化。生态系统服务是直接或间接促进人类福祉的生态功能或过程的产物。实际上它们是自然对家庭、社会和经济的馈赠（Costanza & Daly, 1992）。它们大致可分为 4 类：供应（食物、水、能源）、调节（如防洪、防侵蚀）、文化（如娱乐）和支持（如营养、氧气）。在区域尺度上评估生态系统服务的某些组成部分和模型的终点是最近才开发出来的，尚未在多重压力条件下进行研究。此外，ERAs 很少包括污染物生物效应的机理内容，如基因表达或组织病理学的改变（Hooper et al.，2013）。

由于生态条件将随着全球气候变化而发生不可预测的变化，所以对静态条件和单向变化的简单假设将不再适用。全球气候变化使得我们需要考虑污染物和非污染物压力源之间的相互作用，这可能导致负面或正面的影响。根据 ERA 的重点，在制定问题阶段需要付出大量努力，以确保所有与海洋物种或所关注的栖息地有关的气候变化压力源相互作用可用于评价。由于在预测与全球气候变化相关的风险和确定适当的管理措施方面存在相当大的不确定性，因此一种适应性的管理方法将是必不可少的（Landis et al.，2013）。

Landis et al.（2013）建议生态风险评价要考虑气候变化相应地做出 7 个关键调整：①考虑气候变化的幅度、速度和规模是否会成为特定 ERA 的重要因素，以及后果是否为长期；②根据生态系统服务表达评价终点，为利益相关者提供一个可理解的共通术语；③为气候变化和目标压力源建立因果概念模型；④考虑到对气候变化的响应可能为非线性，对不同的受体有利有弊；⑤考虑区域性和多重压力源的 ERA 方法，例如相对风险模型；⑥确定空间和时间上不确定性的主要驱动因素；⑦为不断变化的环境条件而制定适应性管理计划。

多个压力源之间可能会以叠加、协同或拮抗的方式相互作用。从未来风险评价的角度来看，气候变化的情况下协同作用的污染物效应更大，这将要求对化学污染物制定更严格的环境质量标准。从生态恢复的角度来看，协同作用下消除一个压力源的效益可能大于预期；或拮抗作用下消除一个压力源产生比预期更小的效益。Moe et al.（2013），对气候变化和污染物的多重压力效应的综述中指出，无论是叠加、协同还是拮抗的相互作用，都随着特定的压力组合、物种、营养水平和反应水平（种群和群落）的不同而不同。因此，多重压力源可能会给真实的生态系统带来生态意外，阻碍我们预测 ERAs 影响的能力。

应激反应可以为非线性且具有临界点，因此还需要考虑生物的耐受性（生理适应和遗传适应）。生态系统的恢复能力可能会被一系列的压力源组合所超越。由于生物能够快速适应压力源，理解生物对压力源的进化反应也是 ERAs 的重要组成部分。适应可以在暴露后物种的几代内迅速发生，因此其发生也在 ERAs 的时间线内（Kimberley & Salice，2012）。反应可能首先是耐受，然后是适应、回避，最后是遗传变异（适应潜能）和新性状的出现。然而有证据表明，当同时暴露于多个压力源时，种群对新环境压力源的适应能力会降低，而这种压力源种群的反应具有流动效应。

10.5 展　　望

生态毒理学面临的一个挑战是预测全球变化和污染物在个体水平（如生存、繁殖或生长下降）上的联合效应将如何在种群水平（如丰度）和群落水平（如生物多样性和食物网）表现出来（Stahl et al.，2013）。需要更多层次组织的数据来理解和预测气候变化的影响：在生物层面——生理学、毒性和遗传学；种群水平——丰度；群落水平——物种相互作用和生境；生态系统层面——全球过程（Przeslawski et al.，2008）。需要建立模型和监测方法来解决这个数据缺口。

我们从多重压力源的角度对海洋生态毒理学的理解得益于一系列评价生态系统健康的新工具的开发和应用。表观遗传学、组学和建模方法是可以帮助对全球变化的反应评价的一些新工具，将在下面的章节中进行讨论。

10.5.1 表观遗传学

表观遗传学是一个正在迅速被纳入生态毒理学研究的新兴领域。它研究了由甲基化或组蛋白修饰引起的基因功能或细胞表型改变，而DNA序列没有改变（Connon et al.，2012）。金属、POPs或内分泌干扰物的环境暴露已被证明可以调节环境相关物种（如鱼类或水蚤）的表观遗传标记（Vandegehuchte & Janssen，2011）。表观遗传变化在某些情况下可以传递给后代，即使这些后代不再暴露于引起表观遗传变化的外部因素（Vandegehuchte & Janssen，2011）。

目前表观遗传机制在化学风险评价中未被考虑，也未被用于监测化学物质暴露的影响和环境变化。生物的表观遗传图谱可以识别它们整个生命阶段接触过的化学污染物类别（Mirbahai & Chipman，2014）。表观遗传学在生态毒理学背景下的潜在意义需要进一步的研究，即跨代遗传、化学胁迫诱导的表观遗传变化和表观遗传诱导的应激适应在长期化学暴露下的可能性。暴露于多重压力源后的表观遗传学变化是另一个需要进一步研究的领域（Vandegehuchte & Janssen，2014）。

10.5.2 海洋生态毒理学中的环境基因组学

从多重压力源的角度（如多种污染物与气候变化之间的相互作用）理解海洋生态毒理学的必要性贯穿本章。然而到目前为止，生态毒理学研究都严重偏向于在受控制的实验条件下检测对单一或偶然的二元压力源的反应。虽然这些研究很关键，并能支撑我们对压力源的理解，但在将这些知识扩展到多种压力源对海洋环境的更广泛影响中存在许多局限性：①通常只有相对少量的生物用于生态毒理学分析，而这些生物不成比例地偏向于温带物种；②实验室生物测定无法准确模拟多种污染物的复杂相互作用及其在空间和时间上的暴露变化；③实验室研究提供的污染物如何改变生物相互作用的观点过于简单，包括间接影响，如改变初级生产者和微生物过程可能影响高营养级生物。

虽然可以通过操作性实验（如围隔和移植研究）来实现更真实的环境场景，但可测量的参数数量受到实验设计和逻辑（如重复数量）的严重限制。实地研究（即在自然情景下考察群落）也有其局限性。例如，底栖生物调查可确定底栖生物群落结构与金属之间的紧密关系，虽然两者趋势具有相关性，但其他因素（如未测量的污染物和跨空间的自然变化）也可能是驱动群落变化的因素，因此不能确定是金属引起了这种群落反应。底栖生物调查还有另一个重要但被忽视的局限性，即大型底栖无脊椎动物群落只能代表生态系统中真正多样性中的很小部分。这可能导致一种简单的假设，即微型底栖生物的变化水平反映了生态系统的整体反应。

虽然环境基因组学绝非万能,但它为探索从亚细胞到群落等多个生物组织层次的多重压力源的生态效应提供了新的机遇(Chariton et al.,2015b)。环境基因组学可以广义地定义为对来自环境样本遗传物质(DNA和RNA)的研究,目的是了解生物结构、功能和反应。在较低水平的生物组织中,转录组学从暴露于各种环境的生物中生成RNA转录的基因表达谱,并且有可能根据压力源的作用方式提供独特的信号(如金属与内分泌干扰物),因此其作为早期压力源的研究指标表现出了很大潜力。其他在生物水平的补充"组学"技术包括蛋白质组学和代谢组学,它们的终点分别是蛋白质和代谢物图谱。

如前文所述,最大的挑战之一是理解群落(所有的生物而非传统的代表生物)如何应对复杂的自然和人为压力(如金属和营养)。研究人员通过使用DNA宏条形码技术(即一种针对DNA的分类信息区的高通量测序方法)能够获得潜在的所有生命的生物多样性图谱,提供以前无法获得的生物多样性视图(如硅藻、微生物、真菌和无脊椎动物)。虽然这种方法尚未被生态毒理学界广泛采用,但它已被证明非常适合于研究人类活动对沿海底栖生物系统的影响。例如,Chariton et al.(2015a)使用DNA宏条形码技术,不仅能区分人为活动影响程度不同的河口,还能识别关键的环境压力源如何改变数百个生物种群和群落的个体分布。

由于许多关键的生物地球化学循环是由微生物过程驱动的,因此它们组成水平的改变也会影响生态系统的功能。利用类似DNA宏条形码的方法还可以分析微生物和关键生物地球化学途径中特定步骤相关的基因图谱,例如,nifH和NifK基因通常被用作固氮细菌的分子标记。除用于检测扩增基因的DNA宏条形码方法外,另一种微生物群落分析方法是"鸟枪法测序",其可以检测来自总DNA(宏基因组学)或RNA(环境转录组学)的随机序列片段。这些片段(reads)可以被检测或组装,通过识别蛋白质编码reads并将其与具有已知生物学功能序列的数据库进行比较,从而提供组成图谱和功能图谱。

虽然这些"组学"方法目前在不同的生物组织水平上提供了各种各样的信息数据,但尚不能确定在较低水平组织水平上的表达(如转录组和代谢组)是否能转化为群落功能和结构的改变。鉴于这个主题的广泛性和方法的复杂性,有兴趣的读者可以参阅以下出版物,以便更详细地了解该主题:Chariton et al.,2014;Hook,2010;Paulsen & Holmes,2014;Piña & Barata,2011;Sheehan,2013;van Straalen & Feder,2012;van Straalen & Roelofs,2011。

10.5.3 生态建模

由于收集有关受全球变化影响的环境变量和目标污染物之间的所有潜在相互作用的历史数据不切实际,因此有必要建立预测方法以帮助评价这些相互作用在何时、何地以及如何影响潜在的风险(Hooper et al.,2013)。为了支持预测方法,评估过程中包括机理数据,可通过有害结局路径(adverse outcome pathway,AOP)框架实现(Ankley et al.,2010)。AOP描述了分子起始事件(化学物质与生物靶标的相互作用)与随后跨个体、种群和群落的级联反应之间的联系。Hooper et al.（2013）举例说明气候变化如何影响化学物质的暴露和生物利用度，以及它们与生物体内的毒理动力学和毒物动力学相互作用（Hooper et al.，2013）。

生态模型越来越多地被用于预测全球变化对种群动态、物种分布和生物多样性的影响。生态模型整合了物理和生物过程，并包含了关联的机制（van de Brink et al.，2015）。建模对于场景测试十分有用，它可以识别最可靠的选项（并消除不太可能的选项）和关键的知识缺口。最近的进展包括"全系统"模型的开发，该模型试图包括多种压力源之间的相互作用、替代用途以及生态系统日益增长的压力——不断扩大的人口、新产业、气候变化和气候变异（van den Brink et al.，2015）。

需要将这些模型与生态毒理学更好地结合起来，以考虑和预测污染物与气候压力源、适应和恢复之间的非加性相互作用。例如，基于特征的框架将种群脆弱性定义为外部暴露、内在敏感性和种群可持续性，可以扩展到评估种群对污染物和气候压力的脆弱性（Moe et al.，2013）。

10.6 总　　结

全球变化导致进入河口和海洋生态系统的污染物浓度和类型发生重大变化。此外，气候变化压力导致的生态系统物理化学变化也增加了复杂性。在野外环境中，预测单一污染物的影响已有一定难度，而处理多种污染物和新的压力源时，难度将更大。虽然一个系统可以从单个事件中恢复，但从多个压力源或重复污染事件中恢复则可能导致系统受到损害。变化的速度、频率和幅度是主要的威胁（Przeslawski et al.，2008）。一些物种特别容易受到气候变化的影响，而污染物、病原体、物种入侵、过度捕捞和栖息地破坏等额外的压力放大了影响（Noyes et al.，2009）。

虽然气候变化模型为未来气候变化情景提供了洞见，但它们无法预测气候变化对海洋系统意味着什么。物种必须适应或迁移，否则种群会因无法对气候变化的速度和幅度做出反应而较易灭绝。然而在能够理解这种衰退的原因之前，我们根据地理范围来预测种群对环境变化的脆弱性的能力有限。生态毒理学可以通过提供关于个体生物生理、迁移性和栖息地需求的进一步信息来预测气候变化导致的物种重新分布。

迄今为止，IPCC的报告中尚未考虑污染物，这表明气候变化科学家和生态毒理学家之间需要在研究和政策领域进行更好的交流。有建议指出，一项即时的管理行动应该是监测水和沉积物中污染物的基线浓度（时间序列），并尽可能减少污染物暴露，因为这与比消除气候变化造成的持续压力相比，可能更容易在一开始就解决。然而在压力源间存在拮抗作用的情况下，需要根据具体情况进行仔细的评估，局部干预可能导致无效的、高成本的管理行动，造成管理工作的浪费。

显然，当我们进入一个前所未有的变化时期时，需要付出更大的努力来理解多重压力源，以及如何在地球系统科学更广泛的背景下管理它们。国家和国际污染物环境监管法律和管理框架尚未将全球变化纳入其评估框架。因此，世界各地的决策者和工业界需要开始了解未来变化对海洋环境中化学品对生态系统和人类健康风险的影响，以便能够实施必要对策。

参考文献

ABS (Australian Bureau of Statistics), 2003. Regional population growth, Australia and New Zealand, 2001 – 02. Catalogue number 3218.0. Australian Bureau of Statistics, Canberra, ACT, Australia.

AGARDY T, ALDER J, DAYTON P, et al., 2005. Coastal systems // REID W V (Eds.). Millennium ecosystem assessment: ecosystems and human well-being: current state and trends, vol.1. Island Press, Washington, DC, USA: 513 – 549.

AHNERT A, BOROWSKI C, 2000. Environmental risk assessment of anthropogenic activity in the deep-sea. J. Aquat. Ecosys. Stress Recov., 7: 299 – 315.

AN S, LI H, GUAN B, et al., 2007. China's natural wetlands: past problems current status and future challenges. Ambio, 4: 335 – 342.

ANDERSEN T, CARSTENSEN J, HERNANDEZ-GARCIA E, et al., 2009. Ecological thresholds and regime shifts: approaches to identification. Trends Ecol. Evol., 24: 49 – 57.

ANKLEY G T, BENNETT R S, ERICKSON R J, et al., 2010. Adverse outcome pathways: a conceptual framework to support ecotoxicology research and risk assessment. Environ. Toxicol. Chem., 29: 730 – 741.

BATKER D, SCHMIDT R, 2015. Environmental and social benchmarking analysis of the Nautilus Minerals Inc. Solwara 1 project earth economics report. http://www.nautilusminerals.com/irm/content/pdf/eartheconomics-reports/earth-economics-may-2015.pdf.

BAYLISS P, FREELAND W, 1989. Seasonal distribution and abundance of dugongs in the Western-Gulf-of-Carpentaria. Wildl. Res., 16: 141 – 149.

BELLWOOD D R, HUGHES T P, FOLKE C, et al., 2004. Confronting the coral reef crisis. Nature, 429: 827 – 833.

BJORNDAL K A, 1980. Nutrition and grazing behavior of the green turtle chelonia mydas. Mar. Biol., 56: 147 – 154.

BLASCO J, DURÁN-GRADOS V, HAMPEL M, et al., 2014. Towards an integrated environmental risk assessment of emissions from operating ships. Environ. Int., 66: 44 – 47.

BOESCH D F, BURROUGHS R H, BAKER J E, et al., 2001. Marine pollution in the United States. Technical report. Prepared for the Pew Oceans Commission, Arlington, VA, USA.

BOGDAL C, SCHERINGER M, 2011. Release of POPs to the environment // Climate change and POPs: predicting the impacts. UNEP/AMAP Expert Group Report, Geneva,

Switzerland.

BOUCHARD M F, CHEVRIER J, HARLEY K G, et al., 2011. Prenatal exposure to organophosphate pesticides and IQ in 7 - year old children. Environ. Health Perspect., 119: 1189 - 1195.

BOYER E W, HOWARTH R W, 2008. Nitrogen fluxes from rivers to the coastal oceans // CAPONE D, CARPENTER E J (Eds.). Nitrogen in the marine environment. 2nd ed. Academic Press, San Diego, CA, USA: 1565 - 1587.

BRINK C, VAN GRINSVEN H, JACOBSEN B H, et al., 2011. Costs and benefits of nitrogen in the environment // SUTTON M A, HOWARD C M, ERISMAN J W, et al (Eds.). The european nitrogen assessment: sources, effects and policy perspectives. Cambridge University Press, Cambridge, UK: 513 - 540.

BRODIE J, FABRICIUS K, DE'ATH G, et al., 2005. Are increased nutrient inputs responsible for more outbreaks of crown-of-thorns starfish? An appraisal of the evidence. Mar. Pollut. Bull., 51: 266 - 278.

BROWN L R, RENNER M, FLAVIN C, et al., 1998. Vital signs 1998: the environmental trends that are shaping our future. Worldwatch Institute, W. W. Norton and Company, New York, NY, USA.

VAN DEN BERG H, ZAIM M, YADAV R S, et al., 2012. Global trends in the use of insecticides to control vector-borne diseases. Environ Health Perspect., 120: 577 - 582.

CANTY M N, HAGGER J A, MOORE R T B, et al., 2007. Sublethal impact of short term exposure to the organophosphate pesticide azamethiphos in the marine mollusc *Mytilus edulis*. Mar. Pollut. Bull., 54: 396 - 402.

CARPENTER S R, CARACO N F, CORRELL D L, et al., 1998. Nonpoint pollution of surface waters with phosphorus and nitrogen. Ecol. Appl., 8: 559 - 568.

CASTEL S E, MARTIENSSEN R A, 2013. RNA interference in the nucleus: roles for small RNAs in transcription, epigenetics and beyond. Nat. Rev. Genet., 14: 100 - 112.

CHARITON A A, ROACH A C, SIMPSON S L, et al., 2010. Influence of the choice of physical and chemistry variables on interpreting patterns of sediment contaminants and their relationships with estuarine microbenthic communities. Mar. Freshwater Res., 61: 1109 - 1122.

CHARITON A A, HO K T, PROESTOU D, et al., 2014. A molecular-based approach for examining responses of eukaryotes in microcosms to contaminantspiked estuarine sediments. Environ. Toxicol. Chem., 33: 359 - 369.

CHARITON A A, STEPHENSON S, MORGAN M J, et al., 2015a. Metabarcoding of benthic eukaryote communities predicts the ecological condition of estuaries. Environ. Pollut., 203: 165 - 174.

CHARITON A A, SUN M Y, GIBSON J, et al., 2015b. Emergent technologies and analytical approaches for understanding the effects of multiple stressors in aquatic environments. Mar. Freshwater Res. , http://dx. doi. org/10. 1071/MF15190.

CONNON R E, GEIST J, WERNER I, 2012. Effect-based tools for monitoring and predicting the ecotoxicological effects of chemicals in the aquatic environment. Sensors, 12: 12741 – 12771.

CORRELL D L, 1998. The role of phosphorus in the eutrophication of receiving waters: a review. J. Environ. Qual, 27: 261 – 266.

COSTANZA R, DALY H E, 1992. Natural capital and sustainable development. Conserv. Biol. , 6: 37 – 46.

COSTANZO S D, O'DONOHUE M J, DENNISON W C, et al. , 2001. A new approach for detecting and mapping sewage impacts. Mar. Pollut. Bull. , 42: 149 – 156.

CÓZAR A, ECHEVARRÍA F, GONZÁLEZ-GORDILLO J I, et al. , 2014. Plastic debris in the open ocean. Proc. Natl. Acad. Sci. , USA, 111: 10239 – 10244.

DE'ATH G, FABRICIUS K E, SWEATMAN H, et al. , 2012. The 27 – year decline of coral cover on the Great Barrier Reef and its causes. Proc. Natl. Acad. Sci. , USA, 109 (44): 17995 – 17999.

DUARTE C M, HOLMER M, OLSEN Y, et al. , 2009. Will the oceans help feed humanity? Bioscience, 59: 967 – 976.

DUARTE C M, LENTON T M, WADHAMS P, et al. , 2012a. Abrupt climate change in the Arctic. Nat. Clim. Change, 2: 60 – 62.

DUARTE C M, AGUSTÍ S, WASSMANN P, et al. , 2012b. Tipping elements in the arctic marine ecosystem. Ambio, 41: 44 – 55.

DUARTE C M, 2015. Global change and the future ocean: a grand challenge for marine sciences. Front. Mar. Sci. , 1: 1 – 16.

EASAC (European Academies Science Advisory Council), 2015. Marine sustainability in an age of changing oceans and seas. http://www. academies. fi/wp-content/uploads/2015/06/EASAC-JRC-Marine-Sustainability_Summary. pdf.

ELLIS D, ELLIS K, 1994. Very deep STD. Mar. Pollut. Bull. , 28: 472 – 476.

ELLIS D V, POLING G W, BAER R L, 1995. Submarine tailings disposal (STD) for mines: an introduction. Mar. Geores. Geotechnol. , 13: 3 – 18.

ENELL M, FEJES J, 1995. The nitrogen load to the baltic sea: present situation, acceptable future load and suggested source reduction. Water Air Soil Pollut. , 85: 877 – 882.

ERIKSSON WIKLUND A K, SUNDELIN B, 2004. Sensitivity to temperature and hypoxia: a seven year field study. Mar. Ecol. Progr. Ser. , 274: 209 – 214.

ETTINGER-EPSTEIN P, WHALAN S W, BATTERSHILL C N, et al. , 2007. Temperature cues gametogensis and larval release in a tropical sponge. Mar. Biol. , 153: 171 – 178.

FABRICIUS K E, DE'ATH G, PUOTINEN M L, et al., 2008. Disturbance gradients on inshore and offshore coral reefs caused by severe tropical cyclone. Limnol. Oceanogr., 53: 690 – 704.

FAO (Food and Agriculture Organisation of the United Nations), 2012. The state of the world fisheries and aquaculture 2012. Rome, Italy.

FORTIN M G, COUILLARD C M, PELLERIN J, et al., 2008. Effects of salinity on sublethal toxicity of atrazine to mummichog (*Funduluus heteroclitus*) larvae. Mar. Environ. Res., 65: 158 – 170.

GALANOPOULOU S, VGENOPOULOS A., CONISPOLIATIS N, 2005. DDTs and other chlorinated organic pesticides and polychlorinated biphenyls pollution in the surface sediments of Keratsini Harbour, Saronikos Gulf, Greece. Mar. Pollut. Bull., 50: 520 – 525.

GALLOWAY J N, DENTENER F J, CAPONE D G, et al., 2004. Nitrogen cycles: past, present, and future. Biogeochemistry, 70: 153 – 226.

GARNETT T, GODFRAY C, 2012. Sustainable intensification in agriculture. University of Oxford, Oxford, UK.

GEISZ H N, DICKHUT R M, COCHRAN M A, et al., 2008. Melting glaciers: a probable source of DDT to the Antarctic marine ecosystem. Environ. Sci. Technol., 42: 3958 – 3962.

GLASBY G P, 2000. Lessons learned from deep-sea mining. Science, 289: 551 – 553.

GOUIN T, ARMITAGE J, COUSINS I, et al., 2013. Influence of global climate change on chemical fate and bioaccumulation: the role of multimedia models. Environ. Toxicol. Chem., 32: 20 – 31.

GRAY J S, WU R S S, OR Y Y, 2002. Effects of hypoxia and organic enrichment on the coastal marine environment. Mar. Ecol. Progr. Ser., 238: 249 – 279.

GREENING H, JANICKI A, 2006. Toward reversal of eutrophic conditions in a subtropical estuary: water quality and seagrass response to nitrogen loading reductions in Tampa Bay. Florida USA. Environ. Manag, 38: 163 – 178.

HAJJ – MOHAMAD M., ABOULFADL K, DARWANO H, et al., 2014. Wastewater micropollutants as tracers of sewage contamination: analysis of combined sewer overflow and stream sediments. Environ. Sci. Process Impacts, 16: 2442 – 2450.

HALPERN B S, WALBRIDGE S, SELKOE K A, et al., 2008. A global map of human impact on marine ecosystems. Science, 319: 948 – 952.

HAMM L, STIVE M J F, 2002. Shore nourishment in Europe. Coast. Eng., 47: 79 – 263.

HARRINGTON L, FABRICUS K, RAGLESHAM G, et al., 2005. Synergistic effects of diuron and sedimentation on photosynthesis and survival of crustose coralline algae. Mar. Pollut. Bull., 51: 415 – 427.

HAYNES D, RALPH P, PRANGE J, et al., 2000. The impact of the herbicide diuron on photosynthesis in three species of tropical seagrass. Mar. Pollut. Bull., 41: 288 – 293.

HEDMAN J, 2006. Fate of contaminants in Baltic Sea sediment ecosystems: experimental studies on the effects of macrofaunal bioturbation (Licenthiate thesis). Department of Systems Ecology, Stockholm University, Stockholm, Sweden.

HENNESSY K, 2011. Climate change impacts // CLEUGH H, STAFFORD SMITH M, BATTAGLIA M, et al (Eds.). Climate Change. CSIRO Publishing, Collingwood, VIC, Australia: 45 – 57.

HEUGENS E H W, HENRICK A J, DEKKER T, et al., 2001. A review on the effects of multiple stressors on aquatic organisms and analysis of uncertainty factors for use in risk assessment. Crit. Rev. Toxicol., 31: 247 – 284.

HIGNITE C, AZARNOFF D L, 1977. Drugs and drug metabolites as environmental contaminants: chlorophenoxyisobutyrate and salicylic acid in sewage water effluent. Life Sci., 20: 337 – 341.

HILTON M J, MANNING S S, 1995. Conversion of coastal habitats in Singapore: indications of unsustainable development. Environ. Conserv., 22: 307 – 322.

HOEGH-GULDBERG O, MUMBY P J, HOOTEN A J, et al., 2007. Coral reefs under rapid climate change and ocean acidification. Science, 318: 1737 – 1742.

HOEGH-GULDBERG O, 1999. Climate change, coral bleaching and the future of the world's coral reefs. Mar. Freshwater Res., 50: 839 – 866.

HOEKSEMA R J, 2007. Three stages in the history of land reclamation in the Netherlands. Irrig. Drain., 56: 113 – 126.

HOOK S, 2010. Promise and progress in environmental genomics: a status report on the applications of gene expression-based microarray studies in ecologically relevant fish species. J. Fish Biol., 77: 1999 – 2022.

HOOPER M J, ANKLEY G T, CRISTOL D A, et al., 2013. Interactions between chemical and climate stressors: a role for mechanistic toxicology in assessing climate change risks. Environ. Toxicol. Chem., 32: 32 – 48.

HOWARTH R W, MARINO R, 2006. Nitrogen as the limiting nutrient for eutrophication in coastal marine ecosystems: evolving views over three decades. Limnol. Oceanogr., 51: 364 – 376.

HOWARTH R W, 2008. Coastal nitrogen pollution: a review of sources and trends globally and regionally. Harmful Algae, 8: 14 – 20.

IEA (International Energy Agency), 2012. World energy outlook 2012. Paris, France.

IMO (International Maritime Organisation), 2012. International shipping facts and figures e information resources on trade, safety, security, environment. http://www.imo.org/en/KnowledgeCentre/ShipsAnd ShippingFactsAndFigures/TheRoleandImportanceof InternationalShipping/Documents/International%20 Shipping%20 - %20Facts%20and%20Figures. pdf.

IPCC (International Panel on Climate Change), 2014. Climate change 2014: synthesis report. Contribution of working groups I, II and III to the fifth assessment report of the intergovernmental panel on climate change. Geneva, Switzerland: 151.

JANAKI DEVI V, NAGARANI N, YOKESH BABU M, et al., 2012. Genotoxic effects of profenofos on the marine fish, *Therapon jarbua*. Toxicol. Mech. Methods, 22: 111 – 117.

JONES R, 2005. The ecotoxicological effects of photosystem II herbicides on corals. Mar. Pollut. Bull., 51: 495 – 506.

KEATING B A, CARBERRY P S, BINDRABAN P S, et al., 2010. Eco-efficient agriculture: concepts, challenges, and opportunities. Crop Sci., 50: S109 – S119.

KENNISH M, DE JONGE V, 2011. Chemical introductions to the systems: diffuse and nonpoint source pollution from chemicals (nutrients: eutrophication) // WOLANSKI E, MCLUSKY D (Eds.). Treatise on estuarine and coastal science. Elsevier Science, UK, 8: 113 – 148.

KIMBERLY D A, SALICE C J, 2012. Understanding interactive effects of climate change and toxicants: importance of evolutionary processes. Integr. Environ. Assess. Manag., 8: 385 – 386.

LANDIS W G, DURDA J L, BROOKS M L, et al., 2013. Ecological risk assessment in the context of global climate change. Environ. Toxicol. Chem., 32: 1 – 14.

LAW R J, 2014. An overview of time trends in organic contaminant concentrations in marine mammals: going up or down? Mar. Pollut. Bull., 82: 7 – 10.

LAWRENCE J M, 1996. Mass mortality of echinoderms from abiotic factors // JANGOUX M, LAWRENCE J M (Eds.). Echinoderm studies. Balkema, Rotterdam, Netherlands, 5: 103 – 137.

LI W, GAO K, BEARDALL J, 2013. Interactive effects of ocean acidification and nitrogen limitation on the diatom *Phaeodactylum tricornutum*. PLoS One, 7: e51590.

MACNEALE K H, KIFFNEY P M, SCHOLZ N L, 2010. Pesticides, aquatic food webs, and the conservation of pacific salmon. Front. Ecol. Environ., 8: 475 – 482.

MAGNUSSON M, HEIMANN K, NEGRI A P, 2008. Comparative effects of herbicides on photosynthesis and growth of tropical estuarine microalgae. Mar. Pollut. Bull., 56: 1545 – 1552.

MANNING T, FERRELL D, 2007. Dioxins in fish and other seafood from Sydney Harbour, Australia. Organohalogen Comp., 69: 343 – 346.

MAPSTONE B, 2011. Introduction // CLEUGH H, STAFFORD SMITH M, BATTAGLIA M, et al (Eds.). Climate change. CSIRO Publishing, Collingwood, VIC, Australia: ix – xii.

MARGOLIS S V, BURNS R G, 1976. Pacific deep-sea manganese nodules: their distribution, composition, and origin. Annu. Rev. Earth Planet Sci., 4: 229 - 263.

MARQUES A, NUNES M L, MOORE S K, et al., 2010. Climate change and seafood safety: human health implications. Food Res. Int., 43: 1766 - 1779.

MARUYA K A, SMALLING K L, VETTER W, 2005. Temperature and congener structure affect the enantioselectivity of toxaphene elimination by fish. Environ. Sci. Technol., 39: 3999 - 4004.

MCCOMB A, DAVIS J, 1993. Eutrophic waters of southwestern Australia. Fertil. Res., 36: 105 - 114.

MCCOOK L J, SCHAFFELKE B, APTE S C, et al., 2015. Synthesis of current knowledge of the biophysical impacts of dredging and disposal on the great barrier reef: report of an independent panel of experts. Great Barrier Reef Marine Park Authority, Townsville, Australia.

MCKINNON A, ALLEN J, WOODBURN A, 2015. Development of greener vehicles, aircraft and ships. Chapter 8 // MCKINNON A, BROWNE M, WHITEING A, et al (Eds.). Green logistics: improving the environmental sustainability of logistics. 3rd ed. Kogan Page Ltd, London, UK: 165 - 193.

MIRBAHAI L, CHIPMAN J K, 2014. Epigenetic memory of environmental organisms: a reflection of lifetime stressor exposures. Mutat. Res., 764: 10 - 17.

MOBERG F, 1999. Ecological goods and services of coral reef ecosystems. Ecol. Econ., 29: 215 - 233.

MOE S J, DE SCHAMPHELAERE K, CLEMENTS W H, et al., 2013. Combined and interactive effects of global climate change and toxicants on populations and communities. Environ. Toxicol. Chem., 32: 49 - 61.

MOORE C, 2003. Across the pacific ocean, plastics, plastics, everywhere. Nat. Hist. Mag., New York, NY, USA. http://www.naturalhistorymag.com/htmlsite/master.html? http://www.naturalhistorymag.com/htmlsite/1103/1103_feature.html.

MUMBY P J, HASTINGS A, EDWARDS H J, 2007. Thresholds and the resilience of Caribbean coral reefs. Nature, 450: 98 - 101.

NATH B N, SHARMA R, 2000. Environment and deep-sea mining: a perspective. Mar. Geores. Geotechnol., 18: 285 - 294.

NICHOLLS R J, LEATHERMAN S P, 1995. The implications of accelerated sea-level rise for developing countries: a discussion. J. Coast. Res., 14: 303 - 323.

NICHOLLS R J, WONG P P, BURKETT V R, et al., 2007. Coastal systems and low-lying areas. Contribution of working group II to the fourth assessment report of the intergovernmental panel on climate change // PARRY M L, CANZIANI O F, PALUTIKOF J P, et al (Eds.). Climate change 2007: impacts, adaptation and vulnerability. Cambridge University

Press, Cambridge, UK: 315 – 356.

NORDSTROM K F, 2000. Beaches and dunes of developed coasts. Cambridge University Press, Cambridge, UK.

NOYES P D, MCELWEE M K, MILLER H D, et al., 2009. The toxicology of climate change: environmental contaminants in a warming world. Environ. Int., 35: 971 – 986.

NYBERG E, FAXNELD S, DANIELSSON S, et al., 2015. Temporal and spatial trends of PCBs, DDTs, HCHs, and HCB in Swedish marine biota 1969 – 2012. Ambio, 44: 484 – 497.

OSPAR (Oslo/Paris Convention (for the Protection of the Marine Environment of the North-East Atlantic)), 2004. Environmental impacts to marine species and habitats of dredging for navigational purposes. OSPAR Commission, Publication No. 208/2004, London, UK.

OSPAR, 2008a. Assessment of environmental impact of land reclamation. OSPAR Commission, London, UK.

OSPAR, 2008b. Literature review on the impacts of dredged sediment disposed at sea. OSPAR Commission, Publication No. 362/2008, London, UK.

OZCOASTS, 2010. Economic consequences of acid sulfate soils. OzCoasts Australian Online Coastal Information, Australian Government, Geoscience Australia. http://www.ozcoasts.gov.au/indicators/econ_cons_acid_sulfate_soil.jsp.

PAULSEN I T, HOLMES A J, 2014. Environmental microbiology: methods and protocols. Humana Press, Totowa, NJ, USA.

PELLETIER M C, BURGESS R M, HI K T, et al., 1997. Phototoxicity of individual polycyclic aromatic hydrocarbons and petroleum to marine invertebrate larvae and juveniles. Environ. Toxicol. Chem., 16: 2190 – 2199.

PIMENTEL D, 2009. Pesticides and pest control // PESHIN R, DHAWAN A (Eds.). Integrated pest management: innovation-development process. Springer, Dordrecht, Netherlands: 83 – 87.

PIÑA B, BARATA C, 2011. A genomic and ecotoxicological perspective of DNA array studies in aquatic environmental risk assessment. Aquat. Toxicol., 105: 40 – 49.

PRZESLAWSKI R, AHONG S, BRYNE M, et al., 2008. Beyond corals and fish: the effects of climate change on noncoral benthic invertebrates of tropical reefs. Glob. Change Biol., 14: 2773 – 2795.

REICH K J, WORTHY G A, 2006. An isotopic assessment of the feeding habits of free-ranging manatees. Mar. Ecol. Progr. Ser., 322: 303 – 309.

RIOS L M, JONES P R, MOORE C, et al., 2010. Quantitation of persistent organic pollutants adsorbed on plastic debris from the Northern Pacific Gyre's "eastern garbage patch". J. Environ. Monit., 2010, 12: 2226 – 2236

ROSENBERG R, LINDAHL O, BLANCK H, 1988. Silent spring in the sea. Ambio: 289-290.

SAMMUT J, MELVILLE M D, CALLINAN R B, et al., 1995. Estuarine acidification: impacts on aquatic biota of draining acid sulphate soils. Aust. Geogr. Stud., 33: 89-100.

SANTOS L H, ARAÚJO A N, FACHINI A, et al., 2010. Ecotoxicological aspects related to the presence of pharmaceuticals in the aquatic environment. J. Hazard. Mater., 175: 45-95.

SCHIEDEK D, SUNDELIN B, READMAN J W, et al., 2007. Interactions between climate change and contaminants. Mar. Pollut. Bull., 54: 1845-1856.

SERICANO J L, WADE T L, SWEET S T, et al., 2014. Temporal trends and spatial distribution of DDT in bivalves from the coastal marine environments of the continental United States, 1986-2009. Mar. Pollut. Bull., 81: 303-316.

SHEEHAN D, 2013. Next-generation genome sequencing makes non-model organisms increasingly accessible for proteomic studies: some implications for ecotoxicology. J. Proteom. Bioinform., 6, 10000e21. http://dx.doi.org/10.4172/jpb.10000e21.

SMAYDA T J, 1997. Harmful algal blooms: their ecophysiology and general relevance to phytoplankton blooms in the sea. Limnol. Oceanogr., 42: 1137-1153.

SMITH G, IVES L D, NAGELKERKEN I A, et al., 1996. Caribbean sea fan mortalities. Nature, 383: 487.

SMITH V H, TILMAN G D, NEKOLA J C, 1999. Eutrophication: impacts of excess nutrient inputs on freshwater, marine, and terrestrial ecosystems. Environ. Pollut., 100: 179-196.

SMITH V H, 1998. Cultural eutrophication of inland, estuarine, and coastal waters // PACE M L, GROFFMAN P M (Eds.). Successes, limitations and frontiers in ecosystem science. Springer, New York, NY, USA: 7-49.

SONG M Y, BROWN J J, 1998. Osmotic effects as a factor modifying insecticide toxicity on *Aedes* and *Artemia*. Ecotoxol. Environ. Saf., 41: 195-202.

STAHL R G, HOOPER M L, BALBUS J M, et al., 2013. The influence of global climate change on the scientific foundations and applications of environmental toxicology and chemistry: introduction to a SETAC international workshop. Environ. Toxicol. Chem., 32: 13-19.

State of the climate, 2014. Bureau of Meteorology and Commonwealth Scientific and Industrial Research Organisation, Canberra, Australia.

State of the Environment, 2011. Australia state of the environment 2011. Independent report to the australian government minister for sustainability, environment, water, population and communities. Australian Government, Canberra, Australia.

STAUBER J L, CHARITON A, BINET M, et al., 2008. Water quality screening risk assessment of acid sulfate soil impacts in the lower murray, SA. CSIRO Land and Water Science Report, Sydney, Australia, 45/08: 127.

STAUBER J L, KOOKANA R S, BOXALL A B A, 2012. Contaminants and climate change: multiple stressors in a changing world//Proceedings of Water and Climate: Policy Implementation Challenges National Conference, Canberra, ACT, Australia, 1 - 3 May, 2012: 10.

STIRLING I, LUNN N J, IACOZZA J, 1999. Long term trends in the population ecology of polar bears in Western Hudson Bay in relation to climate change. Arctic, 52: 294 - 306.

VAN STRAALEN N M, FEDER M E, 2012. Ecological and evolutionary functional genomicse-how can it contribute to the risk assessment of chemicals? Environ. Sci. Technol., 46: 3 - 9.

VAN STRAALEN N M, ROELOFS D, 2011. An introduction to ecological genomics. Oxford University Press, Oxford, UK.

TANAKA K, TAKADA H, YAMASHITA R, et al., 2013. Accumulation of plastic - derived chemicals in tissues of seabirds ingesting marine plastics. Mar. Pollut. Bull., 69: 219 - 222.

TEUTEN E L, ROWLAND S J, GALLOWAY T S, et al., 2007. Potential for plastics to transport hydrophobic contaminants. Environ. Sci. Technol., 41: 7759 - 7764.

UNEP (United Nations Environment Programme), 2013. Global chemicals outlook: towards sound management of chemicals. New York, NY, USA.

UNFPA (United Nations Population Fund), 2014. UNFPA state of the world population 2014. United Nations Population Fund. New York, NY, USA.

USGS (United States Geological Survey), 2008. Circum-arctic resource appraisal: estimates of undiscovered oil and gas north of the arctic circle. http://pubs.usgs.gov/fs/2008/3049/fs2008 - 3049.pdf.

VALIELA I, BOWEN J L, YORK J K, 2001. Mangrove forests: one of the world's threatened major tropical environments. Bioscience, 51: 807 - 815.

VALIELA I, 2006. Global coastal change. Wiley-Blackwell, Oxford, UK.

VAN DEN BRINK P, CHOUNG C B, LANDIS W, et al., 2015. New approaches to the ecological risk assessment of multiple stressors. Mar. Freshwater Res., 67: 429 - 439.

VANDEGEHUCHTE M B, JANSSEN C R, 2011. Epigenetics and its implications for ecotoxicology. Ecotoxicology, 20: 607 - 624.

VANDEGEHUCHTE M B, JANSSEN C R, 2014. Epigenetics in an ecotoxicological context. Mutat. Res., 764 - 765: 36 - 45.

VERON J E N, 1995. Corals in space and time: the biogeography and evolution of the scleractinia. Cornell University Press, Ithaca, NY, USA.

VITOUSEK P M, ABER J D, HOWARTH R W, et al., 1997a. Human alterations of the global nitrogen cycle: sources and consequences. Ecol. Appl., 7: 737 – 750.

VITOUSEK P M, ABER J, HOWARTH R W, et al., 1997b. Human alteration of the global nitrogen cycle: causes and consequences. Ecological Society of America, Washington, DC, USA.

VOGT C, 2012. International assessment of marine and riverine disposal of mine tailings. Secretariat, London Convention/London Protocol, International Maritime Organization, London, UK & United Nations Environment Programme-Global Program of Action.

WARWICK R M, 1986. A new method for detecting pollution effects on marine macrobenthic communities. Mar. Biol., 92: 557 – 562.

WEBER M, DE BEER D, LOTT C, et al., 2012. Mechanisms of damage to corals exposed to sedimentation. Proc. Natl. Acad. Sci., U.S.A., 109: 1558 – 1567.

WEF (World Economic Forum), 2014. Global Risk. 9th ed. Geneva, Switzerland.

WILDGUST M A, JONES M B, 1998. Salinity change and the toxicity of the free cadmium ion [$cd^{2+}_{(aq)}$] to *Neomysis integer* (Crustacea: Mysidacea). Aquat. Toxicol., 41 (3): 187 – 192.

World Ocean Review, 2010. Living with the oceans. vol. 1. Maribus, Hamburg, Germany.

World Ocean Review, 2013. Living with the oceans. vol. 2. The future of fish: the fisheries of the future. Maribus, Hamburg, Germany.

World Ocean Review, 2014. Marine resources: opportunities and risks. vol. 3. Maribus, Hamburg, Germany.

WU R S S, 2002. Hypoxia: from molecular responses to ecosystem responses. Mar. Pollut. Bull., 45: 35 – 45.

索　引

A

Abandoning wasteful analysis of variance (ANOVA) procedures　方差分析 2.1
Absorption　吸收 4.1.1
Acetyl cholinesterase (AChE)　胆碱酯酶类 5.2.8
Additive brominatedflame retardants　溴化阻燃剂 1.2.3.2
Adsorption process, definition　吸附过程 1.2.1
Adverse outcome pathway (AOP)　有害结局路径 5.3
δ-Aminolevulinic acid dehydratase (ALA-D) enzyme　δ-氨基乙酰丙酸脱水酶 (δ-ALA-D) 5.2.4

  bivalve species　双壳类 5.2.2.2
  M. barbatus　须鲷 5.2.4
  Methodology　方法 5.2.4
  Prochilodus lineatus　条纹鲮脂鲤 5.2.4
Amphoteric surfactants　表面活性剂 1.2.2.2
Anionic surfactants　阴离子表面活性剂 1.2.2.2
Antioxidant enzyme activity catalase　过氧化氢酶 5.2.2.1
  enzyme activities and methodologies　酶活性反应和定量方法 5.2.2.1
  organic pollutants　有机污染物 5.2.2.1
  oxygen reduction metabolism　氧化还原反应 5.2.2.1
  reactive oxygen species (ROS)　高活性氧（也称为氧自由基）5.2.2.1
  SOD 5.2.2.1
  GPX 5.2.2.1
  trace metals　痕量金属 5.2.2.1
Aquatic biota　水生生物 1.2.2.1
Aristeus antennatus　甲壳动物 5.2.1.1
Assessment factor (AF) method　评价因子方法 2.9
Assimilation　同化 4.1.2
Assimilation efficiency　同化效率 4.1.2

B

Bayesian hierarchical model 贝叶斯分析方法 2.2
Bayesian statistics 贝叶斯统计 2.5
Bias correction factor (bcf) 偏差校正因子 (bias correction factor, bcf) 2.7
 Bioaccumulation 生物富集 4.7
 Application 应用 4.4
 Defined 定义 4
 absorption 吸收 4.1.1
 assimilation 同化作用 4.1.2
 basic illustration 基本图解 4.1.1
 bioconcentration factor (BCF) 生物浓缩系数 4.1.3
 efflux 流入 4.1
 influx 流出 4.1
 kinetic parameters 动力学参数 4.2.2
  assimilation efficiency (AE) 同化效率 4.2.2
  dissolved uptake rate constant ku 吸收速率常数 k_u 4.2.2
  efflux 外排 4.3.3
 modeling 模型 4.4
  equilibrium partitioning model (EqP) 平衡分配模型 4.2.1
  kinetic modeling 动力学模型 4.2.2
Bioaccumulation factor (BAF) 生物富集系数 4.1.3
Bioavailability-based sediment quality guidelines 沉积物质量评价规程 7.2.1.1
Bioavailability 生物可利用率 4
Bioconcentration factor (BCF) 生物浓缩系数 4.1.3
Biomarkers 生物标记物
 δ-aminolevulinic acid dehydratase (ALA-D) enzyme δ-氨基乙酰丙酸脱水酶 5.2.4
  bivalve species 甲壳动物 5.2.3
  Methodology 方法 5.2.4
 biotransformation enzymes 生物转化酶 5.2.1
  phase I enzymes 阶段 I 酶 5.2.1.1
  phase II enzymes 阶段 II 酶 5.2.1.2
  xenobiotics 外源化学物 5.2.1
 description 种类 5.2.1
 environmental risk assessment (ERA) process 环境风险评价 (ERA) 过程 5.1
 future possibilities and probabilities 展望 5.4
 genotoxic parameters 遗传毒性参数 5.2.6
  DNA adducts DNA 加合物 5.2.6

　　　　irreversible genotoxic events　不可逆作用的染色体突变 5.2.6

　　　　mollusks　软体动物 5.2.6

　　　　secondary DNA modifications　二级 DNA 修饰 5.2.6.2

　　　　high-throughput screening techniques　高通量筛选技术 5.3

　　contaminant，in toxicology　毒理学 5.3

　　　　marine genomic resources and examples　海洋基因组资源和实例 5.3.4

　　　　metabolomics　代谢物组学 5.3.3

　　　　omic profiling　组学分析 5.3

　　　　polymerase chain reaction（PCR）　聚合酶链反应（PCR）的转录分析 5.2.7

　　　　proteomics　蛋白质组学 5.3.2

　　　　real-time polymerase chain reaction（RT-PCR）　逆转录聚合酶链反应（RT-PCR）5.3

　　　　serial analysis of gene expression（SAGE）　基因表达系列分析（SAGE）5.3

　　　　transcriptomics　转录组学 5.3.1

　　immunological parameters　免疫学参数 5.2.5

　　　　cellular and humoral levels　细胞和体液免疫水平 5.2.5

　　　　cyclooxygenase（COX）activity　环氧合酶（COX）活性 5.2.5

　　　　lysozymes　溶菌酶 5.2.5

　　　　neutral red retention time（NRTT）　中性红保留时间（NRTT）5.2.5

　　　　nitric oxide　一氧化氮 5.2.5

　　　　phagocytosis　吞噬作用 5.2.5

　　metallothioneins（MTs）　金属硫蛋白（MTs）5.2.3

　　　　characteristics　特性 5.2.3

　　　　classes　种类 5.2.3

　　　　Crassostrea virginica　美洲牡蛎 5.2.3

　　　　differential pulse polarography（DPP）　微分脉冲极谱（DPP）5.2.3

　　　　mollusks/crustaceans　软体动物或甲壳类动物 5.2.3

　　　　Ruditapes decussatus　蛤蜊 5.2.3

　　neurotoxic parameters　神经毒性参数 5.2.8

　　oxidative stress parameters　氧化应激参数 5.2.2

　　　　antioxidant enzyme activity　抗氧化酶活性 5.2.2.1

　　　　lipid peroxidation　脂质过氧化反应 5.2.2.2

　　　　reproductive and endocrine parameters　生殖和内分泌参数 5.2.7

　　　　Endocrine Disruption Chemicals（EDCs）　内分泌干扰物 5.2.7

　　　　Gonadosomatic index（GSI）　性腺指数 5.2.7

　　　　hormone regulation　激素调节 5.2.7

　　　　liver　肝脏 5.2.7

　　　　vitellogenin　卵黄蛋白原 5.2.7

　　　　Vtg concentration　卵黄蛋白原（Vtg）水平 5.2.7

　　　　　Xenobiotics　外源化学物 5.2.7
Biomonitoring　生物监测 4.5
Choice of　选择 4.5.1
　　　diffusion in gels（DGT）　凝胶扩散 4.5
　　　principles　原理 4.6
　　　semipermeable membrane　半透膜 4.5
biotransformation enzymes　生物转化酶 5.2.1
　　　phase Ⅰ enzymes　阶段Ⅰ酶 5.2.1.1
　　　　　Aristeus antennatus　甲壳动物 5.2.1.1
　　　　　CYP　单加氧酶 5.2.1.1
　　　　　CYP1A1　5.2.1.1
　　　　　dibenzylfluorescein dealkylase（DBF）activity　二苯基荧光素活性 5.2.1.1
　　　　　ethoxyresorufin-O-deethylase（EROD）　乙氧基异吩噁唑-O-去乙基酶 5.2.1.1
　　　　　MFOs　混合功能氧化酶 5.2.1.1
　　　　　Mytilus edulis　紫壳菜蛤 5.2.1.1
　　　　　Ruditapes philippinarum　菲律宾蛤仔 5.2.1.1
　　　phase Ⅱ enzymes　阶段Ⅱ酶 5.2.1.2
　　　　　1-chloro-2,4-dinitrobenzene（CDNB）　1-氯-2,4-二硝基苯 5.2.1.2
　　　　　Conjugations　结合作用 5.2.1.2
　　　　　glutathione S-transferase（GST）　谷胱甘肽 S-转移酶 5.2.1.2
　　　xenobiotics　外源化学物 5.2.1
"Black box" animal-based paradigm　以动物为基础的"黑盒"模式 7.4.3
Brominated flame retardants　溴化阻燃剂 1.2.3.2
BurrliOZ　2.8.6
Butyrylcholinesterase（BChE）　丁酰胆碱酯酶 5.2.8

C

Catalase　过氧化氢酶 5.2.2.1
Catchment-land use changes　流域土地利用变化 10.2
　　　acid sulfate soils　酸性硫酸盐土 10.3.8
　　　beach restoration　海滩修复 10.3.7
　　　climate change　气候变化 10.4
　　　coastal industries　沿海工业 10.2.3
　　　coastal waters demands　对沿海水域的要求 10.3.1
　　　dredging　疏浚 10.3.5
　　　emerging contaminants　新兴污染物 10.2.4
　　　fishing and aquaculture　渔业和水产养殖 10.3.9
　　　global water cycles and salinity　全球水循环和盐度 10.4.1.2

hypoxia 低氧 10.4.1.3
land reclamation 填海 10.3.6
methane hydrates 甲烷水合物 10.3.10.2
mine waste disposal 矿山废弃物处理 10.3.10.3
multiple stressors 多重压力源 10.4.2
  bioaccumulated contaminants 生物富集 10.4.2
  climate change stressors and contaminants 气候变化和污染物 10.4.2
  contaminant toxicity 污染物毒性 10.4.2.2
  environmental risk assessment 环境风险评价 10.4.3
  environmental variables 环境变量 10.4.1
  persistent organic pollutants（POPs） 持久性有机污染物 10.2.1
  physiological tolerance 耐性 10.4.2
  species-specific responses 物种特异性 10.4.2
  temperature-dependent chemical fate and bioenergetics models 温度依赖性的化学归宿和生物能模型 10.4.2.1
  toxic effects 毒性 10.4.2
nutrients and eutrophication 营养盐和富营养化 10.2.2
  agricultural intensity 农药 10.2.3
  classification 定义 10.2.2
  coral-eating Crown of Thorns starfish 棘冠海星 10.2.2
  eutrophication 富营养化 10.2.2
  Great Barrier Reef（GBR） 大堡礁 10.2.2
  Peel-Harvey Estuary Peel-Harvey 河口 10.2.2
  persistent contaminants 持久性有机污染物 10.2.1
  Phosphorus 磷 10.2.2
  phytoplankton communities 浮游植物 10.2.2
  synthetic nitrogen fertilizers 合成氮肥料 10.2.2
ocean acidification 海洋酸化 10.4
oil and gas exploration 油气勘探 10.3.3
pesticides 农药 1.2.2.4
  nontarget marine 非目标的海洋 10.2.3.1
  photosynthesizing species 光合作用物种 10.2.3.1
  organochlorines 有机氯农药 10.2.3.1
  organophosphates 有机磷农药 10.2.3
  synthetic pesticides 拟除虫菊酯 10.2.3
  total DDT concentrations DDT 总浓度 10.2.3.1
population growth 人口增长 10.2.1
ports and industry-associated changes 港口和工业相关变化 10.3
Seabed mining 海底采矿 10.3.10.1

  sea-level rise and storm surges 海平面上升和风暴潮 10.4.1.4
  shipping and ports 航运与港口 10.3.4
  temperature 温度 10.4.1.1
Cationic surfactants 阳离子表面活性剂 1.2.2.2
Chemicals of emerging concern（CEC） 新兴化学物质 6.3.10.4
Chemical uptake and effects assumptions roles 化学物质吸收和效应的动态模型 3.1
  building blocks 基础构件 3.2.5
  Calanus finmarchicus 飞马哲水蚤 3.5.3
  closer collaboration 密切的合作 3.8.1
  compartment approach 房室 3.2.5
  complex models 复杂的模型 3.2.3
  confronting models 模型验证 3.2.7
  connecting models 互通的房室 3.2.5
  differential equations 微分方程 3.2.6
  dimension analysis 量纲分析 3.2.6
  embryonic development 胚胎发育 3.8.2
  environmental risk assessment 环境风险评价 10.4.3
  individual-based models 基于个体的模型 3.7.1
  individual life-history traits 个体的生活历史特征 3.1
  "individual tolerance" hypothesis "个体耐受"假说 3.5.1
  intrinsic rate of increase 内禀增长率 3.7.3
  life-history traits 生活史特征 3.7
  marine invertebrates 海洋无脊椎动物 3.1
  matrix models 矩阵模型 3.7.2
  mechanistic vs. descriptive models 机理模型 vs 描述性模型 3.2.4
  model assumptions 模型假设 3.2.6
  model design 模型设计 3.2.1
  nonlinear effect 非线性效应 3.1
  parameter estimation 参数估计 3.2.7
  probability distribution 概率分布 3.5.1
  risk assessment frameworks 风险评价框架 3.8.2
  stochastic death model 随机死亡模型 3.5.2
  sublethal endpoints 亚致死终点 3.6.1
    Capitella teleta 海洋多毛纲海蠕虫 3.6.5
    "DEBtox" model DEBtox 模型 3.6.5
    dynamic energy budget（DEB）theory 动态能量预算理论 2.8
    elimination rate 清除率 3.6.5
    energy-budget model 能量预算模型 3.6.1
    TK linksTK 模型连接 2.6.3

  growth and reproduction data　生长和生殖 3.6.5
  hormesis　毒物兴奋效应 3.6.5
  parameter estimation　参数估计 3.2.7
  physiological mode of action　生理作用模式 3.6.5
  reproduction rate　繁殖率 3.6.5
  TD models TD　模型 3.4.1
  TKTD energy-budget model　TKTD 模型 3.4
 systems and states　系统和状态 3.2.1
 TKTD model parameters　TKTD 模型参数 3.8.1
 toxic effects　毒性 10.4.2
 toxicodynamics（TD）　毒物效应动力学 3.1
 toxicokinetics（TK）　毒物代谢动力学 3.1
1 – Chloro – 2,4 – dinitrobenzene（CDNB）　1 – 氯 – 2,4 – 二硝基苯 5.2.1.2
Cholinesterases（ChE）　胆碱酯酶类 5.2.8
Confounding factors　干扰因素 7.1.4
 abiotic factors　非生物因素 7.1.4
 cultured/field-collected animals　养殖或野外采集动物 7.1.4
 dissolved ammonia and sulfide　氨和硫化物 7.1.4
 long-term exposures　长期暴露 7.1.4
 photoinduced toxicity　光制毒性 7.1.4
 physicochemical conditions　物理化学条件 7.1.4
 tolerance limits　耐受限度 7.1.4
Confronting models　模型验证 3.2.7
Controlled ecosystems populations experiment（CEPEX）　生态系统种群受控实验室 8.3.2
CYP1A1　5.2.1.1
Cytochrome P450 – dependent mixedfunction oxidase（MFO）system（MFOs）　混合功能氧化酶系统 5.2.1.1
Cytochrome P450 monooxygenase enzymes（CYP）　单加氧酶 5.2.1.1

D

"DEBtox" model DEBtox　模型 3.6.5
Dibenzylfluorescein dealkylase（DBF）activity　二苯基荧光素活性 5.2.1.1
Dioxin-like（dl）– PCBs　类二噁英 – PCBs 1.2.2.3
DNA adducts DNA　加合物 5.2.6.1
Dynamic energy budget（DEB）　动态能量预算 2.8
Dynamic energy budget（DEB）theory　动态能量预算理论 2.8

E

Ecological risk assessment (ERA) 生态风险评价 10.4.3
    adverse effects 有害效应 9.2.1
    conceptual model 概念模型 9.2.2
    ecosystem, structure and function 生态系统结构和功能 9.2.1
    effects assessment 效应评价 9.2.2
    environmental stressors 环境应激源 9.5.1
    nonscientific components 应激源 9.2.1
    proactive benchmarks 保护性基准 9.2.1
    proactively predict risk 主动性风险预测 9.2.1
    problem formulation 问题阐述 9.2.1
    rank 排序 9.2.1
    retroactively assess risk 追溯性风险评价 9.2.1
    risk characterization 风险特征 9.2.1
    risk management 风险管理 9.2.1
    risk prioritization 风险排序 9.2.1
    scientific components 科学组成 9.2.1
    uncertainties 不确定性 9.2.1

Emerging organic contaminants 新兴有机污染物 1.2.3
    analytical methods 分析方法 1.2.3
        brominated flame retardants (BFRs) 溴化阻燃剂 1.3.2
        endocrine-disrupting compounds (EDCs) 内分泌干扰物 1.3.3.3
        marine biotoxins 海洋生物毒素 1.2.3.5
        nanomaterials (NMs) 纳米材料 1.2.3.6
        PCPs 1.4.3.4
        per-and polyfluorinated alkyl substances (PFASs) 全氟和多氟烷基物质 1.2.3.1
        pharmaceuticals 药品 1.4.3.4
        polydimethylsiloxanes (PDMS) 聚二甲基硅氧烷 1.2.3.7
    occurrence 赋存
        brominated flame retardants (BFRs) 添加型溴化阻燃剂 1.2.3.2
        endocrine-disrupting compounds (EDCs) 内分泌干扰物 1.3.3.3
        marine biotoxins 海洋生物毒素 1.2.3.5
        nanomaterials (NMs) 纳米材料 1.2.3.6
        per-and polyfluorinated alkyl substances (PFASs) 全氟和多氟烷基物质 1.2.3.1
        personal care products 个人护理产品 1.2.3.7

pharmaceuticals 药品 1.4.3.4
siloxanes 硅氧烷类 1.4.3.7
sources and properties 来源和属性
brominated flame retardants (BFRs) 溴化阻燃剂 1.3.2
endocrine-disrupting compounds (EDCs) 内分泌干扰物 1.3.3.3
marine biotoxins 海洋生物毒素 1.2.3.5
nanomaterials (NMs) 纳米材料 1.2.3.6
per-and polyfluorinated alkyl substances (PFASs) 全氟和多氟烷基物质 1.2.3.1
nal care products 个人护理品 1.2.3.4
euticals, classification 药品，分类 1.2.3.4
polydimethylsiloxanes (PDMS) 聚二甲基硅氧烷 1.2.3.7
sewage effluents 污水 1.2.3.4
Endocrine-disrupting compounds (EDCs) 内分泌干扰物 1.2.3.3
Environmental risk assessment (ERA) 环境风险评价 3.1
Environmental Sensitivity Index 环境敏感度指数 8.2
Equilibrium partitioning model (EqP) 平衡分配模型 4.2.1
Ethoxyresorufin-O-deethylase (EROD) 乙氧基异吩噁唑-O-去乙基酶 5.2.1.1
Evidence assessments 证据权重评价 9
ecological risk assessment (ERA) 生态风险评价 10.4.3
five major environmental stressors 五大环境应激源 9.5.1
Good Ecosystem Status (GES) 良好的生态系统状况 9.5.2
informed decision-making 知情决策 9.1
weight of evidence (WoE) 证据权重 8.2
Exploratory data analysis (EDA) 探索性数据分析 2.4

F

Flame-retarded polymers 阻燃聚合物 1.2.3.2

G

Global change anthropogenic pressures 全球变化 人为压力 10.1
Catchment-land use changes 流域土地利用变化 10.2
IPCC reports IPCC 报告 10.6
multiple environmental stressors 多种环境压力源 10.1
tools 工具 10.5
ecological modeling 生态建模 10.5.3
environmental genomics 海洋生态毒理学中的环境基因组学 10.5.2

　　　　　epigenetics　表观遗传学 10.5.1
　　　　　modeling and monitoring approach　模型和监测方法 10.5
Glutathione S-transferase（GST）　谷胱甘肽 S – 转移酶 5.2.1.2
Good Ecosystem Status（GES）　良好的生态系统状况 9.5.2

H

Halogenated organic flame retardants　卤化有机阻燃剂 1.2.3.2
Heavy metals　重金属 1.2.1
High-throughput screening techniques　高通量筛选技术 5.3
　　contaminant，in toxicology　毒理学 5.3
　　marine genomic resources and examples　海洋基因组资源和实例 5.3.4
　　metabolomics　代谢物组学 5.3.3
　　omic profiling　组学分析 5.3
　　polymerase chain reaction（PCR）　聚合酶链反应 5.2.7
　　proteomics　蛋白质组学 5.3.2
　　real-time polymerase chain reaction（RT-PCR）　逆转录聚合酶链反应 5.3
　　serial analysis of gene expression（SAGE）　基因表达系列分析 5.3
　　transcriptomics　转录组学 5.3.1

I

"Individual tolerance" hypothesis　"个体耐受"假说 3.5.1
Irreversible genotoxic events　不可逆的遗传毒性事件 5.2.6.3

K

Kinetic modeling　动力学模型 4.2.2

L

Lines of evidence（LoE）　独立证据 9.2.2
Lipid peroxidation（LPO）　脂质过氧化反应 5.2.2.2
　　anthropogenic contaminants　源于人类活动的污染物 5.2.2.2
　　malonyldialdehyde（MDA）　丙二醛 5.2.2.2
　　polyunsaturated fatty acids（PUFAs）　多不饱和脂肪酸 5.2.2.2
　　ROS formation　ROS 的形成 5.2.2.2
　　2-thiobarbituric acid（TBA）　β – 硫代巴比妥酸 5.2.2.2

M

Malonyldialdehyde (MDA)　丙二醛 5.2.2.2
Mesocosm and field toxicity testing　围隔与实地毒性试验 8
 advantages　优势 8.1.1
 applications　应用 8.1.1
 artificial stream approach　人工河流方法 8.3.3
 controlled ecosystems populations experiment (CEPEX)　受控生态系统种群实验 8.3.2
 cumulative effects　累积效应 8.6
 definition　定义 8.1.1
 Environmental Sensitivity Index　环境敏感度指数 8.2
 eutrophic conditions downstream　富营养化条件下游 8.2
 field biomonitoring　实地生物监测 8.1.2
 land-based mesocosms　陆基围隔 8.3.1
 large fish　大型鱼类 8.4
  benefits　好处 8.4
  Captive brook trout　捕获的溪红点鲑 8.4
  coldwater salmonids　冷水鲑鱼种 8.4
  dissolved oxygen levels　溶解氧水平 8.4
  filtration unit　过滤装置 8.4
  flow-through/recirculation systems　直流式或再循环系统 8.4
  guidelines　指导方针 8.4
  Osceola State Fish Hatcheries　奥西奥拉州鱼类孵化场 8.4
  physical size　物理尺寸 8.4
  Steelhead trout　虹鳟 8.4
  tank design　水箱的设计 8.4
  water chemistry parameters　水化学参数 8.4
 limnocorrals　湖沼 8.3.2
 littoral and pelagic community　海滨和远洋围隔生物群落 8.4
 littoral enclosures　滨海围隔 8.3.2
 multiple stressor/cumulative effects　多种应激/累积效应 8.2
 National Oceanic and Atmospheric Administration (NOAA)　美国国家海洋和大气管理局 8.2
 ocean acidification　海洋酸化 8.3.2
 ocean temperatures and acidification　海洋温度和海水酸化 8.6
 open systems　开放式系统 8.5
 process/cause-effect relationship　过程或因果关系 8.1.1
 risk assessments　风险评估 8.1.2

single and multiple pesticides　单种和多种农药 8.2
　　single species bioassay tests　单物种生物测定试验 8.2
Metabolomics　代谢物组学 5.3.3
Metallothioneins（MTs）　金属硫蛋白 5.2.3
　　characteristics　特性 5.2.3
　　classes　类 5.2.3
　　Crassostrea virginica　美洲牡蛎 5.2.3
　　differential pulse polarography（DPP）　微分脉冲极谱 5.2.3
　　mollusks/crustaceans　软体动物或甲壳类动物 5.2.3
　　Ruditapes decussatus　蛤蜊 5.2.3
Metals　金属 1.3.1
　　analytical methods　分析方法 1.3.1
　　occurrence　赋存 1.4.1
　　sources and properties　来源和属性 1.2.1
Microplastics（MP）　微塑料 1.5
Mode of Action（MoA）　作用方式 5.3
Mytilus edulis　紫壳菜蛤 5.2.1.1

N

National Oceanic and Atmospheric Administration（NOAA）　美国国家海洋和大气管理局 8.2
Nonionic surfactants　非离子表面活性剂 1.2.2.2
Nontarget analysis　非靶标分析 1.6
No observed effect concentrations（NOECs）　无观察效应浓度 2.1

O

Octanolewater partition coefficient　辛醇－水分配系数 1.2.2.2
Oxidative stress parameters　氧化应激参数 5.2.2
　　antioxidant enzyme activity　抗氧化酶活性 5.2.2.1
　　lipid peroxidation　脂质过氧化反应 5.2.2.2

P

Perfluoroalkanes　全氟烷烃 1.2.3.1
Persistent organic contaminants　持久性有机污染物 1.3.2
　　analytical methods　分析方法 1.3
　　　　dioxins　二噁英 1.3.2.5

pesticides 农药 1.3.2.4
polychlorinated biphenyl (PCBs) 多氯联苯 1.3.2.3
polycyclic aromatic hydrocarbons (PAHs) 多环芳烃 1.3.2.1
surfactants 表面活性剂 1.3.2.2
occurrence 赋存 1.4
dioxins 二噁英 1.4.2.5
pesticides 农药 1.4.2.4
polychlorinated biphenyl (PCBs) 多氯联苯 1.4.2.3
polycyclic aromatic hydrocarbons (PAHs) 多环芳烃 1.4.2.1
surfactants 表面活性剂 1.4.2.2
sources and properties 来源和属性 1.2
dioxins 二噁英 1.2.2.5
pesticides 农药 1.2.2.4
polychlorinated biphenyl (PCBs) 多氯联苯 1.2.2.3
polycyclic aromatic hydrocarbons (PAHs) 多环芳烃 1.2.2.1
surfactants 表面活性剂 1.2.2.2
persistent organic pollutants (POPs) 持久性有机污染物 10.2.1
Pharmaceutically active compounds 药物的活性化合物 1.2.3.4
Phase I enzymes, biotransformation 阶段 I 酶, 生物转化 5.2.1.1
   Aristeus antennatus 甲壳动物 5.2.1.1
   CYP450 细胞色素 P450 5.2.1.1
   CYP1A1 5.2.1.1
   dibenzylfluorescein dealkylase (DBF) activity 二苯基荧光素活性 5.2.1.1
   ethoxyresorufin-O-deethylase (EROD) 乙氧基异吩噁唑－O-去乙基酶 5.2.1.1
   MFOs 5.2.1.1
   Mytilus edulis 紫壳菜蛤 5.2.1.1
   Ruditapes philippinarum 菲律宾蛤仔 5.2.1.1
Phase Ⅱ enzymes, biotransformation 阶段 Ⅱ 酶
   1－chloro-2,4－dinitrobenzene (CDNB) 1－氯－2,4－二硝基苯 5.2.1.2
   conjugations 结合作用 5.2.1.2
   glutathione S-transferase (GST) 谷胱甘肽 S－转移酶 5.2.1.2
Polychlorinated dibenzofurans (PCDFs) 多氯二苯并呋喃 1.2.2.5
Polychlorinated dibenzo-p-dioxins (PCDDs) 多氯代联二苯－对－二噁英 1.2.2.5
Polymerase chain reaction (PCR) 聚合酶链反应 5.3
Polymeric brominated flame retardants 溴化阻燃剂 1.2.3.2
Porphobilinogen synthase (PBGS) 胆色原素合成酶 5.2.4
Potentially affected fraction (PAF) 潜在受影响的比例 2.7
Proteomics 蛋白组学 5.3

R

Reactive oxygen intermediates (ROIs)　活性氧中间体 5.2.2.2
Real-time polymerase chain reaction (RT-PCR)　逆转录聚合酶链反应 5.3
Receptors of potential concern (ROPCs)　潜在关注受体 9.3.2
Ruditapes philippinarum　菲律宾蛤仔 5.2.1.1

S

Saltwater toxicity tests　海水毒性试验 6
 ambient monitoring　环境监测 6.3.10
 ambient toxicity　环境毒性 6.3.10
 anthropogenic contaminants　人源污染物 6.3.10.4
 biotic and abiotic factors　生物和非生物因素 6.3.10
 chemicals of emerging concern (CEC)　新兴化学物质 6.3.10.4
 Cnidarians　刺胞动物 6.3.8
 Copepods　桡足类 6.3.5
 Decapods　十足类 6.3.6
 Echinoderms　棘皮动物 6.3.3
 Fish　鱼类 6.3.1
  atherinid species　6.3.1
  embryo-larval development test　胚胎幼体发育试验 6.3.1
  flow-through conditions　通流条件 6.3.1
  life-cycle test　生命周期试验 6.3.1
  pacific herring embryos　太平洋鲱鱼的胚胎 6.3.1
  polycyclic aromatic hydrocarbons (PAHs)　多环芳烃 6.3.2
  standardized acute and chronic protocols　标准化的急性和慢性毒性试验规程 6.3.1
  Toxicity Identification Evaluation (TIE) procedures　毒性鉴定评估规程 6.3.1
 Guidelines　指南 6.3.1
 Gulf Coast species　墨西哥湾沿岸物种 6.3.10
 legislative requirements　立法要求 6.2
 life stages　生命阶段 6.2
 marine algal toxicity tests　海洋藻类毒性试验 6.3.9
 mollusks　软体动物 6.3.2
  brood stock　亲鱼 6.3.2
  embryo-larval development tests　胚胎幼体发育试验 6.3.2
  metamorphosed juvenile red abalone　新变态稚态红鲍螺 6.3.2
  mussel veliger larvae　地中海贻贝幼体 6.3.2

  nonrenewal tests　零换水的试验 6.3.2

  red abalone　红鲍螺 6.3.2

 mysids　糠虾 6.3.4

 non-contaminant factors　非污染因素 6.3.10

 ocean acidification　海洋酸化 6.3.10.4

 protocol sensitivities　试验方案的敏感性 6.3.10

 rotifers　轮虫 6.3.7

 Society of Environmental Toxicology and Chemistry（SETAC）　环境毒理学和化学学会 6.3.10

  standardized test protocols　标准测试方案 6.3.10

  state and federal guidance　美国联邦和各州的指南 6.3.10

  stormwater monitoring　雨水监测 6.3.10.2

  test species　试验物种 6.3.1

  Toxicity Identification Evaluation and Toxicity Reduction Evaluation（TIE/TRE）procedures　毒性鉴定评价和毒性消减评价规程 6.3.10

  transitional coastal environments　过渡海岸环境 6.1

  transitional environments 6.3.10.3　过渡环境

  tricyclic PAHs　三环 PAHs 6.3.10

Secondary DNA modifications　二级 DNA 修饰 5.2.6.2

Sediment toxicity testing　沉积物毒性试验 7

 adequate planning and good design　充分规划和良好设计 7.1.1

 adverse outcome pathways　有害结局途径 7.4.4.2

 applications　应用 7.1

 benthic species　底栖生物 7.1

 Bioavailability-Based Sediment Quality Guidelines　基于生物利用度的沉积物质量指南 7.4.1

 biomarkers/genomic endpoints　生物标记物或基因组终点 7.1

 "black box" animal-based paradigm　以动物为基础的"黑盒"模式 7.4.3

 Characterization　特性 7.2.1.3

 chemical analyses　化学分析 7.1

 chronic effects　长期影响 7.4.3

 collection and storage　收集与储存 7.2.1.1

 confounding factors　混合因素 7.1.1.5

 control and reference sediments　控制和参考沉积物 7.2.1.2

 ecological-ecotoxicology effects　生态毒理学效应 7.4.2

 ecological factors　生态因素 7.4.3

 ecological risk assessment（ERA）　生态风险评价 7.4

 exposure conditions　暴露条件 7.1.1

 monitoring　监测 7.2.4

exposure routes and concentrations 暴露途径和浓度 7.1.1.3
identification evaluation 鉴定评价 7.3
lab-on-a-chip technology 实验室芯片技术 7.4.4.1
LC50，EC/IC10 thresholds LC50，EC/IC10 阈值 7.2.5.3
mechanistic effect models 机理效应模型 7.4.3.1
noncontaminant stressors 非污染应激源 7.1
organism responses, robust evaluation 生物反应，可靠评价 7.2
response endpoints 反应终点 7.1.1.2
result analysis and reports 结果的分析和汇报 7.1.1.6
statistical analyses and repeatability 统计分析及重复性 7.1.1.4
test acceptability criteriashiyan 试验可接受标准 7.2.5.1

Sediment toxicity testing 沉积物毒性试验 7
    test endpoints 试验终点 7.1.3.1
        adverse outcome pathway (AOP) models 有害结局路径（AOP）模型 7.1.3
        amphipod and copepod species 端足类和桡足类生物 7.1.3
        behavioral test endpoints 行为试验终点 7.1.3.1
        chronic/subchronic effects 慢性或亚慢性效应 7.1.3
        chronic toxicity test 慢性毒性试验 7.1.3
        decision-making 决策 7.1.3
        genotoxic/mutagenic responses 遗传毒性/致突反应 7.1.3.2
        molecular-based biomarkers 基于分子的生物标志物 7.1.3.2
        suborganism-level endpoints 亚生物体水平终点 7.1.3
        time-dependent endpoints 时间依赖性终点 7.1.3.3
    test exposure conditions and setup 试验暴露条件和设置 7.2.3
    test organisms selection 试验生物选择 7.2.2
        bacteria and algae 细菌和藻类 7.1.2.2
        contaminant exposure pathways and sensitivity 污染物暴露途径和敏感性 7.1.2.1
        crustaceans 甲壳类动物 7.1.2.2
        estuarine and marine sediments 河口和海洋沉积物 7.1.2.2
        estuarine and marine whole-sediment toxicity tests 河口和海洋全沉积物毒性试验 7.1.2.2
        larval and juvenile stages 幼虫期和幼年期 7.1.2.2
        organic contaminants 有机污染物 7.1.2.2
        contaminated sediments 污染沉积物 7.1.2
        whole-sediment toxicity test species 全沉积物毒性试验物种 7.1.2.2
    toxicity determination 毒性测定 7.2.5.2
Serial analysis of gene expression (SAGE) 基因表达系列分析 5.3
Society of Environmental Toxicology and Chemistry (SETAC) 环境毒理学和化学学会 6.3.10

Species sensitivity distribution (SSD) modeling 物种敏感性分布法模型 2.7
    Bayesian hierarchical model 贝叶斯层次模型 2.7
    bias correction factor (bcf) 偏差校正因子 2.7
    degree of freedom 自由度 2.7
    "extrapolation" technique 外推法 2.7
    missing dimension 缺失维度 2.7
    potentially affected fraction (PAF) 潜在受影响的比例 2.7
    probability density function 概率密度函数 2.7
    random sampling 随机抽样 2.7
    stomatal response 气孔反应 2.7
    toxicodynamic (TD) model "毒物动力学" TD 模型 2.7
Statistical design and analysis 统计设计与分析 2
    assessment factor (AF) method 评估因子方法 2.9
    BurrliOZ 2.8.5
    Concentration-response modelingadaptive approach 浓度-响应模型自适应方法 2.6.1
        ANOVA techniques ANOVA 方法 2.6
        bioassay modeling 生物测定建模 2.6
        cucumber shoot weight 黄瓜茎重 2.6
        drc packagedrc 包 2.6
        exponential-threshold model 指数阈值模型 2.6
        hormesis and hysteresis 毒物兴奋和滞后 2.6
        idealized C-R curves 理想状态下的 C-R 曲线 2.6
        logistic function 逻辑斯蒂函数 2.6
        odds ratio 步比 2.6
        one-way ANOVA methods 单向方差分析方法 2.6
        parameters 参数 2.6
        posterior densities 后验密度 2.6
        probit analysis 概率分析 2.6
        random variable 随机变量 2.6
        response-generating mechanism 反应-生成机制 2.6
        S-shape curve S 形曲线 2.6
        subjective assessment 主观评价 2.6
        total error, stochastic error and lack-of-fit 总误差、随机误差和失拟误差 2.6
        toxicokinetic-toxicodynamic (TKTD) models 毒代-毒效动力学 2.6
    concentration-response (C-R) phenomenon 浓度-响应现象 2.1
    data processing and handling big data 数据处理 2.4
        data manipulation, R package tidyr 数据操作：tidyr 包 2.4.1
        data visualization, R package ggplot2 数据可视化 ggplot2 包 2.4.2
        data wrangling 数据整理 2.4

    exploratory data analysis (EDA)　探索性数据分析 2.4
    guideline documents　指南 2.4
    unstructured approach　非结构化的方法 2.4
    US EPA ECOTOX database　美国环保局生态数据库 2.4
dynamic energy budget (DEB)　动态能量预算 2.8
ecotoxicological data　生态毒理学数据 2.3
environmental data　环境数据 2.3
estimation and inference　估计和推断 2.5
    Bayesian statistics　贝叶斯统计 2.5
    D-optimality　D-最优性 2.5
    ecotoxicological setting　生态毒理学 2.5
    environmental planning and assessment process　环境规划评估过程 2.5
    estimation and hypothesis testing　参数估计和假设检验 2.5
    fractional factorial design　分式析因设计 2.5
    frequentist statistics　频率统计 2.5
    general linear model　广义线性模型 2.5
    linear regression　线性回归方程 2.5
    nonlinear model　非线性模型 2.5
    off-diagonal entries　非对角项 2.5
    posterior densities　后验概率 2.5
    "protective" concentration　"保护"浓度 2.5
    random variables　随机变量 2.5
    simple regression　简单回归 2.5
    statistical resampling techniques　统计重采样技术 2.5
    subjective probability　主观概率 2.5
    variance-covariance matrix　方差-协方差矩阵 2.5
factors　因素 2.3.1
fractional factorial design　分式析因设计 2.3.1
information and communications technology　信息和通信技术 2.8
judgmental sampling　判断抽样 2.3
laboratory-based toxicity tests　实验室毒性试验 2.3
MOSAIC 2.8.4
natural resource management agencies　自然资源管理机构 2.9
nonconformity　非一致性 2.3
no observed effect concentrations (NOECs)　无观察效应浓度 2.1
orthogonal fractional factorial designs　正交分式析因设计 2.3
package fitdistrplus fitdistrplus　包 2.8.3
probability sampling　概率抽样 2.3
publication-quality graphics　出版质量图形 2.9

purposive sampling 有目的的抽样 2.3
randomness assumption 随机性假设 2.3
R function fitdistr（） 2.8.3
R package drc 2.8.3
R package ggplot2 2.8.2
R package webchem 2.8.1
species sensitivity distribution（SSD）modeling 物种敏感性分布法模型 2.7
time of day（TOD） 一天中的时间 2.3.1
ToxCalc program 2.8
toxicity measures 毒性检测 2.2
    acute effects 急性效应 2.2
    arbitrary scaling and pooling 任意地缩放和组合 2.2
    Bayesian methods 贝叶斯分析方法 2.2
    chronic effects 慢性效应 2.2
    dose-response experiments 剂量-响应实验 2.2
    ecosystem protection 生态系统保护 2.2
    guidelines 指南 2.2
    quantification 量化 2.2
    random variable 随机变量 2.2
    statistical properties 统计特性 2.2
    toxicity, definition 毒性定义 2.2
    waste stream 废水 2.3.1
Stressors of potential concern（SOPCs） 潜在关注应激源 9.3.2

T

2-Thiobarbituric acid（TBA） β-硫代巴比妥酸 5.2.2.2
Toxicity Identification Evaluation and Toxicity Reduction Evaluation（TIE/TRE）procedures 毒性鉴定评价和毒性消减评价规程 6.3.10
Toxicodynamic（TD）model 毒物动力学模型 2.7
    body-residue data 机体残留数据 3.4.2
    damage compartment/receptor kinetics 损坏房室或受体动力学 3.4.3
    dose-response models 剂量-反应模型 3.4
    internal chemical concentration 内部浓度 3.4.1
    linear-with-threshold relationship 线性阈值关系 3.4.1
    TKTD models TKTD 模型 3.4
Toxicokinetics（TK） 毒物动力学 3.2, 4.1.3
    fate modeling 归趋建模 3.3
    mercury 汞 3.5.3

 Mytilus galloprovincialis 地中海贻贝 3.3.3
 one-compartment model, first-order kinetics 基于一级动力学的单室模型 3.3.1
 parameter estimation 参数估算 3.3.3
 simple one-compartment model 简单单室模型 3.3.2
 transport mechanisms 转运机制 3.3
Toxicokinetic-toxicodynamic (TKTD) models 毒代-毒效动力学模型 2.6
 energy-budget model 能量预算模型 3.6.4
 parameters 参数 3.8.1
Trace analytical methods 痕量分析方法 1.6
Trace metals 痕量金属 1.2.1
Transcriptomics 转录组学 5.3.1
Trophically available metal (TAM) 营养用金属 4.3.2
Trophic transfer factor (TTF) 营养转移因子 4.2.2

W

Wastewater treatment plants (WWTPs) 污水处理厂 1.2.2.2
Water Framework Directive 《欧盟水框架指令》1.2.2.3
Water-sediment partition coefficient 水质-沉积物的分配系数 1.2.2.2
Weight of evidence (WoE) 证据权重 9.3
 bioaccumulation assessments 生物富集评价 9.3.2
 biomarkers 效应生物标记物 9.3.2
 decision matrix 决策矩阵 9.3.2
 established and accepted method 公认的方法 9.3.1
 intermittent effects 间歇效应 9.4
 laboratory toxicity studies 室内毒性研究 9.3.2
 lines of evidence (LoE) 独立证据 9.2.2
 logic system 逻辑系统 9.3.1
 receptors of potential concern (ROPCs) 潜在关注受体 9.3.2
 resident community assessments 暴露群落评价 9.3.2
 sediments 沉积物 9.3.2
 semi-quantitative approach 半定量方法 9.3.1
 stressors of potential concern (SOPCs) 潜在关注应激源 9.3.2
 toxicity and bioaccumulation testing 毒性和生物富集试验 9.4
 types 类型 9.3.1

X

Xenobiotics 外源性物质 5.2.1

后　　记

　　海洋与人类生存息息相关，与国家兴衰紧密相连。21世纪人类进入了新一轮大规模开发海洋、利用海洋的历史时期。人类活动所导致的气候和生境变化、入侵/引进物种、富营养化和化学污染物等，正在对海洋造成持续的有害影响。

　　海洋生态毒理学数十年来获得了突飞猛进的发展。在中山大学海洋科学学院的支持下，译者开设了"海洋生态毒理学"本科课程。该课程重点培养学生在科学研究中综合运用相关知识的能力；为认识海洋污染生态危害、可持续开发利用海洋资源以及海洋环境保护等热点问题奠定扎实的理论基础；引导学生深入体会"绿水青山才是金山银山"的意义，激发学生对海洋生态毒理学研究的热情。然而，译者发现目前缺少海洋生态毒理学系统性知识的教材。综合多方面材料，译者选择翻译了Elsevier出版的、由诸多国际知名学者编撰的图书 *Marine ecotoxicology*。希望能让更多的学者、海洋污染防控和管理工作者，以及相关专业的学生较系统地了解目前该领域知识体系及管理方法和经验。此外，借此译书抛砖引玉，希望专家学者协力编写符合中国海洋战略和特色的海洋生态毒理学教材。

　　译者衷心感谢中山大学海洋科学学院领导和老师们对于本书出版的支持。感谢团队吴玉萍教授对本书翻译工作的支持。感谢恩师暨南大学生命科学技术学院杨宇峰教授长期以来对译者工作的关心与帮助。感谢中山大学生态学院刘之威副研究员和海南大学海洋生物与水产学院罗洪添副研究员的校阅。感谢吾妻在本书翻译过程中对译者的协助与支持，使译者可心无旁骛地完成该译书工作。感谢刘亚婷老师、唐丽丽老师、宁曦老师和广州中山大学出版社编辑对于此书出版所作出的努力。

　　译者虽然始终保持严谨细致的态度完成此书翻译工作，但专业水平与能力有限，此次为首次译书，译文存在错漏及不妥之处在所难免。敬请各位读者、专家和同仁，不吝指教，共同使译书臻于完善。

孙　显

2023年12月